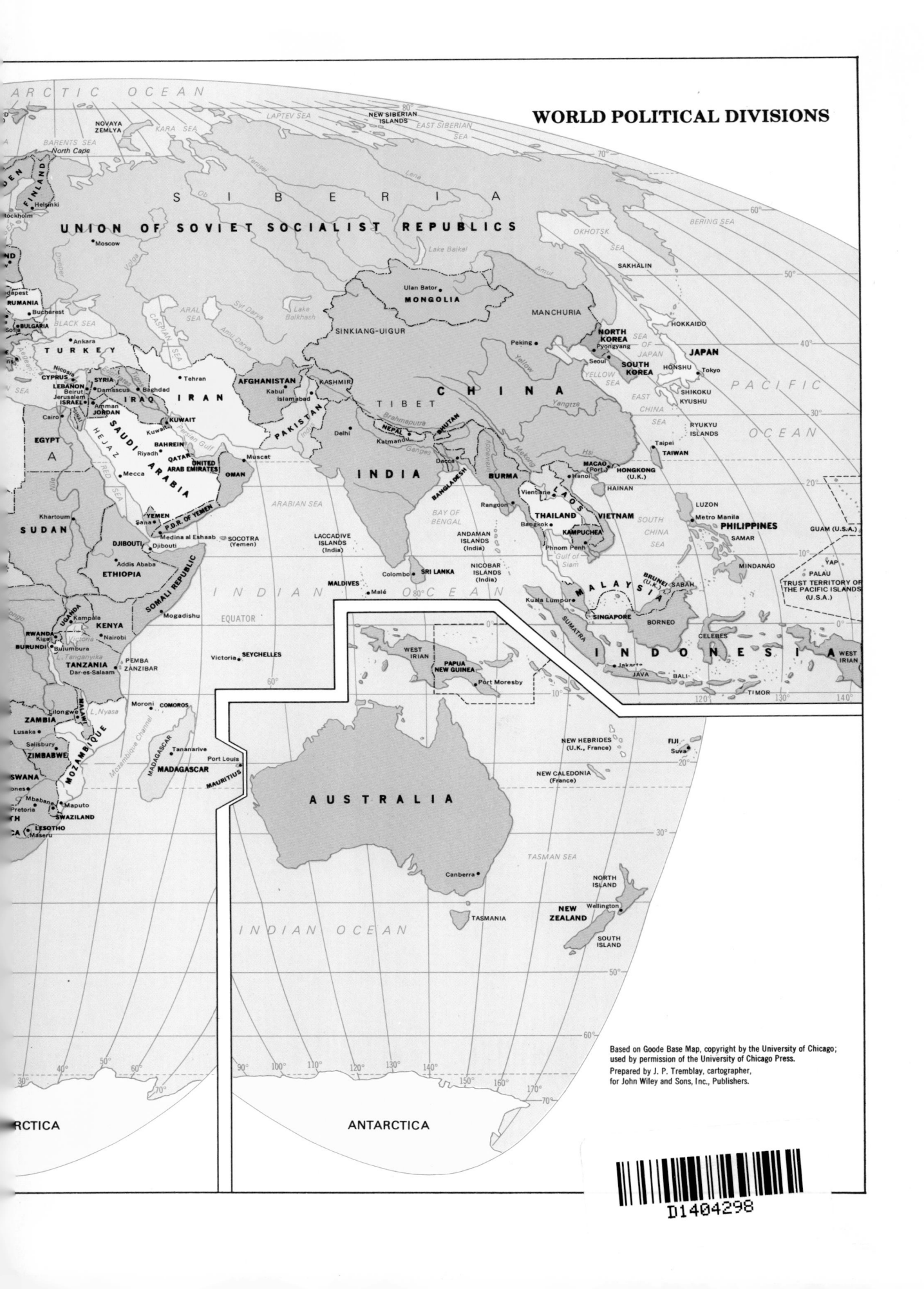

WORLD POLITICAL DIVISIONS

Based on Goode Base Map, copyright by the University of Chicago; used by permission of the University of Chicago Press.

Prepared by J. P. Tremblay, cartographer, for John Wiley and Sons, Inc., Publishers.

D1404298

INTRODUCING
CULTURAL
GEOGRAPHY

J. E. Spencer

University of California
Los Angeles

W. L. Thomas

California State University
Hayward

INTRODUCING CULTURAL GEOGRAPHY

SECOND EDITION

Cartography by
Robert E. Winter

John Wiley & Sons
New York • Santa Barbara
Chichester • Brisbane • Toronto

About the Cover:

The three circles depict (from back cover to front cover):

1. Antiquity — Cave painting, Egyptian art, and Greek architecture.
2. Middle Ages — Renaissance art, architecture, and navigation map.
3. Modern Age — Lunar module and sky scrapers.

Library of Congress Cataloging in Publication Data:

Spencer, Joseph Earle.
  Introducing cultural geography.

  Bibliography: p.
  Includes index.
  1. Anthropo-geography. 2. Social evolution.
3. Ethnology. I. Thomas, William Leroy, 1920–
joint author. II. Title.
GF41.S69 1978   301.2   77-20230
ISBN 0-471-81631-0

Printed in the United States of America

10 9 8 7 6 5 4 3 2

PREFACE

chap 1
read &
answer questions

Introducing Cultural Geography has been written to present a broad view of the cultural development of humankind on an earth that has been modified and reshaped during the period of human occupancy. It assumes no previous study of the subject and is designed for use in introductory courses in geography. An awareness of the viewpoints expressed here is, we believe, important for the geographic education of all persons, intelligent citizens and future geographers alike, whatever their later individual interests and professional perspectives may be.

OBJECTIVES

We have emphasized unifying concepts, features, and processes important to a basic understanding of the role of human beings in changing the surface of the earth as their knowledge and skills have accumulated through time. The subject matter of cultural geography is large and complex for presentation in an introductory course, no matter how it is organized. By "painting with broad strokes," we have focused only upon those elements we consider essential to the achievement of this primary objective. We have attempted to present step-by-step explanations for each concept and each new development introduced as we discuss the cumulative ability of the human race to cope with changes in the physical and biotic worlds. We have provided a glossary of terms in which we have carefully defined those concepts, terms, and processes necessary for even a simple discussion of this historical growth of human culture.

THEMES

The ecological and evolutionary approaches employed in the first edition of 1973 have been retained and amplified. In this volume we outline a series of themes:

1. The ways in which the planet Earth has been modified by physical and biotic processes during the existence of mankind,
2. How human technologies and culture systems have grown during that long time span,
3. How human living systems in separate parts of the earth have developed at varying rates and into different culture systems,
4. How these human technologies have matured to become the dominant agencies in modifying the earth in the modern period of human occupancy, and
5. How human technological systems have brought about the modern overpopulation of the earth.

The discussions of these themes are broadly systematic on an evolutionary time scale that contains no fixed rates of change nor levels of achievement.

ORGANIZATION

We try to cover the whole growth of human cultural development in a compact manner by dividing the material into units that accord roughly with chapters. Chapter 1 presents viewpoints toward the earth, the analytical framework employed, a statement of the basic culture processes, and concepts of time, space, and culture as these affect human living systems anywhere on the earth. Chapters 2 and 3 present the changes in the earth during the era of human occupancy, the beginnings of the human race, and the simplest elements of cultural development that occurred in the distant past. Chapters 4 through 6 review the historically divergent evolution of regionalized culture systems and group technologies, and preview those developments that brought about the ability of human beings to travel over and utilize the whole earth.

Chapters 7 through 10 deal systematically with technological modernization of the earth in differential terms, as communication and transportation systems have fostered world trade and industrialization. In these chapters we present, in addition to an explanation of the growth of population on the earth and the issues of contemporary urbanization, a glimpse of worldwide human conditions at the present time. Chapters 11 and 12 examine the human role in modifying the earth as a physical and biotic ecosystem; they also discuss the ecological problems and alternatives facing modern mankind in the continuance of systems of complex culture.

SPECIAL FEATURES

All the 94 maps and diagrams have been revised for easier reading, and 21 of them are new to this edition. There have been additions to the list of tables, and these now number 21. We would caution that some of our maps are quite tentative and are presented as graphic expressions of ideas aimed to stimulate further thought and inquiry. We have retained the photographic essay as a useful instructional aid, and have expanded the patterns into six separate essays, for which many new photographs have been employed. Additionally, new photographs have been placed at appropriate locations in the text itself. The glossary of terms, presented at the end of the text, has been expanded as an aid to the reader, and the bibliography has been both updated and expanded in range.

TO THE STUDENT

The table of contents lists the primary subdivisions of the separate chapters and the page locations on which these begin. At the beginning of each chapter is a list of key concepts, worded to relate to the primary and secondary headings within the text of the chapter. *Before* you read a chapter, please read the list of key concepts as a guide to what is to be presented. Then be on the lookout for the discussion of each concept, and the related subject discussion and viewpoint, when reading the chapter. Pay particular attention to the figures and tables, since they expand upon what you are reading. At the end of each chapter is a summary statement of objectives for that chapter, with references to the text pages covering each statement or question. These are indications of skills, viewpoints, or knowledge we would like you to acquire. These summaries review the essential points of the chapter to help you recall and relate them to each other. As you review the objectives after reading the chapter, you may check on your own learning process to know how well you can handle the materials presented.

ACKNOWLEDGMENTS

Our revision owes much to those who kindly took the time and effort to present constructive comments about ways in which this volume could be improved. The present text reflects many of those suggestions. There has been rearrangement of chapter topics, the addition of new topics, new maps and tables, changes in the photographic essays, and the elimination of unnecessary text materials. We are particularly grateful to Dr. Mary Lee Nolan and to Dr. William D. Warren for their painstaking efforts at identifying various infelicities and overly complex language, and hope that we have achieved a more smoothly reading text. We are also grateful to Mr. Robert E. Winter as the cartographer who revised or rendered for publication all the maps and diagrams in the volume. Shortcomings remaining in this second edition are, unfortunately, the responsibility of the authors only, who have persisted in the presentation of certain features that we feel make this a distinctive textbook.

Canoga Park, California   *J. E. Spencer*
Hayward, California   *W. L. Thomas*

CONTENTS

LIST OF FIGURES

LIST OF TABLES

CHAPTER 1
WHAT CULTURAL GEOGRAPHY IS ALL ABOUT

KEY CONCEPTS

Culture Involves Traits, Complexes, and Systems
There Are Four Basic Culture Processes
Functional Concepts Have Interlocking
    Relationships

Man/Environment Relations Are Changeable
Differences in Areas Occur Through Time
Geography Is a Single Discipline in Three
    Divisions

This book focuses on the slow growth of human culture on our changing earth. Human culture, in the collective sense, is worldwide and serves to distinguish mankind from other animals. Culture, in the collective sense also, is the sum total of historically learned human behavior and ways of doing things. Human culture did not develop collectively, however, but in separate units put together by different groups of people, each of whom occupied a specific territory on the face of the earth. Every group of people responds to opportunities in the physical and biotic environment in which they live. Each group that has ever occupied a territory works out a way of living in that particular environment. In the simplest early context, that way consisted of finding ways of securing food, shelter, amusement, and "things to do." That way of living, however, also involved working out patterns of individual behavior toward other members of the group, social practices, and sets of rules that operated for the group as a whole. Individuals born into the group learned the practices that provided support and also learned the social patterns of behavior that enabled the whole group to live satisfactorily within their territory.

Each such group of people develops a particular living system that is somewhat different than that of any other group resident nearby or at some far distance. Such a system comes to be a part of the total living environment within which a group operates. The group then finds itself surrounded by a total environment that is a threefold affair, made up of the physical environment, of the biotic environment, and of the cultural environment that the group has created for themselves in their own particular way. This threefold total environment becomes a continuing influence in the way of life of the group thereafter, and in this context the culture system becomes in time increasingly important.

No human group has lived totally alone, without contact with any other group at some time—at least not since the human race grew numerous enough to start spreading out over the earth. And, in time, the living system of no human group remained totally and solely devised by that group alone. Each group both gave to, and received from, other groups some bits of learning in the way of material objects, ideas, and practices. Most of the early history of mankind on the earth was a matter of devising ways and means in many different environments. Over the long span of time, however, the separate developments in particular regions have been pooled to some degree and have come to constitute the collective body of culture possessed by mankind on our earth.

We make no distinction between the bodies of culture that formerly were separated into two parts and labelled material culture and nonmaterial culture. Every human living system inevitably involves social contacts among members of the culture group, along with organized ways of procedure as the group earns its living and seeks its way of life. Our focus is on the local development of human behavior and the impacts on the earth produced by groups of people living together in many different parts of the earth. We are concerned with the exchanges of many items between these local groups, and we are concerned with progress made in particular regions, since people in some regions achieved more rapid development than did people in some other regions. We are interested in the ways in which people have reacted to their local environments, both in terms of the manner of using them, and in respect to the impacts made upon them. Each culture system, however, is a distinct and complex mix of many kinds of ideas and practices, and a culture system must be examined in respect to its different elements.

CULTURE INVOLVES TRAITS, COMPLEXES, AND SYSTEMS

The term culture system is a very broad and all-inclusive one, and it is impossible to examine the whole of a regional culture in one piece. It is necessary, therefore, to break down a culture system into units and parts in order to understand it. The patterned units of a culture system consist of several kinds of smaller parts, and these can be studied in different ways. The smallest unit of culture is termed a *culture trait*, referring to a distinctive individual item, whether this be a behavioral trait or the particular use of a tool. Related single culture traits that go together in practice form a *culture complex*. Culture complexes in different culture systems may contain unequal numbers of culture traits. Keeping cattle is a human culture trait, whatever their use. Milking the cattle, drinking milk and making and consuming butter, yogurt, and cheese is a small but distinctive culture complex. Eating beef, using cattle to pull plows and wagons, and wearing tanned cowhide as shoes and garments is a culture complex related to the dairying complex. A large assemblage of culture complexes fit together into a *culture system* (Fig. 1.1). A specific territory inhabited by a population follow-

Figure 1.1 The Culture Hierarchy, a Schematic Representation. A, B, C, D, and E represent individual culture systems, shown at the top as culture regions. Regions A, B, and C form a culture world, I. System D is linked to E and others in another culture world, II. Culture region A has four subregions, with its culture hearth in subregion 1.

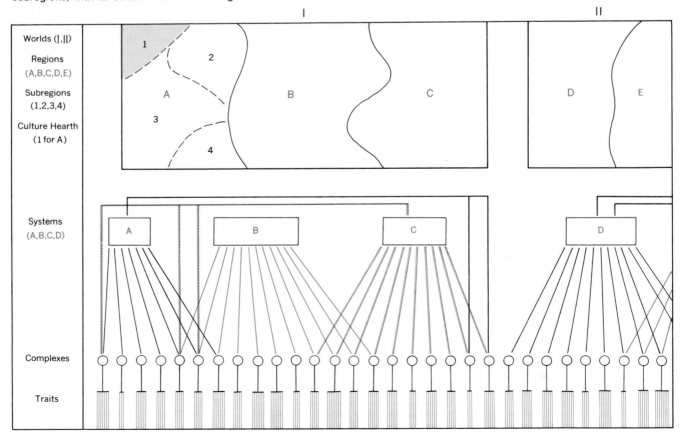

ing a particular culture system forms a _culture region_. Mexico, Canada, and the United States each form culture regions. A group of regions in which related culture systems are practiced is termed a _culture world_ or a _culture realm._ Canada and the United States form a culture realm, whereas Mexico belongs to a Latin American culture realm. In a further example, the two related traits of making and eating Roquefort cheese began in a specific locality in France, and they evolved as part of the dairying complex in what came to be the French culture region. Dairying, as a culture complex, was practiced in all other regions of northwestern Europe, but in each of these regions people came to make and consume their own local varieties of cheeses. That these regions all shared in the early dairying complex is one of the elements in the very large assemblage of culture complexes that developed through time to form the distinctive European culture world.

In the growth of a culture region, many of the core ideas and basic practices were first formulated in a localized part of the larger territory by a closely related group of people who shared in their development. These ideas and practices then spread to nearby groups of people, or were carried by settlers moving from the home locality, so that a larger area came to share in the core ideas and basic practices. The local region in which the initial and most important ideas and practices were formulated is termed the _culture hearth_ for that particular culture system. Even though a large region may come to share in the core ideas and basic practices, there may be variation in related and nonbasic elements in different parts of the larger region. A culture region, therefore, may contain a number of areas, each of which is a _subregion._ Thus, in our example of France, although dairying and cheesemaking are basic traits in the whole culture region, there are many local varieties of cheese produced in France. And, though Canada forms one major culture region, the French-speaking sector of eastern Canada forms a subregion.

The spread of particular complexes from a culture hearth has frequently carried those complexes beyond the specific culture region into territory inhabited by different peoples who then interpret aspects of the new complexes in different ways. These varied ways, plus the locally developed core ideas and basic practices, may in time create a culture system that must be recognized as forming a separate culture region. The shared patterns, however, identify the culture systems and regions as related. In eastern Asia, the core concepts and basic practices formulated in the North China culture hearth eventually spread all over China to create a Chinese culture region. Many of the practices also spread to Korea, to Japan, and to Vietnam. Korea, Japan, and Vietnam each formulated significant ideas and practices of their own so that each forms a separate culture region. Since all four societies (China, Korea, Japan, and Vietnam) share many features in common, not shared by other culture systems, all four can be grouped into one culture world.

Culture traits and complexes are structured in particular ways by different societies, and each trait or complex serves a particular role or function in a given society. Structure has reference to the arrangments of the parts of any whole, and role/function refers to the purpose each part plays in the operations of the whole. Human societies have innumerable ways of arranging culture traits and complexes in combination to form a system, so that each unit operates to achieve its purpose—at least a society may hope that things will work that way. A human society is made up of individuals, each of whom has personal needs in biological, economic, political, social, and spiritual terms. Some of the biological needs, such as the need for oxygen, water, and sleep, do not require much conscious effort in many parts of the earth. But many human needs derive from the fact that the human individual rarely lives a completely solitary life, but lives with others in groups. The personal needs of each become the cumulative needs of the group, and the satisfaction of those needs requires that people adopt common ways of meeting them. For the individual to earn a living, find compatible social contact, develop suitable recreational activities, and satisfy his spiritual needs, depends upon the interrelations worked out among the members of the group in their normal living system. As cultural geographers, we are interested in the different ways in which culture traits, complexes, and systems have been structured. Each of these may play a different role in separate structured combinations. Roles vary according to the perceptions and desires of the separate societies living on the earth through time, as human beings first began to create bits of culture and organize these bits into systems.

To sum up, our concept is that culture is the product of human societies living in particular regions of the earth over extended periods of time. For each society that product is the created and assembled, maintained but steadily readjusted, distinctive tradition of a way of life. Each tradition is comprised of loosely correlated esthetic, social, economic, political, and ethical codes of behavior, and members variably share ideologies, habits, customs, procedures, and technologies. Each society primarily utilizes the resources of its own region, however these may be shared, to establish patterns of daily living, and each society's activities imprint on the surface of the inhabited region a distinctive expression of human occupancy.

THE FOUR BASIC CULTURE PROCESSES

Important to our viewpoint in this book is the way in which change is brought about in producing new culture traits, modifying culture complexes, and changing culture systems. There are a great many cultural "processes" operative through which change occurs. We may speak of the "process of industrialization," the "process of urbanization," or the "process of acculturation." Each of these involves the integrated operation of several basic and elementary processes working together in different ways to achieve a complex set of results. The four most basic elementary processes (each contributing to a complex and interlocking process such as industrialization) are *discovery, invention, evolution, and diffusion* (Fig. 1.2).

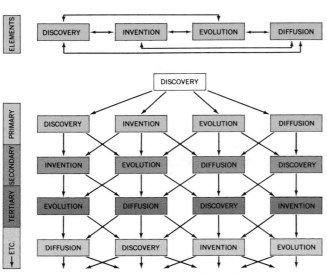

Figure 1.2 The Interlocking Relationships of the Basic Culture Processes. The oversimplified model at the top contains all of the elements. In the lower diagram the interlocking relationships are more fully spelled out.

1. DISCOVERY is the act of finding things that are new to the experience of people but that occur naturally in the physical/biotic environment, such as the black rocks (coal) that could be burned to produce heat, the virus that produces the disease smallpox, electricity that can be artificially produced and controlled, and the presence in the human blood system of many different components. The term is sometimes used too loosely to refer to things that are not properly discoveries. Basically, discovery is the way in which mankind learns about the physical world.

2. INVENTION is the human act of creating something new that does not occur naturally in the physical/biotic world, such as the paper used in printing this book, steel (an artificial compound of metallic minerals), a television set, or the electric cooking range now often found in our kitchens. Most often we tend to think of inventions as material artifacts, but the political state is a human invention, as is the legally incorporated settlement we term a city. The greater share of the things we live with in daily life are human inventions that enable a pattern of living different from that of the animal kingdom. Invention is one particular aspect of the wider creative process now being termed innovation.

3. EVOLUTION is the process of making small changes over time periods in things previously produced or arranged by human inventive processes, thereby making things more complicated (often termed "making things better"). The improvement of domestic housing in advanced societies illustrates the evolution of the simple shelter into the modern home. The annual changes made in the shape, equipment, and furnishings of the automobile demonstrate the evolutionary process by which the first invention of the automobile has been made both more complicated and more satisfactory to our contemporary living systems. The amendments, made and pending, in the Constitution of the United States illustrate the process of evolutionary change at work.

4. DIFFUSION is the process of spreading, distributing, or scattering of something away from its locality of original occurrence over a period of time. The slow spread of the plow tools in the development of early agriculture illustrates the historic process, and the twentieth century spread of the telephone illustrates the process in a modern context. The wheat plant, originally native to a region of Southwest Asia, has been diffused by human agencies into all those agricultural regions of the earth in which the plant will mature its grain, and current research efforts are pushing it both towards the Arctic fringe and into the subtropical margins of the earth. The English language, although not spoken by all peoples everywhere, has diffused to an almost worldwide distribution from an island in northwestern Europe. Diffusion can occur through the unconscious copying of human activities by peoples not originally doing things that way, and it also takes place through purposeful borrowing of ideas, traits, procedures, or artifacts from other societies. The students from other societies studying in the United States are one of the important mechanisms of cultural diffusion. Diffusion has often occurred through forced acceptance under politicomilitary pressure, and it occurs as human beings migrate from one region to another, taking their artifacts and ideas with them. Diffusion can also occur as idea

stimulus, in which the idea of something spreads without the artifact itself being transferred, but often, at a later time, with some comparable artifact being made. The rather recent spread of the concept of industrialization to the so-called nonindustrial societies illustrates idea stimulus.

The basic culture processes operate in interlocking and repetitive fashion, with a "feedback" aspect, for the human mind sometimes perceives that some development through one process may enable a further creative accomplishment that then brings the earlier process back into play again. For example, the discovery of a new raw material that can be substituted for an old material may lead to evolutionary change in the machine employed in fabricating the product; the diffusion of the use of the new raw material may inspire evolutionary changes in the technique of processing the material, which in turn may inspire the invention of a different and new machine employed for fabrication; the diffusion of this new machine back to the original area of raw material discovery may result in further evolutionary change in both processing and fabricating techniques (see Fig. 1.2). 8-24-81

It is impossible to predict accurately the sequence of changes and "progress" in human affairs that operate through the basic processes, but it is clear that all four processes function in complex ways, conditioned both by the surrounding physical and cultural environments and circumstances. The development and elaboration of any advanced culture system in its particular place and time has been possible only because an everchanging combination of culture traits has been assembled from all areas of the world then known to the populace of that culture system. People in such an advancing culture have unconsciously or consciously employed the basic culture processes in the elaboration of their changing culture.

As one example of the above statements, only a few of the elements of the material culture and consumer behavior of people in the United States derive from purely internal, initial discovery and invention. The larger part of our culture system is compounded of ideas, raw materials, products, technologies, and procedures gathered from all over the earth. In the broad sense, the culture system of the United States has enabled and fostered enormous application of the basic culture processes to the elaboration of techniques of using the resources of the whole earth in the improvement of the way of life among our population. That a long-prevalent attitude has been that "there must be a better way to do it" clearly indicates that in the United States the four basic culture processes have been consciously pushed to make all possible discoveries, to make subinventions continuously on older things and ways, to carry on

basic research leading to new product invention, to maintain evolutionary patterns of change, and to gather up all the good ideas from all other peoples of the earth. That the people of the United States have done this too effectively is now one of the complaints of people elsewhere on our earth.

Not only is the use of the basic processes effective for changing the material ways of life on our earth, but that use has been effective in the development of the many varied cultural landscapes of the earth. Every object that owes its presence to human action was either invented, imported by diffusion, or altered from some older pattern by evolution. Each element in a cultural landscape has its own origin, its own history of acceptance with a particular culture system, and its own pattern of modification through time.

In the earlier prehistoric and historic eras, efforts at discovery into physical laws and the nature and utility of earthly resources was not pursued in any organized way, although our remote ancestors did amazingly well at discovery in the eventual sense. The diffusion process operated slowly, spasmodically, and incompletely, since the linkage between parts of the earth was neither complete nor continuous in time. Invention often was an almost accidental occurrence, and it was seldom purposely sought in the ways in which modern science seeks for new developments. Evolutionary changes came slowly during very long time intervals, as indicated by the lack of dramatic change in the life of earlier societies. In the modern period, in contrast, institutionalized agencies in the advanced culture systems carry on conscious and continuous pushing of all four basic culture processes in concerted efforts to expand the ways of life. One of the probable explanations for the slow rates of cultural change in the "ancient world" is that the operation of the basic cultural processes was left almost entirely to chance and intermittent human curiosity, whereas in our modern world those processes have been actively fostered. It is clear, in the historic sense, that the rates of culture change have been most rapid and far-reaching in those societies that have been open to change and that have recognized that the basic culture processes may be employed productively. That, in fact, is what the continuing Industrial Revolution has been all about.

FUNCTIONAL CONCEPTS AND INTERLOCKING RELATIONSHIPS

Members of every living human society are surrounded by a complex physical and biotic world into which has been intruded the whole of their particular culture system. This system composes part of the total environment within which a society carries on its daily life. Most individuals accept this situation with hardly a thought of the interrelating complexities of life. In the simplest classification possible, there are four factors that bear upon the life of a society over a long-term period. These factors work one against another in some proportionate way, as independent variables, to bring pressure to bear upon the continuance and successful life of a society. Out of these four factors there arise six continuing interrelationships comprising the specific sets of forces that influence the increasing or decreasing success of the society. Chiefly, these interrelationships operate over the long-term period, but any of the factors (as a changing element or variable) may become significantly operative in the details of life of the individual on any given day. The four factors, and their interrelationships are depicted in the accompanying figure and spelled out in the tabulation below (Fig. 1.3).

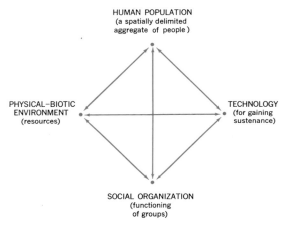

Figure 1.3 Functional Concepts and Interrelationships in Cultural Geography.

The four independent factors are:

1. **Population:** human population groups occupying spatially defined territories. The numbers of groups and/or numbers of people in each territory may be rising or declining.

2. **Physical/Biotic Environments:** spatial entities undergoing changing relationships, but possessing resources that have particular carrying capacities, and that offer potential support in the maintenance of human livelihood.

3. **Social Organization (Cultural Environment):** organizational structures of interdependent units (families, kin groupings of all kinds, associations) permitting integration and functional operations

through adoption of sets of rules, formulations of living systems, and elaborated division of labor.

4. **Technologies:** sets of integrated techniques (abilities, ideas, procedures, tools, and systems of operation) employed by a population to gain sustenance from environments and to carry out cultural objectives.

The six operative interrelationships are:

1. **Population ↔ Environment:** a mutually interacting, dynamic two-way system in which a whole environment affords potentials to a whole population but which undergoes change through resource use by human spatial occupancy. Small populations at low levels of culture (combining variables 3 and 4, above) realize few of the opportunities present in environments and make slight impact on environments; the larger the population and/or the higher the level of culture the greater is the achievement in environmental use and also the higher becomes the human impact on environments.

2. **Population ↔ Social Organization:** interacting relationships between size/distribution of a culture group and the culture forms/structures adopted or devised by organized societies. Small populations normally utilize few forms and structures, whereas elaboration of culture forms and structures is indispensible to large populations. Advancing culture levels generate added and more complex forms and structures. Social organization tends to be functionally persistent and to increase population, since organizational structures enhance probabilities of human survival, but social organizations can decay, producing hazards in living systems and decline in population.

3. **Population ↔ Technology:** an interdependent relationship between manpower and technological system. Increasing population ordinarily enhances technological advancement through an enlarging brain pool and increasing labor specialization, yielding improved levels of living (but see interrelationship 4, below). Very small populations having simple social structures usually exhibit simple technological systems; the larger the population (to some undetermined threshold number) the greater the potential for technological advancement.

4. **Organization ↔ Technology:** social organization ordinarily adapts to the technological toolkit at the command of a population in dynamic terms. Technological levels and creative/permissive patterns of social change are interdependent, *normally* facilitating technological advancement/change derived from internal inventiveness

or inward diffusion of advanced technologies/social systems. Simple organizational structures usually exhibit simple technological systems; the more complex the social structuring, the greater the opportunity for technological complexity. However, there is a factor of variability in acceptance/hindrance of organizational change; culturally conservative societies often decline to permit constructive social change, thus inhibiting technological advancement. Historical lags in acceptance of social change in culturally conservative societies are normal, preserving antiquated technologies and yielding variable ratios of underdevelopment when compared to economic development of the most advanced systems.

5. **Environment ↔ Organization:** individuals and societies must cope with specific conditions of occupied environmental regions. Low-quality environments requiring a full-time quest for food prevent development of organization for other ends. Isolation of small groups in separated regional environments tends to prevent organizational advancement. In fact, necessity seldom mothers invention; food plentitudes afford leisure time, facilitating creativity in diverse elements of cultural expression. Small populations having simple organizational structures can occupy only restricted areas. Organizational improvement offsets environmental hazards and facilitates development, but complex organizations make for stronger impacts on environment.

6. **Environment ↔ Technology:** technologies provide means of utilizing potentials of environments; interrelationship is dynamic under cultural advancement, based on changing perception by members of society of the potentials of an environment. Earliest man was directly dependent on immediately usable resources and was unable to occupy tropical rain forests, deserts, cold steppes, boreal forests, and tundra. Invention/evolution of the food, clothing, and shelter technologies made almost the entire earth habitable. Technologies are not random tool kits but understandings-techniques-procedures-tools devised and integrated in depth to attain comprehended objectives. Advancing technologies made possible usage of added resource potentials in expanding sequences, improving perceived qualities of environments. Technologies create human habitats from wildernesses, sequentially altering environments in interdependent ratios.

To exemplify the ways in which some of the variables and the interrelationships may complicate life on some days, consider the two following "bad" days in the lives of two Americans: (1) An urban dweller awoke to find it raining "again!" That meant skipping breakfast and taking the raincoat, hat, and boots again, since it would be a long walk up the hill from the parking lot. The car almost flooded out crossing that low intersection at the bottom of the hill, and the two moths that had crept into the car overnight kept flitting around to complicate the already poor visibility caused by the worn blades on the windshield wipers. The parking lot attendant was unusually grumpy and more than normally profane because drivers were leaving too many cars in awkward positions on this wet morning. Finally, she made it up the hill, half-soaked, only to discover that the building foyer was jammed since two elevators were out of operation because of electrical short circuits induced by rain leaking into the control room. Several of the office staff were absent with the spreading influenza epidemic, and that threw office routines into disorganized attempts to cope with problems that normally went smoothly. Despite laboring well past the normal quitting time, a backlog of work had accumulated, promising another difficult day tomorrow. (2) The rural farmer awoke in the dim early morning light to the sound of his radio broadcasting a warning that the river had overflowed its banks and that flood conditions were imminent. "Get the cattle out of the bottom pasture" was his first thought. An attempt to get help from two "high-ground" farmers did not work because the phone was dead from downed lines. Hurrying past the chicken sheds on the way down to the lower pasture the farmer saw several dead birds lying about the yard, and there went a fox scooting off carrying a chicken. The supply company was unable to deliver the fencing that had been ordered because the Teamster's Union was on strike. Rounding up the cattle was a full morning's work because parts of the lower pasture were already flooded and some cattle had retreated to a low hill that now was an island. The afternoon was spent in fruitless attempts to locate the veterinarian who had promised to vaccinate the pigs. All local veterinarians were already out on emergency calls to the scattered farms in this rural countryside.

In both examples one can find the "workings" of all three aspects (physical, biotic, and human environments) of the situation in which each of us lives. The physical environment was expressed through bad weather and the physical relief of the two areas that brought trouble. The biotic environment expressed itself in terms of wild life and disease germs. And the human environment showed its own effects through the disorganization of its facilities. On "good days," with the sun shining and temperatures at an agreeable level, most of us forget about the complex interrelations between the sequential patterns of our daily lives and the environmental conditions that may alternate in their influences.

MAN/ENVIRONMENT RELATIONS AS CHANGEABLE

What is the control in man's relations with the earth? Is he "the product of his physical/biotic environment" and thus here by pure sufferance according to some higher law, his patterns of happiness or misery dependent on his chance of birth in a good environment and on whether he interprets well and follows faithfully that higher law? Or can man really choose to make his own future in any physical/biotic environment on the earth according to his vision of the possible and his steadfast pursuit of knowledge about that environment which gives him power over it? Can human culture grow and evolve to the level at which man can make over any regional environment in its ecological totality?

The traditional views in these matters depend very clearly on the levels of culture and the views of those making the generalizations. Simple and small societies stand in awe and in not a little fear of their physical/biotic environmental surroundings, normally seeking to counteract the dangerous forces in the environment through the use of ritual and magic. But modern science views the issues simply as complex problems in the perfection of the technology necessary for controlling the forces of nature. Historically, the conservative has scoffed at the farseeing visionary who would master natural law in the creation of a new technology. In view of the changing technology of the last century, however, who can say for certain what the limit of power controlled by human will may be?

Within the European culture world, affected by Judeo-Christian doctrine, it was long held that mankind was a creature of God, set upon an earth of God's design. In this, the teleological view, mankind *was* the product of his environment, and the earth was a temporary abode testing selective human fitness for an afterlife in Heaven. In the late nineteenth and twentieth centuries there has been the growth of ideas promoted by Science and by the increasing secularization of populations. Secularism holds that there is no God-given control over human destiny and that the growth of human knowledge and technology can in time free man from environmental and other omnipotent controls. Secularism has shared with Science the view that exploitation of the earth through the discovery and application of physical laws can improve levels of human living. At its greatest extreme this view becomes cultural determinism. All geographers, consciously or unconsciously, take positions somewhere between these two extremes. Too often, however, the position taken is a static one, applying the single viewpoint to all of human time and to all peoples at all levels of cultural development.

In this book the position taken toward any form of determinism is conditioned by the level of culture of the group under discussion. The levels of daily living of the very earliest members of the human race were fully controlled by the qualities of their physical and biotic environments, because those early members were naked, they possessed no basic knowledge of physical laws, and they had developed no significant cultural technologies. Through long periods of time the descendants of earliest mankind gradually learned more and more about how to cope with the simpler problems of environment by developing systems of shelter, patterns of clothing, particular techniques for getting food, and simple techniques for making tools and utensils. After several millions of years, the most advanced societies have acquired tremendous resources of knowledge, skill, power, and technology. For modern, highly developed societies the generalization about the influence of environment on human life no longer can be applied in the same terms. A flexible and sliding-scale viewpoint, therefore, is necessary toward the issues of physical environmental versus cultural determinism.

A small contemporary society possessing only a low-level culture system of slight technological power must still live close to nature and accept aspects of control from the physical environment. A large society possessing a high-level culture of great technological power is much more free of the influences of physical and biotic environments. For example, in the United States almost the whole body of human culture may be interposed between the external physical environment and the pattern of daily life, even though we cannot totally disregard that physical environment. In New Guinea, however, but recently out of the Stone Age, population units, levels of culture, and controlled technology do not permit ignoring physical environment to any significant extent. (That not all members of an advanced society live prosperously does not alter this principle. Inequalities within any one culture system are not produced by environmental influences, but result from imperfections in the operation of the culture system itself.)

In the twentieth century it is irrational to take a static view of human relations to environments. In describing or analyzing these relations, it is necessary to establish the levels of culture and technology that are capable of being applied to the problems of living by the group in question. It is equally true that no group is totally controlled by environment (since all living human groups do make choices in using their territories) and that no human group has the power to free itself totally of all environmental influences.

The more advanced the culture system, the more complex are the interrelations among physical/biotic environments and human living systems. Advanced societies have the capacity for altering the environments in ways never open to earlier societal choice. Increasingly it becomes a question of the quality of human organization of the living system as well as a question of the intelligent manipulation of the biotic and the physical conditions within the whole ecological system. In the operation of advanced societies, human activities interact with environmental conditions on many more fronts than in simpler societies. Even the most advanced societies still have facing them many major problems in manipulating the environment.

DIFFERENCES IN AREAS THROUGH TIME

In no two centuries since the human race began has the earth remained exactly the same, and today's earth is like none earlier. In this everchanging physical setting the human race slowly developed many separate ways of doing things. Culture, itself, became an element in the earthly pattern. Groups of people, their territories, and culture systems are now arranged in patterns never known a few thousand years ago. At present the earth is a complex mosaic of regions whose peoples have different time perspectives and different life styles. The relations between regions, culture systems, and living patterns have involved space and time in a great many different ways. At any time within the last ten thousand years, some territories have been small, the numbers of people few, the cultures simple, and time relatively unimportant, since tomorrow would be like today. Other territories have been large, the numbers of inhabitants very large, the cultures complex, and time an important issue, since change was taking place. In general, within the last ten thousand years, societal territories have been growing larger, the numbers of people have been increasing, the levels of culture have been reaching greater complexity, and time has brought change more quickly.

In some parts of the earth today small groups of people still think only in terms of the areas in which they live, but in other places people have learned to think globally in terms of One World. In times past some people were able to think in terms of large regions with large populations, but the One World perspective found among some people in modern societies has developed only in recent centuries. Despite this growth of a One-World, long-term perspective, the act of daily living, for most human beings, normally has a localized territorial limit and is concerned only with the immediate present.

The Familiar Local Area

Strongly built into human consciousness is the concept of "home"—a sense of territoriality, a feeling of security, and a set of psychological reactions oriented to some part of the earth. Home consists in the truly familiar area in which there are no unknowns (Fig. 1.4A). To the small child the home area may be only a few hundred feet square, whereas to the seasoned adult traveler it may encompass a very large area. Every human being has been "away from home" at some time and has felt the relief and pleasure of getting back.

The nature of home varies with the person and with the earth. The individual sense of home is compounded of several elements. The configuration of the land surface is one such element, but others lie in the shape and kind of plants that grow on the land, in the form and placement of buildings, in the nature and type of the decorative material applied to land and buildings, in the sounds, sights, and smells that belong to the area, and in its language, signs, speech, clothing, and the kinds of animals, big and small, that share its space.

At a distance from home lies the "neighborhood," the proximate area within whose range the composition of things on the surface of the earth is also well known (Fig. 1.4B). Here not every detail is precisely memorized. Changes can occur, or surprise elements may appear in time without upsetting equilibrium. To the child, the area involved may not be very large, but to an experienced traveler it may be sizable. As the child matures, the home area and the neighborhood tend to expand, (Fig. 1.4C), and as one travels increasingly, both home area and neighborhood grow, although remaining distinct, strictly local, territories. Home and neighborhood have for each of us a special kind of setting on the surface of the earth. To some, home and neighborhood must be in or on the hills, to others they must be on the flatlands. There are people oriented to the forests, to the open lands, to the watery surfaces, to the arid lands, to the rainy tropics, or to those parts of the earth having marked seasonal changes and snow in winter. The urban dweller who is ill at ease in the quiet rural scene has his counterpart in the unhappy villager in the great city.

A short stay in some new environment can easily induce a new sense of "home" and "neighborhood"; remain a few weeks in a different place and new definitions of "home" and "neighborhood" begin to emerge. One set of definitions may fully replace a prior set. Human beings vary greatly in these respects, from the person who never has more than a single home in his lifetime to the rootless wanderer who seems never to have any home at all.

The neighborhood tapers off into a transitional zone in which familiarity with the scene distinctly

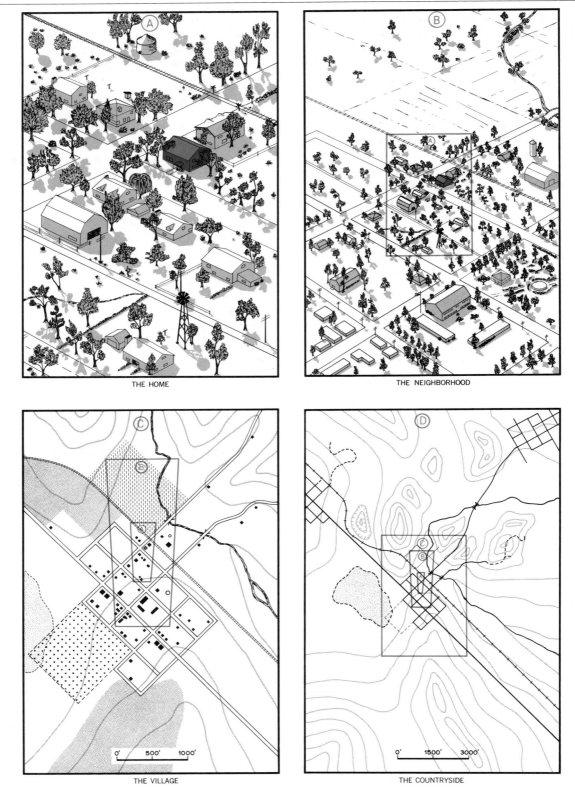

THE HOME

THE NEIGHBORHOOD

THE VILLAGE

0'   500'   1000'

THE COUNTRYSIDE

0'   1500'   3000'

Figure 1.4 Familiar Ground. (A) The Home (scale approximately 1:2000); (B) The Neighborhood (scale approximately 1:6000); (C) The Village (scale 1:16,667); (D) The Countryside (scale 1:50,000).

ABOUT THE MAPS IN THIS BOOK

Maps, tables, diagrams, and photographs are important parts of this book and should be given careful attention. The six clusters of photographs are accompanied by their own textual essays; the maps, tables, and diagrams, however, also must be "read," but they require using a different method that will take a bit more effort. Maps use symbols to represent, at a reduced size, selected aspects or parts of the real world. The information contained on a map must be interpreted according to the placements of and meanings assigned to the symbols. All complete maps include four aspects: title, scale, orientation, and legend.

TITLE

Each map in this book has a title, as stated in the figure caption, that indicates "what the map is all about."

SCALE

Each of the four sections of Fig. 1.4 on the opposite page is at a different scale. The largest scale is (A), the closest to reality in showing detail but portraying the smallest area. The home and the area of (A) also are shown in (B), at one-third the size and hence a smaller scale, which thus covers a larger area, the neighborhood of the home. Scale is the proportion between the map and the part of the earth it represents. Graphic or bar scales are indicated near the lower edge of the two maps, (C) and (D). Proportional relationships are expressed in the fractions set off by parentheses in the figure caption.

ORIENTATION

Customary usage places the direction of north toward the top of most maps, with either an arrow or lines representing the earth's grid of parallels and meridians used to "orient," or indicate directions to, the map user. Figure 2.3 (page 28) is a map of the world on which the earth's grid is drawn as intersecting lines over the oceans numbered at 15-degree intervals. The north pole is represented by the line at the top of the map, and the map reader can easily determine directions.

LEGEND

Symbols on a map are of three kinds: points or dots, lines, and areas. Each can take several forms and thus have multiple meanings. In (C) on the opposite page, the small black squares and rectangles that represent buildings in the village are dot symbols. Lines are used to portray streets and roads, a railroad track, a stream; and (in light gray) contours depict lines that indicate the shape of the land by connecting all points of the same altitude. Areas are covered with dots and dashes to represent orchards, plowed fields, growing crops, or, as at the left margin, the dry bed of a former lake. A legend, absent from the opposite page, is shown for Figure 2.3 (page 28) in which the meanings of the lines and areas, represented by different symbols and colors, are explained along the base of the map, just above the Figure number and title. You should read carefully the legends of all of the many maps in this book.

WORLD MAPS

The base map used throughout this book to depict the world, or its several continents, is standardized, as is Fig. 2.3, but at varying scales. The arrangement of the earth's grid used in this book is known as the Interrupted Flat-Polar Quartic Equal Area Projection. The importance of using such a projection has to do with area. In this book all text maps show areas in their true sizes relative to one another. The important quality of equivalence, meaning equal area, is essential to the indication of quantitative distribution. The end-paper maps are on a different projection but they, too, show areas in true size relative to one another.

lessens (see Fig. 1.4C). This may be the village as a whole, within which there are numerous familiar items: homes of friends, particular trees, stores occasionally visited, a playground, or a movie theater. Within this zone only some of the marks are well established in the mind, and there are unknown and unfamiliar blanks between these points. One is not lost here, ordinarily, but neither is one as thoroughly at home as within the neighborhood.

Beyond the village lies the countryside (Fig. 1.4D). There will be gross familiarity with the general scene in that one has been here before; true recognition and full familiarity, however, often are restricted to a few particular points one has learned to pick out to maintain a sense of location. One can get lost here at night, and changes through time deceive one as to one's precise whereabouts.

The Outside World

The transition zone between the home territory and the outer country (see Figs. 1.4C and 1.4D) is sometimes broad and gradual and sometimes sharp and sudden, but beyond this zone lies "the outside world," a distant, unfamiliar land where things are entirely different (Fig. 1.5). It is another sector of space inhabited by beings with conditions of another sort and quite unfamiliar.

The human being tends to wonder about the outside world, for it must be different from home and neighborhood. How distant it lies depends entirely on the human outlook. To the child it may begin within a few blocks, to the rural villager it can be the next village a few miles away, whereas to the seasoned traveler it may lie thousands of miles off. What makes the outside world different? The very landscape is unfamiliar in that landforms, plant growth, and climate are arranged in combinations unlike those at home, and the fields and crops, the villages and towns, the roads and transport consist of different mixtures. It is also likely that the daily living routine, the diet, the clothing worn, and the songs one hears are dissimilar. The language men speak may differ, and the people may themselves seem strange. The habits of living appear to be novel, the social framework is somewhat unfamiliar, and the political structures and organization of regions by which people live also seem different.

The Slowly Changing Character of Place with Time

When an adult returns to an old childhood home he is often startled to observe how near home was to the street corner or tree where he played as a child, although these are remembered as places that had been at the far edge of his neighborhood. Examination discloses that many things remembered are still there, but their relationships seem very different from the memory of them, and totally unfamiliar items have meanwhile intruded.

This review includes two kinds of changes that occur to areas on the earth's surface. The first is the scale differences that, with the passage of time, humans develop in reaction to areas, and the second is the actual slow changes that come about as areas fill up with human developments, attain growth or suffer depletion in vegetation, and suffer soil erosion as the cultural landscape alters. Both are relatively slow changes in terms of human chronologies but quite rapid in terms of the full history of man on the earth. Were a resident of ancient Athens able to stand today before the Parthenon and survey the city he had known as a child, he might well pick out a large number of familiar items, but contemporary Athens would contain many changes that have emerged slowly during the intervening centuries (see photo p. 301).

The dimensions of space have altered greatly within the limits of human history (in this volume "human history" refers to the whole sequence of human events from the beginning of mankind). The shortening process has not always operated at the same pace, but it has been continuous. People now living are well aware of how effective distance has shrunk within the last generation as modern high-speed transportation and communication have developed.

In the time of man on the earth, the slow process of landscape change has also been continuous, if uneven. As the numbers of people have grown greater, all over the earth, "vacant lots" have been filling in, wild forests have been shrinking, decorative and other planted growth has been expanding, crop fields have been enlarging their coverage, and cultivated landscapes have slowly replaced wild ones (Fig. 1.6).

Under the human hand all parts of the world have been slowly changing. Most of the regions of the earth have come to show the accumulating works in long-compounded patterns, whereas a few areas show accumulation for a period and subsequent abandonment. Such now-abandoned cities as Troy, Mohenjo-Daro, Jarmo, Machu Picchu, and Chaco Canyon show occupance for certain time periods only.

For any one portion of the earth's surface, elements of place and time have been interwoven into the changes that have taken place. For example, an early campsite on a natural levee amid the marshy lands near the mouth of the Mississippi River, where New Orleans now stands, formed a useful bit of dry land to early Indian occupants of the lower river country, but the potential water transport connec-

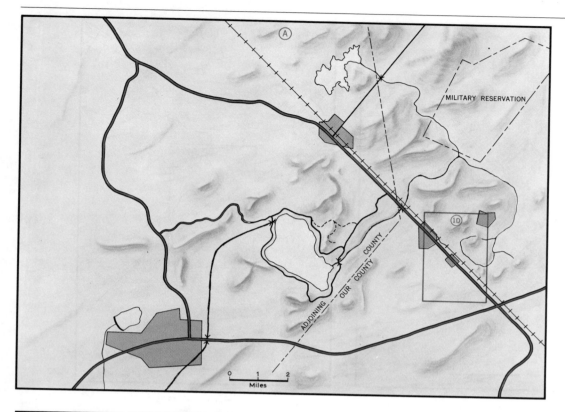

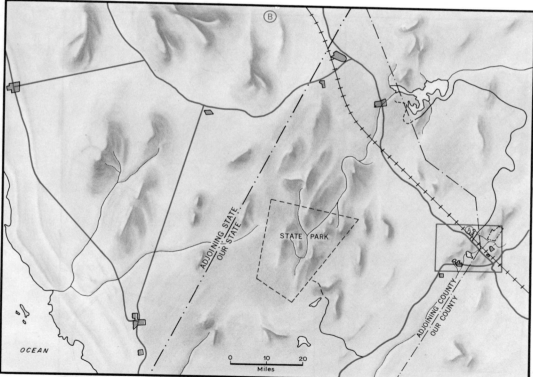

Figure 1.5 The Outside World. (A) The Next County (scale 1:200,000); (B) The Next State (scale 1:1,600,000).

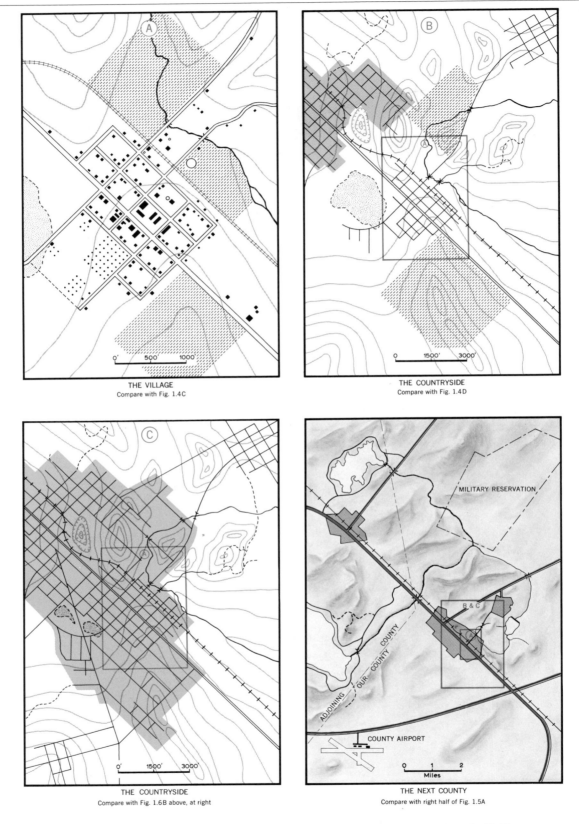

THE VILLAGE
Compare with Fig. 1.4C

THE COUNTRYSIDE
Compare with Fig. 1.4D

THE COUNTRYSIDE
Compare with Fig. 1.6B above, at right

THE NEXT COUNTY
Compare with right half of Fig. 1.5A

Figure 1.6 Change in Time and Place. (A) The Village becomes a Town (scale 1:16,667); (B) The Countryside as Suburb (scale 1:50,000); (C) The Countryside becomes a City (scale 1:50,000); (D) The Next County (scale 1:200,000).

tions with other such sites were of little consequence. Eventually the marshy delta became a refuge to another people, the Acadian French, who also lived along the levees as subsistence cultivators and hunted in the marsh. Later Andrew Jackson fought the British in the delta country, not for the dry levees and the marsh hunting, but for the potential connections of the place with other areas of the world.

The great river has changed its course in detail since that first campsite was used. People enlarged the dry land strip of the levee, drained some of the marshes, planted sugarcane, and progressively changed the use of the site. The river continues to alter the mouths of the delta, and people continue to expand the buildings, roads, canals, levees, and fields. Sites for chemical factories, oil fields, and petroleum refineries replace muskrat-hunting grounds. The connections with St. Louis and Pittsburgh have become important, and so have those with Panama, Singapore, Cape Town, and London. Man's concepts of space and time have been altered since the first Indian looked out over the marsh from a levee.

Abrupt Changes in Time and Space

Not all change on the earth's surface occurs slowly or in great detail. The physical record reflects numerous unexpected sudden changes. Major developments have taken place in the time that man has inhabited the earth, but in the long stream of human occupancy they constitute the exceptional and spectacular rather than the normal and average kind of change.

About A.D. 330 the course of the Tarim River, in central Asia, suddenly shifted, altering for centuries the course of the Great Silk Route. In 1883 two-thirds of the island of Krakatoa, between Java and Sumatra, disappeared in a volcanic blast that filled world skies with clouds of dust for months and created great oceanic waves that drowned thousands of people (such a wave is properly called a seismic wave, or a tsunami, and not a tidal wave.) In 1906 the course of life in San Francisco, California, was rudely jarred by a major shift of earth along the San Andreas Fault. Mankind is restricted in his reactions to such violent expressions of physical force, for little—other than accepting them—can be done. That we now appear to be making some progress in predicting both volcanic eruptions and major fault-line shifts does not alter the necessity for acceptance of such happenings.

Perceptions of Areas,
Places, and Conditions

One aspect of the changing nature of places and conditions derives from the patterns of human perception about the world, its parts, and its conditions. People all over the earth form mental "images" or "pictures" of "what it is like" in other places not visited but about which they may have heard little or much. These images are derived from their own ways of thinking, and the information is interpreted, or "filtered," through their concepts of what things should be like, based upon what they already know about such things. The result is a subjective image. If the information is scanty, out of date, or wrong, the resultant image may be quite different from the actual case, and it may require a long time to develop a correct image. Sometimes, if the information "filter" is colored by a particular pattern of thought, even correct information may be interpreted in a particular way that results in a distorted image. The subjective image may only very slowly be corrected through the receiving of new and more accurate information and the changing of the filter system of thought about what certain things should be like.

Individual human perceptions of distances are extremely variable, as are perceptions of height, area, volume, and numbers. The perceptions of "livability" of particular regions vary tremendously. Ideas about the frequencies and dangers of earthquakes, floods, landslides, tornadoes, and hurricanes vary greatly. In general, it seems the case that many people do not greatly fear, or think extremely dangerous, those conditions with which they live regularly, and they learn to cope with them.

In the first half of the nineteenth century, the first travelers to leave the eastern seaboard of the United States and to visit the high plains of Nebraska, Kansas, and eastern Colorado, perceived the region as the "Great American Desert," and they thought it an entirely useless region. Their perception of "good country" was "filtered" through an image of what good country ought to be: forests, numerous running streams always carrying water, and plants always green in summer. The seasonally dry high plains lacked forests; running water was infrequent; and many of the plants ranged in color from buff to yellow to brown throughout much of the year. The perception of the high plains as useless country took decades to correct. Few contemporary residents of Colorado east of the Rocky Mountains think of their home region as being a desert.

GEOGRAPHY AS A
SINGLE DISCIPLINE

This book deals with the geography of the earth from the viewpoint of cultural geography and, from what has been said so far, it is clear that the viewpoint is broad-ranging, concerned with long spans of time and with how human living systems have grown and developed on our earth. The viewpoint of the cultural geographer, however, is only one particular viewpoint within the discipline of geography as a whole. We close this first chapter with a short discussion concerning the nature of geography and the position of cultural geography within the larger discipline.

Geography deals with the surface of the earth and with the ways in which the human race lives on and uses that earth. The "surface of the earth" is understood to include the lower atmosphere surrounding our planet, the actual surface of the ground on which we walk, and the thin layer of crustal material that forms the continents, islands, and sea basins including the oceanic waters. This narrow zone forms the air-sea-land interface within which all earthly life occurs. Here, conditions differ from place to place, and many different processes have been at work for a very long time to produce the present aspects. Geography is the only body of knowledge whose primary focus is directed at the earth's surface as it has been occupied by mankind. In the larger sense, geography is interested in everything that goes on within the air-sea-land interface, because geography is concerned with the spatial and process aspects of *both* the interface itself *and* with the activities of all forms of life. The above can be shortened to say that geography is concerned with knowledge about the earth and mankind, or to say that geography is a study of the earth as a human habitat.

Knowledge has to do with human learning, all kinds of learning. Some disciplines now deal in specialized ways with the knowledge of material things above, on, or below the surface of the earth. Knowledge about the art and music, or even the agricultural systems, of different peoples cannot be dealt with in the same ways as knowledge about the movement of the earth's crust or the advance and retreat of glacial ice sheets. As more and more knowledge has accumulated, it has been divided into segments, with particular names attached to each. These segments form the many separate disciplines that we customarily group into the physical sciences, the life sciences, the social sciences, and the humanities. Geography, in dealing with the earth itself and with all forms of life on the earth, has parts of its subject matter in each of the four groups of disciplines.

Physical Geography, Biogeography, and Human Geography

Since the earth involves nonliving and living matter, and since mankind has set itself off from other forms of life, the discipline of geography today involves a threefold subdivision, each having its own specific orientation. One of these, *physical geography* (allied to the physical sciences), studies the physical phenomena and processes relating to the air, water, rock, and other materials that can be described, measured, mapped, photographed, and sensed by remote controls in spatial and process terms. The second, *biogeography* (allied to the life sciences), examines all nonhuman animal and plant life forms, and the related physical and biotic processes. The third subdivision, *human geography* (allied to the social sciences and the humanities), studies the living systems of the human race as it occupies the earth.

From the mid-seventeenth century through the early twentieth century, physical geography was considered to be the primary and most important phase and received the greatest share of attention. Biogeography was thought to be an aspect of physical geography. Wild animal life was almost totally ignored, however, and plant life was lumped together as "vegetation," and paid scant heed. Human geography was chiefly concerned with material culture, or the things that people built or laid out on the earth, such as roads, dwellings, field patterns, and boundary systems. Most geographers then considered any unoccupied area as a "natural landscape," the good parts of which could be made productive through the removal of dangerous animals and all wild plant life, and through the replacement by "development" of crop fields and settlements. Most geographers then also considered that the patterns of human occupance, and the levels of life attained, were controlled or at least strongly influenced by physical environmental conditions, chiefly landforms and climate.

Since the early twentieth century, however, there has been increasing recognition of the role of human technologies and the power of human societies to develop occupance patterns according to human choice. There has been a steady increase, too, in the recognition that human societies can change the face of the earth, for good or for detriment. These changing viewpoints have affected each of the subdivisions of geography. The position taken in this book on the question of physical environment versus culture was stated on an earlier page (see page 8).

All three geographic orientations have varying numbers of specializations and regional focuses (Fig. 1.7). Physical geographers divide themselves according to the specialized fields of study, such as geomorphology, weather elements, climatology, soils geography, or water resources. Others focus their studies on the glacial or arid-land aspects of landforms, on bioclimatology, and so on. Biogeographers divide themselves into plant geographers, animal

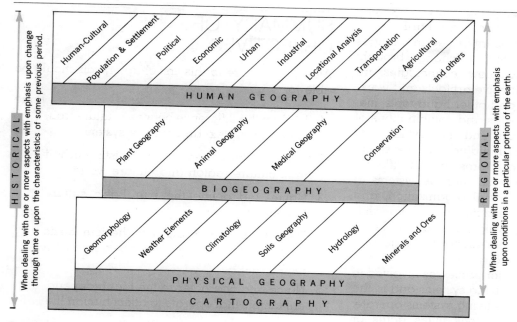

Figure 1.7 The Orientations and Primary Subdivisions within Geography. Basic to all aspects of geography is cartography (mapmaking), which includes all forms of representing graphically all or part of the earth—the physical, living, and human processes and features of its surface.

geographers, medical geographers, or conservationists, and they sometimes further specify a regional interest, as in the humid tropics, the arid lands, the high mountains, or a particular continent. Human geographers divide themselves into more categories than do the others. They may be concerned with human or cultural geography in general, or they may specialize in population, settlement, political, economic, urban, or industrial geography. Transportation, locational analysis, and agricultural geography are examples of other subdivisions, and as the subject matter is more fully studied there are other newer specialties developing.

Each of these subdivisions of geography has its own particular subject matter and precise methodology, just as do the specialties within the broad field of medicine. Each contributes to the broader understanding of how people live upon and utilize the earth and its resources. Although there are all these specializations within geography today, the discipline of geography comprises a single field of study out of its primary focus on the surface of the earth as humanity's home. All geographers share broadly the same outlook toward the earth and all, ultimately, are concerned to varying degrees with the life of *Homo sapiens sapiens* on this earth.

Cultural Geography as a Branch of Human Geography

Within the third orientation, human geography,

there is a subdivision known as *cultural geography*, and this is the specific orientation of this book. In the thinking of some geographers human geography and cultural geography are one and the same. For others, the two are no more synonyms than geomorphology is a synonym for physiography. We view human geography as of much broader scope than cultural geography. We cannot pretend to deal in depth with all aspects of humanity or with all elements of human culture that are dealt with by the specialists within the subject of human geography. Many of the specialists in some subdivision of human geography take a *synchronic* view of human life on the earth (occurring at the same time), a view that often is focused on the contemporary scene. Cultural geographers usually take a *diachronic* view of the development of human activity (occurring in succession through time). They view the changing development of landscapes, culture systems, and technological traditions as requiring long periods of time for growth to maturity.

Cultural geography is concerned specifically with systems of human technology and cultural practices as these are developed by societies in particular regions of the earth through time. Human practices have been altered through time by human inventiveness, and constructs placed on the earth by human builders have had their forms and functions altered over time. Mankind is an active agent on the surface of the earth, operating in regional systems of culture. These facts lead the cultural geographer to some form of evolutionary and ecological systematization of the human occupance of the earth.

The cultural geographer is interested in changes in technology, and in the kinds of change brought into environments in which new technologies are applied. These concerns involve artifacts and working systems such as hoes and other hand tools, mechanical tool systems, species and combinations of crop plants, kinds of looms and the textiles they produce, and the styles of building and the settlements they produce. Also of interest are the differing and changing social elements such as systems of marriage, structurings of social organization, and the kinds of economic organizations. Such specifics are diagnostic criteria by which to measure development, cultural transfers, evolutionary change, and creative skills in the regionally variant culture systems. There is vital concern for processes involved in both the change in human living systems and in the environments in which those living systems operate. There is but little concern for tons, miles, acres, or other sheer vital statistics of contemporary occupance of a landscape, since these belong to the realm of economic geography. Similarly, there is little concern for some of the things that belong in the interpretive toolkit of the specialties of population, political, or urban geography. There is but slight concern for the functional geometry of interpretation of the strictly contemporary scene. There are, instead, recurrent questions as to "What?" "When?" "Where?" "How?" and "Why?" The chief goal of the cultural geographer is the comprehension of the workings of culture systems in spatial units throughout the time span in which humankind has occupied the earth. Our concept of culture was set down on page 3.

With this chapter as background orientation, we turn now to the examination of the world that the earliest men and women entered.

SUMMARY OF OBJECTIVES

After you have read this first chapter, you should be able to:

	Page
1. Explain the difference between culture trait, culture complex, and culture system.	2
2. Define culture hearth, and distinguish it from culture region and culture world.	3
3. Name and define the four basic culture processes.	4
4. Describe and diagram how the four basic culture processes are interlocked in actual occurrences.	5
5. Name and characterize the four independent factors that operate to affect human life.	6
6. Diagram the six interrelationships by which each culture system functions.	6
7. Distinguish between the two points of view, teleological and secular.	8
8. Draw a map of your earliest remembered home area, and then draw the limits of the neighborhood and where the outside world began.	10
9. Illustrate, from your own experience, a perceived image of some area that turned out to be wrong when you visited it.	15
10. Explain what the phrase "surface of the earth" really includes.	16
11. Identify the three subdivisions of geography, and describe what each generally includes.	16
12. Discuss the basis for the cultural geographer's concern for both space and time.	17

CHAPTER 2
PLEISTOCENE HABITATS AND THE RISE OF HUMANKIND

No realistic understanding of humanity's tenure on earth is possible until it is recognized that the earth undergoes change independently of human activity. To provide a background for the understanding of present-day relationships, this chapter deals principally with what happened during most of the several million years that mankind has been on the earth. First, the chapter describes the greatly changed physical conditions that, prior to mankind's appearance, were so very different from those of the present. Second, it discusses the effects of these changing circumstances on the life-forms that are significant to human beings. Third, it explains the evolution of the key attributes of our human ancestors. Finally, it tells how, through the invention of language, toolmaking,

control of fire, the use of shelter and clothing, culture emerged to distinguish mankind forever from other evolving life-forms.

MAJOR CHANGES ON THE EARTH PRECEDED HUMANKIND

The belief that the land is unchanging is a myth of long standing. Study of a globe, or a map in which continents are drawn in their true shapes, will show, however, that some continental slopes (outer edges) can be fitted together as in a huge jigsaw puzzle.

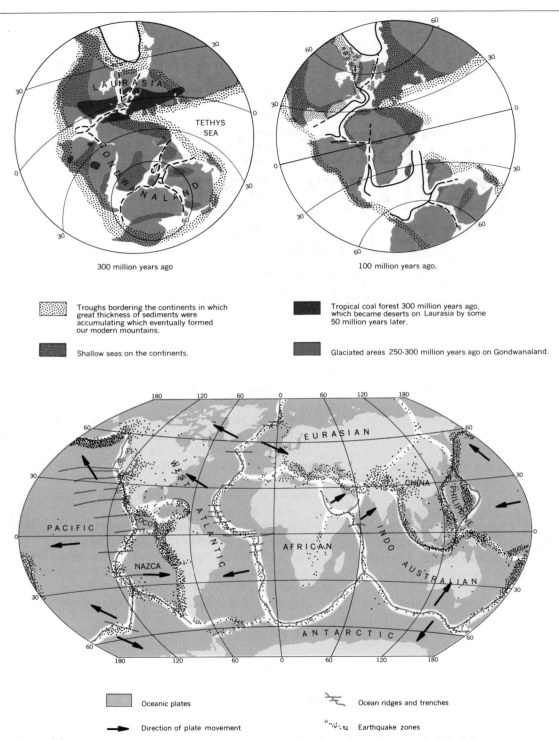

300 million years ago

100 million years ago,

Troughs bordering the continents in which great thickness of sediments were accumulating which eventually formed our modern mountains.

Shallow seas on the continents.

Tropical coal forest 300 million years ago, which became deserts on Laurasia by some 50 million years later.

Glaciated areas 250-300 million years ago on Gondwanaland.

Oceanic plates

Ocean ridges and trenches

Direction of plate movement

Earthquake zones

Figure 2.1   Changing Positions of the Continents. Top: Gondwanaland, in general, drifted from southern subpolar to equatorial regions, and Laurasia gradually drifted north. Bottom: Present location and directions of movement of plates, oceanic ridges, ocean trenches, and earthquake zones.

What we now know from the study of the earth is that the continents were once together, about 230 million years ago, but have since separated very slowly and have drifted apart to form their present arrangement on the globe.

Plate Tectonics Created Continental Drift

Research and debate over the last sixty years have resulted in the concept of continental drift and, more recently, in explanation of crustal movements by means of plate tectonics. The crust of the earth consists of six large portions or plates, and a dozen smaller ones that are in intermittent but very slow motion (Fig. 2.1). These units move or drift a few inches per year. They ride on a semiliquid zone beneath the crust, with the continents locked into some of the plates, so that they are carried along by them. As the plates shift, continents are moved together or torn apart, and ocean basins are enlarged or made smaller. During these events, earthquakes, volcanic eruptions, and the opening of trenches in ocean bottoms occur where and when ruptures exist in the earth's crust.

The crust of the earth is only 3 to 5 miles thick beneath the floors of the ocean basins, but it is from 13 to 35 miles thick beneath the continents. Below the crust the semiliquid zone, from 40 to 150 miles in depth, is weak, and is subject to plastic flow. This is known as the asthenosphere. Here movements take place that very slowly deform the crust and reshape the surface of the earth. Ocean floors are spreading (Fig. 2.2, right), and new oceanic crust is being generated along the "mid-ocean" ridges, which are the cracks opening between the plates as these are being pulled apart. The older parts of the crust are being melted and reabsorbed as they are pushed

down in the deep trenches at the bottoms of the oceans along the edges of the mountain and island arcs that form the edges of plates (Fig. 2.2, left).

The ocean floors are all relatively young in terms of earth history (Table 2.1). None is older than the Upper Jurassic (Age of the Dinosaurs), and all are being constantly recycled. By contrast, rocks on the continents are up to twenty times as old. A worldwide relationship exists on the sea floors, since the newest oceanic crusts occur on either side of the "mid-ocean" ridges. The age of these new units of crust increases away from the center, with the oldest crusts at the farthest distance away. The "mid-ocean" ridges are spreading at rather uniform rates and the new sea floor, as it is being generated, is being imprinted like a magnetic tape whenever the magnetic field of the earth is reversed.

The form of the present system of "mid-ocean" ridges is a nearly complete circular ring through the Antarctic Ocean, with branches northward through the centers of the Atlantic and Indian oceans (see Fig. 2.1, bottom). The Pacific ridge is not in mid-ocean, but occupies an off-center position owing to the westward drift of the American continents. The East Pacific Rise of this ridge in its northern extent has been overrun by westward-moving North America.

Plants and Animals Evolved Differently on Separating Continents

The continents have been moved long distances during the last 200 million years (see Fig. 2.1). They

Figure 2.2 The Concept of Sea-floor Spreading, shown schematically. At Right: New ocean floor is generated along a mid-ocean ridge. At Left: Old sea floor descends beneath a continent or island arc creating an oceanic trench, and a downsloping zone of earthquakes and volcanic activity—the "rim of fire". (After Sullivan, 1974, p. 69)

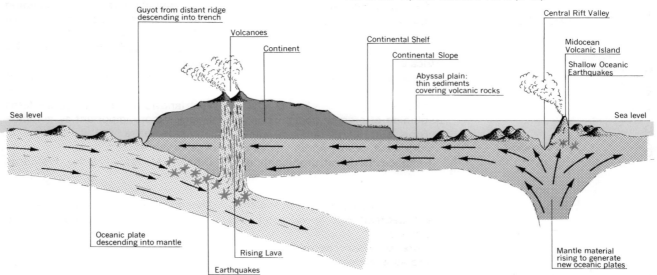

Table 2.1 Time Chart of Earth History

Era	Period (Years before Present) in Millions	Continental Movements	Chief Developments in Life-Forms
CENOZOIC / QUATERNARY	HOLOCENE (Recent)		Humankind dominant on earth; birds in air; fishes in ocean; competing with insects on land and in air
	—0.011—		
	PLEISTOCENE	Latest Ice Age	Evolution of *Homo sapiens;* extinction of some mammals
	—2—		
CENOZOIC / TERTIARY	PLIOCENE	Arctic ice begins forming Major uplift of Rocky Mountains Baja California separates from mainland	Mammals able to migrate between the Americas
	—7 ± 1—	Antarctic ice sheet reaches maximum size	
	MIOCENE	Isthmus of Panama rises from the sea Mediterranean isolated and dries up Rapid growth of Himalayas Red Sea begins to open Africa moves northward producing Alps	Early hominoids spread into Asia and Europe Apes develop in Africa and India
	—26 ± 1—		
	OLIGOCENE	North America overrides Pacific plate, altering mountain formation in west	Rapid rise of mammals
	—37 ± 2—		
	EOCENE	Intense volcanic activity in western North America India begins impinging upon Asia Australia begins drift northward from Antarctica	All main groups of mammals exist
	—58 ± 2—		
	PALEOCENE	Land bridge between the Americas submerges	Rise to dominance by mammals Widespread expansion of flowering plants
	—65 ± 2—		
MESOZOIC	CRETACEOUS	New Zealand separates from Antarctica North America and Europe separate India breaks from Australia and Antarctica Africa and South America separate in breakup of Gondwanaland	Earliest primates, in North America Mass extinction of reptiles, including all 15 dinosaur genera, and two-thirds of all coral genera First birds; mammals small and inconspicuous Angiosperms become very prominent
	—135 ± 5—		
	JURASSIC	Andes forming Beginning of present ocean basins Europe and North America move northward from Africa, splitting Pangaea the great protocontinent	Rich land flora: conifers, cycads, ferns Reptiles more important as amphibians decrease Age of dinosaurs Dinosaurs; first lizards; first true mammals
	—190 ± 10—		
	TRIASSIC	First rifting between Gondwanaland and Laurasia with "pivot point" at Gibraltar Closure of ocean between Europe and Asia; formation of Urals	Mammal-like forms from reptiles Reptiles spread throughout the world
	—230 ± 10—		

Table 2.1 (*Continued*)

Era	Period (Years before Present) in Millions	Continental Movements	Chief Developments in Life-Forms
P A L E O Z O I C	PERMIAN ————280 ± 10————	All continents reassemble to form Pangaea, the great protocontinent, in final closure of ancient Atlantic Second Ice Age occurred on all southern continents (Gondwanaland)	Modern insect groups appear
	PENNSYLVANIAN ————310 ± 10————		Rich swamp flora; first conifers, formation of widespread coal deposits
	MISSISSIPPIAN ————345 ± 10————	Development of the Appalachians	Earliest reptiles Earliest land vertebrates
	DEVONIAN ————400 ± 10————		Conquest of land by plants and animals; first amphibians; first insects Age of fishes
	SILURIAN ————425 ± 10————	Enormous mountain-building	First true fishes Age of reef corals; first land plants
	ORDOVICIAN ————500 ± 10————	Ancient Atlantic begins to shrink	First vertebrates Free oxygen in atmosphere; higher proportion of CO_2
	CAMBRIAN ————600 ± 50————		Atmosphere clears; sunshine reaches earth First shelled invertebrates
	1500———	First Ice Age Several cycles of mountain building	Dawn of life in water First oceans, without life
	PRECAMBRIAN 4.7 billion	Laurentian revolution resulting from thermal convection with heat generated by decay of radioactive elements Accretion from a dust cloud, as a cold body without life	Anhydrous phase; CO_2 + CO added to primitive atmosphere

have been in motion intermittently and have had different relative positions and shapes and sizes at different times. The single landmass of Pangaea began to break up in late Triassic time, splitting into Laurasia in the north and Gondwanaland in the south. Early in the Cretaceous period, Africa and South America separated in the breakup of Gondwanaland and the Andes Mountains began to form. Later in the Cretaceous, India broke apart from Australia and Antarctica, and near the end of the Cretaceous, North America and Europe separated.

Australia and Antarctica were connected until 50 million years ago, in the Eocene epoch. After Australia separated, and as it drifted northward, its climate became increasingly warmer and more hospitable, allowing the evolutionary development of plants and animals that could not have developed in Antarctica. In the mid-Eocene, India began shoving against southern Asia. North America overrode the main Pacific plate in the Oligocene, creating mountains systems in the west. Beginning with the Miocene epoch, the northerly motion of Africa created the European Alps, following which the Red Sea began to expand and the Antarctic ice sheet began to form.

The end of the Miocene saw the spread of early hominoids (animals resembling humankind) into Asia and Europe, the rapid growth of the Himalaya Mountains, the drying up of the Mediterranean Sea, the rise of the Isthmus of Panama from the sea to connect the two American continents, and the maximum development of the Antarctic ice sheet. The

In this photograph of India and Ceylon (lower right) taken from the Gemini XI spacecraft, the earth looks like a globe. There is considerable cloud cover. From this distance, even the Tibetan Highlands, seen on the horizon, produce no obvious bulge in the earth's form.

major uplift of the Rocky Mountain region occurred at the end of the Pliocene, at the same time as Arctic ice began forming.

As the positions of continents and oceans changed, there were considerable alterations of climate. These changes led to major migrations, extinctions, and patterns of accelerated evolution in many plants and animals. As large landmasses pulled apart, a great increase in the length of coastlines resulted. Climates of these landmasses then were modified by the size and location of any nearby water bodies. For example, the increasing closure of the Isthmus of Panama altered the ocean currents, and the great equatorial current across the full width of the Pacific and Atlantic was changed to separate circular systems in each ocean. Warm ocean currents moving poleward created new storm paths, delivering more warm, humid air to central North America and northern Europe. In winter this resulted in the heavier snows and when the accumulation exceeded the summer melt, large continental glaciers developed.

The separation of areas formerly joined led to discontinuous distribution of similar plants and animals. These life-forms became less and less similar as time passed. On areas that were isolated the longest, plant and animal forms peculiar to the region evolved, such as the marsupials, the group of mammals that developed in the isolation of South America and Australia. Today, Antarctica is the only continent completely encircled by spreading ocean floors and is, therefore, the only one from which all others are moving away.

Continents have been brought together as well as carried apart. The uplift of such mountains as the Urals, Alps, and Himalayas are all believed to be the product of continents coming together again. Australia drifted northward against Southeast Asia, just as India "collided" with southern Asia during the past 50-odd million years. Such coming together reduced and eliminated old areas of sea and caused the development of quite different climates. These, in turn, put great biotic pressures upon plants and animals in the merged areas, leading to increased competition among the forms that came into competition with each other. The result has been both the more rapid extinction of some species and accelerated evolution in others. For example, only 20 orders of reptiles evolved in their two original homes, Gondwanaland and Laurasia, during the 200 million years from the Permian through the Cretaceous periods. Yet in the 65 million years since then, mammals have diversified on eight continents to produce 30 orders. Later continental merging resulted in 13 orders of land mammals becoming extinct. This enormous diversity of the mammals was a direct result of the breakup and drifting apart of the two supercontinents. As with other mammals, it is probable that mankind could not have come into existence without this process of continental breakup and merger. Knowledge of continental drift helps decipher the migration and dispersal patterns of plant and animal groups across space and through time.

Angiosperm Plant Life Became Dominant

Among the roughly 350,000 known and classified living species of plants, only about 700 belong to the Gymnospermae, but more than 250,000 belong to the Angiospermae. The gymnosperms include the coniferous forest trees of the earth, those that provide much of the commercial timbers, firewoods, and wood pulps. The angiosperms, or flowering plants, include almost all garden varieties as well as leafy trees and shrubs; they seem to have come into existence rather suddenly in the Cretaceous period about 100 million years ago (see Table 2.1).

Able to adapt quite readily to all aspects of environmental change, the angiosperms also developed superior abilities in both reproduction and wide-ranging distribution. They represent the climax of plant evolution to date. By their particular characteristics, they produced new kinds of vegetation that spread widely over the earth to become the dominant forms in the Cenozoic era of the last 65 million years. The evolution of the angiosperms appears to have taken place in climatically mild regions with seasonal aridity. The equatorial rain forests of today are more a ''museum'' for ancient forms of angiosperms than the ''birth cradle'' for their development, although the issue is still under debate.

Although many angiosperms are woody perennials, large numbers of them developed into biennials, annuals, and herbs. The angiosperms differentiated both aboveground in stem, leaf, and seed features and belowground in root characteristics. Important aboveground features are the deciduous leaf-shedding that occurs with frost or drought and the fruiting-seeding that makes for rapid reproduction. An important belowground feature has been the root modification by which food energy becomes stored in the fleshy ''roots.'' This, in turn, permits reproduction by vegetative propagation, which is achieved by cutting or breaking off a piece of the root or tuber and putting it back in the ground.

The seasonality of the nonwoody plants, the fruiting-seeding, and the root storage features were of primary importance for the development of animal life. The evolution of a wide variety of leafy grasses and herbs, and the many kinds of seeds, fruits, nuts, tubers, and edible flowers made possible the evolution of many new kinds of bird and animal life. The plants of seasonal growth habit, producing nutritious crops of seeds and soft leaves, greatly aided the evolution of many of the ''herd animals'' familiar to us today. These same characteristics provided mammals with a useful and widely distributed food supply, and made it possible for early humans to become gatherers of seeds, fruits, nuts, roots, and tubers.

Mammals Rose to Dominance and Spread Widely.

About 6000 species of mammals, including mankind, exist on the earth. Mammals originated in the Triassic period (see Fig. 2.1), but remained few and insignificant until the beginning of the Tertiary period, about 65 million years ago. The mammals dispersed over the earth and evolved into many species during the Tertiary. Today the northern limits of land mammals are the limits of land at the higher latitudes. Sea mammals, such as the whales and seals, live even farther north. In the south mammals inhabit the southern ends of the primary landmasses. Habitable islands of the continental type always have mammals, but they usually are fewer in kind than those of the adjacent mainlands, since the water barrier has a screening effect on migration.

Most families of mammals are less than worldwide in distribution. Mankind shares worldwide distribution only with three families of bats, some mice, and some rats. All other existing families of land mammals are more limited in distribution, and are strongly influenced by climatic zoning. The tropics contain the most species and have the most widely distributed families. The temperate zones are less diverse, and the Arctic is still less diverse. This diversity is primarily a matter of subtraction: temperate faunas are essentially tropical faunas from which many families have been subtracted; arctic faunas are the result of even more subtraction by climatic influences.

Direction in the dispersal of mammals (the net result of adding and subtracting two-way exchanges) has been from the Eurasian zone of the Old World toward Australia and North America, and from North to South America. Excluding human intervention, only one group (*Phalanger*) has slightly penetrated the Eurasian zone from Australia, and no South American groups have reached the Old World. Since the greatest numbers of mammal faunas are tropical, the primary zone from which mammals dispersed undoubtedly was the African-South Asian section of the Old World.

THE PLEISTOCENE AS A UNIQUE EPOCH

The period during which humans became the dominant species on the face of the earth takes in the last two or three million years. To this epoch the English geologist Sir Charles Lyell in 1839 gave the name *Pleistocene*—most recent (see Table 2.1.). The Rise of Mankind occurred amid profound changes in physical aspects of the earth and amid enormous changes in the distribution of plants and animals. Mankind was witness to extensive and repeated glacial action, with huge ice sheets occupying as much as three times the area occupied by glaciers today. The climates of the great zones over the earth changed markedly in temperature and precipitation: snow lines dropped and lifted; sea levels all over the earth fell and rose more than 500 feet; ground in the northern latitudes froze and thawed. As a result, great lakes grew and died away, rivers altered courses, and whole populations of plants and animals shifted locations. Thus, the long era of human emergence from other biological forms and the great time span of human prehistory is linked with the whole of the Pleistocene record.

An appreciation of the earth's changing character during the Pleistocene epoch is vitally necessary for an understanding of mankind's rise to dominance on the earth. Our present landscapes for the most part are not solely the products of the physical and biotic processes or forces now acting upon them. Rather, much phenomena whose distribution we seek to understand have their own histories of having been molded or shaped or influenced by the changing forces acting during the Pleistocene. The earth that mankind inherited was a Pliocene earth: human development coincided with the changes of the Pleistocene. Thus an examination is necessary of those characteristics of the Pleistocene epoch that were of fundamental significance for the successful emergence of our human ancestors.

The Pleistocene or "near-recent" epoch has, in many respects, been the most unusual period in the history of the earth. During the last two million years, the long, gradual evolution and dispersal of plants and animals was abruptly disrupted by several kinds of related, important factors, such as:

1. *Isolation*, by upheaval of mountain barriers and through climatic change or separation of islands from continents by rising sea levels,
2. *Extinction*, by failure to adapt to changing climatic conditions or failure to migrate with shifting life zones, and
3. *Succession*, the replacement of old groups by the spread of new dominant forms.

For example, the overall evolutionary trend during the Pleistocene was a gradual withdrawal and extinction of large mammals, a process that is still going on. There were many more kinds of mammals in existence during the early Pleistocene than survive today. For example, among those that have become extinct and part of the Pleistocene fossil record are: (in Europe) the straight-tusked elephant, cave bear, cave lion, woolly mammoth, woolly rhinoceros, and saber-toothed cat; (in North America) all the camels, all the horses, all the ground sloths, giant bison, giant beaver, stagmoose, several cats including the saber-toothed, several mammoths, and mastodon. As the predominant form of life in this world, mammals are definitely declining in numbers and varieties through naturally evolutionary processes.

Mountain Building and Climatic Change

One of the unusual features of the present earth is the widespread existence of high mountains. For most of earth history the landforms consisted of plains or low rolling hills, which are thought of as seldom exceeding 1000 feet in altitude. The first outlines of the mountain systems we recognize today began to take shape during the Miocene.

The unusual character of the last two million years was ushered in by a series of mighty upheavals that elevated the previous lowland landscapes by 10,000 feet or more. Whether in New Zealand, Switzerland, the South American Andes, or the Asian Himalayas, the resulting mountain systems are everywhere high, and they were carved into deep, narrow valleys only shortly before the Pleistocene. Too little time has elapsed for these mountains to have been reduced to lower or softer forms.

The elevation of high mountain chains around the earth altered many specific aspects of the climates of the earth, and produced a greater diversity between wet and dry, and hot and cold, regions on the earth's surface. In many locations these mountain systems are sufficiently high to deflect the general sweep of the air masses in the lower atmosphere. The specific paths taken in many regions by near-surface movements, as prevailing winds, is new to the earth. The locations of regions of heavy rains or snowfalls, on the front sides of mountains against which prevailing winds now blow, and the locations of arid regions, behind high mountains, result from the present unusual arrangement of the earth's mountain systems. Similarly, there now are basins newly sheltered from cold polar air masses. The precise locations of many distinctive climatic regions, such as the mild highland zones near the equator, are new to the climatic systems of the earth.

Glacial and Interglacial Stages

In what is termed a glacial period, the annual volume of snow that fell in polar and high mountain regions was greater than the amount of annual summer melt, and as the snow accumulated it solidified into large masses of ice, termed glaciers. During an interglacial stage there was more melting than accumulation, so that the areas covered by ice became smaller. Our present period, the Holocene, is an interglacial stage. Nearly 85 percent of the area now covered by glacial ice (Fig. 2.3) is in Antarctica (4,862,200 sq. mi.) and more than 11 percent (666,200 sq. mi.) is in Greenland. Most of the present glacial ice occupies polar regions where ice persists because of the cold climates. The rest of the earth's present glaciers (see Fig. 2.3) constitute less than 4 percent of the total area of glacial ice.

Following an early and prolonged phase, the Pleistocene culminated in a sequence of three ice advances, glacial stages, of different extent and duration. These stages were separated by longer periods of warmer time, called interglacials, when temperatures were at least as high as at present. During the Pleistocene, glacial expansion took place chiefly in the high middle latitudes, and glaciers accumulated to cover three times their present surface. The largest areas at maximum extent (see Fig. 2.3) were:

1. North America (6,033,400 sq. mi.), where the Laurentide ice sheet merged with the coalescent Cordilleran ice to cover an area larger than present Antarctica and Greenland combined.
2. Antarctica (5,093,700 sq. mi.).
3. Scandinavian ice sheet (2,300,000 sq. mi.), combined with glaciers of British origin.
4. Siberian ice sheet (1,627,000 sq. mi.).
5. Greenland ice sheet (833,500 sq. mi.).

The continental glaciers of the Pleistocene were unequally distributed over the world's continents. Nearly 35 percent by area lay on North America. There was very little glaciation in the tropical or equatorial regions or in the southern continents, except for Antarctica.

In addition to the large ice sheets there were a great many local areas at high altitudes in which small glaciers accumulated. At times an ice sheet filled part of a valley, blocked the normal drainage outlet, and forced the melt water to spill over and cut a new course around the ice sheet. One such temporary diversion of the Columbia River in the state of Washington formed the Grand Coulee. Another example is the formation of the course of the middle and upper Missouri River across the Dakotas, where former eastward-flowing streams were diverted and permanently connected southeastward to become tributary to the Mississippi River.

As the Pleistocene great ice sheets gradually wasted away by melting, there came to be exposed two new kinds of landscapes very significant to the contemporary world. These are the icescoured plains, hills, and lower mountains reduced to pitted (now lake-strewn) hard rock surfaces, and the glacial drift plains, on which the eroded sediments were deposited in a wide variety of minor landforms. Such surfaces make up significant portions of North America and northwestern Europe. Smaller areas of glacial ice locally reproduced both the scoured and drift surfaces in many areas of the earth. The present limit of frozen subsoil or bedrock represents a lingering distribution of a former, more widely spread distribution. Permafrost (permanently frozen subsoil) still accounts, however, for 25 percent of the world's land surface in continental areas with long, severely cold winters and vitally affects many human activities in such areas.

The Pleistocene ice sheets caused unusually strong and continuous winds to blow outward from the ice margins. Fine materials were picked up by these winds and transported variable distances. The coarser sands were deposited in close proximity to the glacial outwash margins, forming sand dunes. Finer particles were carried much farther and deposited in thick beds of loess. Loess is very fine-grained silt and dust transported by wind and deposited in layers of new surface material that cover up the older, underlying landforms (see Fig. 2.3). Huge volumes of loess were deposited in thick beds of 50 to 500 feet during and after the Pleistocene glacial stages. At later dates much of this material was removed and redeposited by streams.

Loess deposits are widespread in central North America, and are scattered over central and eastern Europe. A wide but discontinuous band of sand and loess stretching from the Caspian Sea to northwest China seems related both to glacial outwash and to wind action on the sometimes bare and arid expanses of central Asia. Much of the North China loess has been reworked by streams. In general, loessial soils have been among the more easily worked and more productive soils of the earth, a fact significant to the whole history of crop growing.

Arid and semiarid regions in the interiors of all the continents, except Antarctica, contain many saline lakes and dry basins of numerous extinct lakes. During the growth of the glacial ice sheets the belts of eastward-moving storms shifted progressively toward the equator. Increased storminess produced increased rainfall. Greater cloudiness, combined with lowered temperatures, reduced the evaporation rate. As a result, surface runoff created many large freshwater lakes in previously dry basins. Such temporary cooler and wetter (pluvial) conditions are most evident in the subtropical deserts, as in western North America, north and east Africa, and Southwest Asia. Alternation of pluvial and dry stages occurred repeatedly all over the earth. Pluvial stages are considered to coincide with glacial advances,

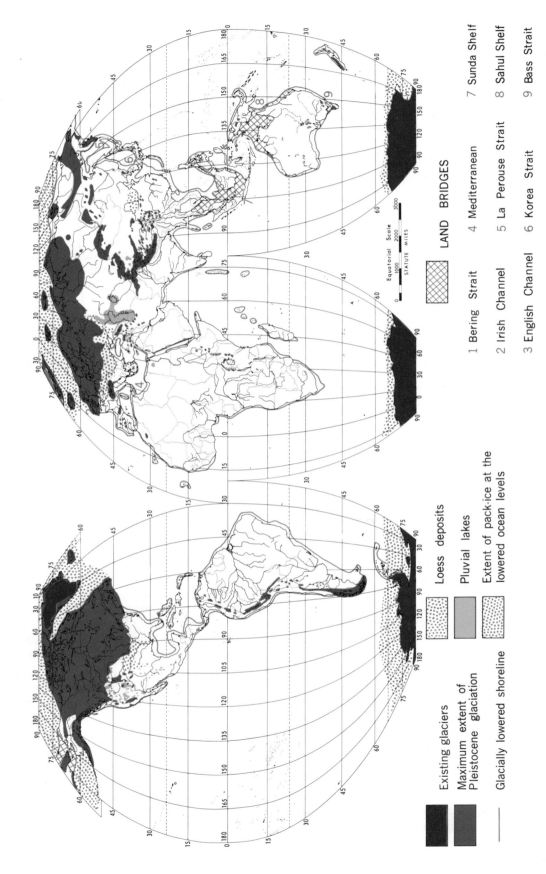

Existing glaciers

Maximum extent of
Pleistocene glaciation

Glacially lowered shoreline

Loess deposits

Pluvial lakes

Extent of pack-ice at the
lowered ocean levels

LAND BRIDGES

1 Bering Strait
2 Irish Channel
3 English Channel
4 Mediterranean
5 La Perouse Strait
6 Korea Strait
7 Sunda Shelf
8 Sahul Shelf
9 Bass Strait

Equatorial Scale
0   1000   2000   3000
STATUTE MILES

Figure 2.3 Existing Glaciers and the Extent of Pleistocene Glaciation.

whereas dry interpluvials occurred during interglacial stages.

The glacial (pluvial) stages consisted of periods that included both expansions of vast ice sheets and widespread growth of lakes on the continents. When so much water had been evaporated from the sea surface and was "locked up" on land in a frozen state or trapped in closed basins, normal runoff of surface waters back to the sea was reduced. Consequently, the level of the sea was conspicuously lowered, resulting in worldwide emergence of continental shelves. These changes in sea level (falling when ice and lakes expanded and rising again as glaciers melted and lakes evaporated) were worldwide, or *eustatic*, since the whole ocean was affected uniformly.

The tentative conclusions that can be made about sea-level changes during the Pleistocene are the following:

1. The last glacial-stage lowering of sea level, relative to present sea level, was at least 300 feet and probably near to 350 feet.
2. The previous interglacial-stage rise of sea level, relative to land, was about 100 feet above present sea level.
3. The total sea-level drop, from the last warm interglacial stage to the subsequent glacial stage minimum position, was at least 400 feet.
4. Repeated interim Pleistocene sea stands approximated present sea level and also contributed toward creation of various present coastal features.
5. During periods of low sea level several important land bridges were exposed for long periods of time.

The general effect on river valleys of the slowly swinging sea level was twofold: as sea level lowered, the faster flowing rivers eroded valleys; then, as it rose, streams dropped their loads of sediments into embayed river mouths. During glacial stages large land areas were added, coastlines changed location, and former islands were connected to the mainland by exposure of narrow straits or land bridges (see Fig. 2.3).

The example of Bering Strait will suffice to indicate the significance of the existence of land bridges. The glacial-stage exposure of Bering Strait created a broad plain 600 miles wide, north to south, connecting Siberia and Alaska. Warmed by an ocean current, the southern margin of this plain became covered with long thick grass, ideal fodder for herbivores. Animal traffic was two-directional: North American camels and possibly the wolf and rodents moved westward to Asia, but there always seems to have been more movement eastbound from Asia to North America, including musk-oxen, antelope, rein-

deer, elephants, bears, cats, moose, cattle (bison), sheep, and humans.

Shifting Life Zones caused Further Biotic Migration

The three ice ages and the interglacial periods of the Pleistocene dramatically disrupted the distribution of all previously evolved life-forms. Although the overall effects were worldwide, Eurasia and North America were particularly affected. Additional disruption was provided by early Pleistocene mountain formation. Northern faunas and floras were dislocated at least four times by the advances and retreats of the ice sheets. They receded toward the equator during glacial stages and then returned toward the pole during interglacial stages. Successive patterns of distribution did not fully duplicate previous ones. Sometimes arctic-type populations moved upward on mountains during interglacial stages, rather than withdrew northward, and survived at high altitudes as isolated relicts.

The post-Pleistocene period of the last 11,000 years—the Holocene epoch—may be considered a partial interglacial period at less than its highest point, since ice sheets still lie on Greenland and Antarctica. The complex northward spreading of readvancing forms of life, following the withdrawal of the last ice sheet, is still going on. Plant migration, in response to these new conditions, has been going on actively during this period, and more speciation has been occurring in addition to the development of some new balance in plant associations. This period, however, has also been one during which the organic agencies, most notably mankind, have also been active in causing changes in the balancing operation.

HUMAN EVOLUTION INVOLVED BIOTIC/ CULTURAL INTERACTION

Life has displayed innumerable variable trends. Mankind was not inevitable, and yet our kind *did* originate after a long continuous sequence of occurrences. For our study that result is of primary significance. Mankind, the most highly endowed form of organized matter that has yet evolved on the earth, is the Pleistocene's principal product.

All living things are related. As in all relationships, some life-forms are more closely related than others. Mankind, for example, is closer in material origin as well as in functional and structural similarities to animals than to plants. The degree of relationship can be carried further and narrowed: humans are vertebrates, not invertebrates; they are mammals, not fishes or amphibians or birds or reptiles. And, among mammals, humans are primates. Humans are the most generalized of all mammals, since they have never depended on any one physical skill at the expense of many others. Perhaps the most satisfactory way of describing mankind's status is to say that humans are unique in having a singular combination of abilities, rather than in possessing any single ability exclusively. As primates, the earliest humans shared in the evolving dietary systems of the group of primates.

The larger primates long ago added some meat to their diets. Vegetable diets contain much roughage and lots of vitamins, but have little protein and add only a few calories per unit of volume. Animals that depend on vegetation for their food must spend much of their time eating, even though that food be readily available. Since meat contains little roughage, is highly nutritious, and is high in calories, meat eaters can spend less of their time getting enough to eat. A large primate that was able to run on its hind legs and to include meat in its diet then had time for play, exploration, and inquisitiveness. Our ancestors did just this, using their upright posture, manipulative hands, evolving brains, and free time for all sorts of activities not previously attempted by other forms of life. From cooperation in the gathering of food, communication through gestures and sounds, and the sharing of food, human culture emerged.

The singularity of human evolution lies in the fact that the human species evolved culture. Among a human being's peculiarities are adaptability and the capacity to learn in order to improve future behavior in light of past experience. By means of culture, mankind profoundly modified its own biological evolution to become so unlike any other life-form that the evolution of humanity cannot be understood only in terms of those causative factors that operate in the biological world.

Homo habilis as an Advanced Hominid

Table 2.2 places in order an interpretation of our human ancestors in a time perspective, as we currently interpret that record. Some interpret the record differently, and all such interpretations are subject to change as research continues and new evidence is found. Our purpose is to emphasize the immensity of that stretch of time—two or more million years—the latter portion of an even longer period during which the gradual evolution of human forms and the accompanying cultural development took place. Half the Pliocene and the first three-quarters of the Pleistocene were spent in coping with the problems of the cultural transition from using tools to making them and the biological consequences of

this change. Table 2.2 also outlines our interpretation of the present consensus that human evolution was a continuum, reconstructed as a series of orderly, discrete stages:

1. *Australopithecus*, which we here accept as an evolutionary form that reached its dead-end extinction in time, but which had its own common ancestor with

2. *Homo habilis*, which became the continuing line of evolutionary change, and who evolved into

3. *Homo erectus*, who in time evolved into

4. *Homo sapiens neanderthalensis*, who evolved into modern mankind, *Homo sapiens sapiens*.

We must, however, guard against presenting too simple an evolutionary picture. In their evolutionary development the earliest humans probably went through somewhat the same kinds of changes as did the other mammals. Some branches of hominids existed side by side for time periods before some of these branches became extinct. A surviving branch began, finally, to evolve at an increasingly rapid rate, giving rise to ourselves.

The earliest known hominids are Ramapithecines, found in northern Pakistan, and its East African counterpart, Kenyapithecines, dated between 10 and 14 million years.* South African and East African finds of fossils of *Australopithecus* offer evidence of further evolutionary progress. At least two species of *Australopithecus* are commonly recognized, *africanus* (a slender, graceful form) and *robustus* (a heavier, more muscular form), the earliest of these now being dated at more than four million years. *Australopithecus africanus* had achieved an upright posture, was about 4 feet tall, weighed about 50 to 80 pounds, and had short, light jaws and small front teeth (Fig. 2.4). The brain averaged between 450 to 500 cubic centimeters, less than half that of modern man.

Operating in environments ranging from parklike savanna to dry grassland, Australopithecines were extremely clever animals. They clearly were tool-users, and tool-modifiers, able to use their hands effectively, but the evidence so far suggests that they were not toolmakers able deliberately to fashion stone tools to a set, regular, and evolving pattern. It seems highly unlikely that *Australopithecus* possessed the mental skills to take the crucial step of using one tool to make another, the critical factor in the development of cultural technology.

Other East African fossil finds have provided evidence of a species now labeled *Homo habilis* (handy man) with dates that extend from 600,000 to

---

*In the singular designation of a species the term is italicized, but in the plural reference it is not.

Table 2.2 The Rise of Mankind During the Last Two Million Years

Time Units	Stage	Years (before present)	Climatic Oscillation	Evolution of Life-Forms	Emergence of Hominid Forms	Years (before present)	Cultural Phase (tool types)
HOLOCENE (Recent)	Postglacial or Interglacial		Warm	Fully modern fauna		9000	NEOLITHIC
		11,000			Homo sapiens sapiens	11,000	"MESOLITHIC"
UPPER PLEISTOCENE	Würm (Wisconsin) glaciation (last major cold climatic phase)		Cold			40,000	UPPER PALEOLITHIC (blades)
	Eem (Sangamon) Interglacial	70,000	Warm	Division of fauna into alternating glacial and interglacial assemblages	Homo sapiens neanderthalensis	100,000	MIDDLE PALEOLITHIC (varieties of tools; regional specializations)
		125,000	Cold				
	Riss (Illinoian) glacial complex { Warthe glacial / Ohe Interstadial / Saale glacial }		Warm / Cold				
MIDDLE PLEISTOCENE	Holstein (Yarmouth) Interglacial	200,000	Warm				LOWER PALEOLITHIC (bifacial tools; hand axes)
	Elster (Kansan) glacial complex (first major continental glaciation) { Elster II / Cortonian / Elster I }	275,000	Cold / Warm / Cold	Largely modern genera			Use of fire (Choukoutien; Escale) Dispersal through Old World
LOWER PLEISTOCENE	Cromerian Interglacial	500,000	Warm		Homo erectus	600,000	
	Kedischem (Nebraskan) complex (great complex of cold and warm phases prior to the first glaciation) { Menaplan (Gunz) glacial / Waalian Interglacial / Eburonian glacial }	1,000,000	Cold / Warm / Cold	Great floral and faunal changes as most Pliocene elements disappear			BASAL PALEOLITHIC
BASAL PLEISTOCENE (or Villafranchian in the broad sense)	Tiglian (first Interglacial)		Warm	Pliocene fauna	Homo habilis and Australopithecus (coexisting since the middle Pliocene)	1,750,000	Crude chopping tools in use
	Villafranchian (in restricted sense)	2,000,000	Mainly Cold			2,000,000	
PLIOCENE			Warm				Crude chopping tools in use

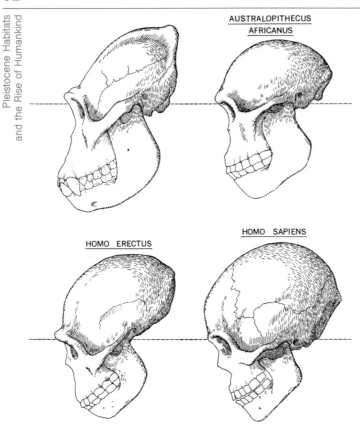

Figure 2.4 Evolution of the Skull. Note how the brain case (above the dashed line) has enlarged while the size of the teeth, and especially the jawbone, has decreased. *Homo sapiens* thus evolved a relatively large brain and a small face. (After Washburn, *Scientific American*, September 1960)

of tools that have survived, from the Basal Pleistocene (Basal Paleolithic) of tropical Africa, are those not of wood, shell, or bone, but of stone. These tools are crudely flaked chopping tools, haphazardly manufactured, but intended for future use. *Homo habilis* was definitely a toolmaker fashioning crude versions of both stabbing and cutting tools, suggesting conceptual foresight in an animal who was intentionally fashioning tools for future use to catch the other lesser animals for food. Such first halting steps in expanding a subsistence technology opened a wholly new way to reinterpret the external environment. *Homo habilis* became embarked on a long-term program in cultural advancement that improved its ability to cope with its surroundings, better equipped it for survival, and led toward the evolution of the physical species *Homo*.

Australopithecus of several varieties lived in Africa for several million years (Fig. 2.5). South Africa, which has produced so many of their remains, was presumably not the place of origin but rather a cul-de-sac, an area in which they could survive later than elsewhere. The long coexistence of *Australopithecus* and *Homo habilis* until the extinction of the former, went hand in hand with their habitation of eastern Africa. Since no similar evidence is available from any other part of the world, Africa must be considered as the place where man emerged and where just about every significant biological and cultural advance took place, at least for the first 2.5

between 4 and 5 million years ago, hence coexisting with *Australopithecus*, but evolving into the later *Homo erectus*. The estimated brain size for the type specimen of *Homo habilis* is about 725 cubic centimeters, falling beyond the upper end of the range for *Australopithecus* but short of the lower end of the range for *Homo erectus* (750 cc) by many interpretations. As the first ancestors that showed brain expansion beyond the *Australopithecus* level, *Homo habilis* represents an evolutionary advance. This was a small human weighing between 90 and 100 pounds and having smaller cheek teeth than *Australopithecus*. Such evolutionary changes as increased brain size and small tooth and jaw size had not progressed very far, yet these changes were sufficient to open the way to further cultural development. The tendency for brain sizes of the *Homo* lineage to increase as these creatures evolved is an indication of increased intelligence. The decrease in size of cheek teeth indicate the increased use of tools for processing plant and animal foods.

It is now thought by many authorities that *Homo habilis* was the earliest maker of tools. The only kind

Figure 2.5 *Australopithecus* Sites in Africa.

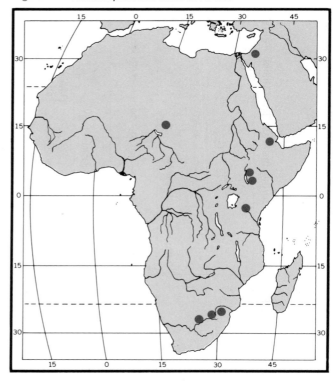

million years. But it would not be surprising if further research turned up an advanced type of *Homo habilis* in Java, where the remains of his successors have been discovered.

Before the end of the Pliocene, our primate ancestors had adapted to habitual tool-using and to erect posture, which resulted in the freeing of the hands; and during the Pleistocene that followed, our hominid ancestors, as their brains increased, adapted to the manufacture of tools to designed and consistent patterns. As cultural toolmaking became the mark of humanity, it also became the method by which to refashion the earth's surface into a human habitat.

Homo erectus of the Lower and Middle Pleistocene

The *Homo habilis* phase of hominid evolution merged imperceptibly with that of the *Homo erectus* phase during the long time span of the Lower Pleistocene. The doubling of cranial capacity during the *Homo habilis* stage enabled *Homo erectus* to have twice the brain size as that of *Australopithecus africanus* (see Fig. 2.4). The logical inference is that *Homo erectus* was more effectively organized than

were all varieties of *Australopithecus* and *Homo habilis* and was thus better able to survive since *Homo erectus* had the capacity for further elaboration of tools.

Homo erectus had to be fully mobile. Tools made for a particular purpose were discarded when the group moved on. It was easier to manufacture new tools as needed than to transport large quantities of stone tools. One or more hand tools, a throwing stick or club, and a wooden spear were the likely tools (weapons) carried by people of the Lower Paleolithic cultural tradition until the control of fire permitted the occupation of caves—as at Choukoutien, China—for shelter, protection, and a semipermanent home. The working of wood must have been greatly eased by the use of fire, including the felling of trees by lighting a fire at the base of the trunk. Easy woodworking in turn provided new methods of hunting, as in the construction of traps and pitfalls. The greatly increased number of sites known from the Middle Pleistocene (end of the Lower Paleolithic phase) indicates a small increase in overall population (Fig. 2.6).

Early people used stone not only directly as a tool but also as a way to make other equipment more

Figure 2.6 Mid-Pleistocene Distribution of *Homo erectus*.

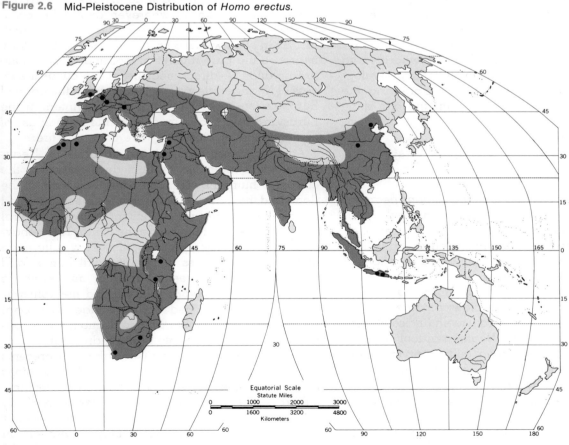

Equatorial Scale
Statute Miles

• Fossil Sites

efficient—in other words, mankind had invented tool-making tools. For example, the making of a sharp point on a digging stick increased the range of foods available to the toolmakers. The stone tools of the Lower and Middle Pleistocene indicate that they were used for special activities that could be performed in a variety of physical/biotic environments. The connection of stone tools with meat eating has probably been overstressed. The most important tools must have been of wood—for example, clubs, throwing sticks, knocking sticks, and pointed sticks (as spears or for digging), and carrying receptacles of bark or rind, or of animal origin, as bladders, stomachs, and skins to contain liquid.

As the name *Homo erectus* indicates, these creatures stood erect. They had an average adult stature of about five feet two inches, with brain sizes ranging between 750 and 1,400 cubic centimeters. Adults had large browridges that lacked recognizable foreheads (see Fig. 2.4) because the frontal lobes of their brains, not yet having evolved and expanded, were distinctly smaller than ours. Short faces, broad noses, and thick skulls are typical of many early types of men.

Discoveries of fossil remains from several other parts of the Old World permit us to infer that *Homo erectus* was widely distributed. Figure 2.6 is based on the *Homo erectus* finds made through 1974.

In Europe the transitional development of *Homo erectus* led to the enlargement of the skull (see Fig. 2.4), which became more rounded at the back although still retaining the heavy browridges and low forehead at the front. The evolutionary development of *Homo erectus* thus led to increasing the brain size before significantly altering the form of the face.

Emergence of *Homo sapiens neanderthalensis*

The evolutionary position of *Homo sapiens neanderthalensis* and the significance of its cultural development required nearly a century to decipher. Neandertals were the creators of Mousterian culture *prior* to the reduction in the form and dimensions of the Middle Pleistocene face.

The Neandertal face was larger than that of modern humans because its larger jaw contained bigger teeth with longer roots. The front teeth (incisors) were a heavily used part of the Neandertal tool kit. They were used for biting, scraping, cutting, prying, peeling of bark, cracking of nuts, and holding objects being manipulated by the hands. As a result of such use, the front teeth were usually worn down to the gums in individuals reaching fully mature ages.

The Neandertal phase of human evolution began about 100,000 years ago (Table 2.3), during the last interglacial stage. This phase continued for about 70,000 years, past the maximum of the main Würm glacial stage, to about 31,000 years ago. The culture of *Homo sapiens neanderthalensis* is identified as Middle Paleolithic. In Europe and Africa the cultural development from the artifacts of *Homo erectus* to the stone tools of the Neandertals proceeded without a break. Skeletal evidence, in the large browridges and large jaws while the brain size was enlarging, confirms the transition. Neandertaloids, over a span of 70,000 years, developed a greater sophistication than *Homo erectus* in toolmaking, as the cruder hand axes were replaced by flake blades and superbly worked smaller hand axes of triangular or heartshaped form.

Neandertal people began with the Old World distribution of *Homo erectus* (see Fig. 2.6), from which they evolved. As a result of improved cultural adaptation, the Neandertal populations extended their range poleward into central Europe and into central Asia. They are the first people we know to have lived in a truly cold climate. That is, before the time of glacial intensity, Neandertaloids were able to penetrate north of the frost-free zone into the midlatitudes that experienced winter cold. Toward the end of the interglacial period—about 70,000 years ago—Neandertaloids effectively occupied Europe. In considerable number, and over a wide expanse of territory, caves were occupied wherever available; occupancy was year round where food resources permitted, as in southwestern France. These sites, in turn, improved the probabilities of tool and human bone preservation. A direct measure of the cultural success of Neandertals was their survival at the edge of Europe's ice sheet and through the winters of the Eurasian midlatitudes, made possible by their use of clothing, shelter, and their sure control of fire.

Another measure of cultural advance by Neandertaloids is indicated by their development of the custom of human burial. The careful disposal of the dead and the inclusion with them of objects intended for use after death are actions indicative of funeral rituals that served to reassure the living. This behavior clearly suggests a more sophisticated culture, an increasing awareness of man's social bonds, and a considerable degree of abstract and symbolic thought pointing toward religion and metaphysics. Neandertaloids reached a crucial stage in personal awareness and a mark of humanization, the recognition of death as the fate of the individual.

The accompanying chart (see Table 2.3) presents the contemporary view of the place of *Homo sapiens neanderthalensis* in Pleistocene time and Paleolithic cultural development. Neandertals were fully *Homo sapiens;* their brains had a capacity equal to our own, although the frontal lobes may not have been so highly developed. Perhaps by 80,000 to 100,000

Table 2.3 Upper Pleistocene Humans in the Time Perspective of the Last 125,000 Years

Years (before present)	Glacial Stages (in Europe)	Climate	Hominid Form	Cultural Phases (in Europe)	Tool Types (in Europe)
	RECENT				
5000				NEOLITHIC	
	8000	Cold Temperate		9000	
	9000				
10,000				MESOLITHIC	
	11,000	Cold, but warming		11,000	Bow and arrow
15,000				12,500	Specially equipped for reindeer hunting Eyed needles, cave art, reindeer-antler artifacts
				17,000	Extraordinary pressure-flaking; willow-leaf and laurel-leaf lance points.
		Very cold	HOMO SAPIENS SAPIENS		
20,000	LATE WÜRM (III)				
				21,000	Smaller knife blades Ivory wedges for splitting wood and bone
25,000				UPPER PALEOLITHIC	
	26,000				
		Cool Temperate		25,000	Most skillful flintworking First true boneworking (pins, awls, light spearheads)
	28,000				
30,000			CRO-MAGNON		
				31,000	Earliest blade cultures Slender, straight, keen-edged flakes
35,000	MAIN WÜRM (II)	Very cold		35,000	
40,000				40,000	
	42,000				
45,000					
	46,000				
50,000	EARLY WÜRM (I) UPPER PLEISTOCENE				
60,000					Wooden hafts added to stone tools
		Cold		MIDDLE PALEOLITHIC MOUSTERIAN	Multifunctional tools, thick, heavy flakes Improved techniques of manufacture rather than functional specialization
70,000					
			HOMO SAPIENS NEANDERTHALENSIS		Two standard types of tools 1. D-shaped sidescrapers 2. Triangular point or knife
80,000	EEM INTERGLACIAL				
90,000					Small, bifacial, hand axes
				100,000	
	UPPER PLEISTOCENE		HOMO ERECTUS	LOWER PALEOLITHIC	Sidescrapers
125,000	MIDDLE PLEISTOCENE				

years ago the human brain reached its present state of development and since then has scarcely changed. The explanation for lack of further increase in brain size is twofold:

1. There was an upper limit to brain size in the newborn. Brain sizes larger than could be passed through the birth canal of the female pelvis were at a reproductive disadvantage for survival of both child and mother.

2. Cultural development, as reached by early Neandertals, made a large and more complex brain unnecessary. As further experiences were shared through language, individuals learned more and more from one another, and increasing knowledge was passed on from one generation to the next. Beyond a critical point of brain complexity sufficient for the individual to be able to speak and to learn, further increases of intelligence conferred no advantages toward survival. There are suggestions that the Neandertals could not yet utter a full range of sounds, so that true language had to await the appearance of *Homo sapiens sapiens*, but this does not alter the basic point concerning the transmission of learning.

During the Middle Paleolithic, the most obvious cultural advances made by Neandertal peoples had to do with the invention and usages of many new kinds of stone tools for cutting and the manipulation of materials. The very large teeth of the Neandertals then lost their significance for survival, since improved tools had taken the place of teeth. Thereafter, the sizes of teeth and jaw were free to vary. The long-term result of random genetic mutations was the natural selection of teeth and jaws of reduced size. Instead of teeth that met together, there developed an overbite, in which the upper teeth extend forward beyond the lower teeth, as do your own for the most part. Thus the interplay of cultural development and natural selection led to the creation of the smaller face of modern *Homo sapiens sapiens* (see Fig. 2.4).

Rise to Dominance of Modern Mankind

Coming into its own as the successor to the Neandertals, the modern species *Homo sapiens sapiens* is a product of the last glacial stage (see Table 2.3). In Europe this succession may have taken as long as 9000 years, from 40,000 to 31,000 years ago. This was a slow process by our present time dimension but rather abrupt in light of the previous four to five million years of earlier human evolution. The end result was the rise to dominance of our modern racial strains of humans, in a variety of forms almost undistinguishable from those of the present day. Culture skills had not evolved at the same rates as human biology in this interval, however, and the first representatives of truly modern mankind lived rather simple lives. We still are uncertain about many aspects of the transition from the Lower Paleolithic to the Upper Paleolithic.

Stone implements of the Upper Paleolithic have been largely interpreted as the tools of hunters. These were weapons for killing, blades for skinning and cutting, and scrapers for preparing hides. As a

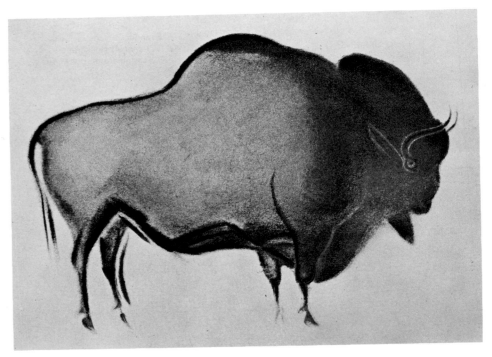

This multicolored painting of a bison found in the cavern of Font-de-Gaume in the Dordogne region, France, shows the skill of Paleolithic artists.

result of dependence on the things best preserved, and then recovered by archaeology, we have perhaps exaggerated the importance of hunting in the life of early peoples. But there is no doubt that Upper Paleolithic humans in Europe during the late Würm glacial stage were skilled hunters of big game. That specialization in Europe was the result of a unique time-and-culture coincidence, in which evolving cultural abilities allowed people to achieve the big-game kills just before and during the period when the large mammals of the Late Pleistocene were becoming extinct. Over most of the Old World Upper Paleolithic people went on using their stone tools for cutting and splitting wood, for cutting bark, and for digging roots.

Further evidence from the Upper Paleolithic reveals the heights of human achievement then attained: the fantastic paintings in reds, yellows, browns, and blacks on the walls and ceilings of great caves. The most expressive cave art was found in southern France and northern Spain, where limestone regions contained a great many natural caverns created by underground solution drainage. Thousands of murals were created, of abstract signs and of animals such as the mammoth, bear, lion, bison, and reindeer—the great animals of the Pleistocene—on whose successful hunting Paleolithic peoples depended. Evidence of a spiritual/ritual life was clearly expressed in the hunter's quest for assurance of a continued human existence, a future supply of game, and of success as the hunter of certain preferred animals.

CULTURE BECAME THE NEW METHOD OF HUMAN DEVELOPMENT

Human evolution thus had two components that are interrelated and interdependent: the biological and the cultural. But, in producing culture, biological evolution transcended itself. The capacity to learn a symbolic language and a variety of cultural forms was added to other characteristics of the species *Homo*, which made possible the ability to profit by experience and adjust one's behavior to the requirements and expectations of one's surroundings.

Humanity's greatest prehistoric accomplishments—(1) dispersal across nearly the whole of the earth; (2) adaptation to nearly the whole enormous variety of the earth's physical/biotic environments; and (3) slow growth in numbers to a population of about five million (an average density of about one person per ten square miles) by 10,000 years ago—resulted from the development of culture. The achievement entailed an immense journey, in both

space and time, since the cumulative increase in cultural ability was extraordinarily slow, by any modern measure. Humans emerged from being wholly animals to become forever more than animals.

Cultural development provides yet another means, beyond the biological method, for adaptation. People achieve adaptation to their environments through learned rather than inborn automatic responses, and most do so successfully. Knowledge enables more individuals to survive by choosing or creating suitable environments for themselves. The development of the human faculty for abstract symbolic and cultural transmission was certainly a radical innovation, transforming the whole of human life, after which humans lived in a new dimension of reality as compared to other life-forms. The unique characteristic of humanity was that cultural traditions, when learned, opened further possibilities for unlimited growth.

Culture is a new emergent in the biology of mankind but, unlike bodily characteristics, it is transmitted socially. Culture is a new form of evolution by means of which the human species has been able to progress more rapidly than any other species. Cultural evolution and biological evolution are not alternatives, but are superimposed upon each other. The evolution of culture has not suspended or taken the place of biological evolution. Cultural evolution, however, has assisted and complemented biological evolution in that it permits an increasing adaptation of the genetic heritage of ethnic groups to multiple environments. This adaptation has given rise to the many different racial strains occupying separate regions of the earth.

Human Tropical Origins: Easier Food-Gathering

Human beings are omnivorous, but, with rare exceptions in special times or places, the great bulk of their food, in calories and nutrients, has been of plant origin. Big-game hunters have attracted much attention, and meat in diets provides better archaeological evidence—through bones of animals, birds, and fish—than do the more perishable organic remains of plant materials. For most of human time, the limited technological skills restricted mankind to unspecialized food-gathering. A great variety of locally available edible roots and plants, as well as many smaller mammals, insects, and birds were used. Flesh of larger animals was probably eaten only on a scavenging basis for much of earliest time, but men learned techniques of hunting in groups and were eventually able to kill the larger animals.

Wild vegetable foods are more abundant and more varied in equatorial and tropical latitudes, where moisture is available (at least seasonally) and year-round warmth permits continuous plant growth. Seasons of cold and drought in any region reduce the volume and variety of plant materials potentially useful to man. Several hundreds of species of plants in a localized area would provide roots, tubers, seeds, nuts, bark, and fruits in quantity for consumption. Unspecialized food-gatherers also consumed such readily available edibles as insects, grubs, lizards, snakes, tortoises, birds, small rodents, bats, eggs, and the helpless young of larger animals. Shrimp, bivalves, snails, crabs, and fish were obtainable from inland waters, and the varied shellfish of the seashore supplied a valuable source of protein and iodine for people of limited technological skill. In sum, the humid and subhumid tropical and subtropical climatic regions, particularly in association with lake, river, or coastline, were optimal areas for early hominids.

Food-gathering was not a process of free wandering. Continuous and assured access to potable water was a first essential, followed by the need for adequate food. The logic of supply indicates the need to keep the expenditure of human effort within the limits of the energy value derived from the food secured. Least effort and continuous food supply were fundamental early principles to be followed for success in perpetuating life. Progress was possible by success in locating and occupying areas where resources suitable for human food were superior, thus giving an initial advantage to those groups who occupied such regions.

Fire: The Great Discovery

The greatest—certainly the bravest—of mankind's early triumphs, and one leading to subsequent successes, was the discovery of the control and utilization of fire. The knowledge of how to make a fire, to keep it burning, and to use it for specific purposes was a most useful achievement. No other animal can do it! Exploiting fire long ago became one of the few universal traits of human culture, and evidences of its use date back at least to *Homo erectus* in several parts of the earth.

Fire for early people was both tool and weapon. Its presence in the form of heat and light at night deterred or drove away other animals; it provided warmth that enabled human beings to extend their range and to survive in hitherto unexploited territories with cooler climates. It also provided a focus of activity for a group of families, a sense of belonging to a common "hearth," for which provision of fuel and keeping the fire or sparks alive were of overwhelming concern.

The small band gathered round its hearth derived a tremendous benefit from the discovery that cooking (initially by roasting) made meat more edible and digestible and that plant starches and tough vegetable fibers are broken down by heat into forms digestible by humans. Possessing fire, people could and did move outward from their "home territory" with greater confidence. No matter what newfound wild plant materials were gathered, they could be rendered palatable and provide sustenance. No matter what new wild animals might be encountered, they could be driven by fire into bogs or traps or pits or over cliffs, thence killed and eaten as a source of meat. Fire made possible an extension of humanity's range over the earth, from the subtropics into the higher latitudes. It also improved the possibilities for slowly increasing the number of people and maintaining kin-group solidarity, thus further differentiating mankind from other animal forms.

Another Cultural Achievement: Shelter

The shelters that people learned to create also greatly aided adaptation to living in a variety of physical environments more extreme than those of the tropics and hence enabled mankind to extend vastly its range over the earth. Human dwellings are exceedingly varied in the raw materials used, in the climatic conditions (winds, temperatures, precipitation, seasonalities) being countered, in their sites, and in their forms and functions (see Photo Essay No. 1, pp. 70–82).

Early human occupancy of caves and rock shelters (formed by overhanging cliffs) is perhaps overemphasized, simply because the archaeological record is so much better preserved in their littered and buried floors and debris mounds than in the open-air living sites along streams and lake shores, where materials were both greatly scattered and exposed, and later more easily removed by running water, waves, and wind.

Food-Gathering to Food-Collecting: The Trend Toward Specialization

The basic connections that early people established with their physical/biotic environments in order to sustain themselves hinged on their methods of obtaining food and securing shelter (settlement) within the territories occupied. Throughout the long duration of the Pleistocene, survival depended on skill in locating and securing food that grew as a result of natural biotic processes.

It is useful to distinguish among several probable types of human existence during the Pleistocene (Fig. 2.7):

1. **Unspecialized Food-Gathering** (example: *Homo habilis*).
   • Naturally determined mammalian subsistence.

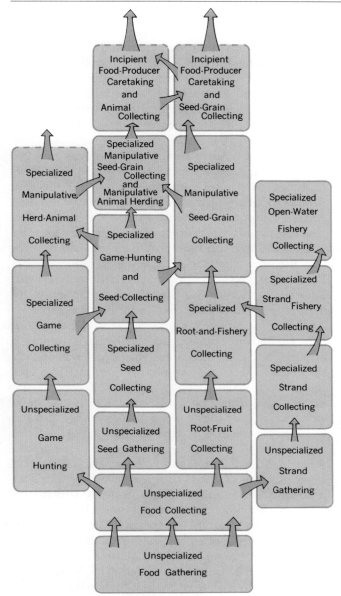

Figure 2.7 From Food-Gathering to Food-Collecting, and to Incipient Food-Production.

- Wandering within limited territories.
- Temporary shifting settlements of one or only a few days' duration.
- Tools fashioned but not yet standardized (e.g., the very crude and variable chopping tools).

2. **Unspecialized Food-Collecting** (example: *Homo erectus*).
   - Selective hunting, strand gathering, and fishing, in particular brief seasons of availability during calving and spawning.
   - Increased skill in harvesting temporary surpluses of food beyond what could be consumed immediately.
   - Restricted wandering, within fixed (known) lim-

its, including identification with, and defense of, a "home territory."
   - Mostly temporary settlements, but occasionally seasonal settlements perhaps of several weeks', or even several months', duration.
   - Variety in standardized tool forms (mixed tool assemblages) within regions, including "tools to make tools."
   - Distinction in space between where food is collected and where consumed; beginning of techniques for preservation of surplus food collected.

3. **Specialized Food-Gathering** (example: *Homo sapiens neanderthalensis* in Europe).
   - Increased skill in hunting of larger animals (herbivores).
   - A greater proportion of meat (high protein content) in the diet.
   - Wandering within somewhat delimited territories, with increased knowledge of the better locations for food-gathering.
   - Temporary shifting settlements of one to several days' duration; less mobility whenever an increased supply of food permitted.
   - Standardized toolmaking traditions for specialized tools; core bifaces, flakes, and choppers with broad distributions for each tool type.
   - Materials mostly consumed where gathered, not much spatial distinction in activities, but more feasts than famines, although a seasonal "starvation period" was possibly an annual occurrence.

4. **Specialized Food-Collecting** (example; some groups of *Homo sapiens sapiens*).
   - Intensified hunting, fishing, and collecting of plant products (seeds, berries, nuts), with emphasis on maximizing food surpluses.
   - Seasonal differentiation in activities between plant collecting and hunting for game, fishing, or both.
   - Trend from restricted wandering toward center-based wandering, that is, toward seasonal settlements of some month's duration, to which human groups returned each year. In regions of especially great food surpluses, as along coastlines with continuous supplies of shellfish or huge seasonal salmon runs, the beginning of semipermanent settlements of some years' duration became possible. Reduced frequency of "starvation periods" became normal.
   - Beginning of plant manipulation (fibers, poisons, drugs) and greater concentration on the taking of fish, fowl, mollusks, and the fleeter mammals (ungulates).
   - Marked regional distributions of tool types, great skill in food preservation (drying, salting,

fermenting) and storage (baskets, bamboo tubes, pit caches).
  • Spatial separation of food collecting and food consumption, the latter including preservation and storage activities at campsites.

5. **Incipient Food-Production** (example: Mesolithic cultural stage that marked the end of the Pleistocene and beginning of the Holocene [Recent]).
  • Experimental manipulation by specialized food-collectors of favored plants and animals within their natural habitats (e.g., watering or weeding of plants, protection of particular animals from other carnivores, burning of grazing areas to improve the food supply and hence an increase in the number of those animals favored in hunting).
  • Semipermanent to near-permanent settlements; less wandering necessary as greater volume and variety of foodstuffs are obtained as a result of increased skills (knowledge of all the useful things to be found in a local region; techniques for acquiring, preserving, and storing food surpluses); fewer famines, but "starvation periods" recurrent among less able cultures or in seasonally poor environments.

These five types of existence were developed in different times and places by varieties of prehistoric people. The differences among them were very real for those who lived by their means. What seem to us, with our advantage of hindsight, to be only very small differences in the conditions for human existence were precisely those that made "all the difference" in those cultural developments that enabled some groups of Paleolithic people to surpass others.

The Significance of Mental Activity

Increased mental capacity coincided with the great expansion of the cerebral cortex, with its deep folds and convolutions, in the brain. Particularly characteristic of modern humans (*Homo sapiens sapiens*) is the vast size of the frontal and temporal lobes of the brain, with their nerve cells used for association and storing of memories. *Within the human brain lies the evolutionary bridge from the biological to the cultural.* Infinite varieties of experience are accumulated as memories, stored in the brain, and there reshaped into a limited number of generalized ideas or concepts. The development of recognizable and meaningful abstract symbols, both graphic and acoustic, enables people to share experiences and transfer ideas from one individual to another. The apparently small step of the sharing of memory and experience made all the difference in human evolution.

The cerebellum, the back part of the brain that is concerned more with skill functions, is just as large as the cortex. With its large areas devoted to manipulation of thumb and hand, the human cerebellum is about three times the size of that of a great ape. From this difference in size and distribution comes the conclusion that the parts of the brain involved with skills had to start developing before humankind became proficient in the use of tools. Along with the development of the brain, however, there also probably resulted natural selection of the most effective tool-users—those who remained alive to reproduce had both the most evolved brains and the best ranges of manipulative skills. This is why millions of years of human evolution were necessary before *Homo sapiens* rose to dominance. It is easy for modern people to learn manipulative skills because the way the brain developed makes such skills possible. For an ape, however, the things that humans do with little effort, such as the precise manipulation of the fingers, are impossible and even inconceivable.

Human self-awareness, which developed along with the cerebral cortex, made possible a number of cultural achievements. One very early activity was synthesis of thought—an improved capacity to visualize, to conceptualize, to reach conclusions, and to act on them. Early people did not live just to make stone tools. They made those tools, whether of wood, shell, bone, or stone, to express inner urges that were regularized by definite patterns of conceptual thought. They transmitted to others not only material objects whose methods of preparation were observable but also idealized forms engraved in human memory and later to be duplicated. The development of conceptual thought, expressed as art, mysticism, and ritual, was progressing all the while people were battering away at cobbles to produce chopping tools and hand axes. The evidence is clearly visible in the many fine tools whose design and workmanship are far more elaborate than was needed for their purely practical functions of cutting, scraping, or pounding. Aesthetics, as a sense of style or of correct form, was an early human achievement, conferring inner satisfactions.

Any delineation of the qualities of humanness must include the considerable ability to communicate with others in elaborate symbolic ways: speech, writing, printing, whistling, drumming, and gestures. Spoken language is a purely mental achievement that leaves behind no material remains. Nevertheless its development was every bit as significant for human evolution as the capability of toolmaking and tool-using. The idea of language—the patterned association of certain sounds with certain meanings—was an invention of the first magnitude. No human group ever known has been without a fully developed, richly symbolic means of verbal communication.

Language, the systematization of signs and words, requires directed thought, but it also permits thought to be directed. Once analytical thought was possible, language expanded rapidly with increasingly complex vocabulary and word order. As language grew, it became a record of the general elaboration of total human culture.

Since all primates are social animals with strong gregarious instincts, humans too can be assumed to have always acted in groups. As the human body and the human mind slowly but continuously evolved from an animal ancestry, so did human society. Without companions of like kind, people could never have developed their mental powers through communication or conceptual thought. The individual is born into a group; the whole process of growing up through a long period of helplessness and immaturity is as important for the society as it is for the survival of the individual. Over a protracted period of childhood, individuals acquire the language of their home and community, learn the socially approved ways of behavior, and gain the necessary skills to manipulate the artifacts of the society—in a word, they become *enculturated*. Society thus insures its survival by constant replenishment and education of its individual members.

MANKIND SLOWLY OCCUPIED THE WHOLE EARTH

Upper Paleolithic humans possessed the cultural combination of fire, shelter, clothing, containers, and weapons that enabled them to compete with carnivores in following the herd animals anywhere, confident that they could feed themselves from the hunted game. The growth in stature of humans placed them among the elite as a large land mammal.

The way was open, through cultural means, for the human species to occupy all the landmasses that could be reached from the Old World (Fig. 2.8). The record indicates that Upper Paleolithic man was the first to invade Australia and the Americas.

During the late Upper Pleistocene people learned to live successfully in all the world's physical/biotic environments except the true deserts, the high plateaus and mountains, and the oceanic islands. By 15,000 years ago humans not only had reached all the world's continents except Antarctica but also had spread to their far extremities, such as Tierra del Fuego, clearly establishing a range of subsistence groups, from specialized food-gatherers to specialized food-collectors, in a host of local territories. No other single animal species has ever accomplished as widespread a dispersal over the earth as have

humans. This astounding accomplishment, which occurred during the peak period of differentiation of modern people into many local racial stocks, was all but completed before the time when civilization (and history) began. It was no mean achievement for small groups with limited skills who migrated without plan or foreknowledge of final destinations. But, significantly for humanity's later development, the whole planet was at last inhabited.

Mankind's Varied Pleistocene Environments

The change from glacial to postglacial conditions differed greatly from place to place. *Deglaciation*— the melting back and wasting away of the great ice sheets—was a process, not an event.

The beginning of the last great deglaciation occurred immediately after the continental ice sheets reached maximum size, that is, from the very peak of the last major ice advance—about 35,000 years ago. The full postglacial stage was not reached until the end of the deglaciation process, which occurred at different times at different places. The continued rise of sea level until almost 7500 years ago caused oceanic waters to spread onto continental shelves to positions approximating recent shorelines. Lower courses of streams were flooded to form estuaries and bays, and coastlines became more irregular. At that time, also, began the formation of present coastal landforms: the erosion of sea cliffs; the building of bars, spits, and beaches; and the filling of lagoons to become salt marshes.

Near the end of this period of the last major deglaciation falls the cultural revolution known as the *Mesolithic*, a coincidence of great environmental and cultural change. The duration of the Mesolithic, from 11,000 to 9000 years ago, encompassed a long period of human trial and error in adapting to a new set of conditions for living, hence to a great change in culture.

For human living on the earth, the period of waning ice sheets created considerable changes in the physical/biotic environment, hence changes in opportunities. The hunting folk disappeared from the more arid parts as they followed the migrating big game. Those who stayed on in drying-out places became gatherer-collectors with less specialized habits, but their habitation sites became fewer and their ranges more restricted because water could be found less readily.

Following gradual disappearance of the ice, millions of square miles of land in the higher latitudes were opened for human colonization. The newly created lakes, ponds, and marshes were replenished by aquatic life and migrant waterfowl. Northern woodlands, wet prairies, and marshes became feeding ranges for such big game as muskox, moose, elk,

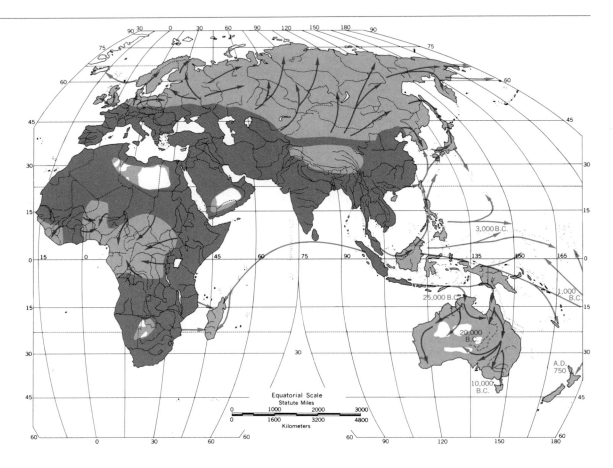

bison, and deer, and the hunting peoples followed these animals northward.

The coastal peoples were crowded back by the rise of the sea. But as river mouths became embayments and shorelines became longer and more diversified, more people could live as strand collectors, distinguishing among the fauna of rocky reefs, gravel beds, sand, and mud flats. The newly formed shallow lagoons and bays offered a favorable setting for experimentation with rafts and boats.

Valleys filled with alluvium, and river floodplains grew to include natural levees, meander scars, oxbow lakes, and backwater swamps. The various grasses, reeds, lilies, and palms all colonized the marshy parts of the floodplains, as new environments took form, developing the physical/biotic conditions on the earth we recently have known.

Characteristic features of the Old World Mesolithic were the following: (1) most stone artifacts were reduced in size; microliths (very small worked-stone tools) became diagnostic; (2) the bow and arrow became widely used, the arrowhead being a form of microlith; (3) fishing gear became varied, as fishhooks, sinkers, and floats were introduced, and use of the harpoon became most advanced; (4) boats came into greater use; (5) woodworking tools were developed; (6) grinding slabs appeared; (7) stone was shaped by deliberate grinding; (8) the domesti-

cated dog was spread widely; (9) pottery began to be made.

Many of the features sketched in this chapter possess profound significance in the development of life in the Holocene epoch on the matured land areas of the earth. Some were critical to the cultures and the numerical growth of humans as occupants of earth. Environmental changes have been intricately bound up with the coming to dominance of *Homo sapiens* as one particular life-form in all parts of the earth.

The surface of the earth in recent time displays a great variety of landforms. In the sheer range of altitudes, the variety of surface features, and the wealth of variation within comparatively small areas, the world of recent time is probably the most complex assemblage of earth forms that has ever existed. The variety of conditions now existing just above the solid crust of the earth, in the lower atmosphere, presents the widest range of weather and climate that has ever obtained for our earth.

Future Conditions on the Earth

The Pleistocene was an unstable period in earth history, with many dramatic physical and biotic changes taking place. The rapidity of these developments has significant implications for the future characteristics of both the physical and biotic envi-

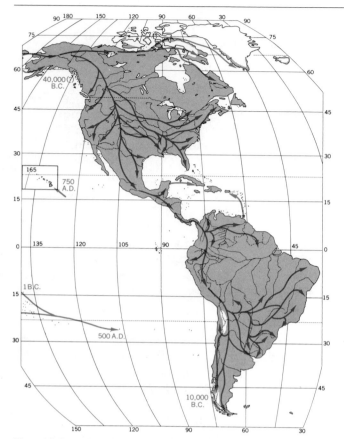

Mid-Pleistocene distribution of Homo erectus (after Fig. 2.6)

Spread of Homo sapiens

Major routes of migration
with dates indicating earliest known arrival

Uninhabited areas

Figure 2.8 The Spread of Mankind over the Earth.

earth undergoing rapid change. In the course of time, measured in geological terms, many of the present forms, conditions, and qualities of regional environments may well be altered quite markedly. Two examples can illustrate this point:

1. The loess deposits scattered about the earth are probably little older than *Homo sapiens*. It is quite possible that late Neandertal peoples experienced the initial beginnings of loess deposition. Almost immediately after the first significant thicknesses of loess were laid down, different kinds of plants were able to colonize these new landscape surfaces. Animals able to subsist on those plant forms then promptly occupied the new regions. Human groups soon found the loess regions to possess useful varieties of plant forms and rich populations of browsing animals, and profitably utilized both. Later peoples certainly found the loess areas easily cultivable, some of which soon became food-producing "breadbaskets." The loess regions are still among the more productive agricultural areas of the earth.

   How long will those loess areas remain on an unstable earth for the use of mankind? Such soft and erodable surface deposits are not permanent elements of the earth's crust in terms of the geological time scale. They will be removed by the physical processes currently at work despite all human effort to conserve them. How long it will take to remove them, under changing physical processes, is not now predictable. As those deposits are removed, however, the populations then occupying such areas—in that limitless future—will begin increasingly to face physical conditions and environmental qualities of landscapes quite different from those that have existed for the recent few tens of thousands of years.

2. The first human inhabitants of the region we today call Japan almost certainly walked overland across the lowland plain that today forms the bottom of the Yellow Sea, sometime within the last thirty thousand years. At that time there was an extensive land mass off the coast of Asia, instead of the island archipelago that exists today. The present Inland Sea of Japan was then a valley region between surrounding highlands. Later, as sea level rose at the end of the last glacial era, the

ronments now present. There has been too little time for physical conditions to reach a level of stability, and for biotic assemblages to come into ecologic balance with those conditions. There is, even, no certainty that a stable era, geologically, is in the offing. These facts have implications for the long-term future of human living systems on an earth that many people now regard as belonging to mankind.

Human beings tend to think of the present physical/biotic world as permanent in forms, characteristics, conditions, and qualities. We also tend to look forward to future human existence as being in perpetuity, that is, forever. This, in a sense, is why so many now are concerned to conserve all resources, and to preserve things the way they are. Despite the experience of almost disastrous short-term droughts, destructive earthquakes, damaging floods, and other sorts of regional catastrophes, every society tends to think of its own future as limitless in future time. In sheer reality, of course, most people do not think in truly long-term periods of time (see p. 9), and the limitless future is a vague, if persistent, concept.

A great many of the present physical and biotic conditions, however, are but temporary aspects of an great lowland plain was flooded, to become the Yellow Sea; those first inhabitants retreated up-slope and became island residents. The inland valley between the highlands became flooded, turning into the Inland Sea, and only the unstable, earthquake suffering mountain tops of the former highland remained as land areas—the present island archipelago. In time other peoples arrived by sea, and the populating of Japan continued. Gradually, a culture system was developed in southern Japan, focused on the attractive local environments around the Inland Sea, and this matured into the modern Japanese living system. Those unstable mountaintops still shake under the pressures exerted by the great Pacific Plate.

What physical changes may geologic processes produce in the Japanese environment in a limitless future? How permanent is the Inland Sea, and how permanent are other features of the contemporary Japanese landscape? What patterns of modification of living system do inhabitants of Japan face in the truly far distant future when there might not be an Inland Sea?

By the workings of evolutionary processes, *Homo sapiens sapiens* came into the occupance of an earth of tremendously varied environmental conditions. Our series of richly divergent culture systems developed during a period in which there occurred some of the most spectacular changes in all geologic time. Better supplied by physical environments and surrounding biotic assemblages than at any other possible time, modern people occupy the most potentially useful resource ranges the earth has ever presented. That physical and biotic world has altered and improved during the very appearance and occupance by mankind. Can we count on further improvements being provided, infinitely, to our earthly home? The physical and biotic worlds will go right on changing, infinitely. We can hope that the future changes do not restrict human occupance, but we must also hope that cultural evolution may keep pace with those environmental changes if the earth is to remain, infinitely, a satisfactory earth for humankind.

SUMMARY OF OBJECTIVES

After you have read this chapter, you should be able to:

CHAPTER 3

THE BEGINNINGS OF CULTURE SYSTEMS

KEY CONCEPTS

Territoriality Is a Basic Culture Trait
Religion Is a Basic Organizing Force
Social Organization Develops Grouping Systems
Tribal Organization Requires a Chief
Resource Systems Evolve into Domestication Systems
 Beachcombers Were the First Fishermen
 Plant Domestication Had a Very Slow Evolution

Shifting Cultivation Was the Basic Neolithic Garden System
Animal Domestication Began with Small Animals
Animal Husbandry Provided an Alternate Economy
Housing Systems Develop Settlement Patterns
Power Technology Was Slow in Evolving

From our present knowledge of the beginnings of mankind, the initial development of culture, and the first occupation of different environments, we cannot draw many significant contrasts between the levels of culture in the separate physical patterns. The Paleolithic Age was more than two million years long, and the results achieved were neither very high nor sharply distinguished. The development of culture started slowly, and it grew by very small additions. The numbers of new developments were very few in any one region during any one time interval. Early Paleolithic peoples only slowly learned basic bits of knowledge about environmental conditions and possible human practices. This slight knowledge promoted only a very slow increase in the numbers of people, but it provided ability to survive in competition with predators. Distinct bodies of knowledge about how to live in different regions developed as peoples in one environment learned to use certain resources that did not occur in other places. Languages began to develop differences as those early groups each named the separate kinds of plants, animals, water bodies, weather conditions, and kinds of landforms found in their particular region of occupance.

Sources of foods, the needs for shelter, the amounts and kinds of clothing, and the materials for making tools all varied slightly in the separate occupied areas. Although the patterns of living differed from place to place, the levels of accomplishment remained both very low and similar in degree. All human groups were engaging in the basic aspects of discovery, and were making very simple inventions of the same general types, but evolution in the level of achievement was very slight, and there were few new features that could spread from one group to another by diffusion.

TERRITORIALITY AS A CULTURE TRAIT

The concepts of home, neighborhood, and the outside world were discussed in Chapter 1. This feeling of an individual for a home territory is related to the broader concept of biological territoriality. It would appear that mankind inherited a sense of territoriality from the animal background that did not disappear during biological evolution. Territoriality in some form is one of the strongest expressions of human reaction, and one to which mankind has made many additions in the creation of systems of culture.

In earliest mankind the sense of territoriality probably took a form similar to that of the higher animals. The tendency to form social groups probably made the basic territorial unit one occupied by a population that could conveniently live and travel together. The area of the unit probably depended quite directly on the food productivity of the general region. The limited technology of earliest mankind undoubtedly resulted in a direct ecological relationship between the territory and the population. Thus one can say that earliest mankind was directly and strongly dependent upon the quality of the inhabited environment.

The first human advances in culture could have had one of several results. The following are suggested:

1. A focus of interest in particular parts of a territorial range considered more productive than others, resulting in a decrease of the group territory.

2. An increase in group mobility in food-gathering, resulting in the enlargement of the territorial area.

3. An increase in efficiency in food-gathering, resulting in the contraction of the territorial area.

4. An increase in the efficiency of defense of the margins, resulting in a tighter hold on the territorial area.

5. An increase in the population that could subsist within the territorial unit, resulting in full control of the occupancy of the territory, or in the creation of pressures for expanding that territory, or both.

Regardless of which result came first, any advance in culture could well have altered the definition of what constituted a group's unit of territoriality. All later additions could have had some impact on the concept of the territorial unit. We must recognize that environmental change took place, during the Pleistocene, as mankind was first refining notions of human territoriality. In part, this refining process can be ascribed to biological evolution of the human species, but increasingly it must be ascribed to increments of culture. If it be held that the human race did adapt biologically to varied physical environments in the earliest periods, it must be accepted that the development of culture in later periods began to lessen mankind's direct ecological dependence on environment.

Whatever early mankind's ability to live in different kinds of environments, the habits of territoriality remained. Having divided up earth space to begin with, human culture groups have been redefining the concept of territoriality with every significant change in culture. The contemporary problem involved in control over the oceans is but the most recent aspect of the problem.

Human attitudes toward space in general have differed greatly. Some groups of humans came to prefer large amounts of geographic space per unit of population, whereas others have preferred smaller units of space per person and a closer pattern of social contact. In the same way there are many different attitudes of preference regarding landscape in general. Some people have sought the coasts; others have preferred the high mountains, the open plains, or the deep forests. Culture groups, just as individuals, develop these preferential attitudes. We cannot be certain how old or how strong may be a particular group's preference for a specific kind of environment. We can only say, in dealing with peoples in historic time, that certain groups have lived for specific periods in particular environments. Many groups have been forcibly ejected from regions of long occupance. Within historic time, many human groups have had to occupy territories that might be termed second, third, or fourth choices, since they were deprived of occupance of territories previously inhabited.

RELIGIOUS BELIEFS AS AN ORGANIZING FORCE

Earliest mankind did not understand all the happenings of Nature, and certain of these occurrences must have been terribly disturbing. In this lack of understanding, fear of the unknown, and belief in the existence of supernatural power, lies the origin of all human religious beliefs and practices. A quest for assurance, and activities aimed at securing favorable treatment by supernatural powers, are involved in all formulations of religious systems. The attribution of power in this sense is both to physical objects and natural phenomena, and to beliefs in beings and things of another world. The magic, ritual, mysticism, and formal ceremonial form a complex continuum designed to deal with the supernatural. The belief in supernatural power enables separate peoples to develop different interpretations of relations between themselves and the physical world. It also provides a concept of, and an explanation for, the universe. This belief establishes a mechanism by which people can appeal to the forces that lie outside the control of both the individual and the culture group. The forms, practices, and creeds developed by culture groups in this broad range of situations vary from the simplest animistic beliefs and practices to the most complex theological and philosophical constructs. *Animism* describes beliefs that good spirits and demons inhabit natural objects. Animism characterized early societies, and is found among some simple societies at present. The organized *religions* are constructs of civilized societies during historic time, but many religions retain many elements of animism. The recent development of *secularism* expresses the thought of the world of Science that everything can be rationally understood and explained without recourse to mystical appeasement of the supernatural.

Regardless of the specific design of a religious system, the human social tendency reinforces the felt need for a belief in a god, or gods, and for systems of interpretation of faith. This same tendency produces a social grouping of mankind into units larger than the natural family. This social grouping by religious systems is one of the basic structural patterns of human life on earth that continues in force. Religious social grouping becomes a regional phenomenon; the basic attitudes involved become strong elements in the formation of culture systems; thus religious patterns become dominant elements in the differentiation of cultures all over the earth.

Among all human groups there have been those persons best able or most interested in interpreting supernatural phenomena for the rest of the group. The shaman, the soothsayer, the fortune-teller, the priest, the minister, the evangelist, the lay preacher, and the religious reformer, all belong to the cult, fraternity, or order that specializes in establishing contact with the supernatural. These agents practice the arts of appeasement and interpretation, developing and maintaining the beliefs and rituals or the ceremonies and creeds by which the group may receive comfort, understanding, and interpretation. Initially their function was formulation, whereas later it became preservation and maintenance of the proper beliefs and doctrines. As such, they oversee the social formulation of culture systems and serve as bulwarks of precedent against untoward change.

Although the initial demand for a system of dealing with the supernatural might seem to have specific limits, every such system developed influences extending throughout the culture, for who could be certain as to what human act, or what failure in human action, annoyed the gods? Taboos against the eating of certain products, the definition of modesty, the rules of marriage, and the prescription of social behavior were initially set down as religious rules in simple societies. The development of the religious system, therefore, extended not only to interpretation of supernatural phenomena, but also to the regulation of the daily life of all members of a group. In these ways religious forces became highly significant in the shaping of early culture systems.

SOCIAL ORGANIZATION, GROUPING, AND LEADERSHIP

The smallest functioning human group is the natural, or nuclear, family, consisting of two parents and immediate offspring. It is unlikely that human occupance of the earth began through the dispersal of single families. It is much more likely that the *band* formed the first living group. As the simplest organizational grouping unit, the band is not properly a political organization but is a loose and changeable social unit. As such it operates as a communal group in the economic sense, sharing the proceeds of the search for food, tools, and weapons. Early Paleolithic populations were small, and their patterns of dispersion were such that political organization was not necessary. Conflicts that could not be resolved within a band resulted in the withdrawal by a dissenting splinter group, which itself functioned thereafter as a separate band. The splintering-off of band members into new bands was a result of good conditions and growing numbers of people. Sizes of bands seem to have been between twenty and one hundred members, and numbers above the latter figure normally resulted in the splintering-off of a group large enough to form a new band.

It can be assumed that early, small societies showed little internal structure, as compared with large modern societies. Internal grouping was chiefly by sex, age groups, and personal characteristics. Kinship patterns, beyond the nuclear family, provided varying kinds of subgroupings that bound members into units. Leadership in the small and simple society was exercised by the more experienced and wise elders who had strong personal qualities. These features remain true for the modern remnants of simple societies still in existence. All early societies developed particular systems of marriage, thereby requiring the development of certain elements of social organization. If marriage outside a given kin group became the custom, the clan system was devised to control marriage, with husbands or wives being chosen from different clans. The *clan* is a social group having unilateral kinship; thus one does not join a clan but is born into it. If marriage within certain kin groups was permitted, then other lineage-line structures were devised. According to the rule followed in a given society, residence location for newly formed nuclear families and inheritance patterns for personal property within families, clans, or lineage groups were variably arranged.

As the populations of the early societies increased beyond simple band limits, then further social organizational elements had to be devised. Territoriality, language, and external relations between societies also began to be elements in the total organizational structure. The clan system began to function beyond the level of marriage control, and the clan leaders took on various administrative roles within a society. Leadership tended to become more formalized, and the privileges, duties, and obligations of particular members of the society became more specific. Simple kinship systems tended to function less well in the life of the society, and the beginnings of social ranking systems began to appear. As these developments took place, however, kinship organization tended to be less effective in the management of society, and separate elements of political organization began to be created by which stronger authority could be exercised than that provided by purely personal qualities.

TRIBAL ORGANIZATION AND THE CHIEFDOM

With significant growth in the number of bands and the total number of people inhabiting a region, problems of competition for territory began to occur, along with problems within groups over the exercise of personal and territorial authority. This condition brought about the evolution of political organization. The tribe and the chiefdom form the first clear marks of political organization. A *tribe* is a group of bands that have entered into a voluntary alliance by which they acknowledge a specific leadership that possesses authority above and beyond the level of the family, clan, or band. Within a tribe all units and members worked together in economic, social, and ceremonial matters, although bands retained a certain share of internal autonomy. All members of a tribe recognized the authority of the *chief*, who became the executive authority in the exercise of external relations and the defense of the tribal territory. In all cases a degree of authority and superior social status marked the role of the tribal chief.

The chiefdom, in early stages of tribal organization, often was an elective status on some basis, and there were distinct limitations on authority. In the later stages of political development within tribal structures, authority increased, and the role tended to become hereditary within a particular family, lineage, or clan line. To the chief always went economic rewards of leadership in the form of gifts of produce or handicrafts, but the chief had the obligations of managing the redistribution of economic

goods throughout the tribal membership in some pattern determined by the social structure of the units of the tribe. Among gatherers, collectors, and hunters the upper limit to effective numbers of a tribe seem to have been in the vicinity of five hundred people. Above this limit, face-to-face contact and language uniformity are difficult to maintain, and the tendency is toward the formation of another tribe.

A primary function of the chiefdom, in external matters, concerned negotiation with other tribes, or leadership in war. The formation of an aggressive entity often resulted in territorial changes and the redistribution of peoples, and sometimes in structuring a larger regional organization through dominance by the stronger tribe and subservience or tributary status of weaker culture groups.

In regions where considerable competition arose among peoples no longer closely related by ethnic or cultural ties, tribal organization sometimes matured into the *confederacy*. Such a grouping is chiefly political and territorial in objective, bringing together tribes that are culture groups discretely separated by many different attributes. The confederacy ordinarily resulted in a linked relationship, in a formal and negotiated way, for a period of time, and for a single or several objectives. Confederacies are bound to dissolve in time; for, the groups that compose them are separate political or cultural units, each of which possesses in its own right, autonomy, specialized interests, specific jealousies, and ambitions that eventually splinter the confederacy.

All the early forms of organization freely permitted the differentiation of early populations into separatist groups for residential location, the growth of bodies of culture, and the development of local regionalisms. It is only with the Middle Neolithic that organizational forms tended to work toward cohesion, aggregation, and unity. Secondary functions of social organization were the establishment and maintenance of friendly relations between culture groups or the alignment of friends and enemies. Other secondary functions relate to control over basic resources and evolving production and distribution systems. In the case of unfriendly relations, organizational groupings sought to establish the bases on which belligerency could be regulated and kept to a minimum.

There is a tendency in the modern world to consider peace as normal and to consider militant competition (war) as abnormal. Competition for living space is, however, a normal state of affairs for all life-forms, reckoned in the concept of territoriality. Throughout human time militant competition for living space has been more usual than has peaceful existence. The bid by human groups for territorial living space leads to the condition described as war. Once human organization developed beyond the

A group of Dinka village chiefs in southern Sudan is being addressed by a senior chief. Each of the village chiefs carries a ceremonial symbol of his rank. The Dinka have only a simple tribal organization.

level of the band, a share of the people, time, resources, and energy of every group, probably, has been devoted either to the acquisition of added territory or to the defense of the existing territory. The warrior has always been honored in a time of crisis in such matters, and some societies have placed high values on war and militarism. More generally, however, humanity has sought peace and has stimulated the search for the means of achievement of that condition in which peace everywhere is the goal. So far, mankind has not been successful in achieving this cultural ideal.

Warfare may result from the strong desire for the conquest of territory or from the determination to defend territory. For a people subjected to strong external pressures and stress, there is another alternative to war—the abandonment of territory, gradual retreat into poorer territory, group dispersion, or refuge in some still unclaimed territory. The hills and mountains in several parts of the earth have often served as zones of retreat, areas of refuge, and regions of dispersion for peoples who either chose not to defend, or could not successfully defend, their original regions of occupance.

Although we view war, today, as largely a political and economic issue, there is ample ground in animal behavior for suggesting that much of early human warfare was marked by ritual performance, ceremonial sham battling, and large amounts of bluffing, and elements of all of these have been present among the remaining simpler societies in recent time. Sheer brutality, of course, has always been with the human race, and there are numerous historic examples of brutal conquests. Warfare in the punitive sense, however, is a product of civilization and culturally advanced technologies.

RESOURCE SYSTEMS AND THE DOMESTICATION PROCESSES

The human race is omnivorous, in contrast to such herbivores as cattle and to such carnivores as the cat family. Biologically, the earliest humans inherited from their animal ancestors certain tolerances and preferences for food items, but all peoples have since gradually developed certain additional patterns of diet preference. Culturally, *taboos* became attached to certain food items through experience, resulting in habits of refusal to consume those items at any time, although the true reasons for the nonedibility were slow in being understood. Knowledge of fruiting seasons, spawning periods, migrating periods, and other life-cycle elements for both plant and animal forms were slowly learned for the different new regions in which peoples settled down to live. In becoming widely distributed over the earth, mankind took the first steps in the evolution of a wide-ranging series of dietary systems based on different patterns of local resources.

Through trial and error in human campsites various simple processings evolved, such as aging, drying, soaking in water, leaching in water, pounding, roasting, and grinding. The inventions in domestic food-processing technology such as boiling, baking, parching, broiling, and frying, were important in making possible the better utilization of many food sources and in crystallizing human food habits and dietary systems.

On the production end of the food supply the knowledge of seasonalities, as in the ripening of fruits and tubers, developed annual harvest patterns for many resources and stimulated varying techniques of acquisition. Hunting and fishing procedures, seed-gathering techniques, and other means of acquisition stimulated the invention of specific tools, utensils, and implements for specific use.

On the processing side, continued experimentation resulted in the devising of a wide range of utensils, implements, tools, and their related procedures. Such processing included means of rendering initially poisonous items palatable, inedible items edible, and momentary surpluses storable. These procedures systematized the acquisition of food in response to sheer biological need, further differentiating human dietary systems from those of animals.

The steady elaboration of this whole complex of regional developments in food supply became surrounded by numerous habitual ideas, taboos, insistences, preferences, and dislikes. In the variance of the natural-resource range lay the basic origins of multiple systems, but in human extension of taboos and preferences lay the cultural sources of variation that led to a great multiplicity of dietary systems in the occupancy of the earth.

Beachcombing, Fishing, and Aquatic Resources

Beachcombing for edible, interesting, curious, and strange things is probably one of the oldest human activities. The tide zone has always yielded a return of something for anyone who searched. As a fishing technique, beachcombing is extremely varied, widely developed, and an ancient procedure. Along seacoasts, lakeshores, in river estuaries, and in shallow lake waters the trapping of marine products is an old technique using different specific procedures and materials. The simplest of trapping procedures is the building of rock-ringed enclosures that, at low water, trap a variable product in a procedure closely related to beachcombing in natural tide pools (Fig. 3.1). At its most advanced level, fish trapping involves some very complex constructions in fairly deep water with ingenious recovery systems.

Fishing, as a technology, displays a bewildering range of procedures dependent on the local environment, the marine biota present in any area, the cultural circumstances, and the level of technology in use. The traditional lore of "where the fish are," "what the fish will bite," and "how to catch a fish" demonstrates that the fisherman usually lives in a world of his own. The element of sport seems always to have been attached to fishing, along with the ever-present need for subsistence.

The variation in means of catching fishery products has been accompanied by a wide variation in their processing, storage, and consumption, around which a rich lore has also developed. Out-of-the-water prompt consumption, both raw and processed in some manner, lies at one extreme; traditional forms of drying, salting, smoking, fermenting, and pickling processes lie at the other. To the latter have been added the modern procedures of canning and freezing. With these modern exceptions, the processing of fishing products represents an ancient set of techniques.

Large portions of the world, of course, allowed little fishing, yet it is only during the era of drying out after the last glacial period that the arid lands of the earth have failed to provide a resource for those who would fish. In more recent time, with the evolving of complex dietary systems motivated by religious principles, fishery products have become either taboo or periodic alternatives to other food sources. With the evolution of crop growing and animal husbandry the fishery resources of the earth became a dietary protein complement utilized in many different ways by crop growers and raisers of animals.

Plant Domestication

Early mankind became skilled in the knowledge of the properties of plants long before people began to domesticate them into purposeful producers. The selection of fuel for fires, the processing of string and cord, the making of particular wooden tools, the securing of poisons for several uses, the use of plant

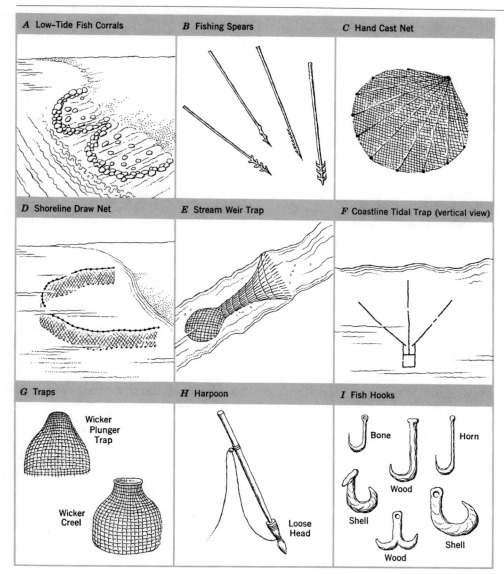

Figure 3.1 The Simple Tools of Fishermen.

A Low–Tide Fish Corrals

B Fishing Spears

C Hand Cast Net

D Shoreline Draw Net

E Stream Weir Trap

F Coastline Tidal Trap (vertical view)

G Traps

Wicker Plunger Trap

Wicker Creel

H Harpoon

Loose Head

I Fish Hooks

Bone

Horn

Wood

Shell

Wood

Shell

extracts as medicines, the building of shelters, the processing of plant materials for clothing and their employment in magic rituals and ceremonials all lie beyond the direct appropriation of plant products for food usage. Simple culture groups living close to their plant world know and differentiate many hundreds of species of herbs, roots, grasses, shrubs, and trees.

There is good cause to ascribe domestication of useful plants to more than one reason and not to attribute the entire development to the process of working with food plants. The interests were significant in fibers, colors, dyes, poisons, glues, spicy flavorings, strange exotics, "magic" plants useful in religious ritual, and plants yielding items of medicinal value. One of the very old traditions attaching to certain domesticated plants is the "sacredness" theme, owing to rather peculiar growth characteris-

tics. Our modern viewpoints, when most of us no longer live in vitally close relationships with the plant world, may well see the whole matter in too secular and too "scientific" a manner.

Prior to purposeful planting and harvesting, early peoples in many different regions of the earth must have affected the qualities and growth habits of large numbers of plants in unplanned and unconscious ways by cutting, picking, burning, and collecting, and scattering of seeds or root stocks. By such actions, early peoples affected both the natural reproduction systems and the distributions of plants they thought useful. And by such actions early peoples affected plants differently, depending on where they lived. In zones where grasses produced large amounts of small seed, the struggles of early peoples to subsist had a different physical/biotic impact than in zones where many of the useful prod-

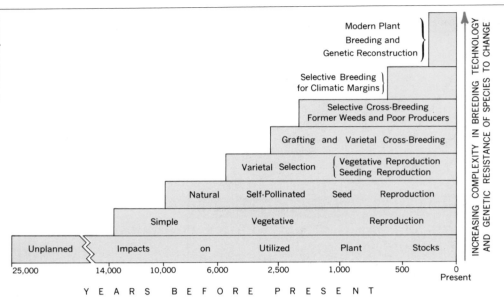

Figure 3.2 Plant Domestication Technologies, a schematic diagram.

ucts were fruits or nuts growing on shrubs and trees, or in regions where climbing vines and large-leafed plants produced the thickened roots dug for food. A distinction can be made between plants that reproduce by seed and those that maintain continuity by vegetative reproduction. The late Pleistocene was a period of rapid change in the intrinsic and highly plastic characters of the angiosperms by purely natural processes, and it is most likely that Paleolithic peoples stimulated the patterns of change for a good many species of plants, particularly in the regions of mild environmental conditions.

The time period during which plant domestication occurred used to be set at about 5000 years ago, and the whole process of domestication was thought to have taken place rather suddenly within a short interval of time. As more and better archaeological work has been done, and as more is learned about plants, the dates for domestication are being pushed back further and further. In the late 1970s the conventional date is placed at about 11,000 years ago, but this date is still thought of as providing too short a time interval. We believe that the very earliest beginnings of "domestication" took place with a few easily adaptable plants in the humid tropics and subtropics (not much affected by the last glacial period) as far back as 25,000 years ago. Very gradually, as people learned more about plants, the domestication process was increasingly effective with plants that were harder to change or in which significant changes took place very slowly. Domestication has been a very slow and long-continuing process, and its effects have been cumulative (Fig. 3.2). Some of our important crop plants have been brought into domesticated form, growth habit, and productivity only in the last two thousand years.

Even today, domestication is a continuing process with such plants as the *Hevea* rubber tree, some of the palms, and some of the forage grasses, and modern foresters are beginning to think about domesticating types of trees good for timber.

Figure 3.3 Regions of Significant Plant Domestication. Numbers correlate with regions listed on Table 3.1 pp. 54–55.

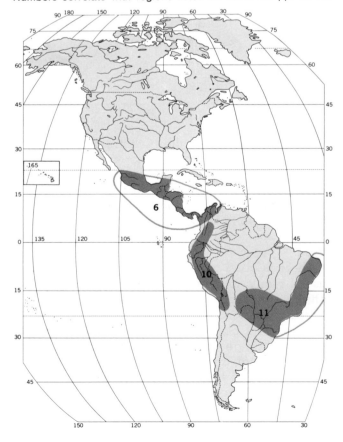

The above does not at all suggest that crop growing and "agriculture" began 25,000 years ago, but it does suggest that the whole process of plant domestication is more complicated than previously thought. Domestication procedures, with crop plants the ultimate result, were the work of sedentary peoples who lived with those plants for long periods of time, and became very familiar with them. The collecting of seeds, fruits, berries, leaves, roots, and tubers, and the eating of them or the other uses made of them, produced preferences for particular varieties among the users, who could watch situations in which cross-pollinations or mutations occurred, to produce new varieties. Natural changes took place in many such plants as the glacial period ended and postglacial conditions developed. Early peoples furthered these patterns of change by their constant disturbance of environmental conditions, and certain plants were favored at the expense of others in this process.

One theory places incipient domestication and the beginnings of crop growing in the humid, lower hill-country of Southwest Asia. This theory holds that around the wheats and the barleys lie the beginnings of all crop cultivation. We do not question this as one of the beginning locations, but how could rootstock and tuber planting derive from these beginnings? It seems more likely that there is more than

one kind of origin to domestication of plants and the beginnings of crop growing, since procedures involving the seeding plants are quite different from those developed around the plants that reproduce by vegetative propagation. Recent evidence points to a conclusion that both domestication and crop growing were multiple developments that took place in several equable regional environments. In such areas, culture groups had been long resident, having extended experience with the plants of those separate regions. The accompanying map and table (Fig. 3.3 and Table 3.1) depict our interpretation of the distribution of domestication and initial crop growing for the more common crop plants.

In regions in which wild seeding plants formed the bases of a gathering economy, we believe that crop planting based on seed-planted species could have developed. Where wild seed-bearing plants were less common, but where roots, tubers, and fleshy growths formed the bases of a gathering economy, it seems likely that crop planting based on vegetative reproduction would have resulted. The two sets of procedures are not interchangeable in the beginnings of crop planting. The seeding grain plants are not native to the tropics and the warmer, more humid subtropics, whereas originally few of the plants yielding large roots and tubers and propagating themselves by vegetative reproduction were at home

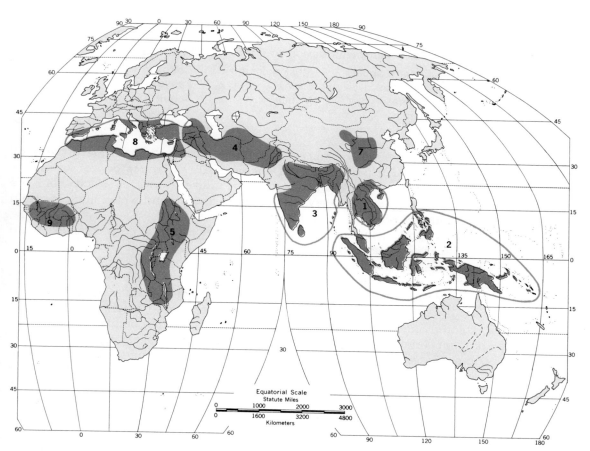

Table 3.1 Chief Source Regions of Important Crop Plant Domestications

A. Primary Regions of Domestications

1. THE UPPER SOUTHEAST ASIAN MAINLAND

Citrus fruits*	Yams*	Eugenias*	Teas
Bananas*	Cabbages*	Job's tears	Tung oils
Bamboos*	Rices*	Lichi	Ramie
Taros*	Beans*	Longan	Water chestnut

2. LOWER SOUTHEAST ASIAN MAINLAND AND MALAYSIA (INCLUDING NEW GUINEA)

Citrus fruits*	Yams*	Sugarcanes	Lanzones	Gingers*	Cardamom
Bananas*	Almonds*	Breadfruits	Durian	Brinjals*	Areca
Bamboos*	Pandanuses	Jackfruits	Rambutan	Nutmeg	Abaca
Taros*	Cucumbers*	Coconuts	Vine peppers*	Clove	

3. EASTERN INDIA AND WESTERN BURMA

Bananas*	Rices*	Peas*	Vine peppers*	Kapok*	Sunn Hemp
Yams*	Amaranths*	Grams	Gingers*	Indigo	Lotus
Taros*	Millets*	Eggplants	Palms*	Safflower	Turmeric
Beans*	Sorghums*	Brinjals*	Mangoes	Jute	

4. SOUTHWESTERN ASIA IN GENERAL (NORTHWEST INDIA–CAUCUSUS)

Soft wheats*	Poppies*	Beets*	Apples	Plums*	Pistachio
Barleys*	Oat*	Spinach	Almonds*	Figs	Walnuts
Lentils*	Rye*	Sesames	Peaches*	Pomegranates	Melons
Beans*	Onions	Flax	Soft Pears*	Grapes*	Tamarind
Peas*	Carrots*	Hemp	Cherries*	Jujubes*	Alfalfa
Oil seeds*	Turnips				

5. ETHIOPIAN AND EAST AFRICAN HIGHLANDS

Hard wheats*	Barleys*	Oil seeds*	Coffees
Millets*	Peas*	Cucumbers*	Castor beans
Sorghums*	Beans*	Melons*	Okras
Rices*	Vetches	Gourds*	Cottons*

6. MESO-AMERICAN REGION (SOUTHERN MEXICO TO NORTHERN VENEZUELA)

Maizes	Sweet potatoes	Custard apples	Muskmelons	Cottons*
Amaranths*	Squashes	Avocados	Palms*	Agaves
Beans*	Tomatoes*	Sapotes	Manioc	Kapok
Taros*	Chili peppers	Plums*		

far outside these two regions. The distinctive wild plant associations of the large territory forming the humid tropics and subtropics meant that a great range of wild plant material lay open to manipulative procedures by early peoples. Culturally, each of these environments contained human groups that had already developed preference systems, food habits, and manipulative systems based on wild plant products by the time domestication can be considered to have yielded assemblages of crop plants.

We cannot be certain of the regional origins of all domesticated plants, but we can form certain broad generalizations about their beginnings (see Fig. 3.3 and Table 3.1). Although we seem to have lots of crop plants, the early historic procedures domesticated only about 1850 distinct species and subspecies of crop plants out of the 350,000 species of higher plants. Meanwhile several hundred species now on the threshold of domestication are grown as

almost wild plants. Many of the forage grasses fall into this category. These figures do not, of course, include the decorative plants, of which there are several thousand and around which there is also a long history.

In the early development of crop complexes and cropping systems, people in different regional sectors of the earth concentrated on particular groups of plants locally available. Thus it is that the basic cropping system of Neolithic southeastern Asia can be said to have been based chiefly on taro, yams, bananas, and a series of other plants that reproduce by vegetative association. Similarly the cropping systems of southwest Asia came early to depend chiefly on wheat, barley, and a series of minor seed crops. In the New World maize, the squashes, and a series of beans became the primary base of the cropping system in Middle America and its northern and southern margins.

Sometimes domestication is spoken of dramatically

Table 3.1 (*Continued*)

B. Secondary Regions of Domestications

7. NORTH CENTRAL CHINA (INCLUDING THE CENTRAL ASIAN CORRIDOR)

Millets*	Cabbages*	Rhubarb	Bush cherries*	Jujubes*
Barleys*	Radishes*	Mulberries	Hard pears*	
Buckwheats	Naked oat*	Persimmons	Apricots	
Soybeans	Mustards	Plums*	Peaches*	

8. MEDITERRANEAN BASIN—CLASSICAL NEAR EASTERN FRINGE

Barleys*	Grapes*	Parsnips	Carrots*
Oats*	Olives	Asparagus	Garlic
Lentils*	Dates	Lettuces	Sugar beet
Peas*	Carobs	Celeries	Leek

9. WESTERN SUDAN HILL LANDS AND THEIR MARGINS

Sorghums*	Fonio	Peas*	Gourds*	Kola nut
Millets*	Yams*	Oil seeds*	Oil palms	
Rices*	Beans*	Melons*	Tamarind*	

10. ANDEAN HIGHLANDS AND THEIR MARGINS

White potatoes	Strawberries	Quinoa	Arrocacha
Pumpkins	Beans*	Oca	Ulluco
Tomatoes*	Papayas	Cubio	

11. EASTERN SOUTH AMERICA (CENTERED ON EASTERN BRAZIL)

Taros*	Pineapples	Cacao	Tobaccos
Beans*	Cashew nut	Passion fruits	
Peanuts	Brazil nut	Cottons*	

*The asterisk indicates domestication of related species or hybridized development of new species during domestication in some other region or regions. Some of these secondary domestications were later than in the original region, but clear evidence of chronologic priority seldom is clear-cut.

The plural rendering of the crop name indicates that several different varieties/species either were involved in initial domestications or followed thereafter.

The term "oil seeds" indicates several varieties/species of small-seeded crop plants grown for the production of edible oils, without further breakdown.

In regions 2 and 3 the brinjals refer to the spicy members of the eggplant group used in curries, whereas in region 3 the eggplants refer to the sweet vegetable members.

None of the regional lists attempts a complete listing of all crop plants/species domesticated within the region.

The table has been compiled from a wide variety of sources.

as an Agricultural Revolution, but it was a long process rather than a sudden event. Manipulative procedures certainly began in the late Paleolithic. The term *Mesolithic* denotes the era of transition away from Paleolithic exclusive dependence on gathering and collecting. The Mesolithic should be regarded as a fairly long interval involving several millennia, during which late Paleolithic man was bringing about a whole group of cultural advances that would enable the Neolithic to be typified by a fundamentally different way of life. By definition the *Neolithic* is an era in which the most advanced peoples utilized production systems rather than appropriation systems (gathering and collecting) only.

CROP-GROWING AND CROPPING SYSTEMS

We do not really know much about how early peoples began the sequence of activities that eventually produced crop growing as a regular procedure. We can presume such an origin as the "rubbish heap" concept, as an illustration, or we can presume any one of several other possible origins. In the rubbish heap notion, there is the presumption that early, repeatedly used human campsites created small clearings in areas of some pattern of wild plant cover. Such clearings created disturbed patches in the vegetative cover, and these may have been enriched by the repeated disposal of camp wastes. Accumulations of such soluble organic compounds created small islands of good soil around campsites. Within these areas camp residents processed collected wild seeds and dropped some of them accidentally on the ground. Alternatively, in other environments, they may have thrown away bits of rootstock or plant portions able to grow by vegetative reproduction, and people may well have thrown away pits of fruit or lost stray nuts that remained unopened. Enriched by the accumulated wastes and favored by the open ground, many kinds of plants would have found such disturbed spots quite acceptable growing sites. Such chance occurrences would have created different kinds of wild plant associations best described under the heading of "wild dooryard garden." Products of such growth would have been collected and used just as would those same items gathered far afield. Perhaps more luxuriant growth and more of the desired product would have resulted. What mental step then was needed to conceive of scattering planting materials purposefully around the open ground of the campsite? This we cannot trace, but if such did occur, the dooryard garden had become a purposeful planting site, and the resulting yield had become a crop. Given such sites, changes in the more adaptable wild plants would have taken place, and the domestication process would have been under way. In this presumption, there is also implicit the first crop-growing procedure, the first garden site, and the beginnings of a long evolutionary pattern.

The rubbish heap origin may not have been the only beginning of crop growing, but in some way Mesolithic peoples did begin to plant and harvest. At the outset planting-harvesting did not totally replace gathering-collecting as the source of all human food supply, since the skills were but little developed. The first returns from plants purposefully grown must have been very like those secured from collecting wild products, and the procedures employed must have been similar. The first "crops" produced cannot have amounted to much in volume, and peoples had to continue gathering and collecting for most of their subsistence.

There were also other problems concerning production systems. Early peoples must have engaged in trial-and-error methods of plant domestication. They knew nothing about the qualities of soils or the means of handling them, and nothing about cultivation practices. They had no specific tools or implements to aid operations, no concepts of weeds and how to control plant growth. Although they knew gathering seasons, they had to learn the planting calendar. Early mankind knew that the gods often produced annual sequences of weather that altered usual patterns of seasonality, resulting in abundances and scarcities. Not understanding these things, early planting groups placated the gods and sought their assistance in many different ways to make efforts at growing crops turn out favorably. These latter efforts often took the form of magic rituals.

Slowly, Mesolithic peoples worked out procedures and combinations of procedures for planting, weeding, cultivation, scaring off wild animals, and for winning the favor of the Gods of Nature. There began the development of tool systems, clearing-planting-weeding systems, selection systems aimed at better crops, and harvest and storage systems (Fig. 3.4). Mesolithic peoples had no preconceptions

of what constituted a good cropping system. The earliest fields were jungle plots in which all kinds of things, including weedy growth, grew in profusion in an approximation of the conditions of nature, modified only a little by the efforts of mankind.

Trial-and-error efforts at plant domestication resulted in the accumulation of a group of domesticated plants that could support a population for a large part of the year with a varied food supply. In each region in which the domestication processes went on, there emerged a group of crop plants more or less ecologically suited to the local environment and suited to the provision of a reasonable diet (see Table 3.1). This involved, first, a primary crop plant that supplied the carbohydrate dietary bulk, around which most crop rituals centered. It also involved a group of secondary plants that yielded edible oils, vegetable greens, varied sorts of sugars, spices, and flavorings, between-meal snacks, and medicinals. Minor elements were catch-crops (substitutes chiefly important only in "bad" years when the main crops failed, to avoid famine), alternate-season substaples, and sacrificial-ritual-magic items. Almost every group grew a few odd plants of peculiar characteristics that appealed to different groups. In the non-food category, in every domestication zone, there developed at least one fiber-crop plant yielding a string-netting-clothing textile. There was at least one beverage crop of nonalcoholic or alcoholic potential. There were a few plants utilized in making dyes, facial cosmetics, and body paints. The old and richly developed domestication zones showed the plant-replacement process, some early domestications becoming relict as a later domestication gained in favor through the yield of a better product.

It is clear that as domestication took place people were gathering the products from some plants that we have long considered to be weeds. A weed, essentially, is any unwanted plant that is growing spontaneously (and usually vigorously) in the midst of a plot of desired plants; thus a cucumber plant may be a weed in a pumpkin patch. The continued association of "weed" species with "crop-plant" species was an important aspect of the development of domestication.

We favor the dooryard garden as the place where cropping was first practiced, whether or not the rubbish heap notion accounts for its beginning. We regard the dooryard garden as the place wherein learning to handle plants under human control matured into cropping procedures. And in this pattern it is very likely that women were the family members who got things started, since they spent most of their time around the campsite processing food materials, working with clothing materials, and caring for the young while the men were loafing, hunting, fishing, or carrying on ritualistic fighting. Useful as the door-

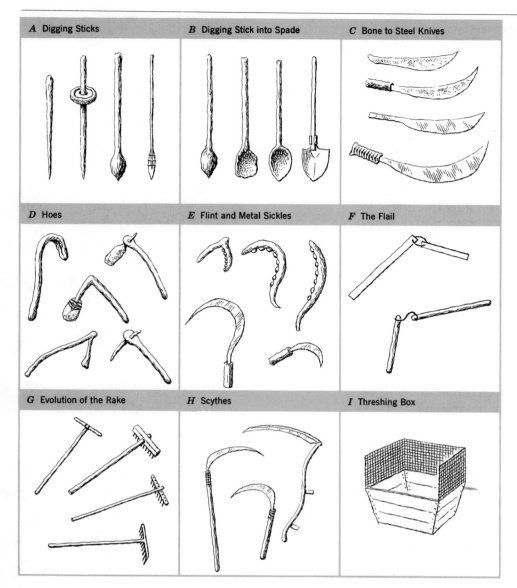

Figure 3.4
The Simple Tools
of Gardeners.

A Digging Sticks

B Digging Stick into Spade

C Bone to Steel Knives

D Hoes

E Flint and Metal Sickles

F The Flail

G Evolution of the Rake

H Scythes

I Threshing Box

yard garden was in the learning process, it could not produce enough food to supply a family regularly, even when complemented by the results of hunting and fishing.

Since gathering and collecting still was necessary, and since women did much of this, it is likely that women out collecting things found other local spots that, with a little clearing away of the wild plant cover, could serve as additional gardens. When such were also planted, they extended the range of production, and one or more such small sites might have helped greatly in providing food supplies. But such spots, lacking the enrichment of the ground around the campsite, would have produced worthwhile "crops" for only a season or two. This might well have led to the repeated search for other small planting sites that could be planted for a year or two. And out of this double kind of practice—maintenance of the campsite dooryard garden, and the selection

of additional garden spots in the neighborhood—there may have evolved the beginnings of the cropping system known as shifting cultivation. This, admittedly, is a somewhat speculative accounting for the origin of shifting cultivation, the cropping "system" by which new plots of ground were selected each planting season, cleared, and cropped for a year or two.

Shifting Cultivation

Whatever its Mesolithic origins, shifting cultivation became the basic cropping system of the Neolithic period. Working with small plots, seldom over an acre per laborer, this is really a form of gardening, since very simple tools and hand labor are involved. It is a natural rotation system in which patches of ground are cleared (chiefly by cutting and burning), planted with little preparation of the ground, given

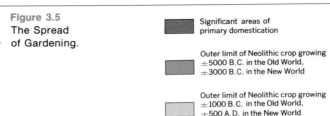

Figure 3.5
The Spread
of Gardening.

Significant areas of
primary domestication

Outer limit of Neolithic crop growing
±5000 B.C. in the Old World,
±3000 B.C. in the New World

Outer limit of Neolithic crop growing
±1000 B.C. in the Old World,
±500 A.D. in the New World

some weeding, assisted by some magic performance, and then hand-harvested. The ash residues from the burning provide a dressing of soluble nutrients to assist productivity. After a few croppings and heavy weed growth, a new plot is chosen and the routine repeated. Good yields normally accompany such operations; the vegetative association rapidly regenerates on plots no longer cropped; and little harm results to the landscape when the cultivator can move at his own discretion in a territorial range not heavily overpopulated. As a cropping system, shifting cultivation operates on the principle of *usufruct:* the cultivator does not own the land but holds rights to the clearing, cropping, and harvest of the yield of land, whereas the social group to which he belongs (lineage, clan, culture group) holds ownership-in-common for the land within the territorial range.

Shifting cultivation, as perhaps the earliest basic cropping system, spread gradually over the inhabited earth in which Neolithic man learned to grow crops and could cope with environmental conditions. Crop plants were moved from their home regions to different ones, and new trial-and-error procedures were needed to adapt them to new conditions; new crop varieties resulted from selection procedures, and again new practices were developed to accommodate to the new sets of conditions. Shifting cultivators always continued to collect particular wild plant foods from the open ranges around them, and also engaged in hunting or fishing or both to round out the dietary system. They also very often retained the dooryard garden as a place for trying out new varieties and for growing special plants. The two forms of gardening spread very widely over the world during Neolithic time (Fig. 3.5).

Shifting cultivation is still practiced among many of the simpler culture systems of the earth. Today its distribution is chiefly restricted to the tropics and subtropics where it may be, in fact, one of the best possible land-use systems. About two hundred million people are currently supported by shifting cultivation. It represents a holdover from the Neolithic period of an earlier era, and it will probably disappear in time as pressures of world population demands for food require the permanent cultivation of the remaining cultivable lands of the earth. Several

kinds of site-shift, and other accompanying practices, are used today, each of which is traditional among its users. In Figure 3.6, the random shift (top) indicates a specific sequence that might be followed in areas of rural dispersal of settlement and low population densities. The linear shift (center) involves the periodic relocation of hamlets or villages when distances from garden plots become great. The cyclic shift (bottom) usually permits permanent village settlement. Working systems, tool systems, commodity exchange patterns, and the recultivation of particular plots of ground after periods of fallow all form a part of a group's system of operating shifting cultivation.

Animal Domestication

It is uncertain just how animal domestication came about, but calculated and consciously purposeful action for economic ends does not seem to describe the development. Such action is not in accord with the other very early kinds of cultural development. More likely, domestication began as an unconscious development involving the symbiotic relationships between man and the small birds and animals who both scavenged his campsites and found in them a refuge from their own predators. It is likely that the

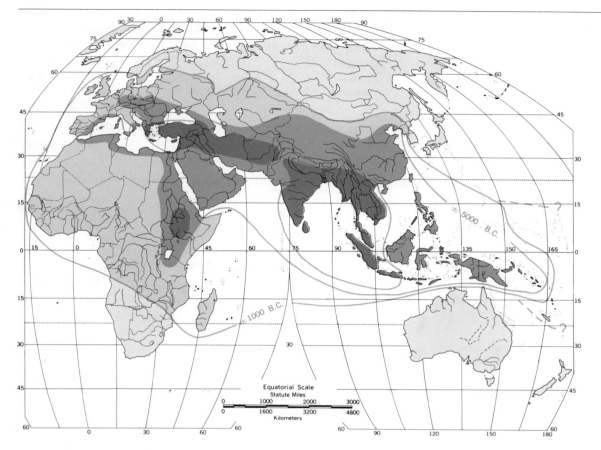

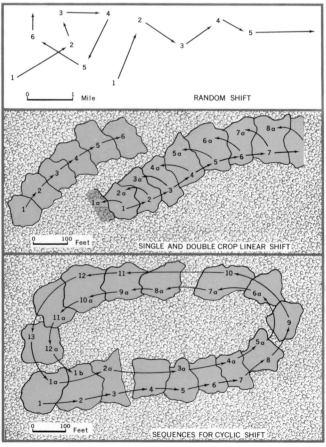

initial aspects of animal domestication took place prior to the significant onset of crop growing. There may have been the taming of very young animals and birds, secured from litters and nests close to the time of psychological imprinting of the young.

Very important to the domestication process were the symbiotic relationships between the small animals and people in the development of playmates—pets who were cared for. At some uncertain point in time there came into operation the practice of considering certain animals as sacred, to be killed only for important religious rituals. Undoubtedly, simple economic utility of domesticated animals grew in importance once such animals became common. Early mankind ate all the small wild animals as part of a regular diet, and this must have been an additional factor in the whole process.

Most likely the dog, the cat, the pig, the jungle chicken, the pheasant, and perhaps some of the pigeons and ducks were among the first animals to enter into the domestic relationship (Table 3.2). Archaeological evidence seems to suggest that the goat and the sheep also came early into a similar symbiotic relationship. Goats and sheep were large

Figure 3.6 Systems of Field Shift in Shifting Cultivation.

Table 3.2 Chief Source Regions of Important Animal Domestications

1. SOUTHERN AND EASTERN ASIA, SOUTH OF THE HIMALAYAS–YANGTZE VALLEY

Pigs*	Chicken*	Ducks*	Zebu cattle[a]
Dogs*	Pheasant*	Geese*	Water buffalo
Cats*	Peacock	Silkworm	Mithan
			Banteng (?)

2. INTERIOR EURASIA, NORTH OF THE HIMALAYAS AND WEST OF CHINA PROPER

Goats*	Horses*	Yak	Taurus cattle*[a]
Sheep*	Camels*	Reindeer	

3. SOUTHWEST ASIA, MEDITERRANEAN BASIN, NORTHEAST AFRICA

Pigs*	Ducks*	Donkey*	Sheep*
Dogs*	Geese*	Horses*	Rabbit*
Cats*	Pigeons*	Goats*	White rat
	Guinea fowl*		

4. MESO-AMERICA, INCLUDING THE ANDES

Llama	Muscovy duck	Turkey	Guinea pig
Alpaca			

enough and strong enough, when well domesticated and trained to the operation, to be of utility to man in carrying pack loads, although we cannot be certain how early such usage really began.

The chronology of domestications of larger animals is quite uncertain, for the dates have been moved progressively back into earlier time as archaeological research continues. A suggestive sequence of domestication would put cattle, the donkey, the horse, the reindeer, the camel, and the elephant in that order, with the first domestic cattle dating from as early as 7000 B.C. and the elephant from perhaps 2500 B.C.

The locations of domestication of all the animals are still uncertain (Fig. 3.7). Once the processes began, there apparently was diffusion of the concept, and numerous secondary domestications presumably occurred from southeastern Asia to the Atlantic Ocean. The accompanying chart (see Table 3.2) presents an attempt to categorize all of the animal domestications, but many problems remain surrounding many species on the list. It is notable that animal domestication succeeded far better in the Old World than in the New, for which the list is both small and not significant.

Prior to domestication Paleolithic man apparently had learned to drive, round up, or direct the movements of wild herd animals sufficiently to be able to exert some regional control over their seasonal movements. Many of the small animals and some of the birds are not strongly migratory but instead alternate seasonally between local regional ecological situations. Evidently Paleolithic man could derive considerable food support from many of the animals in a semicontrolled situation. It is likely that large herd animals often threatened to overrun the

Figure 3.7 Regions of Significant Animal Domestication. The regional names correlate with those used on Table 3.2, pp. 60–61.

Table 3.2 (*continued*)

5. MISCELLANEOUS AND MARGINAL NOTATIONS

Swan—appears to relate to northwest Europe at a late date.

Elephant—often considered domesticated, the Indian elephant seldom breeds in captivity and cannot be considered fully domesticated.

Onager—sometimes considered a historic domestication, extinction makes this doubtful.

Honeybee
Stingless bee   Although kept in captivity in production for man, it can be argued that
Cochineal   these forms are only manipulated.
Lac insect

Many of the modern laboratory animals, the fur-farm animals, and the zoo-bred animals comprise a large group of animals to which the older definition of domestication (change in bone structure, shape, or size from the wild species) does not fully apply. Current manipulation of many "wild" animal forms may eventually produce new breeds and new conditions of animal living systems that will require redefinition of the term *domestication* as applied to animals.

---

*The use of the asterisk suggests that several wild species or subspecies probably entered into the evolution of the domesticated breeds.

The use of the plural in the name indicates domestication in more than one regional environment.

^a The interbreeding of Zebu and Taurus cattle in both eastern and western Eurasia, both very early and more recently, has produced many mixed breeds of cattle under both wild and domesticated conditions.

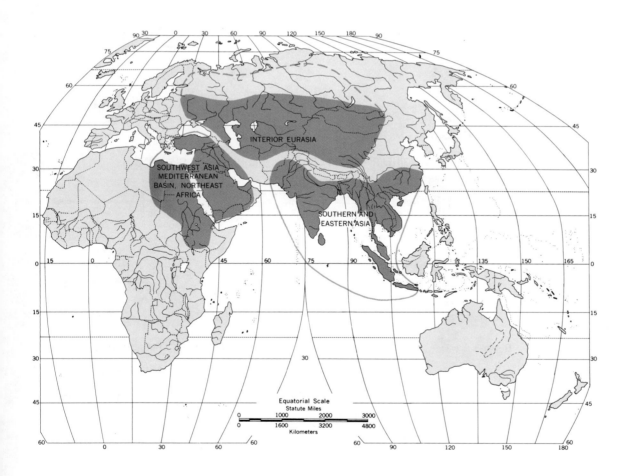

patches of crop plantings when these early farming patterns lay near the herd ranges; driving off the animals could be combined with hunting.

BEGINNING ANIMAL HUSBANDRY

The initial domestication of the small animals gave Mesolithic peoples a complementary aspect to their slowly developing activities in crop domestication. The two separate kinds of domestication broadened the range of economic possibilities. A common theme in the older social science literature refers to the evolution of hunting peoples into pastoralists and then into agriculturists. There is little truth in this theory of straight-line development as the standard of cultural evolution. Most probably the domestication of the Old World small animals such as the dog,

the cat, the pig, and some of the birds was achieved by some of the same culture groups that were also making progress in domesticating different varieties of plants. As the first animals and plants came within the limits of domestication, it is likely that both were used as food supplies by the same culture groups without specialization in either.

Pastoralism was not born mature, but was worked out slowly, as peoples learned how to care for grazing animals such as sheep and the goats. The other large animals that characterize mature pastoralism, such as cattle, camels, and horses, came into domestication rather late, and were added only after pastoralism had begun its development. In the early stages there must have been much experimentation in grazing patterns, in the kinds of country in which each animal could succeed, and in the kinds of care that were required by the different animals. There is, therefore, a slow evolutionary development of pastoral economy, whatever its precise nature.

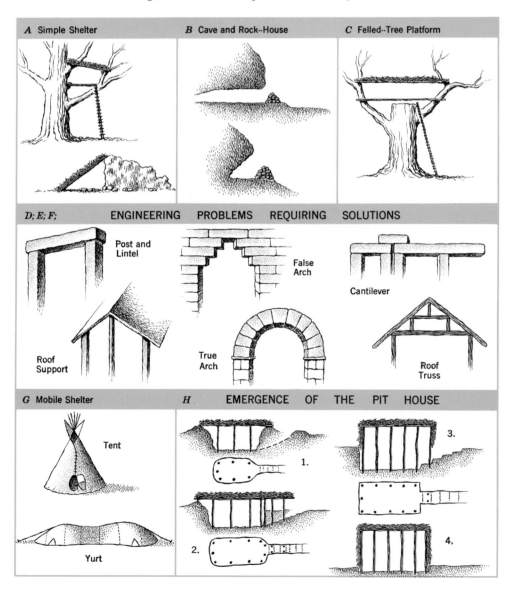

Figure 3.8
Elements of
Housing Systems.

It is notable that most of the animal domestications took place in the Old World, against a very small number in the New World (see Table 3.2). In both realms, however, small animals were among the first domestications so that the beginnings of animal husbandry were present at an early point in both hemispheres. Later, large-animal domestications in the Old World made forms of pastoralism possible. The New World lacked good large-animal potential domesticates and, therefore did not develop early forms of pastoralism. The kinds of pastoralism developed in the New World came about only after the post-Columbian transfer of Old World animals to the New World.

In many of the early environments in which culture groups were working at the general domestication processes, the kind of animal husbandry that slowly evolved into pastoralism probably formed an alternative to crop growing. In the heavily forested equatorial tropics and humid subtropics, early peoples could make clearings on which to grow patches of crops, and they could keep such small animals as could find food in those environments and around human campsites. They could not keep many large grazing animals owing to the scarcity of grass in forested regions. In the increasingly dry territories on the climatic dry margins, early peoples could begin to work with grazing animals such as sheep and goats, but could not grow crops except at oases. In the early stages of learning to grow crops and maintain animals, only restricted territories possessed environmental qualities that would make both forms of economy possible. It seems likely, then, that as shifting cultivation began to be shaped into a crop-growing system of economy in the mild and humid lands, the alternative of pastoralism also began to be shaped into a specific form of economy in the dry zones of the Old World. These two separate forms of economy led peoples into the development of two quite different human living systems.

Figure 3.8 *Continued*

HOUSING SYSTEMS AND SETTLEMENT PATTERNS

In the phrase "the caveman era" there is allusion to the cave as the earliest human habitation. People have always used caves when they were at hand, and it is likely that good caves formed the best shelters on the poleward margins during the last glacial advance. Having to spend nights and winters in the open in most parts of the earth where there were no caves, early peoples built shelters of several sorts as one of the early technological achievements.

There is no way of knowing fully what the earliest human shelters were like, for they must have been built of perishable materials, subject to rapid collapse and early decay (Fig. 3.8, and see Photo Essay No. 1, pp. 70–82). The earliest archaeological evidence of housing suggests oval-to-near-rectangular shallow pits dug into the ground, entered by a ramp at one end, and presumably roofed with some perishable plant materials. There seem to be evolutionary forms of the pit dwelling, which grew larger in floor plan and suggest the use of vertical posts for lifting the roof. The pit dwelling seems to have been a northern-hemisphere form that may have begun during a cold period in the Paleolithic Age. In other parts of the earth there must have been other elementary shelter styles also, but there is little evidence of what they were like.

By the early Neolithic Age, marking the onset of food-producing as opposed to food-collecting, a notable expansion in domestic building had taken place, although its technological and regional origins are quite unclear. Both the circular and rectangular floor plans were in use. Construction involved wattle and daub (reeds, twigs, and small branches interwoven and daubed with mud plaster), sun-dried bricks, poles, crude log slabs, and stone, as evidenced by archaeological sites. Building in more perishable materials also must have continued, but evidence does not remain. Structures ranged from small beehive units to large and complicated units of multiple stories. There is disagreement over the timing and roles of fortification, palace, storehouse, factory, and temple building, but all of the forms are part of the Neolithic record. Neolithic architecture blossomed quite fully, and many regions present a wide variety of architectural types and technological systems.

Along with domestic residential forms there came a variety of auxiliary buildings for purposes of storage, penning, sheltering of animals, and perhaps the development of the workshop. There is considerable evidence that very early in the development of domestic shelter building, religious motifs appear in the siting, directional facing, inclusion of shrines, focusing on sacred objects, and disposal of the dead. Among some peoples the cooking fire was inside the primary structure, but among others it became excluded from the living quarters, thereby requiring an attached auxiliary structure.

Once the primary technological elements of architecture were invented and diffused widely, there developed regions of specialization in particular forms and with particular traditions. These traditions, however, become the expressions of regional cultures. Throughout historic time the technologies and styles of architecture increased, as did the materials employed, becoming elaborated into a bewildering variety and wealth of cultural forms in which regional diversity became extremely marked.

There has been a common assumption that earliest man was nomadic, and that permanent settlements necessarily awaited the beginnings of cultivation. There is good reason to believe that this was not the case. The size of some of the kitchen-midden shell mounds suggest rather sedentary occupancy of a fine site for fish, oysters, or clams. In the primordial environments of the Paleolithic Age, when man had not yet highly disturbed the plant and animal associations that he found, small bands of people could have found many attractive resource ranges in many parts of the world that would have provided ample resources within rather small areas. It is possible, therefore, to suggest that sedentary living patterns began in many parts of the Paleolithic world rather promptly after the elaborating of basic subsistence systems. Earliest man must have tried many kinds of ways of working out these subsistence systems, and the elaboration of the periodic travel route through a given territorial resource range, as a mobile system, must be reckoned as only one of many alternatives.

Taking the fair assumption that early human economic subsistence systems may have ranged from quite sedentary to quite mobile, it is possible to suggest that all the basic forms of settlement, short of the town-city, really date from the Paleolithic Age Fig. 3.9 and Photo Essay No. 2, pp. 114–123). Among large numbers of working bands of human beings subject to different solutions to problems of human organization, the evolution of social structure certainly varied greatly in early times, as it continued to do later. A given regional population living in loose contact could have varied in its living patterns from full dispersal of nuclear families throughout a peaceful zone to relatively tight concentrations of bands in given favored sites.

It is doubtful that warfare significantly affected either architecture or the settlement pattern during most of the Paleolithic Age. By the Neolithic the techniques of both aggressive and defensive war-

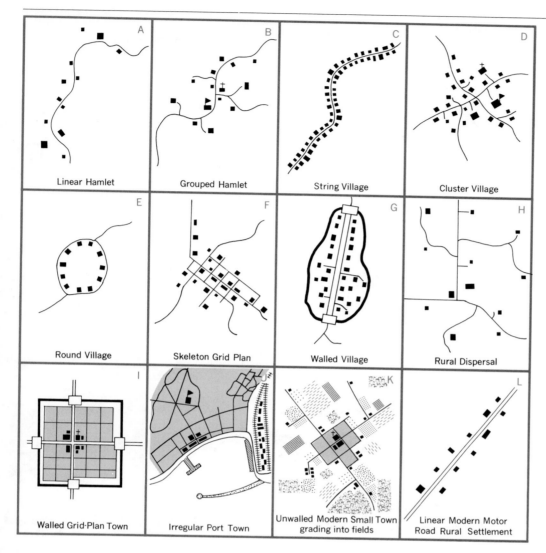

Figure 3.9 Basic Settlement Forms.

A — Linear Hamlet
B — Grouped Hamlet
C — String Village
D — Cluster Village
E — Round Village
F — Skeleton Grid Plan
G — Walled Village
H — Rural Dispersal
I — Walled Grid-Plan Town
J — Irregular Port Town
K — Unwalled Modern Small Town grading into fields
L — Linear Modern Motor Road Rural Settlement

fare had become sufficiently developed to have implications for both architecture and settlement. Certainly the construction of formal fortifications derives from the onset of the Neolithic, and the diffusion of fortifying devices probably spread rapidly. The selection of defensible sites for settlements had an effect both on settlement location and on the architecture of fortification. From the early Neolithic into historic and modern time, war has been an element in both fortification architecture and settlement patterns.

THE EVOLUTION OF POWER TECHNOLOGY

During most of human history, the tools that people could create and employ were small ones whose utilization depended heavily on the application of human muscle. Until a very late date, a mass-productive task required large numbers of human laborers over long periods of time.

Once the stone axe was well developed a man could cut down a very large tree in a surprisingly short time, but the task of doing anything with the felled trunk was a very different story. The spear-thrower added considerable leverage to the arm of a skilled and strong man, but the device made it possible to fell game, or enemies, at only a slightly greater distance (Fig. 3.10). Once people learned to use fire as a specific tool, as in burning the heartwood out of a log in shaping a dugout canoe, skill in manipulating fire could speed up certain kinds of operations. And hunting game by firing the grass cover of grazing ranges could help produce huge volumes of fresh meat at the bottom of the cliff off which the game was driven.

The devising of various agents to extend the power of human muscle has ranged through a wide variety of props and aids. The carrying basket, for shoulder, back, or head load, is one of a series of devices

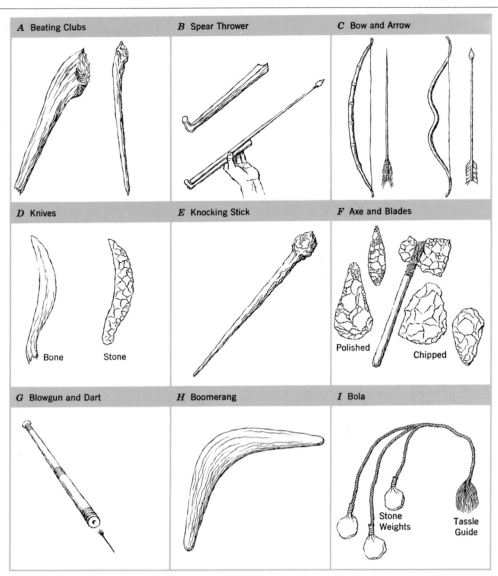

A Beating Clubs

B Spear Thrower

C Bow and Arrow

D Knives

Bone   Stone

E Knocking Stick

F Axe and Blades

Polished   Chipped

G Blowgun and Dart

H Boomerang

I Bola

Stone Weights   Tassle Guide

Figure 3.10
The Early Simple
Hunting Tools.

(Fig. 3.11). Skid boards, rollers, and various systems of "greasing" a surface played their separate roles in various areas at different times. Learning to package loads into sizes suitable to human strength was an early skill.

Once animal domestication began, mankind acquired another work agent of varying ability. Even dogs, goats, and sheep were put to work carrying, pulling, and grinding in parts of the earth in which they were available. As larger animals were domesticated, the process of power application was extended in many different ways in distinct regional patterns. In regions without domesticable animals, the delivery of power remained dependent on human muscle. The travois, the sled, and skid road, and finally the wheel were means of utilizing animal power more efficiently (see Fig. 3.11). Invention of the plow, seeding tube, cultivator, leveler, and scraper

using animal power provided another range of applications. The pack train of groups of animals, according to their availability, was still another variant. In different parts of the earth people employ most of the traditional methods of utilizing animal power, according to their skills and availability of animals of particular kinds, thus representing a continuation of late Neolithic systems.

The harnessing of wind in the development of power production was a slow matter that did not appear on land for a very long time. The development of sail systems on watercraft came much earlier for varying portions of the earth and probably relates to certain wind zones. Such systems were integrally linked to the skills of building the craft themselves. The use of wind in power production reached its productive climax in the windmill. Although the origin of the windmill may lie in a device involving a prayer wheel, its development in ancient Persia as a source of economic power was linked to the grinding of grain. The Arabs adapted it to pumping water, and in this usage it spread almost around the world in later time.

The use of water in power development was also slow and dependent on other kinds of inventions. The simple waterwheel for grinding grain, and later for lifting water, seems to have originated somewhere in southwest Asia (Fig. 3.12). Its spread involved variations in both construction and usage in developing power for mankind, but the inadequacy of engineering and construction knowledge prevented fruitful development. Not until the seventeenth century did real advances in the development of waterpower come about, and the modern evolution of hydroelectric power is the climax in this series.

In a sense, the learning to make and use charcoal in firing operations is a special development of power. However, the early operations that involved firing different kinds of fuels never matured into the production of power useful in separate applications until the invention of the steam engine and other forms of engines burning some kind of fuel.

Throughout early human history, therefore, people were severely restricted in the kinds of operations they could conduct. The wonder is, of course, that early mankind occasionally could have accomplished so much in this manner, but these results did not come during the Paleolithic. The huge mounds, the pyramids, the excavations, the long canals, the walls, and the other massive human works came only during the late Neolithic and protohistoric periods, at which time social and political organization finally permitted the massing of human energies for a given task.

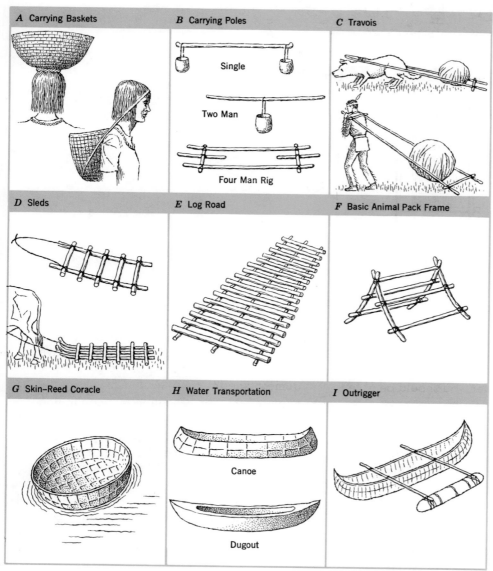

Figure 3.11
The Tools of
Simple Transport.

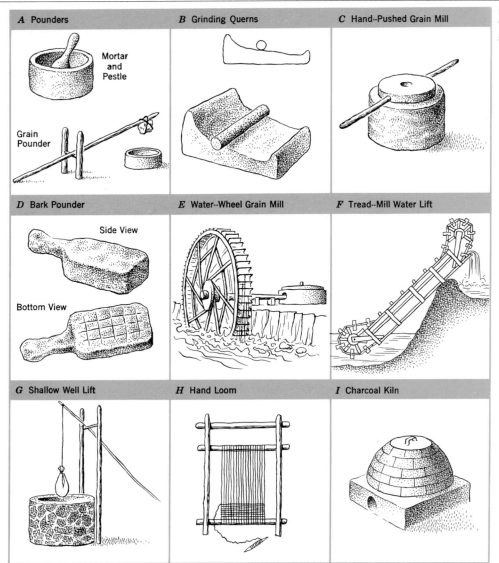

Figure 3.12
The Tools for
Simple Processing.

A Pounders — Mortar and Pestle — Grain Pounder

B Grinding Querns

C Hand--Pushed Grain Mill

D Bark Pounder — Side View — Bottom View

E Water--Wheel Grain Mill

F Tread--Mill Water Lift

G Shallow Well Lift

H Hand Loom

I Charcoal Kiln

SUMMARY OF OBJECTIVES

After you have read this chapter, you should be able to:

Page

1. Discuss why territoriality is a cultural trait. 45
2. Explain why systems for dealing with the supernatural eventually include rules about everyday human behavior. 47
3. Define *band*, *clan*, *tribe*, and explain why *chiefs* become necessary. 47
4. Understand the significance of plant domestication for human cultural development, and describe the differences between seed (sexual) reproduction and vegetative (asexual) reproduction of plants. In what parts of the earth were each likely to have been the dominant form of the plants first domesticated? 53
5. Explain why domestication is a continuing process rather than a prehistoric event. 55
6. Describe the regional origins of particular dietary complexes (i.e., for each major region of domestication, list the principal animals and the plants providing carbohydrate dietary bulk). 54
7. Describe the forms of shifting cultivation, the advantages and disadvantages, and the significance of this cropping system. 57

These windmills, lining a canal in Holland, serve to pump water out of the hinterland into the canal.

8. Explain the advantages and disadvantages of animal husbandry.   62

9. Describe (after studying Fig. 3.8, p. 62) the features of housing systems that are included in your present residence.   62

10. Discuss the ways in which you have ever employed any of the items depicted in Figure 3.10 (hunting tools), Figure 3.11 (transport tools), and Figure 3.12 (processing tools). For what purpose could each item depicted be used? Which are more likely to be used by women?   66

HOUSING PATTERNS

HOUSING SYSTEMS

Mankind's early systems of shelters were responses to the biological need for protected and private sleeping areas. As cultures developed, the concept of the shelter was extended from simple physical protection from cold, dampness, wind, and sun toward an increasingly complex set of cultural traits centered around the growing idea of the "family home." By the late Paleolithic, the shelter had also begun to serve as a storage place for various bits of "property." During the Neolithic, the shelter began to be separated into various functional parts. This separation has continued through the historic period until now there are often separate compartments for cooking, eating, bathing, sleeping, working, lounging, and playing. People also began to decorate their shelters. Although sometimes it served to indicate social or economic status, such decoration often had no functional purpose other than displaying various forms of artistic expression by the occupants. Everywhere on earth, mankind has always built shelters of some kind. The saying, "A man's home is his castle," indicates the wide range of psychological and cultural drives that have influenced the development of the original simple shelters into the great variety of family homes.

The early development of housing systems was very slow. A number of engineering problems, such as ways of building using different kinds of materials, had to be worked out before the simple shelter could become a house (see Fig. 3.8, pages 62 and 63).

The earliest shelters seem to have been dwellings in common for small groups of people. For most of mankind, however, the maturing idea of housing became family centered with each family coming to occupy an individual shelter of some kind. Many of these were grouped close together. Various historic examples include the compartmented "long house" of Southeast Asia and the equally compartmented compound house, resembling the modern apartment building, which developed in several different places around the world (Photo 10 depicts one variant).

During the Neolithic, as crop growing and animal husbandry began to evolve, space requirements within the shelter increased. The need developed for the working space that became known as the threshing floor. Areas were also required to store crops and shelter animals. In most culture groups, these new areas were eventually separated from the human shelter, giving rise to systems and patterns of crop-storage sheds, animal stables, barns, and other types of auxiliary buildings. The use of the house was generally restricted to people, selected pet animals and birds, and decorative plants. In the modern age, however, we often find the automobile taken back into the house structure via the attached garage.

Photo 1. These tents are the movable homes of a herding unit of pastoralists in Iran. For the most part, they are made of wool and wood poles and use wool rugs for flooring.

Photo 2. The Mongol *yurt,* here seen in winter quarters, uses felt, lambskins, short wood poles, and ropes to form a sturdy, windproof movable shelter.

Photo 3. The extended-family residence compound in Togo, West Africa, is normally built as a fenced-in series of units. Each unit is a round room with a thatched cone-roof and walls of plastered reeds or matting.

Photo 4. The traditional pile dwelling of Southeast Asia is normally set well off the ground. This flimsy Cambodian example is built of bamboo poles with walls of split bamboo matting and a roof of thin bamboo thatching. A ladder stairway gives access to the veranda.

Photo 5. This Yugoslavian log house is one of the early forerunners of the North American "log cabin." Built of squared logs, it is set on a stone foundation and is roofed with thin board shakes.

Photo 6. In the villages of highland Peru such as this one north of Cuzco, housing is often built of adobe brick, stone, or a combination of both and has a thatched roof. A modern road now cuts through the village.

The earliest human patterns of mobility probably involved building fresh shelters at each overnight stopping place. Later, mobile groups began to carry certain special elements with them on their travels. As Old World pastoralism slowly developed, peoples solved the problems of housing in areas lacking good building materials by creating several forms of mobile shelters, such as tents, yurts, and tepees, (see Fig. 3.8, lower section, page 62, and Photos 1 and 2). New World Indians who were mobile seasonally did much the same thing. Lacking large pack animals, they devised a form of travois that could be pulled by either a dog or a person to transport materials for dwellings (see Fig. 3.11, unit C). Historic desires for some mobility in housing among nonpastoralists were met by various kinds of "covered wagons," but the mobile-home problem was never efficiently solved until the development of the modern automobile trailer (Photo 19). "Mobile homes" have also come to include a form of prefabricated housing for permanent use on a particular site.

The "family house" has a great many variants in size, arrangement, and physical disposition on the homestead site. The single-room structure houses many of the world's families. Within these modest units, the small nuclear family can still arrange functional disposition of the enclosed space. Among extended families or families practicing some form of polygamy there is need for a more extensive amount of shelter. Often this is provided through the construction of multiple shelter units that are individually relatively easy to erect (Photo 3). Separate units are then devoted to sleeping, cooking, eating, visiting, or storage according to the culture-group's living system. Most of the world's houses, however, are built for the basic nuclear family. Whatever the grouping provided for, the house has gradually come to be a complex structure. Style, materials, shape, decoration, use, and other characteristics are determined by climate, custom, culture systems, and specific family needs (Photos 4 through 9, and 11 through 16).

Photo 8. Caves dug into hillsides can provide secure housing. These homes in Granada, Spain have smokeshaft ventilators. The fronts are designed to prevent erosion and hillside collapse.

Photo 7. Older homes in the Philippines are often built of wood in a two-story pattern with living quarters on the second floor. This farmhouse belongs to a village leader of modest means. It has sliding windows with panes made of shell. The galvanized iron roof is guttered to lead rainwater to a storage cistern.

Photo 9. The dome-shaped house is an old style in North Africa and the eastern Mediterranean lands. These village houses in Aleppo, Syria, are built of adobe block covered with a clay plaster.

Photo 10. This Pueblo Indian village in New Mexico is one of several forms of early apartment-dwelling complexes. It is built of plastered adobe blocks with roof beam supports. The outer units are normally used for living space, while interior and lower rooms are used for crop storage.

Photo 11. Built of fieldstone in thick mortared walls, with fireplaces and steep roofs of heavily thatched straw, the "cottage" is a traditional style in northwestern Europe. This example is in Dorset, England.

Photo 12. American colonial houses were ordinarily built of wood and followed the northwestern European tradition of the compact structure with a steep roof and one or more fireplaces. This is the Powell-Waller house in Williamsburg, Virginia. It is an example of one of several basic styles, the "Williamsburg colonial."

Historically, the building materials available in an area along with variations in style have created recognizable likenesses in the houses of different culture groups. Thus, one is able to speak of Chinese, English, or Burmese housing styles. Peoples colonizing new lands often take their house types with them. The domestic architecture of the United States and Canada exhibits patterns from many parts of the world; there are also broad zonal similarities in building materials and constructional principles that are determined by environmental conditions. Flat roofs evolved in the dry lands of the Old World, whereas steeply pitched roofs are found in the lands of heavy winter snow. The use of logs, either squared or round, developed in forested Europe where protection from cold and winter winds was sought by builders. The North American log cabin had many European antecedents (Photo 5). In the tropics and subtropics, where a mild climate makes protection from cold winter winds unnecessary, the use of leaves, matting, and simple thatching for walls and roofs has apparently always been common (Photos 3 and 4). Dwellings set off the ground on pilings became the accustomed style in Southeastern Asia and other regions where natural flooding or watery conditions would not allow groundlevel housing (Photo 4). Many areas of the world have caves dug into hillsides to form well-protected shelters, with the fronts finished off in the regional styles of the particular culture (Photo 8). Where suitable wood is scarce, both clay and stone have often been used as building materials, providing still another distinctive style of housing (Photos 6, 9, 10, and 11). Individualism, however, has always been present among home builders. Although styles may have some regional or cultural similarities, there are usually strong contrasts among domestic homes (Photos 12 and 13). The modern period, of course, has seen house construction and styling spread to regions far beyond the area of origin. Thus, one may find a "California ranch house" in eastern Canada or an Arabic flat roof in British Columbia.

Photo 13. The "southern colonial" mansion could be found on many cotton plantations in the southern United States during the nineteenth century. This is "Dunleith" in Mississippi.

Photo 14. This modern, Spanish Mediterranean style, town house is in Manila, Philippines. It is built of stuccoed brick and has a tile roof as well as wrought iron window screens and gates.

Photo 15. In nineteenth-century European cities, the town house was often a tall, three- or four-story structure one room wide. Many had projecting front-window bays to catch the light, as shown in this London street scene.

Photo 16. In old and well-established villages in central China, such as this one near Nanking, the houses are built of whitewashed fired brick and roofed with tile. Chinese village houses front directly on the street, and many have barred windows or solid wood shutters. Not all dogs had been eliminated by 1966, when this photo was taken.

Photo 17. In the solidly built-up portions of many large cities, housing becomes multi-story and consists chiefly of apartments (the British call them ''flats'') erected in variable but repetitive patterns, as in this view of an area in the Westminster-Chelsea district of London.

The habit of placing shelters close together in clusters may well date back to Paleolithic origins of the hamlet or village. The development of crop growing during the Neolithic may have brought about the scattering of these dwellings out on to the land to begin the tradition of the rural farm homestead. In turn, as villages once again became common and grew into towns and cities during historic time, rural housing styles underwent various changes as they were adapted for use in towns. Many houses remained single dwellings set apart on their own lots and built in the various styles of their cultural systems (Photos 5, 11, 12, and 13). Others, however, were sometimes built in rows so as to make use of common walls between the units. Patterns of this sort turn up in all parts of the world (Photos 15, 16, and 17). They are similar in construction and differ mainly in style and decoration as determined by the region where they are found and its culture. In a similar way, pressures of population concentration and cost factors coupled with modern engineering knowledge have allowed the traditional compartmentalized structure to be developed into the modern urban apartment building. Today, these are found in cities all over the world (Photo 10 compared with Photos 18 and 20).

Photo 18. A modern urban architect, working with steel, cement, and glass, conceived this apartment block. Corbusier House, Berlin.

Photo 19.   Spatial mobility in the United States is often very great and takes on elements of comfort and spacious living, as seen in this mobile-home park near Alexandria, Virginia.

Photo 20.   High-rise apartment buildings are widely spaced and have park-like settings in this new and modern suburb of Farst, south of Stockholm, Sweden.

CHAPTER 4
GROWTH AND DIVERGENCE OF CULTURE PATTERNS

By the beginning of the Neolithic period many of the regional culture groups dispersed around the world had learned to find, process, and use many kinds of plant materials, and they had learned a great deal about working with rock and stone materials. They knew what to do with most bird and animal materials, and they understood the basic properties of most forms of water. This is to say that the most basic technologies useful in processing raw materials into useful agents and end products had been developed. No culture group had yet solved the problems around those minerals that yield metals, but they had learned to work with water-soluble minerals. Basic production and processing traditions had been established, and many of the dietary traditions were already fairly well fixed and operative. Basic handicraft systems were in use for textilemaking,

pottery manufacture, stone-tool fabrication, and food-processing. The key long-range exchange routes were beginning to function in the Old World and, to a lesser extent, in the New World.

The Neolithic period did not begin, of course, at a fixed chronological date. Instead, it was a level of cultural achievement reached at different dates by specific culture groups. Achievement of this level has been measured primarily in material artifacts that can be compared and dated, relatively. As the Neolithic culture level was reached by leading societies in some parts of the earth, other societies had not yet advanced so far. Those peoples who had moved into isolated or poor environments, or who had been conservative about taking on new ways, had not yet developed all the bodies of knowledge and skills mentioned above. In contemporary terms, these latter groups were lagging behind, and they showed the beginnings of what we now call underdevelopment.

During the Neolithic period technologies in the processing and production of materials for human use flowered, since many of these materials sprang from production rather than remained dependent on the bounty of nature. Processing technology advanced rapidly, for there were large quantities of materials to be processed. Consumption patterns also increased and became stabilized through cultural preferences. Many of the basic regional cultural preferences of the world thus date from this time. It is in the Neolithic, for example, that wheat products came to be the staple food from southwest Asia northwestward across Europe and northeastward into North China.

THE MATURING OF PASTORALISM

In Chapter 3 animal husbandry was described as a possible alternative to crop growing in the dry margins and the rough lands of the Eurasian part of the Old World. The earliest elements of animal husbandry must have involved simple herding patterns that required mobility for both the herders and their animals, during which the culture groups learned the nature of grazing grounds both as to seasonal food supplies and water sources. We are uncertain precisely where or when a simple animal husbandry began to evolve into a system of pastoralism. *Pastoralism* may be defined as a system of economic production in which seasonally mobile herds of animals convert grassy and shrubby plant growth into products useable by human beings. These products comprise meat, blood, milk, tallow, hide, skin, dung, wool, hair, horn, and bone.

The term pastoral nomad is a common one, but pastoralism and nomadism are not the same thing. Nomadism is aimless wandering or endless roving, and pastoralists do neither. A pastoral group driven from its home range sometimes became nomadic in the search for a new home, but that nomadism ceased when a new territorial grazing range could be claimed and held. True nomads, on the other hand, rarely became true pastoralists.

Early culture groups may have tried both to grow crops and pasture animals. Crop fields require human labor and attention particularly in spring and autumn periods, the very times at which animal herds must be moved either to summer or winter pastures. The two kinds of activity could not be carried on by the same group of people. Only small numbers of herd animals could be maintained on

pastures in areas good for crop production, and water supplies in such areas were often too slight for large herds. As crop growing developed its specialized characteristics, pastoralism did the same, and the two economic systems became increasingly incompatible. Crop growers could keep small numbers of animals as part of their system, and some pastoralists found it possible to grow a few crops, but the specializations of each system led into uses of different environments and different annual calendar patterns.

Early pastoralism was a relatively simple economic system in which a herd of animals supported a group of people. The earliest patterns involved humans following their animals on foot, since the riding animals had not yet been domesticated and included into the system. It is not clear when dairying developed, but it is doubtful that it was extremely early. The seasonal shift of pasturage is the critical element and is known as *transhumance*. In the spring a herd or flock is shifted from its winter quarters to a summer grazing range, and in the fall it is moved back again to its winter range. As human groups worked out these sequences, the seasonal patterns of mobility became both cyclic and very specific, according to the grazing ecology of the herds and flocks involved. Sheep require chiefly grassy browse, relatively mild winters, and do not do well in rough country. Goats will browse coarser plant materials, can take more severe winters, and do well in rough country. As cattle, camels, and horses were gradually added to pastoral animal stocks, their grazing and water demands had to be met by groups using them. Competition among the early pastoralists required staking out territorial ranges that included provisions of browse and water for whatever types of animals different pastoral groups elected to maintain.

There never has been a single precise definition of pastoralism, but critical to all early forms was a specific pattern of transhumance. There also developed forms of specializations in animals kept. Some versions of pastoralism included minor summer crop growing. Some pastoralists used riding animals as integral elements in their system, but others did not. Common to all pastoralists has been the symbiotic relationship with sedentary crop-growing peoples, in which exchanges of surpluses and necessities take place. These symbiotic relations need not be always peaceful, and often they were not. Many pastoralists engaged in militant struggles to maintain grazing ranges in areas in which crop growers sought expansion during humid intervals. Some pastoralists engaged in militant raiding activities into agricultural regions. Almost all pastoralists who adopted camels and horses participated in overland trading operations along one of the great trade routes of the

This flock of karakul sheep is moving across a grassy range in northern Afghanistan. They produce wool such as that shown in the market scene in Photo 8 of Photo Cluster 3.

Old World. In many pastoral zones herds became too large during humid periods of good grazing conditions; these same regions then faced declining fortunes during some later series of dry years.

As the conditions of the last glacial era shifted into the climatic conditions of the last few thousands of years, both mobility and the competitive struggle for territory increased among the pastoral peoples of the Old World. As the last of the highland ice fields melted away, the lessening of runoff dried up piedmont plains, shrank rivers, reduced the extent of oases, and failed to replenish some of the general reserves of groundwater. The changing fortunes of many agricultural peoples, on whom pastoralists depended, disturbed the symbiotic relationships between the settled and the mobile. Out of these competitive struggles perhaps came the earliest patterns of invasion into North China, northwest India, Mesopotamia, the Levant, eastern Europe, the Nile Valley, the Mediterranean fringe of Africa, and the southern Saharan fringe throughout central Africa.

As the struggles between the mobile pastoralists and the sedentary folk of the agricultural zones continued, history records the evolution of new forms of military tactics. Pastoral cavalry tactics improved steadily through time, to be capped by the organizational system evolved among the Mongols in the tenth century, a system that made the militant Mongol pastoralists a scourge from North China to southwestern Asia and well within eastern Europe for a period of over two centuries.

FROM GARDENING INTO AGRICULTURE

The principle of usufruct developed for shifting cultivation was quite satisfactory for the earliest system employed by crop-growing societies, which also kept small animals. There gradually developed the practice of cultivating certain regularly productive kinds of lands on a permanent basis, although it is uncertain how such practices came into operation, or in how many regions this pattern began. *Sedentary gardening* introduced the concept of long-term control over land as a productive resource by individual cultivators as the permanent-field crop garden. Permanent control over particular plots of garden land in effect amounted to private ownership of land. Some developing ideas about the quality of soils, the maintenance of fertility, and the value of continuing care for a plot of ground were necessary. Although hand labor, simple tools, and small-scale operations were still the rule, the cultivator who held a particular plot on a permanent basis had an incentive to both care for his plot and enlarge it, since the land itself now represented a source of wealth and productivity.

The new cropping system produced changes in social organization, in labor systems, and in local exchange patterns by placing an emphasis upon individual family activity rather than the local community group. It is probable that the development of sedentary gardening is related to such features of improvement as systems of terracing, irrigation, drainage, and fertilization, all of which could have helped increase productivity. It is also probable that sedentary gardening brought about the beginnings of dispersed settlement patterns, whereby some families chose to reside on their own lands permanently, thus becoming locally and independently self-subsistent. Certainly the compact village settlement pattern continued, but sedentary gardening turned some of the periodically shifting villages into permanently sited villages.

Among culture groups who preferred compact village settlement to dispersed rural settlement, and who had learned to carry on sedentary gardening, the close social contact probably could have enabled them to increase their productivity by the cooperative development of terraces, irrigation, and fertilization and by concentration upon a particular group of crop plants. Such local territorial increases in production could have accumulated relatively large surpluses of crop commodities beyond the consumption requirements of the local populace. We

still understand very little about such beginnings in productivity levels and specialization patterns, since these were prehistoric developments. It is likely, however, that such local surpluses formed one of the bases for the very considerable trade activity in certain commodities that does seem to have taken place in prehistoric times. Sedentary gardening developed on such a local regional level could have been the basis for the first "commercial" production of crop commodities. We designate this intensive pattern as *formative commercial gardening.* By reason of its small scale, its detailed operations, its hand labor, and its simple tool patterns, this was still a form of gardening. There is considerable evidence of such patterns, in both production and exchange systems, in Indonesia, India, China, the Mediterranean Basin, and in both the Middle American and Andean core areas of the New World.

When land is freshly cleared of its vegetative cover, the underlying soils are loose and soft, and no real cultivation is required prior to planting, a fact the shifting cultivator had long known. But when sedentary gardening began, cultivators found that soils exposed to the sun and rain became compacted and developed a hard crust, so that cultivation was necessary prior to planting. Sedentary gardeners must have devised the basic hand tools still employed by gardeners, such as the spade, the heavy hoe, and the digging rake. But these tools in human hands can cultivate only small plots, and something better was needed by which to enlarge the annually cropped area to support larger families.

It was in Southwest Asia, sometime prior to 3000

This Jordanian farmer is using a primitive plow very much like the one invented some 5,000 years ago.

B.C., where a new group of tools were invented. These were the two-wheeled cart, the plow, the harrow, and the seeding tube (Fig. 4.1). Under regional variation in local conditions in the hill-and-lowland country of Southwest Asia animal keepers and crop growers often lived near each other, and some local populations both grew crops and kept animals. The domestication of cattle provided large animals having the strength to pull tools such as the plow and the harrow. The introduction of draft power and the new tools made possible the cultivation of plots of ground larger than could be worked by human labor alone. *Primitive plow cultivation*, then, represented a new system of production, one that gave the cultivator the capacity to enlarge his garden plot into a farm on which he could produce both animal and plant products. This new system spread over much of the Old World at varying rates, and developed regional variations (Fig. 4.2). We take the position that *the development of animal-powered cultivating tools turned gardening into agriculture*. The new system altered the whole basic structure of crop growing and animal husbandry. Once this system became developed and regionally accepted, farmers in productive areas under favorable conditions could produce significant surpluses of particular commodities. Thereafter the repeated introduction of technological changes in agriculture further improved productivity, and several new systems were later devised, all based on nonhuman power sources (Fig. 4.3). Much of the significant increase in productivity, however, has come only in the last few centuries. These developments (items 8–13 at the top of Fig. 4.3) will be dealt with in later chapters.

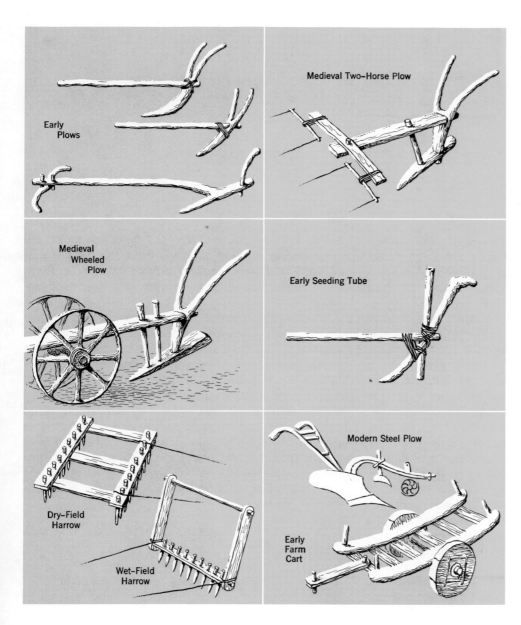

Figure 4.1
The Basic Tools of Agriculture.

Early Plows

Medieval Two-Horse Plow

Medieval Wheeled Plow

Early Seeding Tube

Dry-Field Harrow

Wet-Field Harrow

Modern Steel Plow

Early Farm Cart

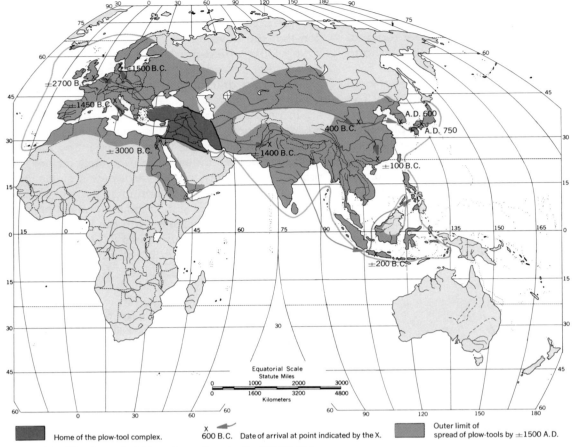

Figure 4.2   The Spread of Plow Agriculture in the Old World.

The development of a plow-cultivation system brought a new complexity into crop growing by integrating plants and animals into a system of production. Farmers now had to provide fodder for their animals if the system was to achieve a higher level of yield, but they could manage larger units of land with the draft labor provided. Plowing systems also altered the shape of the cultivated landscape, as the plots of ground tended to develop into rectangular fields instead of small irregular gardens. In some regions the animals came to be the more important product, and the European development of pastures and hay fields is the best illustration of this. In some other areas the animals remained minor products; pasturage was not provided; and the agricultural landscape clearly reflected this, as in North China.

The early concept of the crop plot as a permanent production site, as developed by the sedentary gardeners, was carried further by the developers of the plow-cultivation system. Permanent control over units of land slowly evolved into the general concept of private ownership of all kinds of land. Private ownership of productive land caused the tribute from the crop field to turn into the land tax that became levied by government. With the land tax, too, there evolved the concept of title to land, the mortgaging of land itself, and around private ownership there evolved systems of land tenure and concepts of tenancy.

The development of plow cultivation into a more productive agricultural system than gardening had important social and economic consequences among societies that adopted the new system. Populations were no longer restricted to the production of their own foodstuffs and could now be supported by surpluses from the new type of farms, which made possible the gathering of people in towns. In towns people could be occupied in many other kinds of activities that contribute to increasing the level, and the enlargement, of culture systems. Perhaps these first shifts of population began the rural-to-town migration pattern.

MIGRATION SYSTEMS

With population increases during the Neolithic, added forms of human movement over the earth

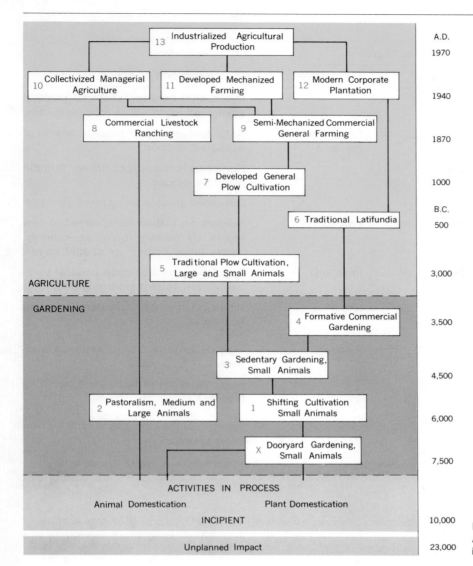

Figure 4.3 Speculative Development of Agricultural Systems. Note that time scale is not uniform.

began to take place. Here, we deal briefly with the whole topic of migration systems. All human movement over the earth accomplishes diffusion of culture, but each is also an element in the differentiation of regions. Our concern, however, is for the various bases for, and the forms of, movement. For convenience, these forms may be grouped under the headings of *exploration*, *migration*, and *regional conquest*.

Exploration often occurs as the result of curiosity, and throughout human time people have been led far afield by curiosity into unknown and far countries. Exploration has also been caused by environmental or cultural processes as searches for some less difficult or more peaceful territory. Fifteenth century Europeans began making a profession of exploration. In our time explorers continue to hunt the earth over for plant, insect, and mineral resources.

Migration properly refers to the transfer of people from one territory to another, divided into two distinct

forms: seasonal movement and one-way permanent movement. Seasonal migration is a form of *transhumance* such as takes place among pastoral people alternating between winter and summer grazing ranges, or as among labor groups moving toward regions needing seasonal labor volumes, or as among vacationers moving to and from winter or summer resorts. One-way permanent movement is the more common and voluminous form of migration. Migration can be either innovative, in seeking new ways of life, or conservative, in trying to preserve known and preferred living systems. Permanent one-way migration can be subdivided into four broad classes: *ecological*, *pressured socially*, *free*, and *mass* (Table 4.1).

The end results of human movement have been the colonization of empty or partially occupied regional territories of the earth by peoples born and raised in other culture regions. Colonization in early times was spontaneous in nature, but in more recent

Table 4.1 Classes and Types of Migrations

Class of Migration	Type of Migration		Description of Action Involved
	Conservative	Innovative	
ECOLOGICAL	Transhumance		Cyclic shift between seasonal pastoral animal ranges to utilize food resources.
		Flight from the land	Rural folk moving to cities in search of economic support.
	Ranging		Hunter-collector cyclic travel through resource ranges.
	Wandering		Displaced group seeking new homeland.
PRESSURED SOCIALLY	Forced or	Forced*	Subject population transplanted to new region by military force, as minority, "slave labor corps," or chattel slaves.
	Impelled or	Impelled	Population transfer upon political partition; term-contract labor migration.
FREE	Group or	Group	Whole-community transfer to new homeland.
		Pioneer	Individuals venturing new life on frontiers.
MASS	Settlement		Regional shifts of population following pioneers.
		Urbanization	Flight from the land and search for new living systems.

*Some migrations can be either conservative or innovative, dependent upon motives of the activating agents. The table was modified from William Petersen, *The Politics of Population*, Doubleday & Company, 1964.

times it has become formalized and has operated under political controls. In all their forms, and throughout human history, migration and colonization have been critical factors creating differentiation among human ways of living on the earth.

Regional conquest combines aspects of exploration, migration, and colonization with political and military expression of power by a strong group against a weaker one. It generally results in altered ways of living, both among the conquered and among the conquerors. In very early times regional conquest may have killed off hunting bands and small culture groups, but the appearance of the political state began to institutionalize territorialisms as formal regions with political boundary lines.

Exploration, migration, colonization, and regional conquest have all been elements of human differentiation of the face of the earth. The complex interworkings of the four specific items have been going on since at least the late Paleolithic and have been repeated again and again. The numbers of people involved in the early patterns were small, but as the population of the earth has grown, the numbers of explorers, migrants, and colonists have grown, too. The patterns of human movement, therefore, became

a cultural tool, constantly adding to the variety of ways of living that have come to characterize the earth.

THE PROCESSING AND SUPPLY OF MATERIALS

The processing of raw materials into usable agents and products has a very wide range, since it includes practically everything that early groups of people could put their hands on. In this section we do not try to discuss the many ways of processing materials, but limit our discussion to the processing of textiles and minerals as two categories of primary importance in the development of advancing culture systems.

Textile Processing

The processes of creating textile materials for clothing, bedding, and other uses was well advanced by the end of the Neolithic in most parts of the earth in which protection from cold weather was needed. Spinning and weaving by then had a worldwide

This Peruvian woman is spinning wool yarn as she walks down the street. She holds a mass of wool under her left arm, twists the fibers together with her left hand, and keeps the spinning whorl suspended and rotating with her right hand. The spindle holds the spun thread, and the weight at the bottom serves to keep it in motion. Similar methods of spinning thread were almost universal until the modern era of the machine.

distribution, although there were culture groups that did not employ these processes.

By the end of the Neolithic, there were well developed patterns of use for abaca, cotton, flax, hemp, hibiscus, jute, ramie, silk, sisal-henequen, wool-hair, and many other plant and animal fibers. Spinning of the "thread" could be done in many different ways, but the spinning whorl is one of the common archaeological artifacts. Weaving was the most widespread of the methods of preparing a textile, and there are many varieties of weaving systems.

In addition to the normal textiles, there are other sources and methods of preparing sheet materials for clothing, bedding, and housing purposes. In areas of cold winters, the curing and tanning of skins, leathers, and furs was highly developed and widespread. In warmer regions there was a widespread pattern of using bast, the soft fibers of the inner bark of many kinds of trees for bark cloth, which is made by several procedures of soaking in water and pounding.

The weaving of thread into a textile, and the pounding of bark into a soft fiber sheet, were not the only processes of producing a sheet of "textile", although they were the most common. Felting, the matting together of fibers of hair and wool, appeared rather late in central Asia as a source of both clothing and housing materials. The making of a floss from raw silk fibers to serve as a liner for a quilted garment or piece of bedding is Chinese, just as the making of a floss of cotton appears to be derived from East Africa-northern India. Knitting is little studied as a technique and seldom reported, but it is known from ancient southwest Asia and became widely distributed in western but not eastern Eurasia. Knitting came to be highly developed in the European culture world. In modern time knitting has diffused all over the earth and, of course, machine knitting is today a major aspect of textile manufacturing.

The wearing of garments is but one aspect of what here is thought of as clothing systems. A better term for the whole complex is *adornment*, for the complex includes a wide range of elements from tattooing and scarification to the application of body paints and cosmetics, the wearing of jewelry, the styling of the hair, and the wearing of garments (see the photos in Photo Essay No. 6, pp. 344–354 and the latter half of the accompanying text).

The use of "textile" materials in housing developed chiefly in areas of scarce vegetative cover and among the mobile culture groups who needed light and transportable materials. Skins, leathers, and furs were widely used in the northern hemisphere either the year around or during summer periods. The American Indian tepee was normally made of such materials. Central Asian mobile peoples often used felt, as in the Mongol yurt. In regions of mild winters, different forms of tents usually were made of strips of woven cloth sewn together (Photo Essay No. 1, photos 1 and 2).

The Mining of Ores and the Smelting of Metals

By definition, the end of the Neolithic is dated to the first conscious metallurgy, that is, the first mining, smelting, and fabricating of metals from ores rather than the use of free copper, silver, or gold. Metallurgy is commonly thought to have derived from the smelting of copper-oxide ores into metallic copper. The date is still conventionally placed at about 3700 B.C., and the location as somewhere in Southwest

Asia, very possibly in what today is northwestern Iran. Recent finds of copper and bronze implements in northeast Thailand, however, raise questions about the dating and the place of origin, since the Thailand finds have been dated tentatively as between 5000 B.C. and 4000 B.C. (Fig. 4.4).

Once the basic technology of smelting was devised, mining became a new kind of operation that went through considerable technological improvement. The details of mining and smelting vary regionally and by specific mineral, but the two technologies are roughly similar for all the common minerals. Early volumes of metal smelted in all cases were but a few pounds per furnace per day, and metals did not rapidly come into common use by everyone.

In contrast to the rather slow diffusion of crop growing over the Old World, the spread of metal smelting moved very rapidly, and by 2500 B.C. bronze products were in use from Britain to North China. The smelting of iron ores may date as late as 1500 B.C. In the New World metal smelting began with gold and silver and was followed by copper

Figure 4.4 The Diffusion of Mining and Smelting.

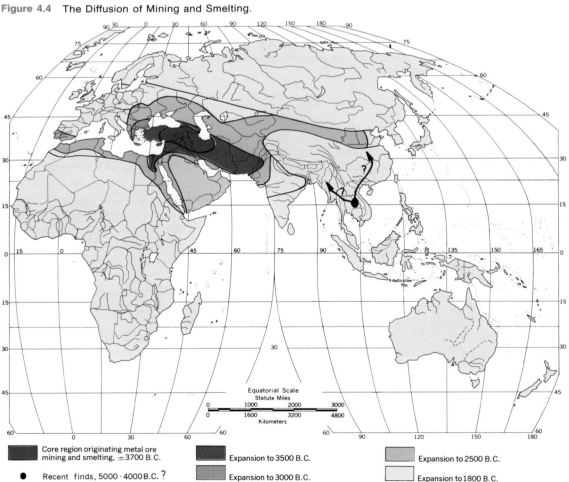

| | Core region originating metal ore mining and smelting, ±3700 B.C. | | Expansion to 3500 B.C. | | Expansion to 2500 B.C. |

● Recent finds, 5000 - 4000 B.C. ?   Expansion to 3000 B.C.   Expansion to 1800 B.C.

and bronze. It is usually thought to be an independent invention that occurred sometime just after the beginning of the Christian Era.

There are two kinds of mining: *surface mining* (strip mining) and *lode mining*. The modern strip miner working coal beds or copper deposits employs exactly the same techniques as the Paleolithic jade miners. Both pick the product from the surface of an area. But, of course, the advantage lies with the modern operator who can pick up a huge volume at a single grasp of his enormous machine. Lode mining did not properly begin until well into the Neolithic, when miners began to dig deep tunnels into rock veins carrying free gold, copper, or precious stones. The essential techniques of tunneling, filling exhausted sections with rubble, leaving pillars of natural stone to support the roof, and opening ventilating shafts are all Neolithic developments. Water was a more difficult problem, and excess water in mines was not properly handled until the invention of pumps in the early modern era. By the time metal smelting was devised, however, miners had learned to work ore bodies to almost any desired depth provided the mines were not troubled with water.

The coming of metallurgy had a tremendous effect on economic technology in all parts of the world to which it spread. Metals went into agricultural tools and into the fabrication of tools of all sorts for cutting, grinding, filing, and shaping. Metals also went into consumer goods of a wide range, from decorative jewelry to cooking pots. Many of our common implements of today date from an inventive application of metallurgy dating as far back as 2500 B.C.

By about 2000 B.C. the properties of most common metals had been learned, their ores located, the smelting and basic fabricating techniques worked out, and most of the basic alloy compositions approximated. From about 2000 B.C. to the modern era, the history of metallurgy is chiefly one of diffusion of basic knowledge and techniques, with each regional culture group reacting slightly differently to the whole complex. Although the knowledge of metals had developed soundly, the cultural inventiveness in how to use them in complex ways had to await the modern era.

This is a photograph of a uranium mine in Ontario, Canada. The miner on the left is using an air-powered drill to prepare holes for the explosives used to blast the ore free. Electric lights, pumps for circulating air and moving water, and powered tools are modern improvements, but the tunneling techniques are still very similar to those developed by Neolithic miners.

THE MEANS OF TRANSPORTATION

A human being can walk longer and farther than he or she can swim, and the inference for transportation is that land transportation is older than water transportation. Peoples in all parts of the earth first devised ways of carrying things themselves before they found other ways of moving volumes of things about. And though human porterage has remained common in some parts of the earth right into the twentieth century, people early worked hard at devising means of moving things so that they did not have to carry them themselves. Short-distance transportation was not a critical problem, but trying to transport raw materials or finished products over long distances was troublesome. The lengthening of transportation on land and sea, however, is a story of the developing of technologies in the packaging of materials and the utilization of motive power other than human muscle. The two types involve different mechanisms and require separate treatment.

Land Transportation

There are five physical aspects to the development of an efficient system of land transportation: the route surface, the motive power, the mechanism of carriage, the packaging of materials, and the terminals at which materials are handled. Man has been slow to coordinate all these separate facets, and only in the twentieth century was a truly effective solution reached in a few places on the earth. Attention was first attracted to the mechanism of carriage, with the wheel as the first significant transport invention, but the formal roadway developed slowly (Fig. 4.5).

The evidence is very scanty until about 1200 B.C., long after the invention of the wheel. The great "commercial roads" of ancient Southwest Asia were chiefly pack-train routes for moving goods, but the road surface itself was often unimproved. Even the famous roads of the Roman Empire were often unusable for wheeled traffic above the flat lowlands. The use of stone paving for the "imperial roads" of both the Persian and Chinese road systems helped the movement of pedestrians, cavalry, mounted messengers, and pack trains. Such road systems often contained steps in the steeper sections, and were not designed for wheeled traffic. Not until the mid-nineteenth century was effective attention given to the nature of the road surface itself for the routes between cities and towns, and land transportation in the hinterlands remained costly, troublesome, and hazardous.

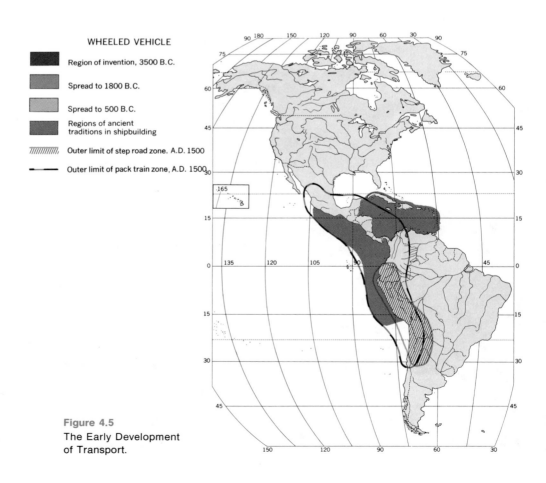

WHEELED VEHICLE

- Region of invention, 3500 B.C.
- Spread to 1800 B.C.
- Spread to 500 B.C.
- Regions of ancient traditions in shipbuilding
- Outer limit of step road zone, A.D. 1500
- Outer limit of pack train zone, A.D. 1500

Figure 4.5
The Early Development of Transport.

We do not know the full story of pack-animal evolution, but in this process the Old World, with its several kinds of large animals in the different regional environments, had a very distinct advantage over the New World where only the llama/alpaca were available. For motive power, although many animals were tried, the ox proved most satisfactory and became the most widespread source of draft power in the Old World. Harnessing the horse as effective motive power was a very late achievement. There were numerous variations in animal power, and harnessing and teaming show much variation in time in different parts of the Old World. Many of the old patterns persisted from about 2500 B.C. well into the nineteenth century, with only gradual improvement through better breeding (stronger animals), better teaming, and better harnessing devices.

The simple travois, sled, skid boat, mud way and other elements of simple transport (see Fig. 3.11, p. 67) by push-pull-sliding were in existence in varied forms from the tropics to the Arctic prior to the domesticated draft animal. Animal power pulling wheeled vehicles was more efficient. Cart and wagon styles varied regionally over the Old World, but the two-wheeled oxcart remained standard in many regions. The development of the four-wheeled wagon appears to relate to northwestern Europe

and the powerful breeds of horses developed there, but this came only during the Middle Ages.

The packaging of materials made little progress so long as the vehicles remained small and motive power was limited. Hand loading and unloading at terminals, fords, and other transfer points, meant that man-manageable parcels were the largest possible units for most materials. The terminals required little elaboration other than what was necessary to serve the needs of travelers and their animals so long as other elements of land transportation were not developed.

The history of land transportation appears to have made significant progress at a rather early date and then to have stood still, technologically, until well into the nineteenth century. Nevertheless, in the diffusion of animals, pack-train systems, carts, wagons, animal harness, and other related features the intervening period is one of marked spread of the basic elements of land transportation over the Old World.

Water Transportation

We know very little about the timing of the development of the earliest watercraft, but many different traditions of constructing distinctive types of watercraft evolved along the shores of oceans, lakes, and

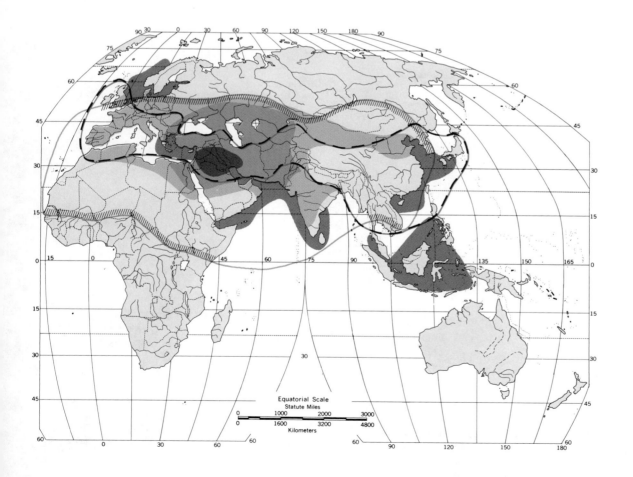

rivers in separate regional centers. Similarly, in the evolution of sails, paddles, oars, steering arrangements, and other boating devices there are both several traditions and rough similarities in far parts of the waters of the earth. Watercraft evolution, in general, shows more localized regional evolution to fit particular conditions than does land transportation (see Fig. 4.5).

Boatbuilding had developed a degree of maturity in several parts of the earth perhaps as early as 3000 B.C. Water transportation shows the same variation in aspects as pertains to land transportation. The medium of navigation, the motive power, the mechanism of carriage, the packaging of materials, and the terminal are significant to the efficient development of the system. The boatbuilding traditions in some regions concentrated chiefly on the mechanism of transport, so that we see many kinds of boats in terms of style, build, material, and waterproofing technology.

Only in a relatively few places did watercraft for large-volume cargo or passenger traffic develop at a really early date. The Mediterranean Basin, the Arabian Sea, Southeast Asia, and the China coast appear to be the more important regions of water transportation in the true sense. The southwest Pacific Ocean is perhaps another such region; its boat-

building tradition is, however, linked to Southeast Asia.

Although a water surface is not directly comparable to land surface in terms of a specific route, there are distinct problems in navigation and the application of power. Nile River craft could not work the Arabian Sea, so that rivercraft, oceangoing craft, and coasting craft had different problems concerning the routes they followed.

In handling cargo the limitations of manpower, lifting devices, and cargo storage meant that only small units could be taken aboard ship except in special circumstances, such as those surrounding the Nile River barges used for hauling massive stones used in constructing the pyramids.

For watercraft the shore forms a terminal point in cargo-handling, and what today is termed a port was seldom needed and only occasionally developed until relatively recently. Safe harbors in the event of a storm became a necessity after ships were built too large to be beached easily or safely.

The history of water transportation showed marked progress very early in particular regions in boatbuilding, in sail rigging, and in evolving special patterns to fit special environmental conditions. As with land transport, the history of water transportation suggests limited accomplishment in geograph-

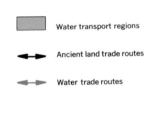

   Water transport regions

   Ancient land trade routes

   Water trade routes

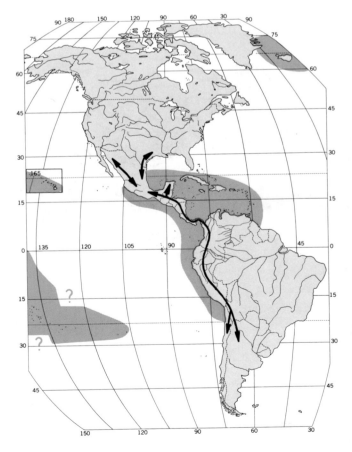

Figure 4.6

The Linkages between Transport Regions in the Ancient World.

ical diffusion of early technology and in the linking of distant regions to form transportation regions. No great advance was made on the whole assemblage of water-transport technology until the modern era, and the rather small size of ships employed on the open ocean by the fifteenth-century European navigator-explorers attest to human courage.

The Linking of Transportation Regions

The gradual extension of transportation routes on land and on water, took place around local regional centers (Fig. 4.6). That is, land transportation very early linked together the several portions of the Fertile Crescent, and the lower Nile Valley was linked by water transportation to Upper Egypt. Similarly, one can see other early regional linkages: North China was linked by land transportation to Chinese Central Asia, the China coast and the lower Yangtze Valley formed a water-transportation cell, the island-peninsular world of Southeast Asia was linked by water, the shores of the Arabian Sea were linked by coastal water contact, the ''amber trade'' linked the Baltic (by land and river) with the Mediterranean, and parts of the New World tropics were tied into a water-transportation cell. Slowly, during historic time, several of these separate transportation cells were interlinked by extensions of the land and water cells. The interlinkage of the Old World was effected, in fact, several times only to be broken apart repeatedly. The final interlinking of the transportation cells of the whole earth came permanently only after the Columbian Discoveries.

EXCHANGE, TRADE, AND VALUE SYSTEMS

The word *trade*, in all Indo-European languages, has references to routes, to products, to occupations, and to human movement. The concept of trade, however, takes in a large number of different but interrelated elements that have had built into them a group of rather complex ideas. Today we normally think of trade as determined by economic and political rules. The modern money-as-the-medium-of-exchange systems express the abstract mental workings of complex and highly cultured societies. In the simpler conditions of earlier periods, trade was less an exchange of products at particular prices than it was the cementing of friendly relations and the fulfillment of social obligations. Product exchange does serve needs of the moment, but it may also have long-run purposes that have nothing to do with material

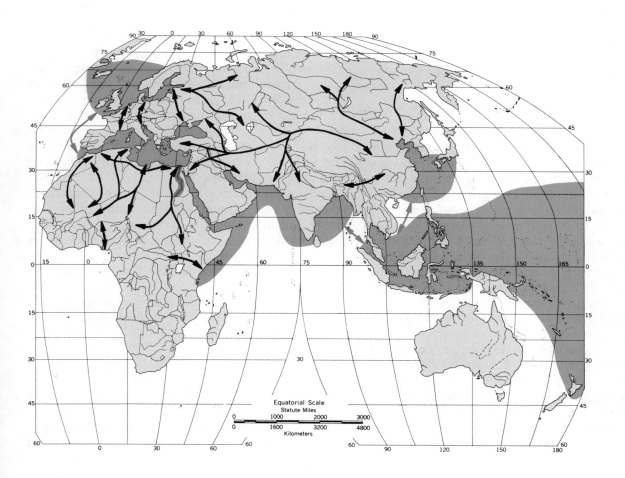

goods. In earlier periods goods were often bartered item against item between persons or groups, and the "profitable" exchange was one in which each side felt that social prestige resulted.

Gift giving is a basic human trait stemming from very early time. It operates between friends to maintain mutually respectful relations and social indebtedness. Gift giving does create an exchange of products, to serve the purposes of trade, but one does not give gifts to those with whom one has no personal relationship. Gift giving was, and is, normally conditional upon the return presentation of a gift having the approximate social value. Gift giving often became institutionalized in a particular society within a distinctive social framework in which it was surrounded by ceremony and perhaps tied to particular calendar intervals.

Strictly *secular trade* rests on social interaction without necessarily involving friendship. Any and all kinds of products and commodities may be involved, according to the level of culture. The pattern of demand and supply for secular trade serves the purpose of economic redistribution. Such trade early began to acquire some notions of fair levels of exchange, leading eventually to the formulation of value systems. Secular trade varies in volume according to the level of the culture system. Volumes of exchange are small in the simple society living close to nature, since the regular level of living may go on without very much exchange. In the advanced society that has developed high degrees of occupational specialization, a large range of raw materials, and a great variety of manufacturing technology, even the simplest daily requirements are met through exchange of money for products.

There are two basic varieties of exchange that have apparently been present from Paleolithic time. One is the exchange of raw or natural products that are the monopoly of particular regions. These are traded outward to all other regions, to be taken in exchange by all kinds of peoples of all levels of culture. Salt, tool stone, free gold, jewel stone, and exotic aromatic plant and animal extracts long enjoyed such monopoly status. Of this list only stone for common tools has dropped out, even today. Every people holding such a natural monopoly has valued it once aware of its worth. Those peoples not possessing that resource have sought to break the monopoly control over the product by developing alternative sources of supply.

The second variety of exchange occurs in manufactured goods and commodities that are known for their utility, purpose, and appeal. Extensive trade in technologically complex and expensive manufactured goods operates chiefly between peoples of comparably advanced culture traits. For example, one can sell automobiles only to peoples who have roads and who know and use wheeled vehicles. Conversely, the variety and amount of trade goods are very limited between peoples whose culture levels are very different. The spread of trade in manufactured goods must be preceded by the spread of patterns of culture within which the goods have meaning.

Several kinds of exchange system are utilized in carrying on trade of all kinds. Versions of *silent barter* may operate between enemies, cultural unequals, or peoples who cannot communicate with each other. Simple *barter* of commodity for commodity on a bargained basis may serve all the requisites of trade between friendly groups, acquainted peoples, or unfriendly peoples in agreed-upon circumstances. *Tribute* between politically superior and inferior groups, or between militarily unequal groups, may not supplant all needed trade but may form the most significant exchange between the particular groups. *Normal commerce*, what we customarily think of as peaceful trade, is more common between culturally comparable peoples, and between friendly peoples occupying different environments. Commerce takes place between distant regions, but normal commerce is an attribute of a culturally advanced set of peoples who have, make, and produce large volumes of commodities ranging from raw products to complex manufactures.

In the long history of the growth of exchange there developed a variety of aids, props, and organizational facets that have facilitated trade. Such features as the agreement on the place at which exchange occurs has developed the marketplace. The timing for exchange, be it seasonal, annual, or at short-term intervals, takes numerous forms. Systems for the arrangement of credit are many. *Regular trade* involves the guaranteed availability of commodities at all or known times, so that the organization of supply systems came about in different ways among varied societies. What we know today in the American supermarket as brand-name certification is actually very old and is the guarantee of the producer that his product is of a specific quality. Diffusion of these cultural systems has been irregular but continuing, both in time and in place. As the Neolithic level of living became dominant over the world, only to be replaced in turn by the Bronze and Iron Ages, the systems of trading also spread, increasing the regional variety in trade systems.

THE CIVILIZING PROCESSES: THE IMPORT AND EXPORT OF CULTURE COMPLEXES

The Neolithic was a period during which the cultural processes matured sufficiently so that man could begin to make his own choices in ways of living and places of residence. Active migrations and regional shifts were man's responses to environmental change. Many of the primary choices were those of powerful ethnic and cultural groups, making it necessary for smaller, less technologically able, and weaker groups to accept the best of second, third, or fourth choices of location and procedure.

In an inhabited world in a state of physical and cultural change the processes of invention, discovery, evolution, and diffusion of culture traits and complexes were tremendously furthered by the very patterns of human migration, regional contact, competitive struggle, splintering-off and rejoining, alliance and separation, and conquest and rebellion. Particularly notable must have been the processes of diffusion, by which most of the then known cultural complexes were spread throughout broad regions (Fig. 4.7 illustrates two examples).

Crossroad location (a central place or node in a transport system) provided a very great advantage to the human occupants of a particular regional site. Cultural contacts, friendly or belligerent, afforded such occupants the maximum exposure to human knowledge, technological skills, materials, and procedures. The converse, of course, held true: the secluded refuge site far from a crossroad position could remain untouched by almost all cultural excitement and human contact. Our concern here is for the progressive elements of change rather than those of stagnation.

There is little doubt that the whole of southwest Asia from northwest India to Transcaspia to the eastern Mediterranean Basin (the classical Near East) formed the key Old World zone within which cultural excitement occurred at particular regional crossroads. Moving through this zone (or available to peoples in this zone) were the wide ranges of Old World plants, animals, technologies, economic procedures, ideas, social constructs, political concepts, religious beliefs, tools, exotic materials, labor resources (including people in some form of social bondage), and forces of leadership. This was a primary meeting ground for the diffusionary currents of features developed all over the Old World. Over thousands of years groups of people moved into Southwest Asia and shifted about, searching for the kind of homeland they desired. The key sector within Southwest Asia was the Fertile Crescent, a strip of hilly country and adjoining piedmont plains stretching from the southern part of what today is Israel, northward to the hilly zone of eastern Turkey, then eastward to the front of the Zagros Mountains of western Iran, and southward along that mountain piedmont. Local regional contrasts in the characteristics and qualities of environment were such that not every local region was equally useful to each invad-

ing culture group. There was continuing conflict over local regions, and periodically new invaders conquered local populations and took over regional controls. Progressive culture groups could turn the best local regions into areas of outstanding development—for its time.

There were other crossroad zones of lesser importance than Southwest Asia, and each of these played a significant role in regional development in its own area. Such a zone was the eastern end of the central Asian corridor, in northwest China., for here several routeways running across eastern Asia came together. The western end of the Mediterranean Basin formed a crossroad zone for African-European contacts. Southern Mexico-Guatemala was a significant crossroads zone for the whole of the New World.

WRITING, URBANISM, AND THE POLITICAL STATE

As gardening and simple animal husbandry matured into agriculture, a share of the population in a region achieving these advances was freed from its daily routine and released to the realm of leisure or creative activity. In this newfound freedom, rather than the sheer necessity of want, lay the seed of creative cultural development. Some people could begin to investigate the world around them in any number of ways. We have referred to the growth of processing technologies, to the development of trade, and to the growth of the means of transportation. There were also several other kinds of efforts that yielded cultural advances that resulted, finally, in the maturing of a higher level of culture, here referred to as civilization.

Civilization involves additional elements of culture, such as systems of writing, number systems and mathematics, some aspects of the rise of science, the development of urbanism, the maturing of political organization into the concept of the state, the formal organization of religious systems, and the development of advanced military defense systems. The list should also include such features as the emergence of transportation, trade, manufacturing; the growth of metallurgy to include iron; and the growth of specific systems of mores, customs, and social patterns.

Protohistoric peoples in different parts of the earth developed several different systems for recording speech, records, orders, warnings, and claims. There are four basic kinds of true writing: picture writing, ideographs, scripts transitional between these, and phonetic writing. *Ideographs* are highly developed and conventionalized symbols with accepted definitions of meaning for things or ideas; the characters in written Chinese are ideographic. *Phonetic writing*

SUGAR CANE

▨ Home region of domestication

→ A.D.600 Diffusion route and date of arrival

⌇ Current outer limits of cane-growing

▨ Current chief production regions

SUGAR BEET

▨ Home region of domestication

→ Route of diffusion

⌇ Current outer limits of beet-growing

▨ Current chief production regions

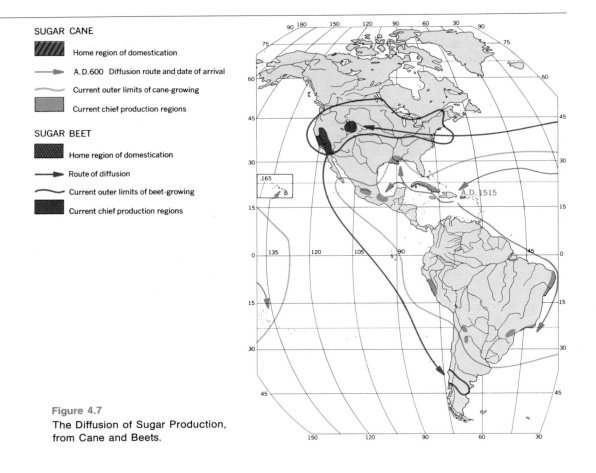

Figure 4.7
**The Diffusion of Sugar Production,
from Cane and Beets.**

uses symbols to represent sound, not things or ideas. Most of the world's writing systems are phonetic.

Three ancient phonetic systems, with minor variants, are known to have been developed: the *syllabic*, the *hieroglyphic*, and the *alphabetic* (Fig. 4.8). Syllabic writing evolved from pictographic symbol systems. Groups of signs stood for syllables of sound, and these were compounded to make words. The cuneiform writing of Sumer in ancient Mesopotamia is our best known example. Although syllabic writing was in use as early as 4000 B.C., it died out by the beginning of the Christian Era and has no known descendants. Hieroglyphic writing also is an early form, which died out by A.D. 600. Ancient Egyptians, who used one form on monuments and a cursive form for business and private needs, provide the chief example of hieroglyphic writing. The most advanced writing systems are alphabetic, in which a single symbol identifies a single sound. Only a small number of symbols are needed (various alphabets use from 21 to 40) because they can be combined in various ways to meet phonetic needs. The idea of an alphabet was devised somewhere in the classical Near East prior to 2000 B.C. and spread outward by migration and by stimulus diffusion (Fig. 4.9).

Many simple systems of numbering rose above the finger-counting level and, in time, related systems were distributed from the shores of the Indian Ocean to the Mediterranean Basin and to the Yellow Sea. The early elements of mathematics, geometry, and astronomy spread throughout that broad region. Early achievements in science shared the same general distribution in the Old World: Egypt, Mesopotamia, Persia, North India, Transcaspia, and North China.

The growth of urbanism resulted from the maturing of agriculture and animal husbandry as the bases for supporting people who no longer produced their own food supplies. What is often termed the "urban revolution" involved a number of creative developments and innovations that greatly expanded cultural patterns and ways of living. These are the results, rather than the causes, of urbanism. Urbanism required integration of political and social practices and organization, and the redefinition of settlement and housing concepts. Urbanism also involved the development of new kinds of architectural structures, including advanced defensive features ranging from city walls and battlements to the garrison barracks. The development of large numbers of structures on specific sites also led to the formalization of the plan of the urban settlement. The most notable of these

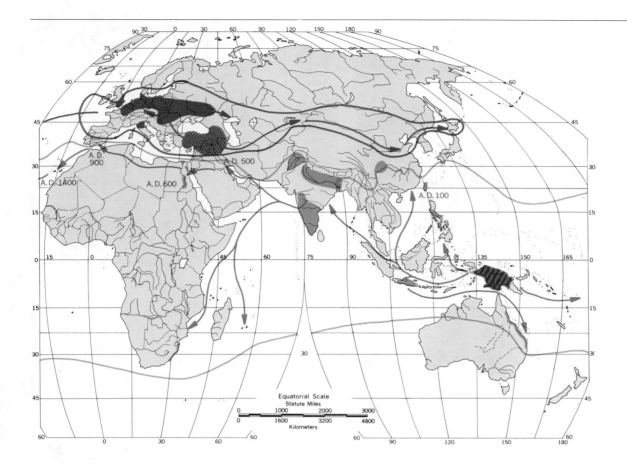

plans was the grid-pattern layout of streets and drains. This plan was widely adopted over the earth in later cities. Not all urban settlements have used the grid-pattern layout, but some formal arrangement had to be adopted to permit access. The growth of social habits around such features as the inn, the restaurant, the factory, the shop, and the temple directly accompanied the development of the physical structures housing such activities.

With the appearance of urbanism, a redefinition of concepts of territoriality required new means of control over surrounding rural lands that supported the urban populations. The evolution of the state as a formal territorialism led to the development of formalized rules governing rights to land and to its cultivation. The old simple systems of giving portions of crop yields as tribute to leaders, shamans, and tribal chiefs was transformed into formal systems of taxation exacted by the territorial ruler.

The formal appearance of the political state as a territorial unit was accompanied by the appearance of the superchief as a political officer in command and control. This was marked in the Near-Middle East, perhaps most clearly related to Egypt, by the figure of the God-King who was all-powerful and the holder of both temporal and spiritual power. The God-King's residence was both a palace and a temple. As a palace it became both the treasury and the seat of political administration. As a temple it became the sacred ritual place and the center of pilgrimage. Both as palace and temple the seat of the God-King required a secular and religious bureaucracy.

In the growth of the political state, the achievement of power to control both the territory and the human population was a primary objective. The developing political institutions were structured to put power in the hands of individuals, a social class, or a military fraternity. The power of a leader achieving the status of a God-King became inheritable by blood-relative descendants.

Implicit in the formalization of the political state was the development of military defense systems. Three aspects characterize this pattern. One element is the military force itself, the growth of the war party into the army and navy. Second is the development of weapons and tactics. Third is the development of fortification. Among different peoples in different regions, the three elements developed variably, with first one and then another taking the lead. The urban settlement could provide increased manpower, it could devise tactics and manufacture

Figure 4.8 Forms of Writing.

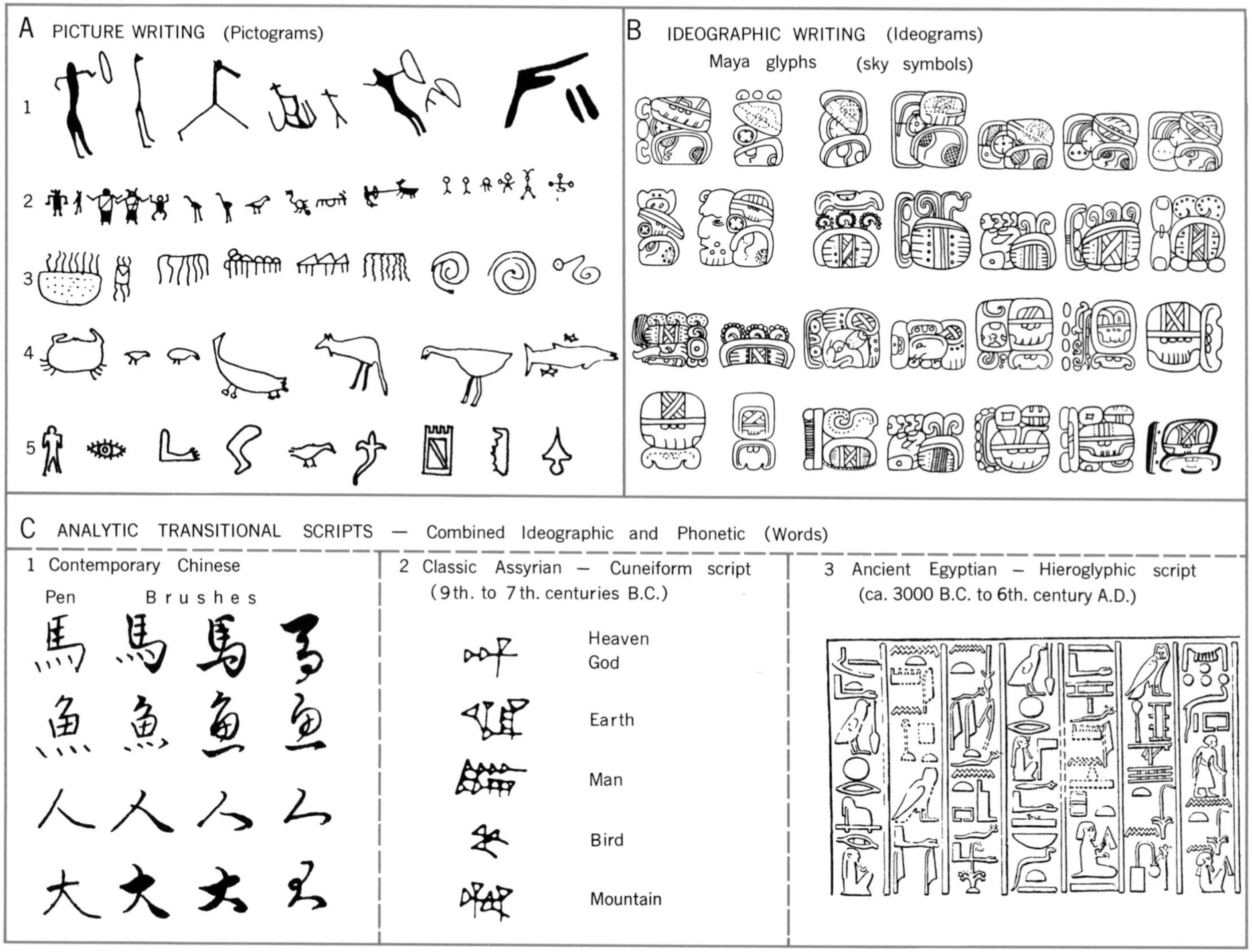
A PICTURE WRITING (Pictograms)
1
2
3
4
5
B IDEOGRAPHIC WRITING (Ideograms)
Maya glyphs (sky symbols)
C ANALYTIC TRANSITIONAL SCRIPTS — Combined Ideographic and Phonetic (Words)
1 Contemporary Chinese
Pen
Brushes
馬 馬 馬
魚 魚 魚 鱼
人 人
大 大 大
2 Classic Assyrian — Cuneiform script
(9th. to 7th. centuries B.C.)
Heaven
God
Earth
Man
Bird
Mountain
3 Ancient Egyptian — Hieroglyphic script
(ca. 3000 B.C. to 6th. century A.D.)

D PHONETIC SYSTEMS

1 Syllabic (syllables)

A Japanese (Katakana and Hiragana)

B Ancient Cyprus

Vowels

2 Alphabetic (letters)

A Mediterranean developments

B Developments in India (since 3rd. century B.C.)

Sources: (A) From *Writing* (Ancient Peoples and Places Series) by David Diringer (1962), by permission of Frederick A. Praeger, Inc. (B) From *Maya Hieroglyphic Writing* by J. Eric S. Thompson (1960), copyright by the University of Oklahoma Press. (C) Item 1 from *Beginning Chinese* by John De Francis (1963), by permission of Yale University Press; Item 2 from *Writing* (Ancient Peoples and Places Series) by David Diringer (1962), by permission of Frederick A. Praeger, Inc.; Item 3 adapted from *The Tomb of the Vizier Ramose* by N. de Garis Davies (1941), by permission of the Egypt Exploration Society. (D) From *Writing* (Ancient Peoples and Places Series) by David Diringer (1962), by permission of Frederick A. Praeger, Inc.

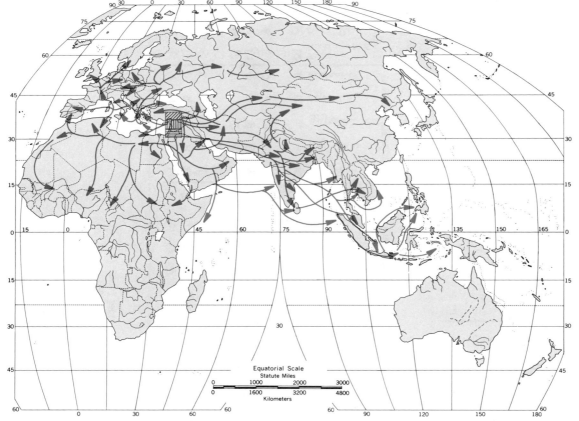

Figure 4.9 Diffusion of the Idea and Use of Alphabetic Writing.

equipment, and it turned into a variably fortified settlement that could resist the siege of the attacker. The military defense system was suitable either to defense or to aggressive attack, and it became formalized into an instrument of the political state.

Urbanism settled on the town (the true city came later) the wealth, power, culture, and initiative-in-development for the territorial unit that became the state. To the extent that the town drew in the produce of the surrounding regions and became the center of administration, it came to hold sway over rural territory and to have the power of creative orientation. To the degree that a town's population was aggressive and creative, it flourished, and with it the territory round about. Whereas the first feeble Paleolithic cultural advances had depended on small groups in relative isolation without the integrative stimulus of interested neighbors, the town became a cultural center in which close associations of peoples with different interests had a stimulating effect on one another. Henceforth the town became the seat and source of cultural advancement.

Figure 4.10 The Centers of Early Civilizations.

THE FORMATION AND APPEARANCE OF CIVILIZATIONS

Our English word *civilization* implies an organized society in a political state of some advanced level. There is no clear distinction between the culture of a noncivilized group and that of a civilized people except that there are more numerous culture complexes among the latter. For all that, it is common to speak of civilization as possessing certain attributes that are greater than those of the so-called simple culture. The distinctions defining civilization are subjective in the minds of the civilized, who then regard certain other peoples and cultures as noncivilized, uncivilized, or barbaric. How, then, do we specify the maturing of relatively simple cultures into civilizations, and what are the attributes, culture components, and culture complexes that signify civilization?

Into every explanation comes the town/city, the political state, organized religion, the social kinship system matured into the citizenship system, the effect of agriculture-animal husbandry, the growth of handicraft manufacturing, metallurgy, systems of writing, and the elements of science. There is neither full agreement on the dating of the maturing of civi-

lization, nor on the specific elements that distinguish civilization from noncivilization.

How many regional examples of early civilization were there? Again there is no full agreement, and the number of regional culture centers achieving civilization ranges from two to nine (Fig. 4.10). Mesopotamia and the Nile Valley appear on every list, and Crete is sometimes added. Northwest India and North China are usually included. There is evidence of two areas in Africa, the Ethiopian Highland and West Africa south of the Sahara, but these patterns are not yet well studied and are often still omitted. The Middle American and Andean Highland regions are included by scholars who take a worldwide view of the growth of human accomplishment. The beginnings are spread no wider than these nine regions, other units being considered offshoots of one or another of them.

Important points about all these regions are that, except for part of the Middle American territory, they lie outside the humid tropical lowlands. None of them lies in a truly formidable climatic region. They are all regions having relatively good physical situations affording human contact with surrounding regions. Each was relatively close to a botanical zone affording numerous plants amenable to domestication as crop plants. All areas possessed tracts of land easily cultivable by the simple early technol-

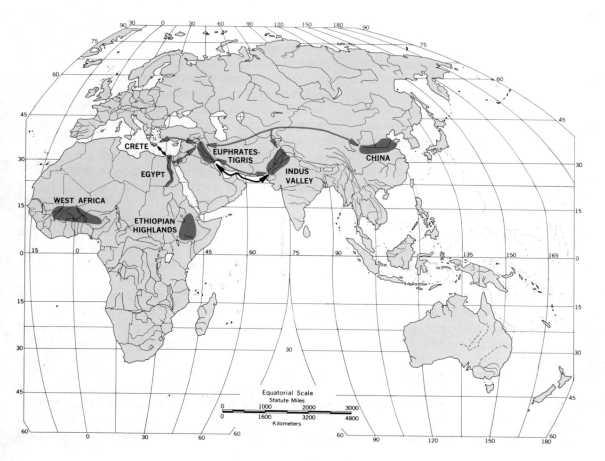

ogies. The Old World areas were all in situations that, in early times, afforded contact with others of the group of developing centers. The two New World areas were in contact with each other. These core regions were the places in which critically important Neolithic culture complexes were brought together.

THE EXPANSION OF THE STATE, AND REGIONALISM

The critical element in the growth of civilization was the ability to organize territory into effective combinations. The genesis of this process took place in the appearance of the town and in the rise of the concept of the God-King. Around each town there was territory utilized for the production of food, fuel, animals, water, and such other natural resources as were conceived important to the particular group. During the period that it took to organize the regional zone into an operating economic and political entity, the maturing of the cultural structure took place, to the end that the town became a city and the territorial holding became a state. These city-states were at first relatively small in area and in population. There was competition between them for control of rural territory, for control of the routeways to distant regions and, in some cases, for control over water supplies. The sequence of development, both spatially and chronologically, was not the same in all areas.

The expansion of a city-state, the dominance of its God-King, and the effectiveness of its administrative bureaucracy, depended in good part on the appeal and the dynamic style of the religious system, the particular sociopolitical construct, and the ethnic composition of the region at large. It also depended on the accessibility of the separate units of the regional environment from the control center. The growth and expansion of the city-state, depended, also, on elements of human creativeness expressed in dynamic terms. Given a reasonably good environment possessing some useful resources to begin with, creative human energy and ambition enlarged the small city-state into a larger regional state controlling rural territory.

Once the essential framework of the regional state was put together as a combination of area, people, and organizational system, creative activity seemed to slacken. There then followed a pattern of calculated, opportunistic, and competitive enlargement. Leadership provided by invaders, rebels, ambitious younger sons, and powerful families visualized new ways of utilizing socioeconomic and politicomilitary combinations to gain and enlarge power. Sargon of Akkad, in an ancient country north of Babylonia,

may well have been a leader of this sort (ca. 2300 B.C.) who both rebelled and enlarged his city-state. His procedures for control over a large territory were significantly personal. By the time of Hammurabi of Babylon (ca. 1900 B.C.) both political structuring and military organization had been perfected to the point that an administrative system could command allegiance at a distance, and a code of laws became operative for at least some social levels of citizenry. The leaders of the Chou (ca. 1100 B.C.) set up their dynastic reign over a large part of North China by capitalizing on political and military constructs to take over and enlarge the previous town-state system in force under the Shang.

These early culture regions had numerous bases in both physical geography and cultural patterns of peoples. The physical geography of the earth was, by about 3000 B.C., fairly well stabilized in its postglacial patterns of climate, glacial wastage, vegetation, animal distribution, and the like. Cultural regionalism, and particularly the political regionalism of the state, became a matter that man could arrange on his own terms.

In the large zone from northern India to North China to the Mediterranean Basin and in the zone from southern Mexico to northern Chile we see similar kinds of constructs, each varying in detail. The specific civil administrative, military, and tax systems differed; the specific gods and their worship systems operated variably. These sociopolitical systems were characterized by differing numbers of classes having different obligations and privileges, by cropping systems varying according to the crop staples, and by other economic constructs showing separation. The broad outlines of each political state, however, displayed notable similarities. Political states could not be self-contained and isolated for two reasons. First, each state was dependent on resources beyond its borders. Second, there almost always were challenges to the patterns of power, either from within the state or by peoples from the surrounding regions wishing to share in, or take over, the state. The state system had developed as a human system of organizing people, territory, and ways of life, but since it was a human construct, the state system was subject to decline, rearrangement, and reconstitution.

As the territorial political state matured, one of its more distinctive aspects was the extension of political control over the rural countrysides to include the crop-producing hinterlands and their rural inhabitants. When rural populations were mobile and employed systems of shifting cultivation or pastoralism, control over populations remained incomplete, although control over territory might be achieved. However, in rural regions in which permanent-field gardening systems, primitive plow cultivation, and sedentary residential systems operated, control over both territory and population could produce a double benefit. Rural manpower and production resources could be made subordinate to the urban political control center, thus establishing a regional economic system. The rural segments were thus incorporated into the early territorial political state. This occurred at a time when the culture patterns of the urban townsmen/city dwellers were maturing into distinctive styles. The rural populations were also developing distinctive life styles under the imposition of political authority from the urban control centers—those styles that have denoted "peasantry." Rural peasant primary production systems and urban handicraft-manufacturing systems developed further in symbiotic relationships as the political state became an integrated functional system. In historic time among different regional cultures of the earth the precise structure of the rural peasantry and its relations to the urban political structure have varied greatly, but there are common functional relations.

THE REGIONAL CULTURE HEARTHS

Human impact on many long-occupied regional landscapes has been so heavy for several thousands of years that it is sometimes hard to conceive how any people ever could have succeeded in building and operating culture systems on some of the regional sites that today impress us as barren, rocky, and inhospitable. The precise chronology of each of the regions to be discussed in this section is still in question. The roots of comparative development lie prior to and behind the written historical record, and archaeological research has not yet completed a full reconstruction of regional settlement of the prehistoric sequence. The dates, and the specific kinds of cultural development, therefore, are subject to continuing change as to interpretation.

In the early clustering of culture groups in particular regional landscapes there were the successes in which technologies, latent resources, and human effort became combined. Success resulted in population growth of two kinds—the natural increase within the region and the in-migration by others from localities in which peoples did less well. Such regions became centers of human traditions and creators of the successful ways of living. They became *culture hearths*, in which cultural traditions became stabilized and in which the human way of living defined the standard of what human living ought to be among the peoples of that region. Out of these hearths diffusion spread the rules of living, the tech-

nologies, the traditions, and the human systems that denoted the civilized society.

Within the broad Old World zone there were several major and minor culture hearths significant to the later growth and spread of Old World culture, and within the New World there were two. In this section we review the growth and crystallization of the most significant of the Old World hearths (Fig. 4.11).

Mesopotamia

The term Mesopotamia refers to the valleys of the Euphrates and the Tigris rivers, the filled estuary of the two, the surrounding fringes of hill country on the east and north, and the margins of the Arabian Desert on the west and south. There is a great deal of variety in the landforms, local climate, and plant cover of the region, and the zone of the river valleys varies markedly from north to south.

The region is situated on perhaps the most frequented crossroads of them all. Its cultural-trading relationships stretch out along the routeways to include the eastern shores of the Mediterranean Sea, the hill country of Asia Minor, the hill margins of Iran, and the littoral of the Persian Gulf. This larger zone is the region termed the Fertile Crescent. Although not all of its portions shared in the early experiences of Mesopotamia, the cultural and historic linkages throughout the larger region are almost inseparable from the patterns within Mesopotamia itself. By its very accessibility the Mesopotamian lowland received a steady stream of militant inmigrants during the fifth and fourth millennia B.C.

There were many centers of local settlement, and an active period of competitive struggle between the city-states delayed the achievement of dominance by any one. As a result, the large regional state did not appear here quite so early as it did in the Nile Valley, although there seems little doubt that cultural diffusion from this region through Palestine to the Lower Nile River country was one of the elements in the maturing of Egyptian civilization. By about 3000 B.C. the Mesopotamian region had become a bright and shining light of culture to an outer world of barbarism.

Accessibility of the Mesopotamian region to nearby peoples was a two-way proposition. Diffusion of culture traits and complexes took place outward in all directions from this regional center. Over the centuries Mesopotamia received many in-migrating peoples bringing new culture traits. Some came as conquerors, some as settlers seeking a home, some as slaves to do the work of society, and others merely as traders. The barbarian in-migrants absorbed the culture of the hearth and became part of it.

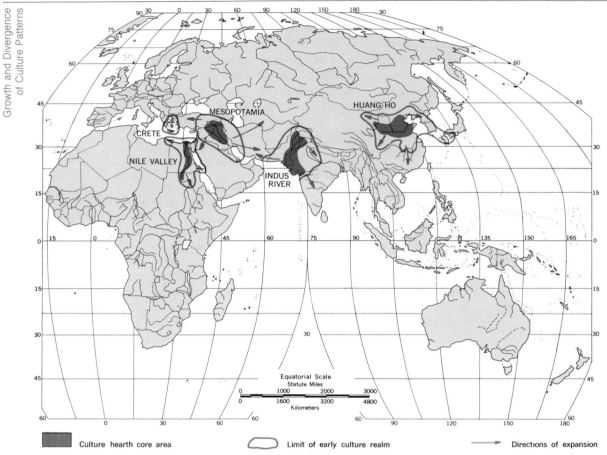

Figure 4.11 The Early Old-World Culture Hearths and Culture Realms.

The professional priesthood, the professional army, the civil bureaucracy, and the merchant traders evolved their own social structures and their own technologies suited to the land and the changing circumstances. These developments insured a continuity to life against continued politicomilitary struggles by many groups for control of the rich lowland.

The earliest archaeological evidence suggests rather sparse and scattered occupancy by sedentary folk in linear patterns along the piedmonts, according to the directions of flow in the water courses. The later archaeological record suggests increases in population and shifts in locality over time. The appearance and growth of formal irrigation facilities was gradual. A peak in both areal occupancy and population patterns in some areas seems to relate to the era of Sargon-Hammurabi (ca. 2300–1900 B.C.). The distribution of abandoned canal systems and the ruins of settlement suggest considerable variation in occupancy, irrigation, and population through time. The peak in historic population density may have occurred about A.D. 1000–1200, well after the importance of the Mesopotamian hearth had declined.

The conquest of Mesopotamia by the rising power of the Assyrian and Persian kings after about 800 B.C. moved the centers of political power away from the heart of Mesopotamia. Mesopotamia, then, as merely one sector in a series of new empires, suffered from poor administration. The ancient culture was passed on; new evolutionary forms of human organization and new technologies were rising. Barbarian lands were becoming civilized, and the creative cultural styles of the old hearth lapsed into a long period of decadence.

The Nile Valley

The Nile Valley is a watered strip of shallow canyon extending through, and standing out in sharp contrast to, the bleak, near-barren desert on either side. In the fifth millennium B.C. the Nile Valley had a gallery forest along its margin. Paleolithic peoples had lived within the flood margins of the river for a long time, since the seasonal flood waters supported an abundant plant cover, a wide range of animals, birds, and freshwater fish life. The upper and middle sectors of the valley enjoyed relative solitude. Near the head of the delta and along the coast the region remained open to access, and there has been a greater history of contact there than farther south. Into this delta region there were brought from Southwest Asia most of the domesticated crop plants and animals that permitted the growth of Neolithic levels of economy for the Nile Valley as a whole.

Within the region known as Egypt, centering on the Nile River Valley, the beginnings of civilization appear to relate to the lower, or northern, region between Cairo and Aswan. A regional political state matured there even before that event took place in Mesopotamia. There were fewer contending city-states, so that the local struggle for control was simpler. The culture hearth of the Nile system lay within the delta zone, although the evidence for much of this must lie buried under the sediments of the delta. The union of the long and narrow region into one political state brought into focus a series of culture patterns in a burst of creative brilliance.

As distinctive and as spectacular as it was, the Nile Valley culture hearth never reached the heights achieved in Mesopotamia. In early times relatively few Nile Valley culture complexes spread far beyond the eastern shores of the Mediterranean Sea. The peoples of the Nile Valley enjoyed independent solitude in the early period but, after about 1700 B.C., numerous invasions brought many new peoples into Egypt.

We know very little about the evolution of control over land and water in the Nile Valley. There always has been irregularity in the volume of flow of the river. A heavy runoff could have flooded the whole plain, cut new channels, and filled shallow depressions. Such a flood would have left behind new silt fills in some areas and temporary ponds and marshy tracts. A scanty runoff might never have wet down the whole floodplain, and might have affected only the margins of the main stream. Such variation was critical to agricultural productivity.

Earliest occupancy must have been patchy, scattered, and centered on those localities having the best year-long conditions. In time, crop growers could have devised techniques of bunding, of building levees, of leveling land, and of channeling water, techniques that could have extended productivity into semicontrollable areas, depending on variations in runoff patterns. Extensive leveling of the land, clearing of the gallery forest, and reducing of the seasonal marshland must have gone hand in hand with increased bunding, water-channeling, and seasonal measurement of flood levels for purposes of religious rituals. However, real control over Nile waters never developed sufficiently to protect against heavy floods or to compensate for scanty runoffs.

Much of the distinctive Egyptian culture system

gradually became so stylized that people in the later centuries sought chiefly to maintain the traditions and forms of the past. When colonies of Greeks, Hebrews, and Romans were resident in Egypt, those aliens learned what they could understand of an aging culture that had no spirit left. To this extent the Nile Valley culture hearth served as a source for the diffusion of many things. Prior to the beginning of the Christian Era, however, Egypt became a political pawn fought over by outsiders. Often subject to misrule, Egyptian civilization lost its vitality and the Nile Valley became a land of agrarian villages living out a routine annual cycle of life framed around the seasonal flow of the Nile River.

The Indus Valley

The Indus River region today seems well separated from Mesopotamia by the high and rugged country of Baluchistan-Afghanistan-Hindu Kush. On its other fronts the Indus Valley is quite open to the east and south. In earlier time, however, the highlands were well populated. The passes were well known, and the patterns of contact were more frequent than they have been in the last few centuries. The archaeological record suggests a Neolithic continuity of occupance from the Indus lowland across to the Mesopotamian lowland, with active and productive economies widely distributed.

The Indus Valley was another great alluvial lowland heavily flooded by runoff of melting glacial ice remaining from the Pleistocene. The lowland and valley bottoms were not easily occupied by the Neolithic hill country folk to the west or by the Neolithic occupants of peninsular India to the southeast. About 3000 B.C. the Indus country was apparently taken over rather suddenly by in-migrant peoples from the north and west. They were well acquainted with culture habits such as dwelling in towns, the city-state, and other advanced aspects of the Mesopotamian culture system. Trade contact between Mesopotamia and the Indus Valley was active. We know little of the social history for the region except for what can be interpolated from the presence of fortified palace-citadels, a precise grid-plan layout, ritual sites, and the apparent sizes of settlements. If their "seal writing" was truly a written language, it cannot yet be read properly.

A second invasion of peoples occurred about the middle of the third millennium B.C., which destroyed the leading cities, such as Mohenjo-Daro. Evidence suggests a rather steady decline in the level of Indus culture in the main core areas after about 2400 B.C. There were maintained on the southern margins of Indus culture, in Kathiawar, many if not most of the elements of classic Indus culture as late as 1000 B.C. The record does not indicate severe depopulation, but rather a decline to a rather crude level more characteristic of towns and villages than of the creative city.

The Indus cities suggest the passion for town planning and the practical functioning of a large settlement. These may have been the chief contribution, for the grid-plan city became more characteristic of north Indian cities, in time, than it did farther west. The impact of Indus culture lay primarily to the east and south. As the later outlines of Vedic Indian culture took shape, there were incorporated many elements developed in the Indus Valley. Although similar in many respects to that of Mesopotamia, Indus Valley culture exhibited a regional specialization of its own.

North China

The main features of the Huang Ho Basin (Yellow River) include a great river, several series of hill and mountain ranges, a huge piedmont alluvial plain, and a region carrying a soft-earth cover of loess. The basin is less clearly an alluvial lowland surrounded by physical barriers than is the case for Mesopotamia, the Nile Valley, and the Indus region. The loess cover is in highland country that has been etched by erosional process and also by soil erosion induced by long human occupance. The regional subdivisions, therefore, present distinctly local conditions in different sectors.

Throughout the Pleistocene and during the closing glacial period, the Huang Basin suffered very heavy flooding. These floods picked up volumes of glacial scour products from the headwaters zone, gathered large amounts of loess in middle reaches, and dumped the whole load into the shallow waters of the Yellow Sea. In its lower course the Huang often shifted its course, thereby spreading the fill over a wide area to create the North China Plain.

The pressure of human occupance has rested heavily on North China, and it is difficult to separate natural landscape conditions from those culturally produced. The strong variation in local environmental conditions offered early peoples many different choices in living situations. That the central Asian corridor opens into this regional complex was important during the drying-out period in which early human populations were on the move. North China presented a set of situations and local conditions favorable to the trial-and-error growth patterns that made the Huang Basin a region useful to early culture groups.

The archaeological record indicates the long occupancy of eastern Asia. There are hundreds of well-studied Neolithic sites within the Huang basin. The region constitutes a zone in which village-dwelling, crop-domesticating, animal-using, and culture-developing patterns matured. The nuclear region appears to have been the narrow and uppermost western margin of the North China Plain and the tributary valleys on both banks of the Huang Ho, in what now is northern Honan Province. The record of the earliest cultural expansion is not yet completely understood, but it is evident that the resulting culture region expanded southward from the hearth area across the North China Plain and westward to touch the terminal end of the central Asian corridor. On both margins local peoples established contact with other culture sources that set in motion further internal development in North China.

Shang China (sixteenth century B.C.) shows a beginning growth of the town-state system, and the emergence of a regional variant of the God-King. There was the functional development of domestic and palace architecture, the inventive-evolutionary development of manufacturing, and the growth of other culture complexes. All these are comparable with those systems and group complexes now well catalogued for Mesopotamia, the Nile, and the Indus regions. The chronology appears to be a little later than for the Mesopotamian culture hearth.

The Chou dynasty's attempt to create the large regional state in North China (eleventh century B.C.) followed similar attempts in western centers. The multiregional environment proved too large for mastery by the organizational concepts employed, and in time North China slipped back to a situation dominated chiefly by small regional states and city-states. It was only with the inward diffusion of added new culture complexes in the several centuries B.C. that the large regional state concept was successful.

Unlike the cases of Mesopotamia, Egypt, and the Indus Valley, there is clear historical continuity between the North China culture hearth and the later Chinese regional culture area, and between early Chinese culture systems and those of historic China. The Chinese culture hearth remained the center of culture growth and the center of regional power as the Chinese state expanded.

The Minor Hearths

Neither archaeology nor history yet provide a clear story of certain other regional developments of advanced culture in the Old World. Here we only make brief reference to those minor regional patterns that form part of the larger picture. These include the insular pattern within the eastern Mediterranean Sea and the regional units within Asia Minor-Syria-Palestine. Once the basic Neolithic production systems were established across Southwest Asia, their more local regional application through diffusion and their local development became possible.

The Island of Crete is such a local regional entity, and the record begins as early as 4000 B.C. with early Neolithic crop-growing clusters of villages on the small plains having good soil, local drainage basins, and other requisites. Both Egyptian and Mesopotamian culture complexes influenced Crete, but a local blending process is also evident, creating a local evolutionary sequence. On Crete the intermontane plains villages were complemented by towns built along the coast to accommodate a significant population increase. Whereas the beginnings of higher Minoan civilization date from shortly after 3000 B.C., the high period comes about 1800 to 1500 B.C., a time when both Mesopotamia and Egypt were disrupted by invasion. The centralization of political, religious, and administrative power turned the island and nearby Mediterranean waters into one functioning unit. Sea trading and the spread of Minoan culture forms from Crete to Aegean islands are part of the record and the means of transferring Minoan culture to the earliest of the Greek peoples (Mycenaeans) coming into the Aegean Peninsula. Around 1450 B.C. Crete lost its power to the Mycenaean Greeks in what probably was militant contact. Crete then became chiefly a region of villages and petty sea trading, exerting little influence on the conditions of Mediterranean life.

The eastern shores of the Mediterranean are backed by a strip of hill country that tapers off in the south but joins the uplands of Asia Minor in the north and that serves as a western border to the Mesopotamian lowlands. Here good lands of winter rain, forested uplands, small river valleys, and a coastal plain form a set of local regions within which Neolithic crop-growing, village-dwelling populations were numerous. As the drying-out process continued, there pushed into this region several ethnic groups of diverse origins. Each carried elements of culture derived from the Mesopotamian hearth blended into its own basic patterns. Competitive struggles for territory were frequent in all parts of the region among Hebrews, Hurrians, Aramaeans, Phoenicians, Hittites, and others from late Neolithic times onward.

Between about 2000 B.C. and the beginning of the Christian Era, this region witnessed a succession of regional cultures. Each shaped some specific construct in such territory as it could take and hold. Repeated alteration of territorial units never totally disturbed the flow of ideas and trade goods, the evolving of specific culture systems, and the broad diffusionary processes by which culture spreads. It was within this region that many discoveries, inventions, and creative formations of technologies took place. For all its turbulence, the zone played a significant part in the development and westward spread of Old World culture complexes.

THE DOMINANCE OF EARLY REGIONAL CULTURE REALMS

The impact of each early culture hearth became felt over a wide range of surrounding territory. This came about through diffusion of its complexes to other peoples as the hearth populations settled adjoining territories. As new peoples moved into the orbit of a hearth, they acquired elements of its culture. This wider geographical region, within which the culture system matured from the stimulus of a culture hearth, we have termed the culture realm. At least four primary and two minor culture hearths in the Old World were centers of the growth of regional civilizations. From these, three primary culture realms matured: the Mediterranean, the Indian, and the Chinese.

The primary diffusionary movements from the Mesopotamian culture hearth were westward to the marginal territories of the Fertile Crescent. They reached into Egypt on the south and the fringes of southeastern Europe on the north. The primary diffusionary movements from the Egyptian hearth were into the lands around the eastern Mediterranean Basin in the earlier period, and westward through the Mediterranean at a later date. In and around the Mediterranean, Mesopotamian and Egyptian cultures became repeatedly intermixed and synthesized.

In western Europe, it long ago became customary to refer to the Greek origin of European culture in general. The Greeks, however, were an immigrant people who moved into the Aegean Peninsula shortly after 2000 B.C. from a location in the barbarian world, still maintaining their pastoral culture on the economic level and a tribal variety of political structure.

The evolutionary translation of culture elements from Mesopotamia, Egypt, Crete, and the eastern shores of the Mediterranean into Greek culture created a new cultural system. This system spread throughout the Mediterranean Basin and back into Southwest Asia. It became a system of culture broad enough and varied enough, to characterize a whole realm rather than a regional culture hearth.

The Vedic pastoral immigrants overwhelmed the populace of the decadent Indus culture hearth about 1500 B.C. Territorial expansion eastward and southward was combined with the gradual absorption of Indus culture, first, and peninsular Indian culture somewhat later. The eventual spread of Indus-Vedic culture southward to Ceylon, and then eastward around the shores of Southeast Asia to Indonesia, broadened its regional impact in the same way that the southwest Asian culture systems spread westward via the Mediterranean.

Southward migrations of peoples out of the North China culture hearth began taking place before the expansionist thrust of the political state came about. The two processes went hand in hand thereafter. The Chinese culture system was more unified and specific in definition than that of either the Mediterranean or Indian, and the Chinese were more aggressive in spreading it as a specific way of life. The southern barbarians either became Chinese in culture or they had to fight for their way of life. Some of the stronger fought for periods of time. Many of the weaker culture groups began emigrating southward into Burma, upper Thailand, and the Laotian high country, a process that still continues.

The Chinese hearth also exerted its influence northeastward into southern Manchuria and Korea, and, a little later, Chinese culture began to spread to the Japanese islands. Northwestward and westward Chinese culture went only where it could go in terms of village settlement and the agricultural way of life. This choice prohibited its expansion into the Tibetan Highland and limited its entry into Mongolia and central Asia to those oases locations in which the settled life was possible.

We have here suggested the regional formation of three culture realms, which largely dominated the Old World by the start of the Christian Era. Around the fringes on all sides lay lands of the "barbarians"—those peoples who continued Paleolithic or Neolithic cultural practices lacking in the "civilized" traditions of each of the realms. Barbarian traits and complexes diffused into the several culture realms to be adopted as intrinsic features. Here and there, on the margins, a people formerly "barbarian" took on enough of the elements of civilization to form regional offshoots and satellite entities. Sometimes putting together something of political organization and military structure, such regional units were able to create sizable and significant regional entities. More common, however, was the forceful assault on the border regions of a particular realm by a group of "barbarians" either seeking entry into the territory of the realm or seeking to acquire the goods and commodities of its culture.

THE COURSE OF CULTURAL EVOLUTION

In this chapter we have discussed the ways in which human culture systems slowly advanced during the

Table 4.2 Cultural Evolution since the Late Paleolithic.

HISTORIC ERA	1975	Mechanization of Agriculture
		Industrial Revolution
	1250	Columbian Linkage of the Earth
		Beginnings of Occidental Science
	A.D.	Founding of Islam
	0	Concept of National State
	B.C.	Founding of Christianity
		Birth of Buddhism and Taoism
		Mature Pastoralism
METAL ERA		Smelting of Iron
		Political State Became Territorial
	2500	Alphabetic Writing Systems
		Concept of the State and God/King, Urbanism, Organized Religion
		Primitive Plow Cultivation
		Metal Smelting in Copper, then Bronze
NEOLITHIC ERA	5,000	Fired Pottery
		Beginning Writing Systems
		Clan Structure and Tribal Organization
		Sedentary Gardening
		Polished Fine Stone Tools
		Elementary Pastoralism
	10,000	Simple Shifting Cultivation
MESOLITHIC ERA		Animal/Plant Domestication
		Maturing of Kinship Systems and Simple Social Organization
LATE PALEOLITHIC ERA		Textile Weaving, Shaped Stone Tools
	20,000	Diverging Spoken Languages

earlier periods in different parts of the Old World. We tried to identify the regional centers of those advancing patterns. The more important aspects were subdivided into topical discussions and then put back together again, in a sense, into regional combinations. The basic cultural processes of discovery, invention, evolution, and diffusion were again and again brought into the account of the patterns of advancement. There is no full agreement among scholars on precisely what features of cultural development were the critical ones in producing civilization out of the simpler culture systems of preceding eras. It is obvious, however, that cultural progress involves the piling up of many different kinds of knowledge and skills, and then working them all into culture systems.

It is clear that not all parts and all peoples of the Old World made important contributions toward the development of civilization. Peoples in some regions were chiefly the receivers of advances made in other areas, and sometimes they did not quickly or easily accept the new advances. But sometimes, also, local developments were not very useful in distant regions in which environmental situations were quite different from those in which particular advances had been made.

The long-term advancement of human culture has been a continuing struggle to find ways in which to utilize the many kinds of resources present in the earthly environment. Human understanding and insight have sometimes been very slow in coming, but mankind has been persistent in searching for better ways of doing things. Once started, however, for the Old World as a whole the process did become cumulative, and the modern New World has shared in the benefits. It is this cumulative aspect that stands out most strongly as the long past is reviewed (Table 4.2).

The tabulation shows only a few items during long time spans in the earlier periods, and relatively more numerous items in recent centuries. The early features were elementary but basic, whereas many of the more recent developments are refinements of old aspects. Taken in total, the record indicates a continued insight into the nature of physical processes, into the qualities of the earthly environment, and it also establishes the persistent ability of the human race to utilize the resources of the earth to its own ends.

SUMMARY OF OBJECTIVES

After you have read this chapter, you should be able to:

Page

SETTLEMENT SYSTEMS

The development of settlement systems is usually discussed under four headings—rural dispersal, villages, towns, and cities. Rural dispersal is often considered the oldest system and the city the youngest. However, it is more likely that the earliest humans camped together, moved together, and lived in groups. This would suggest that the hamlet or village campsite is actually the oldest system of living. As more specialized economic methods evolved, the temporary campsite may have become a permanent settlement.

The rural dispersal of residence appears to have been a response to some cultural factor. Perhaps, it originated during the Neolithic as crop-growing and animal husbandry introduced new relationships between human beings and the land. Rural dispersal is found in all parts of the world among all kinds of ethnic communities where agriculture is the primary means of earning a living. However, it is no longer the dominant settlement system in most large regions. The normal rural dispersal begins with the unlimited but unsystematic choice of land plots. Latecomers find increasingly limited choices available until the process comes to a climax with a fully occupied landscape of irregular-sized plots (Photo 2). This system was common in Europe and can be seen in rural sections of the eastern United States. By contrast, those American regions that were settled after the adoption of the rectangular land survey system have a rather precise and neatly arranged fieldscape accompanying a thinly distributed rural homestead pattern (Photo 4 and Photo 2 in Photo Essay No. 3).

The town is usually larger in size and population than the village. It has a more complex form, and, in addition to serving as a place of residence, it is characterized by a number of secondary functions. The town is of more recent origin than the village, but it may well have developed toward the end of the Neolithic in more advanced regions. Earlier towns normally had some degree of fortification, but few towns originating in the modern era have them. Many old towns have suburbs lying beyond the old cores. Some towns began as campsites, fortified strong points, residential villages, or centers of political administration. They developed in response to the demands of increasing rural populations for secondary services. In the modern era many towns have been founded as such. Recently, the "new town" has been developed as the result of formal regional planning in an attempt to lessen the human pressure on crowded cities.

Photo 1. In the gently undulating Iowa plain, with its maturing woodlots and contour plow lines, the rectangular land-survey lines must be searched for. The large landholdings mean that family homesteads are somewhat regularly and rather widely dispersed. Note that a modern highway cuts across the established road grid.

Photo 2. In western Kenya (near Kisii, not far from Lake Victoria) rural dispersal is by small patchwork holdings so that houses are scattered but close together. The crops shown here are chiefly maize, pyrethrum, and bananas.

Photo 3. When the village residence system is used and the landscape has become fully occupied, the farmlands and the villages create contrasting patterns, as, in this view, is shown of Madagascar.

Photo 4. The grid pattern of the rectangular land survey shows clearly in this view of a dispersed rural settlement in Kansas where the farmsteads are widely scattered. The modern highway cutting across the grid pattern and the town itself offer sharp contrast to the older patterns of the landscape.

Photo 5. The village of St. Martin, Switzerland, is perched on the northern shoulder of the Dent Blanche, and villagers have a view into the valley of the Rhone River. Terraced crop fields lie around and below the village.

There are many kinds of towns. Some occur in areas having rural settlements and thus present sharp contrasts with the landscape (Photo 4). Since towns occur in almost all parts of the world and among almost all culture groups, there is great variety in their sizes and appearances (Photos 7 and 9). Towns reflect the kinds of secondary functions carried on in them. Therefore, they may often occupy sites in which specialized activities are the controlling factor. Some examples are the march site, the agricultural market town, and the port town. Size, morphologic character, leading economic functions, patterns of political administration, the presence of a cathedral or an educational institution, or a particular variety of manufacturing are among the factors influencing the nature and appearance of towns. The populations and functions of towns appear to be related to broad patterns of economic growth and development. Central-place-system studies have advanced conceptual hierarchical structurings to account for the size, rank, function, location, and distribution of many kinds of towns.

The city is an imposing form that, in one sense, caps the landscape developed under human occupation of the regions of the earth. As cities have been culture centers since they were first developed, they not only exhibit the specific culture of their realm, but they also epitomize all of the political, social, and economic aspects of that culture. The strictly hierarchical rank, size, and function approach to cities neglects the consideration of the cultural factors that really differentiate the cities of the earth. Cities differ greatly in their morphologies since so many of them began as residential villages or regional market towns. As formal planning, industrial technology, modern transport-communication, and urban facilities increase, however, the large cities of the earth are coming more and more to resemble each other as intercultural expressions of modern man.

Photo 7. The town of Rockport, Massachusetts, is built around its old fishing harbor. Today, the town functions chiefly as a boating and recreational center.

Photo 8. These new townhouse units near Reston, Virginia, are a variation of the apartment living pattern. Built on an old wooded farm landscape, the "settlement" is an effort to combine a rural atmosphere with urban comfort.

Photo 6. (Opposite) The wooded margins of the Val de Ruz, Switzerland enclose a group of villages and their surrounding fields. The patchwork of crop fields indicates a mature cultural landscape, and the wooded rough areas are all that remain of the original forest cover.

Photo 9. In the Algerian Sahara, settlements occupy the dry localities in compact clusters. Areas with underground water are reserved for date palm groves.

Photo 10. This view of the Montmartre section of Paris shows the older urban tradition in which multistory (but not high-rise) structures create solid blocks of buildings along rather narrow streets and alleys. Such a pattern normally indicates a high population density.

Photo 11. This view of downtown Salt Lake City, photographed about 1971, reflects the basic Mormon town plan with its wide streets and large blocks. Blocks were originally divided into large lots to provide room for houses, gardens, and animal shelters. As the original agricultural settlement became more commercial and residential, some of the blocks were subdivided.

The most common settlement system is the agrarian village. Of various sizes and shapes, these villages are usually closely spaced and are surrounded by small fields of irregular shape and arrangement. Traditionally, they are simply clusters of residences gathered according to some order. Although the agrarian village is the most common type, villages occupy all kinds of sites and accompany many different forms of economic endeavor. Sometimes, settlements similar to the village may result from modern industrial exploitative technology. Today, many regions still have traditional villages. In others, however, the old patterns are breaking down as home handicrafts are replaced by mass-produced consumer goods and stores selling these goods move into the villages.

Nevertheless, the cities of the earth are not really alike. There are some with historic old sections of multistory buildings and canyon-like streets (Photo 10). Others present sprawling patterns of relatively low buildings in which the various functions of a city are somewhat intermixed (Photo 11). A number of cities may also be characterized by their great dormitory function, as displayed by residential sections interspersed with shopping centers, schools, and other community facilities (Photo 12). In some cities, there is a cultural lag accompanying changes in the use to which land is put. A particular area may remain residential even after areas surrounding it have completely changed over to quite different uses (Photo 11, left fringe). The tall skyline of the central business districts of many great modern cities often forms a landscape in sharp contrast to those facilities that need to be spread out over large amounts of space (Photos 13-16). In a few of the great metropolitan centers of the world, all of the cultural functions are present and repetitive, but sometimes they are not visually separated. In these cases, the urban complex presents a massive form and shape which dominates the regional landscape (Photos 14-16). The "flight to the suburbs" occurring around some of the larger urban centers tends to carry urban patterns into formerly rural areas in an effort to escape the problems of the cities while preserving the comforts of urbanism (Photo 8).

Photo 12. The dormitory function of many modern cities takes the form of the single-family home set on its own plot of land, as shown here in Levittown, Long Island, New York. There is a complex street pattern with a scattering of commercial, religious, educational, and recreational facilities.

Photo 13. This view of Melbourne, Australia, presents a contrast of three urban patterns: the broad horizontal spread of the rail yards, warehouses, docks, and factories of the port sector; the high-rise buildings of the business district in the center; and the sprawling residential areas lying in the far distance.

Photo 14. New York City and environs. The massive reach of modern urbanism has spread far beyond the tip of Manhattan Island (left center), the founding point for the little settlement of New Amsterdam. The whole New York region is one great urban complex (see Fig. 10.11) in which all aspects of the modern city are to be seen on the landscape. Brooklyn lies in the lower center, and Queens is in the right upper center. Governor's Island is at the far left center, with the Hudson River beyond. West of the Hudson, across from lower Manhattan, lies Jersey City, with Hoboken to the north. Near the center horizon are Yonkers and Mt. Vernon, and at the far upper right, north of the East River, is the Bronx. The trunk highway leading from the lower right corner across Brooklyn goes under the East River through the Brooklyn-Battery tunnel.

Photo 15. Houston is a relatively new city in one of the fastest growing states in the United States. Since the city is situated on a flat coastal plain, the downtown towers completely dominate the urban landscape.

Photo 16. Vancouver, British Columbia, has also grown rapidly in recent decades. Unlike Houston, the city sits between sea and mountains which provide a dramatic background to the skyline.

CHAPTER 5
DISCONTINUOUS EXPANSION IN CULTURE REALMS

_____KEY CONCEPTS_____

Some Culture Hearths Were Late in Development *There Were Two Primary New World Hearths* *There Were Four Secondary Old World Hearths* *There Were Marginal Satellites to Culture Realms* The European Culture Hearth Exploded	World-wide *The Cultural Background of Europe Produced Militant Competitiveness* *European Explorers and Missionaries Were Followed by Traders* *Spreading Europeans Produced Cultural Reactions*

In the previous chapter we emphasized cultural advances of mankind in the general sense as elements of human creative growth. We brought those developments together in a discussion of the early formation of a few progressive regional centers of civilization in the Old World. We pointed out, but did not emphasize, the fact that the advances in human culture did not equally affect all the world's ethnic groups at the same time in all locations.

In this chapter we deal with three themes:

1. There were many slowly maturing societies and areas that lagged behind the more rapidly developing societies and areas.
2. The late maturing of the European culture realm involved a series of cultural subregions without one primary hearth.
3. The European culture realm expanded to link the earth into one worldwide operating system despite the fact that there continue strong differences among the separate societies of the world in levels of technology, levels of culture, and levels of living.

In the human activities going on around the earth since mankind rose to dominance among the living forms of life, there has always been a steady interplay between three different forces:

1. *People*, as small and large groups, each group living together as a separate society,
2. *Culture*, as that particular combination of material and social technologies, modes of behavior, and values put together by each society, and
3. *Region*, as the territorial unit occupied by each separate society.

The earliest population units of human beings split off into distinct and separate groups, each of which chose to do things somewhat differently. Each group also chose a home territory in which to live, or the group was pushed by more powerful groups into a territory it had to accept. In this selective occupation of home territories, mankind came to inhabit almost the whole of the earth many thousands of years ago. In the earliest periods the sum total of the cultural technologies possessed by each group did not differ greatly, for none of the groups was very advanced. Those human groups that were fortunate in being able to choose and hold a climatically comfortable homeland, rich in easily used products, obviously lived better than other groups that made poorer choices or had to accept poorer homelands. At a primitive level of technology, the quality of the homeland environment was obviously more important than the quality of the culture system.

In this volume a concern is to demonstrate that a societal culture system is extremely important in the long run, as mankind continues to determine ways in which the resources of environment can be made to satisfy human needs and desires. As a society invents new elements of technology, or adapts those technologies invented by some other society, its capacity to manipulate its environment increases. At the highly advanced levels of culture achieved by some societies in more recent human history, the less obvious qualities and more difficult resources of an environment can be utilized to support and expand a pattern of living. At these advanced levels the sociocultural system and the outlook of the population become more important than the simple quality of the environment.

The degree of importance of any one force in the interplay between people, culture, and region has always differed greatly from society to society. In some cases a poor environment presented so few opportunities that the problems of coping with even the maintenance of a simple pattern of livelihood never gave opportunity for the creative spark of internal advancement. If located in an isolated region, off the routes of diffusionary patterns, such a society might fail to receive any new ideas and would long continue its simple system. Conservative societies often chose to move into out-of-the-way locations off the routes of diffusionary currents in order to maintain their traditional culture systems. A passive or conservative society that continued to occupy a good region would begin to lag behind when it ceased to attempt to invent new technologies or when it became unwilling to accept new culture complexes diffusing from other societies. In order to discourage internal inventiveness, a society that had made many developmental steps forward in cultural levels might gradually adopt such conservative viewpoints and refuse to accept new features that later spread along the diffusion routes. But, when a society maintained a strong interest in improving the collective level of living, then the creative accomplishments of its members, and its eagerness to adopt new ways developed by others enabled that society to forge ahead and to become a leader in many patterns of cultural development.

In the cultural advancement of mankind, not every separate society progressed at the same rate. Many of the early simple societies never reached a highly advanced level of cultural development through independent growth. Some societies advanced from the Paleolithic to the Neolithic level, but then did not progress beyond that level, continuing to lead a Neolithic pattern of livelihood. Many of the simpler societies, particularly the smaller ones, have continued to live at either the Paleolithic or the Neolithic levels until modern time. Not all the great steps in evolutionary cultural development took place at the same time. In a few cases there was a very slow beginning of creative advancement, but then a long-continued period of evolutionary growth; hence, when a flowering of a cultural hearth did occur, it would come only much later than the appearance of the earliest culture hearths in other areas. In the previous chapter, we dealt with the earliest of the culture hearths; here we turn to consider the late-flowering cases.

THE LATE-BLOOMING CULTURE HEARTHS

In the appearance of the later culture hearths, two different patterns of development are discernible. First is the culture hearth that evolved in a large region in or near which there had been no earlier primary development of advanced culture. Our list of developing centers of civilization has two cases of this kind, the two New World systems achieved by American Indian populations. Secondly, there is the case of the secondary hearth in which delayed regional achievement takes place on the margins of an older cultural realm. There are four cases of this sort, the African Sudan, the Near East Islamic hearth, the central Asian Mongol hearth, and the peculiarly late development in western Europe. All four of these came to their period of growth at different times. We have set the European case apart from the others to be dealt with as the second theme of this chapter; let us first discuss the two New World developments and the three Old World cases.

Primary New World Hearths

Recent archaeological work has disproved the older theories of the extreme recency of mankind in the New World. There remains, however, a strong and emotional disagreement on how early any people arrived in the New World. Similarly, there is marked disagreement on what Old World culture complexes were diffused to the New World and at what dates. Related to these focal points for differences of opinion is another one: How much independent invention of basic culture traits and core complexes took place in the New World?

The Meso-American Zone. Central-southern Mexico, Yucatan, and Guatemala form a region in which change took place during the late Pleistocene and the early Holocene. The results were somewhat similar to those of the Old World in that a drying-out took place, with changes in plant/animal ecology, water availability, and human living conditions (Fig. 5.1). The whole area is extremely varied in landform units, local environments, living situations, and local conditions. The area lacks a unifying river system such as the Nile or the Indus. Instead, there are regional basins, smaller valleys, karst platforms, and alluvial piedmonts. Many of these are separated from one another by hill-country tracts and mountain systems. There are two long strips of coastal lowlands. The climatic conditions of the whole region range from rainy tropics to mountain alpine to desert fringe. The north and west, where precipitation is seasonal, is a land of summer rains. The whole region is rich in plants of the sort useful to Paleolithic peoples, and it had a varied animal population, although one lacking the kinds of large animals that were domesticated in the Old World.

Meso-America shows the sequence of development familiar in the Old World—Paleolithic, Mesolithic, and Neolithic. The Neolithic now seems to have begun about 7000 B.C. Prior to 1000 B.C., complex culture patterns began to appear in the spread of the community center/temple complex. The slightly later appearance of systems of writing, astronomy calendars, complex irrigation works, the kingship and the concept of the state, and other features unmistakably denote the rise of civilization. The form of the city and its street alignments do not show up in quite the same as in the Old World, but there were some very large urban settlements. Contacts with northern South America are evident in the appearance of metallurgy, but there are regional differences in many of the culture complexes. By about 500 B.C., individual regional culture systems throughout the area had arrived at the equivalent of the city-state level of the Old World. There followed in the centuries prior to the Columbian Discoveries a succession of regional shifts in power, invasions, elaborations of territorial empires, and growth of trade contacts on interregional bases. These developments parallel those of the Old World.

As a hearth region the influences of Meso-America were felt chiefly to the north, with cultural diffusion extending as far as what today is the Southwest and the eastern woodland of the United States.

The Peruvian Coast-Andean Highland Zone. The Peruvian zone of the west coast of South America, ranging from the coast into the highlands, comprises still another region in which human develop-

Figure 5.1 The New-World Culture Hearths and Culture Realms.

The Temple of the Moon originally stood on the top of this great pyramid at Teotihuacan,
northeast of modern Mexico City. It served as a companion to the Temple of the Sun. The two
were constructed about A.D. 200 as the central focal points of the early Toltec capital.

ment progressed strongly beyond the patterns of
Neolithic crop growing and the development of settled life. This is physically a complex zone ranging
from desert on the coast to equatorial temperate
highland and mountain alpine climatic conditions. It
ranges from a narrow coastal plain to intermontane
valleys, basins, and plateaus separated from each
other. Cutting across the zone are snow-fed rivers
whose drainage basins possess tracts of moist and
cultivable soils. Ecological environments vary markedly with elevation and location. Although there is
no great region of uniformity, there is sufficient par-
allelism of environmental conditions between stream
basins located relatively close to each other to provide considerable areas of similarity.

On the coast a cropping economy dates the Neolithic at just prior to 3000 B.C. The growth of settlements and regional political organization came
about 1700 B.C. Irrigation perhaps predates 1400 B.C.
Metallurgy was in evidence by 800 B.C., followed
rapidly by specialization in manufacturing. Highland occupation was only slightly later than lowland
development. Between 1500 and about 200 B.C. there
came the maturing of urban civilization, integration
of interregional trade, and the rise of the regional

political state. Political empires appeared in the
early centuries of our era as the two highland states
of Huari and Tiahuanaco preceded the Inca by
some centuries.

In Peruvian life there was no development of systems of writing to provide a specific chronology of
record. There was almost no development of astron-
omy-calendar systems, and none of "science" in the
Old World sense. At the time of the arrival of the
Spanish, the Inca Empire was, however, a large and
complex political region. It had a high degree of
interregional productive economy, population concentration, and economic wealth, and a high level of
living. There were clearly connections with Meso-
America by land, and possibly some by water. There
had occurred the outward spread of cultural influences both southward and northward along the Andean Highlands.

Secondary Old World Hearths

The history of the diffusion of culture indicates that
discrete populations in particular regions never retain total monopolies over human ways and technologies. Depending on the social value systems of

different societies, there are variations in rates of adoption of newer ways. Around the primary hearths and within and around the primary realms, there was operative the process of creating secondary hearths and culture realms by groups of peoples not active in the life of the primary hearths and realms. With variation in both time and in territorial appearance (Fig. 5.2), these secondary manifestations were combinations of complexes diffused from primary realms and complexes of territorially localized origin. Secondary hearths may involve populations not originally concerned or may include descendants of primary realm populations. However they come about, wherever they are located, and whatever culture systems ensue, such secondary activities are part and parcel of the human tendency to organize earth space for the use of mankind.

The African Sudan. South of the Sahara Desert, a broad zone stretches across Africa ranging from steppe to tropical forest and from flat lowland through hilly upland into rough mountain country. In this large region many different local environments possess markedly different plant, animal, soil, and water resources. This whole area has undergone changes since the height of the last glacial period, but the Pleistocene-Holocene physical record is little more understood than the human record. At least two nuclear areas of significant human creativity can be identified (see Fig. 5.2). One is the Ethiopian highland region in northeastern Africa, and the second is located in central western Africa between the Sahara Desert and the coastal fringe.

The Ethiopian high country was one of the key regions of plant domestication. Diffusion of technolo-

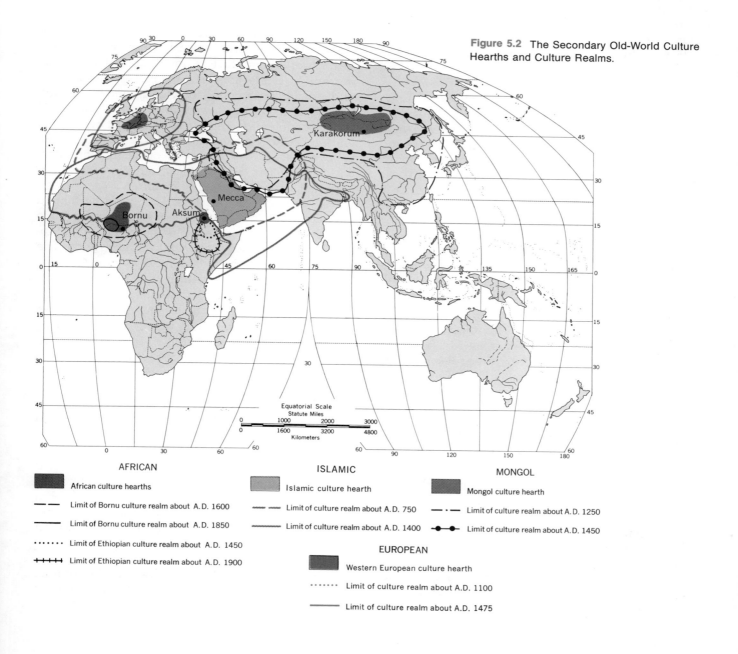

Figure 5.2 The Secondary Old-World Culture Hearths and Culture Realms.

AFRICAN

■ African culture hearths

--- Limit of Bornu culture realm about A.D. 1600

─── Limit of Bornu culture realm about A.D. 1850

······· Limit of Ethiopian culture realm about A.D. 1450

+++++ Limit of Ethiopian culture realm about A.D. 1900

ISLAMIC

■ Islamic culture hearth

─ ─ Limit of culture realm about A.D. 750

∿∿ Limit of culture realm about A.D. 1400

●─●─● Limit of culture realm about A.D. 1450

MONGOL

■ Mongol culture hearth

─·─ Limit of culture realm about A.D. 1250

●─●─● Limit of culture realm about A.D. 1450

EUROPEAN

■ Western European culture hearth

······· Limit of culture realm about A.D. 1100

─── Limit of culture realm about A.D. 1475

gies may be significant, but by about 4000 B.C. crop domestication was under way in some parts of the zone. Village-dwelling crop growers in a Neolithic context must date not far from 1800 B.C. in at least the Ethiopian sector.

The whole African Sudan is too large, too varied, and too little known to permit inferences of uniformity throughout, and there must have been regional and local contrasts. The two nuclear areas may well have been subcenters of local dispersion. One can only set a tentative date as just prior to the start of the Christian Era for the late Neolithic across Africa, and this date marks the kinds of changes that advance beyond the Neolithic for all of sub-Saharan Africa.

Little known archaeologically as yet is the growth of towns, but the process appears to be late—about the start of the Christian Era for eastern Africa and somewhat later in the west. Progress beyond the early Neolithic came slowly to Africa south of the Sahara. Many of its elements rested upon inward diffusion. In the early period no primary culture hearth evolved, and no culture realm surrounded this zone as a positive force.

The further development of sub-Saharan Africa moved more rapidly, based chiefly on the diffusion of political and economic organization derived from north of the Sahara or from Arabia. In the last centuries B.C. a series of trading towns appeared across the northern margins of the savanna belt. These towns evolved into city-states and then into territorial kingdoms, perhaps first organized by outsiders from the north but soon dominated by local leaders and upper classes. Through the centuries the shifts in regional control in this broad zone continued to take place according to the importance of trade routes. The widespread introduction of the camel after about A.D. 400 led to expansion of the trans-Saharan trade routes. Later this led to the introduction of Islam and the southward diffusion of Arab cultural influences. In the Ethiopian sector this was balanced by the influx of southern Arabian peoples and culture, the Rome-India trade routes, and the adoption of Christianity, followed by the introduction of Islam. There gradually matured a distinctive Arabized Negro civilization in the central-to-western Sudan and an Arabized Amharic civilization in Ethiopia. Although there were steady shifts in political controls, social culture, and regional influences, this sub-Saharan zone constituted a culture realm quite separate from that north of the Sahara.

The Islamic Realm. Islam found its early adherents chiefly among populations neither fully part of the Mediterranean realm nor converted by Christian missionaries. Islam originated as a religious revival among and for Arabs, in the seventh century A.D. The new faith incorporated many Judaic religious ideas, and Semitic culture traits. Islam rapidly assumed authority beyond the religious creed alone, to take on a militant political and economic role. Military and political union of the Arabs resulted, but it was based in part on dislikes and grew stronger through militant economic expansion into non-Arab territories. Islam derived from a regional territory neither highly developed in agricultural terms nor possessed of other outstanding resources. Islam derived much of its social strength from its tolerance of minorities and its granting of special privileges to those who joined its ranks. The leadership of Islam came to be centered on a charismatic Caliph who embodied the prophet-force of Muhammad and the absolute temporal power of the king, in a revival of the concept of the God-King.

Within two centuries the territorial spread of the Islamic Arabs had carried them to political dominance in lowland southwest Asia, across northern Africa into Spain, and northeastward into the Transcaspian lands. During this early period of expansion Arabs were busy, first in formulating the Qur'ān (Koran) from the spoken sermons of Muhammad, and, later, in interpreting the Qur'ān and in translating its statements into a workable body of law, social code, economic doctrine, and religious system. So encompassing were these codes of religious law that they came to govern all aspects of life, from architecture to economic activity, thus becoming the handbooks for a whole system of culture.

The later spread of Islamic conquest continued northeastward into central Asia, southeastward into the islands of the Indies and the southern Philippines, and southward into sub-Saharan Africa. The later spread of Islam was carried by non-Arabs in some cases. Arabs followed the militant conquerors as priests (imam) to teach and interpret the Qur'ān, even as far as Mindanao in the southern Philippines.

The great spread of Islam was complemented by the establishment of the tradition of the pilgrimage (hadj) to Mecca and by the practice of daily prayers to Allah while facing toward Mecca. The widespread travel-trading range of the Arab and Persian members of the Muslim realm; their interests in science, technology, and the arts; and their role in preserving and diffusing early knowledge were all highly significant to mankind in most parts of the Old World. But, lastly, the rise of Islam was the addition of a permanent culture realm to the regional cultures on the face of the earth. This was entirely a human product created whole in the mind of man and not the product of a set of nature's resources enclosed in a superior region.

The Mongol Realm. The growth of the Mongol culture realm represented the sudden fashioning of significant political and military bonds among the pastoral peoples of central Asia by the leaders of the Mongols, originating in the genius of Genghis Khan (see Fig. 5.2). The fashioning of superior elements in military strategy and operational horse cavalry tactics resulted in military power that could be focused on particular objectives. This power was far superior to the quality of any of the defensive strategies and tactics then employed in the surrounding lands. Genghis Khan himself participated in operations from North China to Afghanistan, A.D. 1211–1225. Mongol power swept to conquest from China to eastern Europe and to the banks of the Indus, resulting in political and military control over a large share of Asia. The appeal of religious fervor attached to a charismatic leader had little to do with the case, since Genghis Khan remained an animist and religious activities never became significant in Mongol dominance.

Whereas military conquest resulted from superior strategy and tactics, later political control rested on sound military administration based on the procedures and bureaucracies of the conquered territories, with the Mongols holding themselves above and apart from the conquered. Genghis Khan's concept of the state was a simple one, consisting of the pastoral, political elite in the steppe drawing tribute from all surrounding lands, in which each outer region continued its own ways of living in tune with its physical and cultural environment.

In bringing political control over a huge territory under one leadership, Mongol power necessarily operated an "open-door" empire. Ideas, procedures, commodities, and people were free to move from one margin to another. That an Italian, Marco Polo, could become a civilian government official in China under a Mongol emperor is expressive of a tolerance of culture seldom seen in the tight social and regional controls within the agricultural hearths. The rather elementary political concept underlying the political state emerging from the Mongol cultural ascendancy proved only temporarily effective, since little change and development took place in the pastoral homeland in Mongol culture itself. However, the cultural results for the Old World were significant, since the appearance of the short-lived Mongol state produced diffusionary effects perhaps even greater than those of most tightly controlled culture hearths. That Marco Polo, to cite him again as an example, could later spread in Europe the knowledge of how the Chinese lived is the sort of diffusion that promoted tremendous changes in the ways of life on earth.

The Marginal Satellites

We have now reviewed the kinds of culture hearths and the centers of their origins in the development of

several quite different living systems in both the Old World and the New. We have indicated the expansion of these living systems into culture realms, affecting large territorial sectors of the earth. We have not, of course, completed a discussion of all inhabited portions of the earth or of all the different ways of living thus far devised.

Our account may seem incomplete in that we have omitted any discussion of the Roman, Khmer, Javanese, and Japanese empires and have not mentioned the Pueblo culture region of the southwestern United States. From the standpoint of deriving new systems of human living in units of earth space, none of these regional combinations originated basically new systems. Each took culture complexes devised elsewhere and applied them to a region. In that application each improved on many of the earlier applications to produce living systems of greater complexity, greater human population, greater wealth and prosperity, or greater artistic heights than obtained in the original culture hearths. From this point of view the Roman Empire was a practical application of the living systems of two culture hearths—Mesopotamia and Egypt—as joined together by the Greeks in the eastern Mediterranean. The Roman Empire thereby reached heights, population totals, territorial limits, and levels of prosperity not achieved by either Mesopotamia or Egypt.

The regional application of hearth-devised living systems to distant and lightly developed environments has been occurring throughout the human occupation of the earth. As historic time passed and diffusion carried regional cultures to far territories, the total variety of human living systems increased steadily. Within the historic time span, however, there has been regional dominance by the culture systems of a relatively few hearths. Each expanded its influence to affect a sector of the earth. Whereas at 3000 B.C. we could clearly itemize only a very few hearths and culture realms, by A.D. 1400 the number of culture realms had increased, and their zones of influence had been extended to a relatively large portion of the earth.

THE WORLDWIDE EXPLOSION OF THE EUROPEAN CULTURE REALM

This section presents the second and third themes of this chapter, the maturing of the European culture realm, and the interlinkage of the whole earth. The maturing of Europe came slowly and rather late, in a zone far removed from the areas in which outstanding advances in human culture created the earliest culture hearths. The interlinkage of the earth into one system came spectacularly.

The Cultural Background of Europe

Along with northern Eurasia, northwestern Europe underwent marked physical change as the final Pleistocene glacial period turned it into a very cold and wet zone in which life was difficult for the small populations of Paleolithic hunters who lived along the icy southern fringes. As glacial conditions disappeared, plants and animals migrated northward to produce regional patterns described in historical times. The resulting environment varied greatly in local regions, so that northwestern Europe offered many different kinds of conditions. As the separate regions became repopulated by human migrants, there resulted very marked differences in ethnic patterns, economies, and local culture systems.

Between roughly 4000 B.C. and about 1000 B.C., small groups of village-dwelling peoples established themselves in the separate areas. Paleolithic hunters and fishers came first, but then came Neolithic groups who kept animals for meat and milk products, and who carried on a little crop growing and pasturage for their animals. Many of the crop plants were those originally domesticated in the Mediterranean Basin and the Near East, plants not well adapted to the cool and moist conditions of the north country. As selection of crop varieties adaptable to the cool and moist zone gradually succeeded, the local economies of northwestern Europe came to combine animal keeping with restricted variants of crop growing. The new inhabitants of northwestern Europe, therefore, were establishing their basic Neolithic systems of territory and economy at the time the cultural advances of the Near East were achieving the breakthrough from the late Neolithic to the level of civilization.

Some of the major formative factors in the cultural development of northwestern Europe were:

1. The great early migrations of ethnic groups,
2. The rise and fall of the Roman Empire,
3. The differentiation of Europe into nation-states,
4. The role of the North Italian merchant cities during the Middle Ages, and
5. The beginnings of an agricultural "revolution."

The first of these factors resulted, before 500 B.C., in the differentiation by language of the peoples of northern Europe: Celtic in the west, Germanic in the north-central zone, and Slavic to the east. Celts settled in the Iberian peninsula and occupied France, where they became known as Gauls. Others crossed to the British Isles, where surviving Celtic languages are Irish, Gaelic, Welsh, and Manx. Some Germanic-speaking peoples moved north into Scandinavia whereas others fought the Romans along the Rhine and Danube frontiers and later spread west, south, and east: as Vandals in north Africa; Visigoths and Suevi in Iberia; Ostrogoths and Lombards in Italy; Franks and Burgundians in France; and Angles, Saxons, and Jutes in England. Slavic speakers, the ancestors of Poles, Czechs, and Slovaks, moved northwest toward Germany. The ancestors of the Yugoslavs migrated south toward the Balkans, whereas those going northeast were the ancestors of the Russians.

The rise of Rome to military and political dominance began about 500 B.C. among the Latins in the Italian peninsula, the culture hearth of the Romance-speaking peoples, who had been influenced by Greek colonists in southern Italy, and by the Etruscans to the north. Consolidation of the peninsula was followed by the defeat of Carthage, a city-state founded by the Phoenicians on the north African coast, then by conquest of all western Mediterranean lands: Greece, Asia Minor, and the Near East. By 200 A.D. Rome controlled an empire extending from Mesopotamia and Egypt to Scotland, from the Sahara desert to the Rhine and the Danube. During this era, northwestern Europe became acculturated to the traditions of the Mediterranean culture realm. Examples of traits that spread from Rome were such nonmaterial elements as the Latin language, alphabet, and form of writing; Roman principles of law and government; and the Christian religion. Roman engineers improved upon the arch by the use of masonry, and spread the use of the masonry arch and dome, features that were later developed into Romanesque architecture. They also introduced the techniques of masonry and concrete construction employed in aqueducts, bridges, stadiums, public baths, and temples. They constructed thousands of miles of paved highways across the rural landscapes of northwestern Europe and founded hundreds of towns and cities in many parts of the region. These rural landscapes were both opened up and developed to a much higher level than before as forests were cleared, wet lands drained, and fields plowed to supply Rome with grain, wool, horses, and mules. Roman additions to north European agriculture were principally fruits (cherry, plum, peach, and quince) and vegetables (onion, leek, carrot, lettuce, and radish), all of whose names derive from Latin. The Romans carried the cultivation of grapes northward and introduced the making and drinking of grape wine to what is now France and western Germany. It is through this kind of diffusion from the Mediterranean that Europeans look to Rome and Greece as the sources of their cultural heritage.

The "barbarian invasions" of the Early Christian Era disrupted the political, social, and economic structure of much of the Roman Empire, and this cut short the full-scale acculturation of the northern regions to the Mediterranean culture system. The invasions brought in peoples who slowly settled down to sedentary and agricultural ways of living. The newcomers brought cultural traditions of their own, and European life took on a more militaristic and competitive pattern as struggles for control of the many local regions of the area brought repeated change in political terms. Northwestern Europe became more completely Christian in religion, but there was increasing separation between western and eastern Christianity in the zone from France to Turkey. The strong competition for control of local regions in northwestern Europe involved both religious and political elements, producing the feudalism that marked medieval centuries.

Underneath the political and religious struggles for control of territory, rural village life steadily recovered its stability and agricultural productivity after the sixth century A.D. The barbarian invasions may not have reduced local populations significantly in most regions, but in many areas farmlands were abandoned and there was a continued increase in the forest cover for a period of time. As the recovery from the invasions began, population increases occurred, and the regional landscapes returned to a cultivated condition. As local and regional population increases began to take place, linguistic and cultural regionalism developed, and medieval Europe showed increased variety in its cultural patterns (Fig. 5.3). The European zone was not developing into a single culture region dominated by one culture hearth, but into several culture subregions. All these local regions shared in certain broad trends, but each of them was maturing different internal patterns that were distinctly different from one another. A new kind of culture realm was in active formation.

Between about 800 and 1200 A.D., the struggles for regional control in the European world continued, but there gradually appeared patterns of continuity and stability. Regions that had been territories marked chiefly by variations in language, social systems, religious patterns, local agricultural patterns, and local trade systems grew into political states in which a kind of national, cultural sentiment evolved. Just as a zone of Frankish peoples slowly matured into the political state of France, there emerged an England, a Sweden, and a Holland. None of these represented a state that contained a single culture group in the strict sense, for too many things were mixed together without being integrated thoroughly. The physical geography of Europe did not help promote one very large cultural and political region, but neither did it make easy the grouping of small regions into tightly unified political states. However, even though an English king, Henry II, inherited title in the twelfth century to much of what is today France, the peoples of his continental domains were developing a loyalty to their culture system and supported the cause of France.

The numerous kingdoms of northwest Europe and

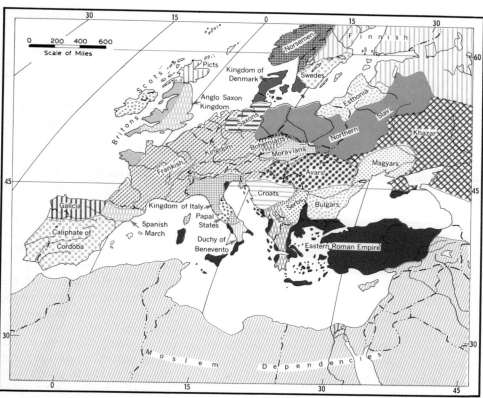

Figure 5.3 The Early European Regionalisms, about A.D. 800.

_{|||||||||||} Tributary limit of Frankish Kingdom

their developing commercial cities provided a new market for trade north of the Alps, in which the city-states of northern Italy flourished as middlemen during the twelfth to fifteenth centuries. Venice, Genoa, Florence, and Milan traded by sea from the eastern Mediterranean and the Black Sea, thence over the Alpine passes into northern Europe. In so doing they transmitted culture traits such as the use of sugar, spices, tea, and coffee in the dietary pattern. They also transmitted the arts of paper and glass manufacturing, and introduced the magnetic compass, the astrolabe, and better sea navigation charts. Northern Italy itself became a great culture center in which large and splendid brick and stone architecture matured. Skills were greatly advanced in weaving and dyeing, in glassmaking, in gold- and silver-smithing, and in jewelry and art forms. Shipbuilding flourished, as did the training of navigators (Columbus, Verrazano, and Cabot were all north Italians, later to sail for Spain, France, and England, respectively). Banking, double-entry bookkeeping, and marine insurance as techniques of modern capitalism originated in the north Italian financial centers and spread northward. Intellectual interests were reawakened, and there was great development of science, philosophy, art, and literature. Universities, as institutions of higher learning, began with the Renaissance in north Italian cities and spread into northwestern Europe.

The most common and widespread farming method in northern Europe during the Middle Ages was the manorial system. Its change, by a process commonly called an agricultural "revolution," began in the thirteenth century in present-day northern France and Belgium. More correctly, the process of change was an agricultural evolution, for it is still going on in Europe. The results of change were important for the development of Europe, because the new agricultural surpluses made possible the population growth that provided both an emergent urban-based labor force and the flood of migrants that were to go overseas a little later.

The manorial system was part of the feudal system of land tenure, which involved a chain of personal relationships. Land was not owned outright, but was held by members of the peasantry in return for allegiance and service to an overlord. The manor and its group of peasant villagers were the smallest local unit in the system. A typical manor occupied a forest clearing of three to ten square miles. It consisted of open, unfenced cropland and common pasture, plus the necessary buildings. These latter usually included the manor house and barn of the manor lord, a church or chapel and a parsonage, a blacksmith shop, a carpenter shop, and a flour mill, along with one or two clusters of cottages for the peasant cultivators. Attached to each cottage was a small shelter for a pig or two and a few cattle. Each cottage also had a small family vegetable garden. All around the clearing were woods, and the local roads were few and often only footpaths through the forest. In many areas trade was meager except for salt and a little iron, and few villagers traveled far. The manor served as a self-sufficient community, providing food and shelter for a few hundred people.

The manor, most frequently, was farmed by dividing the cropland into three major field units, raising crops on two each year and allowing the third to lie fallow for the year. Thus, by rotation each year, each field unit was cropped two years out of three. There were regional variations in this system, ranging from two-field to four-field patterns. Each unit was subdivided into sections separated into narrow strips, each of which was held by a village family. Each family held the right to a number of strips totalling from 10 to 20 acres in all, but the strips were scattered over the field unit among the strips held by other families. All cropland was open, not fenced, and cultivated in common. Some strips were farmed for the manor lord, and others for the support of the church. Every family plowed, planted, and harvested the same crops at the same time. All the village animals were pastured on the commons and on the stubble in the fallow field unit, but they also used the forests for grazing and foraging during the summer.

The manorial system was adequate in times of peace and in years when crops were good, but there was little surplus food for storage or for marketing. Famines occurred when crops failed or when the grainfields were destroyed during times of war. Most critical was the lack of livestock feed to keep animals well fed during winters. Hence animals were of poor quality, and the herds were not very large in numbers. Manure for the fields was not plentiful and crop yields were low, ranging from five to ten bushels of wheat per acre. Since everyone worked together on the same unfenced fields and raised the same crops, there was scant opportunity for experimentation with new crops or agricultural methods.

The breakdown of the manorial system began as early as the late thirteenth century in some parts of Europe. The breakdown gradually produced an agricultural transformation of Europe as one phase in a number of concurrent changes. As cities grew, the merchant class rose in importance. When they supported the king or central government, political power became realigned at the expense of the feudal lords. The agricultural changes supported the emergence of the nation-states and the breakdown of the feudal system. A principal factor was the introduction and development of new fodder crops: clover, alfalfa, vetches, turnip, carrot, rutabaga, and beet. More and better fodder made it possible to keep more and better animals, which in turn made more manure available for the fields. More forest land was cleared, placed under cultivation, and handled more efficiently. Average yields increased dramatically. The system of land tenure was changed as well. Ownership of land was redistributed so that a farmer had all his land in one place which was then enclosed by fencing. Enclosures began as early as the late fourteenth century, a change that freed individual farmers from the rigidity of the manorial system and allowed experimentation with new crops and methods. Land redistribution resulted in some people gaining land and others losing theirs. A large landless rural population came into existence, and many villages became deserted as the landless fled to the towns and cities to seek new means of existence.

Enclosures of land began to change the appearance of the rural landscape in those parts of Europe in which the decline of the manorial system occurred, between the fourteenth and late eighteenth centuries. Previously, the manorial farms had a somewhat open and empty appearance, since there were few roads and no fences, hedges, or rows of trees marking property lines. There were no houses in the fields because people lived in clustered villages. As the enclosure movement progressed, many houses became scattered out on the separate farms, and roads began to be built connecting the farmsteads and manor villages. Hedgerows were planted along property lines, and rows of trees were planted along roads. With the further decline of the manorial system, there began the expansion of the cultivated landscape as large areas of forest became cleared for farmland.

There were interruptions in the expansion pattern that followed the decline of the manorial system. In 1326 the Ottoman Turks began invading eastern Europe. There began in 1334, in consequence, an epidemic of plague that spread from Constantinople westward, that struck western Europe as the "Black Death." During the next twenty years continuing epidemics took the lives of almost 25 million Europeans. Another variety of interruption was the Hundred Years' War (1337–1453) between England and France, from which the latter emerged "victorious," but with a land and people devastated by famine, plague, and marauders.

By the fifteenth century much of the territorial sorting-out process had reached the point that the political and cultural regionalisms of Europe were maturing, based upon the growth of population and the improvement of the material economy (Fig. 5.4). Nation-states were taking permanent shape, and a pattern of cultural bonding was taking place among regional groups of peoples, even though some of this was based as much on dislikes as upon preferences. For example, the Dutch peoples of today are descended from several Germanic tribes, and they

Figure 5.4 The Rise of European Nation-States.

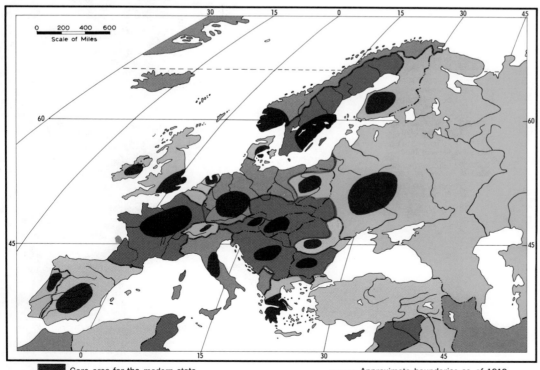

Core area for the modern state ——— Approximate boundaries as of 1913

have preserved some of those cultural differences right down into the present era. The territory they occupied was often controlled and ruled by outsiders over the centuries. The common strong dislike for external control has been an important factor in bringing the Dutch peoples together in a common determination to rule themselves and develop their own distinctive culture system within the broad framework of European cultural trends.

Northwestern Europe, in earlier centuries, had received many immigrants, and its peoples were then concerned chiefly with working out their own internal, regional living systems. There had been towns and interregional trade in Roman times, but these patterns had declined severely. By the twelfth century, however, the growth of towns, trade, and handicraft manufacturing had surpassed the earlier patterns. In the eleventh and twelfth centuries the Christian Crusades into the Holy Land, the Catholic Church missions into central Asia, and the activities of the Mediterranean trading cities in dealing with the East began to open the eyes of Europeans toward other parts of the world and other kinds of commodities. Northern Europeans began to learn a great deal from the study of Arabic scholarly writings by traveling to Spain and Sicily, and educated people were recovering from the traditionalism of monastic Christianity. By the fifteenth century the era of

a quietly rural, village-dwelling, agricultural Europe was coming to an end.

Out of several sources of stimulation, the imaginations of educated Europeans were being awakened, and acquisitive moods were being enlivened, to the end that Europeans were becoming dissatisfied with their own cultural world. The lure of travel, exploration, and new knowledge was affecting many, and as a result they felt impelled to go out to find what lay beyond their own realm. The Era of World Exploration was being born.

European World Exploration and the Beginnings of World Trade

Getting to China from western Europe by walking and riding a camel was slow, although many persisted in probing the land routes eastward across Eurasia. The Mediterranean-Red Sea-Indian Ocean routes were not always open to Europeans. The small boats and primitive ships of the thirteenth century were unsuited for navigation of the open oceans. During the fifteenth and sixteenth centuries the art of shipbuilding developed conspicuously along the coasts of Europe. Harbors evolved into ports as Europe turned its chief attention to the sea routes (Fig. 5.5). By the middle of the sixteenth cen-

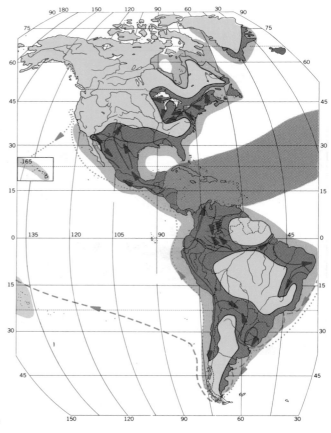

Figure 5.5 The Course of Exploration, 1100–1650.

tury tough and seaworthy (if uncomfortable and small) ships were coming off the ways. Almost from the start these sound, high-freeboard European sailing ships were built as floating cannon platforms. The knowledge of navigation, of world wind systems, and of the physical geography of the seas was extremely slight, and it was by raw courage, ignorance, and avarice that Europeans set out upon the open oceans to find ways to the far countries.

In the somewhat innocently smug belief in the superiority of Christendom, sixteenth-century Europeans almost universally looked down on other peoples as heathen, undeserving of Christian treatment. The militarism of Europe, possessing the tradition of knighthood and the zeal of the Crusades, ranked foremost among the peoples of the world at the time. The acquisitiveness of Europeans had sharpened into an avarice for goods and products unequaled elsewhere. It was the Columbian Discoveries and the consequent explorations that once and for all time brought the whole world within the realm of the trader, the religious missionary, the traveler, the migrant, and the territorial conqueror.

With the map of the lands and seas of the earth established by the explorer-cartographers, and the routes worked out by the sailing ships, Europeans began a pattern of traveling from their homelands more varied than any pattern of human migration

since the early Paleolithic spread of man over the earth. What were they looking for? Anything they did not know about. The European curiosity about the earth mounted steadily until it became an infatuation with exploration, and the profession of "explorer" became an honored one. By the opening of the twentieth century only the interiors of the two islands of New Guinea and Borneo, and the vastness of Antarctica, remained relatively unknown to the Europeans.

Once the Europeans had found their way around the world they set out to acquire portions of the goods of the earth. Northwestern Europe of that day did not produce many commodities useful to the civilized parts of the world. Its own manufacturing technologies were only slightly developed, and its products found no ready sale in the markets of India, China, the Indies, or even the Islamic realm. Europe's rather simple manufactures did appeal to the simpler cultures of the world. The Europeans gradually worked out procedures of trading that substituted regional products for other products until suitable patterns of exchange could work in most places. It was not long before British and French traders, for example, regularly began looting Spanish treasure ships to gain the gold and silver with which to buy pepper in the Indies. The pepper was sold to Germans and Scandinavians in exchange for

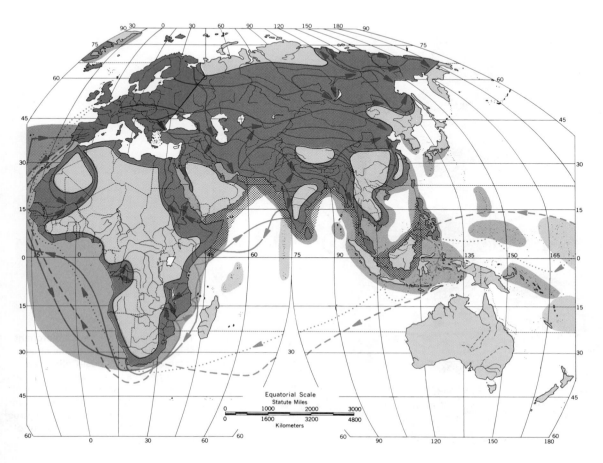

fish, which could then be sold to Spaniards for more gold with which to buy pepper. A little later, after the colonial settlement of what is now New England had taken place, ginseng and furs could be carried from North America to China in exchange for silks, tea, and porcelainware. Most of the silks and porcelain and a share of the tea were taken home to England and France, and the surplus of tea was pushed off on the American colonists for more ginseng and furs.

This interpretation is perhaps not suitable to current economic theorists, but it expresses the inter-regionalism of early modern trade. European mercantilist economic theory held that a country had to sell more than it bought to prosper. Until the manufactures of Europe became acceptable around the world, a bartering process had to obtain. In response to changing economic theory countries of northwestern Europe began to renovate their manufacturing activities to suit their trading programs, and the Industrial Revolution finally yielded the kinds and grades of products acceptable in all far countries. Every ship bringing back stores of goods from a far country stimulated other merchant traders to attempt the same. Armed ships for cargo-hauling came to be built better, bigger, and faster than those built for exploring. Competition for the spoils was militant. The traders of northwestern Europe staked out competitive domains of peoples and their territories in far countries for the enrichment of the homelands. And they competed for control of the seas with floating gunnery platforms.

The Spreading Europeans and Cultural Interaction

Explorers, missionaries, and traders were the first northwest Europeans to venture abroad. It was not long, however, before individuals and groups began emigrating from Europe to become colonists and settlers in lands abroad. Europe of the fifteenth to eighteenth centuries was undergoing growth and expansion, but below the level of the upper classes there were many tight restrictions on life in most European societies. The emigration of small groups seeking social, religious, and economic freedom in some of the newly discovered lands established a pattern that continued to grow (Fig. 5.6).

In the sixteenth century almost all transoceanic travel involved people from northwestern Europe. In the seventeenth century there began a small pattern of migration from northwestern Europe to North America. Also, English, French, Dutch, Portuguese, and Spanish began moving African black peoples to the New World as slave labor on plantations and in mines. In the eighteenth century the flood of European emigrants began their movement around the

This painting of an armed merchant ship on its way to the Indies shows some of the early post-Columbian developments in shipbuilding. These ships were seaworthy and could well defend themselves against attack. In this example, both the human figures and the flying fish are slightly out of proportion.

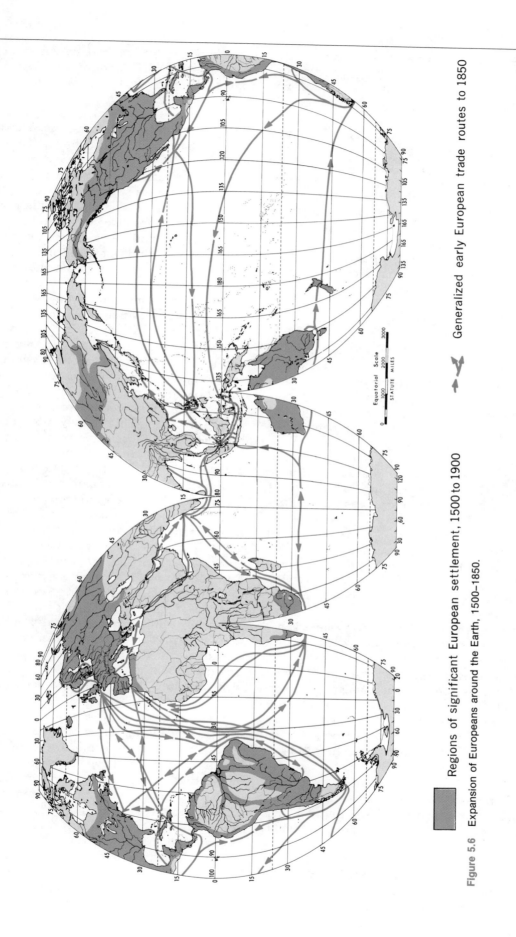

Regions of significant European settlement, 1500 to 1900

Generalized early European trade routes to 1850

Expansion of Europeans around the Earth, 1500–1850.

Equatorial Scale

1000   2000   3000

STATUTE MILES

Figure 5.6   Expansion of Europeans around the Earth, 1500–1850.

141

world. In the same century Chinese began moving into Southeast Asia to carry on agricultural, mining, and trading activities. In the nineteenth century peoples from many lands began to cross the oceans between continents in an increasing flow. For example, the so-called Chinatown turned up in London, Paris, San Francisco, New York, and Havana, and before the end of the century most of the world's large cities held colonies of nationals from all over the earth. Before the end of the nineteenth century, the tourist traffic began to move around the earth as curious people began travelling to "see the sights." As transportation technology has improved in the twentieth century, human mobility has increased by leaps and bounds until in the last decade it almost seems that people from everywhere are going everywhere.

Following the explorers and the first small groups seeking refuge in the freedom of a new homeland, there began to move around the earth a flood of adventurers, colonists, settlers, refugees, soldiers of fortune, mining prospectors, fur-trappers, timber-cutters, Christian missionaries, itinerant traders, commercial travelers, company agents, and inquiring scholars. Each class of migrant was intent on its own purposes, but at the same time each became an agent of cultural diffusion. Many patterns of cultural interaction followed. In moving out to new lands, Europeans often insisted on bringing their own ways and practicing them. This pattern exposed local inhabitants to new customs, practices, and technologies. One of the main purposes of Europeans going abroad was to acquire new kinds of goods. The goods and the technologies that produced them spread over Europe and, in turn, exposed Europe to new patterns of culture. This was cultural diffusion on a grand scale.

The diffusion of elements of culture systems is not directly dependent on transportation systems or purposeful programs. Europeans did seek to spread Christianity as they first traveled the earth, but in no other aspect did they deliberately attempt to spread European culture. The very nature of human beings is to watch other people, the things those people do, and the technologies they use to produce what goods. The human interest factor is the largest motivation in diffusion. Ever since Europeans first ventured abroad, other people have been examining aspects of European culture, often to incorporate something of it into their own system. And, for their part, Europeans were quick to incorporate new technologies seen abroad into their own system, thereby to produce the kinds of goods found around the world. For the English language, the best proof of the absorption of elements of culture from other lands is to examine the sources of words in our unabridged dictionaries.

Culture hearths are creative centers in which curiosity, imagination, initiative, and persistence combine to develop new human ways and change some old ones into better or more complex ways. To the degree that culture hearths become linked to other populated regions, diffusion of aspects and artifacts of culture spread out of the culture hearth. The spread of Old World culture out of both the early and the secondary hearths was relatively slow. The friction of distance reduces the rapidity and the volume of the spread of culture, so that few elements reach from the creative centers to the far corners of the earth, and it takes more time for this to happen. And through their own social controls, a society may reject outside elements or be highly selective of those things it accepts.

Once Europeans started on their world travels, however, many of the old patterns changed. In a very short period, European contacts became worldwide; and broader patterns of diffusion than had occurred in the past were established. Europeans were aggressive in their contacts, and few societies were able to prevent the cross-cultural contacts. Insofar as Europeans became residents settled permanently around the earth, they also became permanent agents of diffusion. Whatever turned up as a new feature, almost anywhere, could be spread all over the earth in the course of time. As the means of transportation and communication have improved, the time lag has been reduced.

Not all societies around the earth have approved of wholesale patterns of diffusion, nor have they willingly accepted the available ways, means, and technologies. Particularly in those places where the Europeans were most strongly aggressive has this been the case. Laying aside, for the present, the question of the justice of European contacts with the rest of the earth, it remains true that the European cultural explosion reached the ends of the earth and linked forever the hundreds of culture regions into one great operating system that has been progressively brought closer and closer together ever since the beginning of the eighteenth century.

The civilizations of western Europe were probably less advanced than those of India or China in 1500, measured in certain subjective kinds of terms. The western Europeans, however, were well practiced in naval and military tactics, and in the competitive skills of dealing with problem situations. All these had been sharpened by the centuries of conflict and territorial struggle involved in the maturing of the nation-states of Europe. They were fortified by their deep belief in the superiority of Christendom over other ways of life; they displayed the aggressive ability to make the most of an opportunity; and they had the courage (perhaps better stated as foolhardiness) to take on the rest of the world. By the end of the sixteenth century, western Europeans had forged ahead of the rest of the world in aggressive initiative and the will to learn about the nature of the physical world. This gave them an advantage in all later contacts.

At the end of the sixteenth century, the hundreds of large and small societies around the earth probably presented a wider range in the levels of civilization/culture than had obtained at any time in the whole of the human past. In the comparative sense, the range lay from the mid-Paleolithic through the Neolithic and the traditionally civilized, into the the civilized-on-the-verge-of-scientific. This comparison can be stated, also, in terms of modern economic theory: the range lay from the totally undeveloped through the barely developing and the underdeveloped into the strongly developing. Patterns of underdevelopment were not created by nineteenth-century Europeans. Degrees of underdevelopment/nondevelopment have been present since the middle of the Neolithic period. The creation and expansion of culture systems, and the spread of their complexes, have resulted in regionally varied patterns, and there have long been those on the advancing frontier and those that have lagged behind.

SUMMARY OF OBJECTIVES

After you have read this chapter, you should be able to:

Page

1. Explain, with examples, why all during historic time some societies have been more advanced, whereas others have lagged behind. 128

2. Discuss why the depiction of the characteristics of any nation must include its peoples, its culture, and its region. 127

3. Locate, name, and contrasts the cultural achievements of the two primary culture hearths in the Americas. 129

4. Describe those parts of the earth that were organized as secondary culture hearths and realms and explain why these appeared later in time than those in the Mediterranean, India, and China. 130

5. Explain why Europe is given disproportionate space in this chapter, compared with other culture realms. 134

6. Discuss the significance for European cultural evolution of any one or more of the five major formative factors. 134

7. Describe the most significant changes in the appearance of the landscape of Europe as a result of the agricultural transformation after the Middle Ages. 136

8. Suggest some reasons why it was Europe rather than another cultural realm that became the first to expand on a worldwide basis. 136

9. Following a review of Table 4.1 "Classes and Types of Migrations," discuss which classes and types were most used by Europeans in their migrations during the sixteenth and seventeenth centuries. In the nineteenth and twentieth centuries. 140

10. Explain the sentence: "At the end of the sixteenth century, the hundreds of large and small societies around the earth probably presented a wider range in the levels of civilization/culture than had obtained at any time in the whole of the human past." 143

CHAPTER 6

THE RECENT CONVERGENCE OF CULTURE PATTERNS

KEY CONCEPTS

Global Diffusion by Europeans Established Worldwide Contacts

Europeans Staked Out Colonial Empires

The Linkage of the Earth Forever Ended Local Isolation

European Settlement Transformed Midlatitude Forest and Grasslands

European Expansion Redistributed Life-Forms Around the Earth

Human Biological Mixing Began to Produce Variable Populations

The Blending of Cultures Has Increased the Rates of Change

The Spread of Technological Civilization Was Outward from Europe

Intercommunication Increases Cultural Inter-

change

European Political Organization Produced the Nation-State

There Are Several Legal Systems in Use Around the Earth

Religious Systems Have Roles and Take Spatial Patterns

Language, Writing, and Systems of Education Are Important Cultural Tools

European Alphabets Are Religious in Origin

Educational Systems Have Different Social Philosophies

The Trend Among Societies Is Toward a Universal Community

Mankind Has Completed an Immense Journey Around the Earth

GLOBAL DIFFUSION BY EUROPEANS

The explosive spread of Europeans over the whole of the known world in the four centuries after about A.D. 1500 was a historical shock to the rest of mankind. As an event it was both unforeseen and without precedent. The rest of the world was not in any way prepared to react to, accomodate to, or resist such a phenomenon. The Europeans carried a complex of ideas, processes, and goods that set in motion a series of irreversible social, political, and cultural changes. Colonization, Christianization, political nationalism, capitalistic industrialization, new forms of militarism, and new aspects of urbanism were all European derivatives. So also were wholly new concepts in public health and preventive medicine, public education, the emancipation of slaves and women, and new political systems such as democracy, socialism, communism, and internationalism. European expansion, with all its ramifications, exploded around an earth that had experienced only the slow and restricted diffusion of civilizing influences from the old, early culture hearths.

This chapter focuses upon what happened to the earth and its peoples during the recent centuries of European ascendancy and dominance, with particular attention to the results achieved by the end of the nineteenth century.

Colonial Empires

Exploration, discovery, armed conquest, the annexation of territory, and colonization by Europeans were processes used in varying combinations by European groups in extending their sway over most of the rest of the world. The establishment of colonial empires was an uneven and haphazard enterprise. As Table 6.1 indicates, the late fifteenth and the sixteenth centuries were epochs of successful Portuguese and Spanish expansion. The Dutch, English, and French contended for supremacy during the seventeenth and eighteenth centuries in North America, in the Caribbean, in South Africa, and in South and Southeast Asia. The fifty years between 1775 and 1825 saw the reduction of the Spanish and Portuguese colonial holdings, particularly in the New World, as most of the colonies gained their independence. England also lost her North American seaboard colonies in this same period.

The greatest era of colonialism was the nineteenth century, as the British turned their attentions to Canada, southern Asia, Africa, and Australia-New Zealand; the French to Africa, Southeast Asia, and the South Pacific; the Portuguese to Africa in a renewal effort; and the Russians to Central Asia, Siberia, and Alaska. In the earlier centuries the struggle had been less for territory than for trade monopolies. Except for Portugal and Spain, the usual early agents of European penetration were the chartered companies that were granted monopoly trading rights by their respective governments. Later, these were bought out by the governments, the territories formally annexed, and colonial government systems established. So open and direct were the nineteenth-century efforts to expand territorial claims that scattered colonies became parts of great, far-flung empires. Germany, Belgium, Italy, Japan, and the United States entered the "Great Colonial Rush" very late indeed, between 1870 and 1898, and Russia completed her expansion in the arid zones of Central Asia only in the last half of the nineteenth century. As late as 1875, not more than one-tenth of Africa had become European possessions. Then 11 million square miles were annexed in 30 years. By 1905 much of the world had been divided up among the European political powers, including Russia.

The Linkage of the Whole Earth

During the nineteenth century European influence became so solidified around the earth that it apparently fixed the habits of globalism for foreseeable historical time. Exploration became almost terminal in searching out the remaining possibly productive parts of the earth. Settlement rounded out occupancy of such sectors as it could find useful. Political control almost reached its final expansion (Fig. 6.1); trade procedures solidified in all parts of the earth; industry turned out products superior to those elsewhere that were in good demand everywhere; and the European technological revolution was running at an increasing pace. Facilitating the diffusion of the new technologies, societies outside the European culture realm reacted progressively to the systems of European culture. After two centuries of self-imposed isolation, Japan started on a program of transforming its Chinese-style agrarian society into a Western-style industrial society including colonial territory and extensive foreign trade. Elsewhere numerous countries shifted their economies to the provision of many commodities useful only in foreign trade, and European industrial technologies invaded many cities around the world.

Prior to the establishment of worldwide linkages, the collapse of life in a culture hearth meant decline for a very large region. In the nineteenth century the decline of China meant the lessening of satisfactory ways of living for Chinese, but it meant less in comparative terms for the progress of life over the earth, since other regions could replace China as a producer, as a market, and as a contributor to ways of living everywhere.

By 1900 European culture systems were so widely spread around the earth, with so many regions contributing to production, trade, cultural expansion, and general productive growth, that never again could the earth relapse into separate regionalisms of a lesser sort. Evolutionary changes had begun to take place in almost every culture that were only to grow in rate of change, depth of penetration, and acceptance by peoples. So many local culture traits and complexes had been diffused around the earth, and were so interrelated in evolutionary change, that particular foods, particular clothing, specific tools, and operating procedures had become worldwide in distribution. It could be said that, in some respects, the peoples of the world were beginning to converge in their ways of living. The humanized earth had become an integrated globe.

In this chapter we also examine the proposition that a multifluctuating system of variant, isolated cultures is no longer possible. Rather, as a result of Europe's demonstration that the entire earth has become mankind's resource base and one intercommunicating system, a *trend* toward cultural convergence set in during the nineteenth century, and intensified during the twentieth century. People in every continent became engaged in establishing, maintaining, and expanding a vast urban-industrial society of extraordinary complexity. As a result, the many traditional cultures—those enormously varied patterns for human living that previously evolved in partial isolation from one another—began to disinte-

Table 6.1   The Growth of Colonialism, to A.D. 1480 to 1900.

Period	PORTUGAL	SPAIN	NETHERLANDS	ENGLAND / GR. BRITAIN	FRANCE	GERMANY	RUSSIA	UNITED STATES	JAPAN	OTHERS
~1480–1500	Cape of Good Hope Treaty of Tordesillas Da Gama to India	Columbus' 4 voyages to West Indies		Newfoundland as first discovered territory						Greenland discovered in 10th century
1500–1600	Brazil claimed Mozambique Goa Malacca Angola Brazil colonized Macau settled Rio de Janeiro founded	Cuba, Puerto Rico, Jamaica Florida Mexico Guam, Philippines Guatemala, Salvador Incas conquered in Peru St. Augustine, Florida founded Manila founded Philippines colonized Buenos Aires founded	Dutch in East Indies				Conquest of Kazan and Astrakan Penetration of Siberia begun			
1600–1700	Formosa (Taiwan) reached	Jesuit state in Paraguay Jamaica taken by British	East Indies Co. (V.O.C.) chartered Spice Islands seized Batavia founded Guyana settled by W. India Co. Ceylon occupied Formosa controlled Tasmania and New Zealand discovered Capetown founded Ejection from Brazil New Amsterdam and Delaware lost to British	East India Co. chartered Jamestown, Va. settled Factories in Madras, Surat, & Bengal Pilgrims at Plymouth, Mass. Barbados settled Puritans at Mass. Bay Colony Jamaica taken from Spanish New Amsterdam and Delaware from Dutch Hudson's Bay Co Bombay as Hdq. of E. India Co. Windward and Leeward Is. Br. Honduras Landing in N.W. Australia	Quebec founded Base on Madagascar Mississippi R. explored; empire established from Quebec to New Orleans Senegal as first possession in Africa	Begin colonization of West Africa	Omsk founded Lena River reached Yakutsk built Amur River reached Okhotsk founded Irkutsk built Ukraine incorporated into Russia Discovery of Kamchatka			Danes take St. Thomas
1700–1800		Revolt in Paraguay Trinidad lost to British	V.O.C. dissolved Ceylon lost to British	Gibraltar captured War in North America; British dominant thereafter Empire in India begun Australia claimed Tahiti, Hawaii visited American Revolution Penang Is. leased Freetown, Sierra Leone British Guiana and Trinidad	Santo Domingo liberated			U.S.A. established		
1800–1900	Brazil independent Republic of Brazil established	Independence gained by Paraguay, Venezuela, Argentina, Chile, Mexico, Colombia, Peru, Bolivia Cuba independent	Cape Province annexed by British Boers trek north from Cape Province, establish Republics of Transvaal, Natal, & Orange Free State	Ceylon as Br. Crown Colony Tasmania taken Cape Province annexed Sierra Leone Crown Colony Malta annexed Singapore acq'd by E. India Co. First Burma War Hong Kong acq'd Maoris surrender New Zealand sovereignty Natal annexed Orange Free State annexed Lagos, Nigeria Canada as Confederation Suez Canal Opened Fiji Islands taken Cyprus ceded by Turks Kenya protectorate Uganda protectorate Burma conquest completed Nyasaland annexed Rhodesia conq'd Sudan as Condominium Boer War in South Africa	Louisiana Territory sold Begin conquest of Algeria Tahiti annexed Libreville founded New Caledonia Cochin-China & Cambodia protectorate Brazzaville founded Tunis as Protectorate Ivory Coast acquired Indo-China formed Madagascar annexed Laos acquired	French in Mexico until overthrown Protectorates over Togo, Tanganyika, Cameroons, S.W. Africa	Finland incorporated as Grand Duchy Kazakhstan acquired Vladivostok concession Tashkent occupied Russian Alaska sold Sakhalin occupied Khiva occupied	Louisiana Purchase Alaska purchased Hawaii annexed Puerto Rico, Guam, and Philippines acquired Samoa	Kurile Islands occupied Chinese defeated; Taiwan (Formosa) acquired	Leopold II founds Congo Free State Italians in Ethiopia

PORTUGAL	SPAIN	NETHERLANDS	UNITED KINGDOM	FRANCE	GERMANY	RUSSIA	UNITED STATES	JAPAN	OTHERS

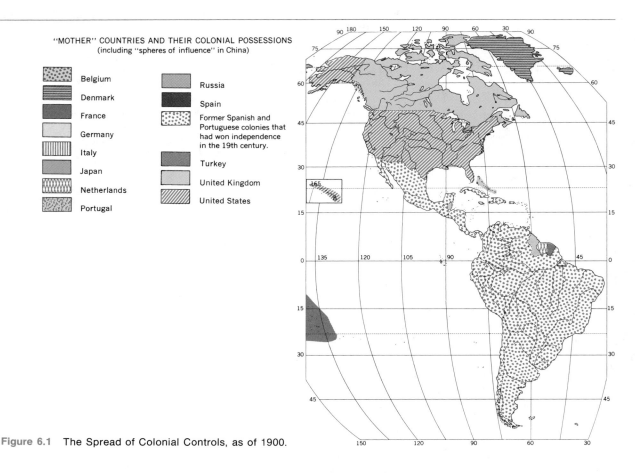

Figure 6.1 The Spread of Colonial Controls, as of 1900.

grate under the impact of the modern urban-based industrial society. The clear trend was that those who lived in a large city or its suburbs anywhere, whether it be Cadiz, Cairo, Calcutta, Canton, Caracas, Chicago, or Christchruch, (and despite certain functional and social differences) began to have more in common in their ways of life than in difference. The faith and hope of those striving to achieve in the worldwide urban-industrial society was that they, too, coud share in demonstrated benefits of increased longevity and improved material well-being.

Transformation of the Midlatitude Forests and Grasslands

The European system of occupancy evolved in a midlatitude region of humid climatic conditions, where forest growth (with perennial grasses interspersed throughout old clearings) and perennial water supplies were everywhere available. As the European explosion carried Europeans all over the earth, they succeeded best in those midlatitude forest lands that had previously been lightly occupied by gathering and collecting cultures or by hunters-gatherers who practiced a minimum of crop growing (see Fig. 5.6, p. 141). Around the earth this meant

chiefly the broad territories of the present United States and southern Canada and eastern Russia-western Siberia, for only small sectors of the world's midlatitude forest lands lie in the southern hemisphere. Only relatively late in time, during the nineteenth century, did European settlement perfect effective technologies for significant occupancy of the warmer subtropics.

Essentially, European colonial settlement occupied lands of small population and slight economic development, using occupancy patterns standard to Europe of the time, that of nearly self-sufficient farming that combined crop growing and animal husbandry. As pioneer settlement took place, the participants at first produced only small surpluses, were not engaged in much secondary economic activity, and remained dependent on Europe for manufactured goods. Settlement became extensive, spreading throughout the range of territory occupiable by the technologies in use. This continued to be true for Russian Siberia until the appearance of the Communist political reorganization of Russian life in the early twentieth century.

In the Canadian-United States zone of the New World, life followed a different sequence. From the seaboard starting point, small-farm settlement ran out of forest country as it went westward into the drier grassed country. Here northern Europeans at

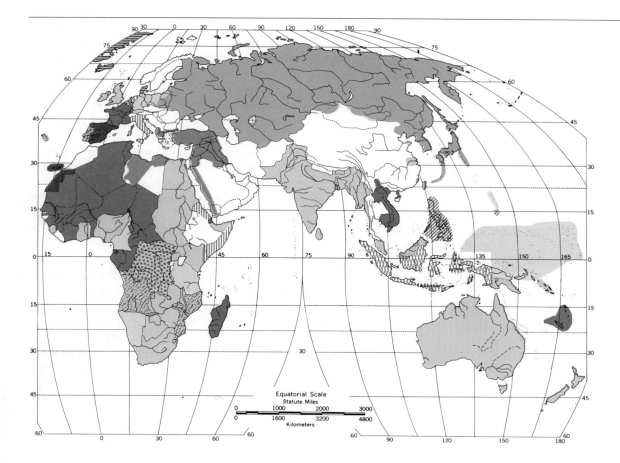

Equatorial Scale
Statute Miles

first were at a loss for suitable technologies, since their traditional concepts of good agricultural land were attuned to humid, forested landscapes. The first uses for the dry west by northern Europeans involved the adoption of cattle-ranching techniques on the open range. These were first developed by the Spanish on the Meseta of central Spain and brought to the New World for application in the drier sections of Mexico. These Spanish techniques diffused northward as far as the edges of southern Canada, and included the "long drive," which became the beloved theme of "Western" movies. Only slowly did cattle ranching in the older tradition yield to a new form of cattle farming on fenced lands on which hay and grain replaced the open-range browse.

Gradually, in the moister sections of the Midwest, the prairie country, a grain-farming/cattle-raising system evolved to replace open-range cattle ranching, although it, too, depended initially on a "long drive" to get the cattle to eastern markets. Some sedentary farmers leapfrogged to the West Coast in both Canada and the United States, but slowly the techniques developed on the moist prairies were applied to the drier parts of the Great Plains. Cattle ranching gave way to the new technologies that employed extensive land-use procedures, animal-powered mechanization, and the new rail-transport facilities to market both cattle and grain.

The integration of grain farming, livestock raising, increasing mechanization, and the commercial exploitation of distant markets by rurally dispersed families on farms of a few hundred acres each, developed a new kind of agricultural system. These techniques also created a new form of cultural landscape by turning prairies and plainslands into farmlands. The cattle and sheep ranchers progressively retreated to the westernmost drier and rougher sectors of the Great Plains, in which crop cultivation was hazardous under the existing technologies.

Faced with constant labor shortages in its ever-expanding settlements, the North American Midwest started the European cultural realm—and eventually the whole earth—on another agricultural revolution, that of mechanization and the application of inanimate power in large volumes to the primary production of food commodities. Although the specific inventions came from several parts of the European cultural realm, it was the American-Canadian farmer who turned crop growing into an industrial pursuit. This particular revolution has not yet run its course in its own homeland.

The European occupancy of the midlatitude forest and grassland country began humbly, but, coupled with inventive technological evolution, continued occupancy added a totally new dimension to settlement of the earth by productive mankind. The European home region has been variably caught up in this developmental procedure, which, in the twentieth century, has been changing the traditional occupancy of the Soviet zone. Elsewhere around the earth in the forest-margin/drier-grassland ranges something of the same sort has begun to take place, chiefly in the twentieth century. Australia notably began in the late nineteenth century to develop the mechanized grain-farming system on the dry margins, with cattle and sheep raising in the drier country beyond and the integration of the economic systems taking place. The Argentine pampa shows a similar, but not identical and not strictly contemporary, kind of evolutionary economic development. And, somewhat slowly in our terms of time, adaptations of the mechanization and industrialization of agriculture are diffusing to other ecological and cultural realms around the world.

The Redistribution of Life-Forms

As Europeans traveled the whole world over and initiated the great patterns of resettlement, they simultaneously set in motion a vast series of transplantations of all kinds of life-forms. Europeans not only took along their decorative-crop plants and domestic animals, but also the common weed plants and the animal pests. An even more numerous class of life-forms was the microorganisms that travel about with humans. Mankind is a disturber of natural and established ecological conditions, creating situations in which other life-forms are provided with fresh opportunities. The plants and animals that follow people around the earth do not necessarily prefer the physical conditions of humanly-created, open, disturbed, ecological niches, but often they can tolerate and exploit them. Purposeful or accidental introduction of an organism into a new environment through human action is often possible only because the existing habitat has been greatly altered in terms of its previous ecological balance. This pattern refers not only to the crop plants and the domestic animals, and to the weeds and the animal pests, but also to the microorganisms that provoke disease among all other forms of life. Although these changes had, of course, been an integral part of the systems of ecological change on the earth ever since life first began, the volume and rapidity of change that began with the Columbian Explorations and have taken place during the last 500 years exceeded in rate and scope the changes of any previous human period.

Disease. The spread of humans has served to disperse mankind's own parasites, whether through infectious diseases or diseases transmitted by vectors or alternate hosts. Nearly all the more deadly epidemic diseases were transported from the Old World to the New World following the Columbian Discoveries. There had been no devastating smallpox, measles, malaria, or yellow fever, perhaps no typhus or typhoid, almost no tuberculosis. From the Old World came the greatest transport of disease to new and susceptible peoples, the most striking example of the influence of disease on history.

The worldwide distribution of contagious diseases is clearly a consequence of post-Columbian intercontinental contacts. For example, yellow fever and malaria were both spread from the Old World to the New. The common vector of yellow fever, the mosquito *Aedes aegypti*, presumably traveled with the slave ships from tropical West Africa across the Atlantic to the Americas. The malarial parasite was not present in the Americas until the arrival of the Spaniards, who brought it from the Mediterranean region. Thus the unhealthiness of the lowland tropics of the New World is a post-Columbian phenomenon.

The unconscious dispersion of mankind's own viruses had a catastrophic effect in contact situations, especially between Europeans and peoples who lacked either immunity or cultural experience in methods of treatment. The decimation of the New World populations was significantly a result of introduced diseases. Throughout the Pacific Basin and in Australia both Old World and New World diseases took terrible tolls during, and immediately after, voyages of exploration or the initiation of settlement programs.

Plants. Accidental introductions of plants probably outnumber purposeful introductions. Dispersal, especially of seeds, can be effected by:

1. Adhesion to moving objects such as a person's clothing, mud on cartwheels or boots, dust in cars or trains.
2. Inclusion of other seeds in crop seeds or in other plant materials, used for fodder or for packing, and in fecal matter.
3. Inclusion in mineral matter, such as soil, ship ballast, or road metal.
4. Carriage of seed intended for purposes other than planting (e.g. seed for food or medical use) that somehow escapes, becomes established, and reproduces.

The plants purposefully dispersed by humans are the crop plants (major and minor), the ornamentals, and the landscape modifiers (as used in reforestation and erosion prevention). The major crop plants have been very drastically changed by the domestication process, for the most part being completely dependent on man for dispersal and reproduction. In some cases, in which they are then known as *cultigens,* they no longer have the capacity to produce viable seeds and thus depend on human-controlled propagation. Most cultigens simply cannot compete with wild vegetation. They do not escape from cultivation successfully to form part of the wild vegetation and so disappear when people cease to intervene on their behalf. Among the crop plants and the ornamentals, however, some readily escape to join with the weeds or the native wild vegetation. Plants used for reforestation and erosion protection frequently include species alien to a locality.

The changes in flora and fauna in many parts of the world, which resulted from the spread of Europeans overseas, reflected the replacement of native plants and animals by exotics. Europeans sometimes greatly transformed landscapes, as in parts of New Zealand, Australia, South Africa, and North and South America, by creating human habitats remarkably identical—in ornamentals, domestic animals, weeds, even wild birds—with the landscapes of their European homelands. Since the end of the fifteenth century, mankind has been the greatest single force for extending the ranges of plants.

Animals. With human habitations have gone the animals who live there (the *commensals*): household mice and rats; roaches, silverfish, and ants; fleas, lice, and bedbugs; and two mosquito forms—*Aedes aegypti* and *Culex fatigans-pipiens*—that, outside of their native Old World habitats, breed only in artificial accumulations of water. As the predominant terrestrial vertebrate, humans have domesticated and spread only a very few of the millions of species of the dominant terrestrial invertebrates, the insects. Wild insects, however, are constantly being moved about by mankind's automobiles, trains, ships, and airplanes. The airplane, with its speed, is the most potent of new agencies for insect dispersal. Fortunately, the establishment of an insect in a new region is not always easy, and only a small fraction survive. Nevertheless, nearly half of the major insect pests of any particular region will be found to be species accidentally introduced by humans.

Mankind's major role as an agent of dispersal of invertebrates has been primarily with the thousands of species that have been accidentally transported with personal travels and shipping, such as the transoceanic introduction of the American oyster drill into oyster beds in England. Humans have also been involved in the dispersal of such other invertebrates as ticks, mites, household spiders, earthworms, and other soil inhabitants. There have been several deliberate introductions of crustaceans for food purposes, such as Chinese crabs and American crayfish into Europe. Mollusks have been moved about a great deal, mostly accidentally, but occasionally introduced into new regions as human food.

The most notorious case of the latter type was the dispersal of the African snail *Achatina* to many tropical Pacific islands by the Japanese.

In the spread of the vertebrate animals, mankind's role almost always has been deliberate and purposeful; the only vertebrates accidentally dispersed are probably rats and mice, lizards and burrowing snakes, and fish—the latter as a consequence of canal digging between previously unconnected watersheds. The deliberate introductions of vertebrates have been for one of four reasons: as domestic animals; for sport, food, or fur; for control of a pest, or for sentimental reasons. Although the domestic animals, in general, have become directly dependent on people for survival, there are striking examples of domesticates escaping and going wild. Such animals include dogs, cats, swine, goats, horses, camels, and cattle. The grazing domesticates have been a very powerful instrument for landscape change in many parts of the world, the goat being an example in its worldwide transfer from the Mediterranean.

Transfers of wild animals for sport are usually attempts to recreate the familiar sport of the remembered homeland and represent human reluctance to change established sporting habits to include a new fauna. Successful introduction of exotic animals may bring unprecedented, unfortunate consequences, if in the absence of natural enemies the new animals increase overly rapidly and become nuisances or destructive agents. One is reminded of the cases of carp or starlings in the United States, deer in New Zealand, or rabbits in Australia.

Humanity Returns to Biological Mixing

In the past few centuries, more people have moved, or have been moved, across longer distances than ever in human history. One of the predictable results of contacts among human groups is that, given the opportunity, interbreeding will take place. Human breeding populations have always been open systems, but they have been rendered far more open over the past few centuries than heretofore. The modern "melting pot" centers such as the United States, Canada, Singapore, and Brazil clearly attest to this point. The trend toward biological convergence has been furthered in the modern world by the vast increase of travel for business, education, recreation, and military operations.

The trend in recent human evolution has been an increasing amalgamation. Civilization enables race convergence caused by genetic exchange to outrun race divergence. The earlier pattern of a world divided into relatively small, isolated populations, each undergoing separate biological and cultural evolution, is rapidly breaking down. Boundaries among most breeding populations are not so much blurred as in flux, shifting in both time and space.

During the past several centuries whole tribes and peoples have disappeared after their lands were invaded by Europeans and other culturally dominant strangers. The native Tasmanians are gone, and so are the Indians of Baja California, the ten tribes of the Great Andaman Islands, and the Yamana of Tierra del Fuego. The Ridan Kubu of Sumatra were decimated by two epidemics of smallpox in 1905 and 1908. Many others are on their way out. These sad cases of ethnic oblivion give us a feeling that human history is a long record of utter extinctions, but this is not true. Just as Neandertal man in Europe was not so much exterminated by *Homo sapiens sapiens* as absorbed by him, so the Tasmanians were absorbed by the Europeans, who replaced them on their islands. A mixed Tasmanian-European population survives today, although the last individual Tasmanian culture-bearer died in 1876.

Eventual elimination of a distinctive localized population most commonly implies only the gradual absorption of a small population's gene pool into that of a larger one. The merging of gene pools from varying populations is producing larger populations different from earlier, more localized ones by showing more individual variations. This is the process now occurring in many complicated ways at varying speeds all over the world. Given sufficient time, many locally identifiable populations will disappear as such, yet each affected locality will have a population with greater variability among its members than now exists.

Biological mixing among humans is considered as advantageous as it is among domesticated plants and animals. The resultant offspring exhibit "hybrid vigor" and, given appropriate cultural circumstances, tend to exceed their parental groups in vitality, stature, and fertility. Human evolution becomes more and more irreversible as the different forms that it takes become increasingly dependent on the conscious actions of people in creating new biological possibilities. The most important thing about the biological convergence of the modern era is that it serves to maintain the unspecialized character of *Homo sapiens sapiens.* If the genetic systems called *races* had ever been isolated long enough to become closed, mankind would have become biologically divergent into different species. The modern convergence of humanity, by keeping open the channels for gene exchange among all human races, has been a great force for human unity. The reality is that no one racial group is in the ascendancy; each is outnumbered by all the others. In the world as a whole we are all minorities.

THE BLENDING OF CULTURES

The simplicities of the ways of life of our ancestors of thousands of years ago are mostly gone; we have been born into the complexities of a quite different age. The remaining thousands of preliterate, noncultivating peoples are relics of the past, surviving by sufferance, almost by accident, in refuge areas not yet preempted by civilization. Despite their apparent simplicity, these cultures differ from one another as a result of long and varying cultural developments, of adjustments to the environment, and of contacts with other peoples. Present-day gathering and collecting (hunting) groups do not necessarily live as did early man. They are, of course, the closest contemporary analogy, but they are not living fossils of the Paleolithic. For example, there are not many groups left today where stone is still being used for toolmaking. One must be very careful of the conclusions about prehistoric life drawn from the study of living gatherers and collectors, because the latter have also been changing over the long span of time. After all, they are our contemporaries, with a longevity equal to our own.

Prior to European occupation of Aboriginal Australia there were perhaps 500 relatively separate social and political units, or "tribes." Some have disappeared; some that are few in membership are now but remnants. Those that are left range between two extremes—the traditionally oriented or the mixed Australian-European. No aboriginal group remains uninfluenced by the urban-industrial society of modern Australia. The Northern Territory, in the Kimberleys, and the central desert are almost the only areas in Australia where some traditional life is still continuing. In another generation or two most of it will be gone, its disappearance accelerated by the impact of white settlement, mining exploitation, pastoral expansion, the effects of governmental policies of assimilation, and the general downgrading of what has come to be called the "old way." All Aborigines are moving in the direction of closer identification with the dominant society, and typically aboriginal traits are receding. The experience of alien impact has not been uniform among the societies; in some urban, suburban, and country areas part-Aborigines have lost their aboriginal identity and become completely absorbed. The direction of cultural change is clear; the only variety in the occurring changes is in the differing rates.

More important than the extinction of cultures has been the process of *acculturation.* Whereas significant evolutionary changes in human biology require time intervals of a much greater length than the span of a human life, cultural changes can be effected ever so much faster. Through learning, people can acquire the characteristics of several or many cultures in their single lifetime. The economic opportunities that the Europeans opened up for the modern world provided plentiful opportunities for cultural contact—on plantations, in mines, in farm settlements, and in commerce and industry. The commercial frontier, especially, brought together people whose tribal and kinship identifications had been largely destroyed and provided opportunities for new cultural amalgams. Conditions of potential economic equality were fostered between participating groups, as the marketplace and the town life around it were relatively freer of moral and political restraint. Urban commercial centers have been focal points for worldwide cultural change in that they afford greater ranges of socioeconomic opportunities and greater freedom for individuals to achieve social mobility.

Whereas the European-established plantations and the mines began as economic systems using slave and other forms of compulsory labor, they evolved patterns of contract and free labor based on a caste, class, or racial social structure. Modern industrial societies are developing both biracialism and cultural pluralism to accommodate peoples of different racial and cultural origins. But both systems have tended to break down under the increasing mobilities of populations. Where peoples of different cultural origins enjoy complete social mobility, blending has developed into a common national heritage, as in Brazil, where a new culture is evident in the fusion of European, American, and African elements.

The cultural impact of the European was the strong impetus for increasing the rate of cultural change around the world. Modifications of existing cultures took place everywhere, with a clear trend toward incorporation of European traits, and were often referred to as the "modernization process." The traditional cultural patterns were uprooted to greater or lesser degree depending in part on the population ratios of Europeans to nonEuropeans, whether Europeans were settlers or sojourners, and the receptivity of the nonEuropeans to the new ways. Selectivity is the keynote in the acculturation processes; some European ways were deemed acceptable, others modified, still others rejected. For example, modern China has ejected Christian missionaries, yet adopted Marxian ideology, political nationalism, and nuclear weapons—all Western creations—in order to become a world power.

On a worldwide basis cultural blending has been taking place. The emergent cosmopolitan culture is neither European nor American, neither Asian nor African, but a fusion of elements in varying propor-

tions. Most cultural studies have been analytical, sorting out the component parts of the new cultures to understand their derivations; more important is the conception that the phenomenon of acculturation has been global and that new values and ideas and personalities are emergent. Hawaii and Japan are avid proponents of the new vision of an open, cooperative world, whereas Rhodesia and South Africa have adopted antiacculturation policies and, through restrictive legislation and police power, sought to stem the tide.

THE SPREAD OF TECHNOLOGICAL CIVILIZATION

The shift from the method of trial-and-error to the application of science originated in western Europe, in a zone reaching from northern Italy to western Poland to Scotland. It began as a revival of learning based on the body of ancient science, but it was greatly stimulated by the Geographical Discoveries and the resulting production of new knowledge, new materials, and new equipment gathered from the ends of the earth. The imagination and creative thinking of Europeans took a new turn that involved a different attitude toward the earth and its resources. This new pattern saw the earth as a source of improvement in human life.

The place where these new developments took shape was Europe, but the ideas, materials, and techniques were from the world at large, as the Industrial Revolution began its development. The coal-powered cotton mill, first put together in 1785, was situated in England. Coal as a fuel derived its ultimate technology from China; elements of the spinning and weaving devices came from Egypt and China; cotton was from the Middle East. The cotton textiles desired were those of India (madras, calico, muslin); the steam engine was European in experimental motivation if English in final perfection. The labor was English, and the market that made the whole thing go was world-scattered. A spiraling advance in material cultural development had begun in which every success led to two more new beginnings.

The ideas and techniques spread from the first cotton mills into all other aspects of extracting raw materials and converting them into useful products, to the end that a great increase in productivity in mining, timber processing, grain milling, and metals manufacture began to occur. And the means and patterns of transportation began to improve. Despite the facts that in earlier eras Italian porcelains had moved to India, that Indian textiles had reached China, that Scottish woolens were worn in Germany, and that Greek olive oil was consumed in southern Slavic lands, these small-volume trade movements had to remain small because of the low rates of productivity in resource conversion and the poor transportation facilities. What the new ideas and technological improvements brought to European economic life was increased volume in goods produced and improved means of moving those goods about. Technological civilization emerged from the eighteenth-century industrialization of northwest Europe, based originally on a factory system that was coordinated by clocks and powered by the burning of coal in steam engines. This system manufactured products with interchangeable parts and employed labor that flocked to the factory towns and adopted an urban way of life.

Although these beginnings remained centered in western Europe, for the most part, until after 1800, it was the European contacts around the world that both helped them along and started the pattern of spread of the new technologies to other parts of the earth. The spread of European settlers to the colonial lands of the earth planted the same attitudes held in Europe. European technology of the new variety was applied to the extraction of raw materials in many parts of the earth to feed the development of industrial production in the home countries.

These new economic patterns spread out over most of Europe during the nineteenth century at the same time that they were spreading around the earth as a whole. The development of general labor skills gradually put Europeans well ahead of the rest of the world in respect to productivity in every form of economic endeavor. Thus European emigrants often became the agents of the diffusion of the Industrial Revolution in the lands to which they went.

The American system, built on that of northwest Europe, broke the components of manufacture into a sequence of standardized tasks that an unskilled laborer could easily be trained to do. Imagination and talent were invested in inventing improved machinery and managing the organized tasks and men. Wage labor was supplied by European immigrants, most of whom had been peasant-based and tradition-oriented. As new Americans, their families rather abruptly became urbanized, factory-organized, market-oriented, and literate as a result of public schooling. The process spread slowly, taking firm hold only after 1865 and gaining momentum by 1880. But the result was that, during the late nineteenth century, the United States was completely transformed from a regional, agrarian-based society into an urban technological civilization.

After 1870 the new civilization spread, in some form, to other lands so that, today, all lands and peoples have felt the impact of technological civilization. Invention of machinery, the harnessing of inanimate energy, and residence in cities are only parts of the story; also involved have been changes in human thought and habit: in concept of efficiency, in social organization, and in personal self-confidence that made possible greater productivity.

We here take issue with the commonly expressed belief that traditional cultures must disappear or remain relics of the past in order for a technological civilization to develop. The most successful growths of industrialization and urbanization have occurred in precisely those areas where industrialization long coexisted effectively with traditional culture and the blending of cultures has been harmonious rather than disruptive. Japan is the outstanding example, proving that acceptance of the new technological civilization has no one-to-one correlation with either Caucasian racial stocks or European cultural values. For all their modernization, however, the Japanese have not surrendered their own cultural mores, as any advertisement for Japan Airlines will indicate, and the Japanese are synthesizing world culture while blending it into their own culture system. To societies of traditional culture, but without some development of material technological traditions, the transition to modern technological civilization requires a little more time and a greater amount of experimental practice.

Intercommunication and Cultural Interchange

The spread of technological civilization has been enhanced by its own products, the new facilities for transport and communication that increased speed and quantity with greater reliability at lower costs. These improvements have made possible greater worldwide intercommunication and trade, and have established a permanent network for cultural interchange. The world's peoples in the twentieth century are tied into a single interlocking communications system of shipping routes, postal services, telegraph cables, telephone lines, railroad lines, newspaper and magazine publications, airline routes, radio and television networks, and communication satellites.

Two brief examples illustrate that the cooperative, coordinated nature of communication in our modern world has created wholly new circumstances.

The International Telecommunication Union (ITU) has existed since 1865. It began in Europe to make possible the sending of telegraph messages across national boundaries (for many years telegrams had to be handed across political frontiers). As new technologies in communication were developed, in telephone calls, cables, radio broadcasts, television, communication satellites, and messages through outer space, the responsibility for their international coordination also became tasks for the ITU.

International cooperation in postal matters also had its origin in Europe during the 1860s. The Universal Postal Union (UPU) was established in 1874, and now its work is all but taken for granted since the world has become a single postal territory for the reciprocal exchange of correspondence. One postal law governs the exchange of mail among all of the UPU's member countries. The routine postal arrangements have established some important matters of principle: freedom of transit by the best means available is guaranteed for mail exchange between countries; each country retains the charges collected on "letter mail," thus enormously simplifying accounting operations.

In the seventeenth and eighteenth centuries, government offices (foreign offices, defense departments, and intelligence groups) were the principal agencies concerned with world events. Today, current knowledge of world affairs is vital to international news agencies, banks, and industrial corporations with far-flung branches and affiliates; entrepreneurs in sports and the arts; scientists and scholars; foreign students; and travelers of all nations. The flow of knowledge is stupendous, whether measured in the bulk of international mail, the numbers of simultaneous channels used in cable transmissions, or in the numbers of overseas publications received in a sizable and active library.

Cyclic growth and decay in cultures, as in the lives of individuals, had a kind of accidental appropriateness in the past when empires, differentially involved in the governing of less organized groups, were highly dependent on resources and a technology they alone commanded. In previous ages of slow transport, natural disasters were suffered locally, recurring famines were impossible to relieve, and wars could be won or lost before becoming known to others beyond the arena of battle.

In the modern intercommunicating world, however, the analogy of growth and decay of individual cultures ceases to be relevant or productive. Admittedly, not all peoples of the earth have yet learned the comprehensive use of all our modern systems to the end that mankind does not yet collectively act to prevent extensive suffering in a particular region of the world during a specific time period. There remain political, social, and economic inhibitions which, so far, prevent full-scale and worldwide application of technologies reducing human need and suffering. This suggests that there is a widespread cultural lag in the applications of technological systems and that material technologies have developed more rapidly and extensively than have political, social, and economic systems. In the practical sense, however, each society is now clearly part of a world-wide network, a single intercommunicating system, that could be self-renewing and self-perpetuating, even though we may feel less than confident that it is reliably self-preservative.

No cultural group that has access to the totality of knowledge of all other human groups, across the chasms of space and past human time, can ever again be truly isolated. Much of the demand for increasing the speed of cultural change has come from the underdeveloped parts of the world, where people have come to know that they need to know more; and, through various programs of scientific exchange and technical assistance, they have discovered the ways to ask.

EUROPEAN POLITICAL AND ADMINISTRATIVE ORGANIZATION

The most cosmopolitan institutions in late medieval Europe were its universities—at Bologna, Paris, Montpellier, and Oxford—each of which attracted students and masters from all the Western world. Europeans studied the past with eagerness in the search for new wisdom, and did not stand forever in awe of its commands. They used the present to exercise choices for action rather than remain compliant with preordained ways of life. They imagined the future as a time of promise and achievement rather than the unfolding of a preestablished certainty. The foundations of the modern secular nation-state were set down in the reformation of Roman law by the medieval law schools at the universities of Europe, and many aspects of conducting international relations derive from the principles used for governing these academic communities, as self-governing faculties divided initially into recognized national groups.

The great contribution of Europe to political organization is the concept of the nation-state (see Fig. 5.3, p. 135; and then front end-paper) Nationalism arose in the later Middle Ages in those kingdoms that were unified within defined and defensible borders and enclosed people who spoke the same language and identified their interests with that of the political territory they inhabited. Ethnic consciousness among a group, as a stimulus to solidarity, was a vital part of the nation-state idea. England was among the earliest of the consolidated states, followed by France and others, as rival kings created national realms, each out of fear of the others. Nationalism, as expressed in monarchy, was the unifying force that created Europe's fragmented territorial units, most of which have persisted into modern times.

Each new state, as it emerged in Europe, was built upon local custom and institutions, but within the framework of a common legal system modified from Roman law. Most early states of the thirteenth and fourteenth centuries were the work of law-minded kings and citizens. The cortes, diets, parliaments, and estates-general all were representative assemblies. The concepts of despotic rule and representative government are contradictory, but what evolved in Europe was the modification of absolutism into constitutionalism supported by new legal institutions. The idea of the state shifted from its initial meaning, that of an extension of the ruler's personality and interests, and came to stand for a conglomeration of individual people who had political rights and obligations. The transformation of sixteenth- and seventeenth-century Europe into a group of separate sovereignties was established in the Treaties of Westphalia (1648), which recognized that states, as distinguished from governments, were members of the society of nations, that states were secular political organisms, and that all states were sovereign, independent, and equal (see Fig. 5.4, p. 137).

Western Europe, at the end of the fifteenth century, no longer possessed any vestige of an international state, such as the Roman Empire had been for its known world. Instead, Europe was an international community of separate and equal states. In their voyages of exploration, Europeans were amazed to discover that there existed ways for solving the problem of administering a territory and its people of which they had not yet dreamed. In its areas of different size, shape, population, and character, the world was found to be even more diverse than the fragmented Europe of the sixteenth century.

It must be remembered that all of men's ways for governing other men derive from a time when almost all the world's peoples were illiterate and were expected to remain so. Except for the Mongols, the great imperial designs in existence in the sixteenth century—Chinese, Indian, Ottoman (Islamic), Turk—were linked inextricably to a religious (the God-King) or quasireligious-ethical (Confucian) orientation that had scant attraction for a secular Europe that was beginning to cope at home with the impact of the printing press and the rise of science.

The discovery that there were other continents beyond the Atlantic and that their pagan inhabitants were rational human beings, as recorded by the Dominican friar, Bartolomé de Las Casas, put a great strain on Europe's image of a world community composed solely of European national states. From sixteenth-century Europe there emerged the ideas of a "collective will of mankind" or "law of nations"; the idea that a world community existed, composed of lands and peoples that differed from one another; the idea that a single superstate, created by military conquest, was impossible of achievement; the idea that the world could not be held together by any single creed in which all peoples would believe.

The logic of these ideas that emerged from the European discoveries was not immediately comprehended. The fervor of saving the world for Western Christendom sent many Spanish and Portuguese overseas to spread the gospel and the territorial extent of European empires. Periodically the dream of world empire burned fiercely, as among the French under Napoleon, the English during the "Pax Britannica" of the nineteenth century, the Germans under Bismarck and Hitler.

The twentieth century's worldwide political structure bears the unmistakable imprint of its European origin and the diffusion therefrom of the idea of nationalism. Most of the new states that have arisen in the Americas, Asia, Africa, and the Pacific have been formed in the image of a European model, or at least use a European political vocabulary: self-determination, nationalism, independence, self-government, human rights, democracy. Much guidance toward the choice of a European model for government derived from tutelage while the new states were colonial territories of European powers.

The World's Legal Systems

What we commonly term law is really codified custom of behavior. Each human society has developed a body of normative principles designed to govern interpersonal relationships as a means of social control, backed by sanctions that enforce punishments to those who act in defiance of accepted social custom. In simple societies there usually has been little separation of the body of custom into sectors, whereas in the advanced society the separation of the whole body of law into specific categories has been normal, wherein each category pertains to a fairly specific body of custom. Such categories as criminal law, civil law, commercial law, and maritime law are exemplary. Any body of law includes two aspects, the substantive (which defines the norms of behavior) and the adjective (which prescribes the procedures and the agents of enforcement). Said in another way, there are three elements to any body of law: (1) the regularizing of behavior in the specific customs, laws, and statutes; (2) the elements of official authority that interpret the laws; and (3) the applications of sanctions against transgressions. In any society, therefore, one can distinguish the law as a body of custom, the institutional forms developed for the interpretation and adjudication of the law, and the institutional forms and systems of punishment developed to carry out the law.

The evolution of the political state produced the

evolution of the concept that transgressions of custom were breaches of the God-King's peace and laws. In the organized political state there occurred the codification of laws, along with the development of agencies that interpreted the law and administered the sanctions. In time the latter agencies became the legislators, judges, courts, assessors, prosecutors, record keepers, bailiffs, sheriffs, police, and jailers, and others of the government bureaucracy who administer the law in all its varied forms in the advanced society. Around this whole set of constructs, for the whole body of world law, each society has developed its own specific legal institutions designed to make the law, to interpret the law, and to enforce the law.

As private property in land became an accepted principle, for example, there not only was the simple expression of law to this effect, but there also developed the institutional forms of the deed recorder, the tax assessor, the tax collector, the official treasurer, the financial controller, and the title guarantor. As a body of mineral law came into being there developed not only the laws concerning rights to minerals but also the institutional forms and agencies designed to interpret mineral law, administer mineral law, and effect sanctions for transgressions. In the modern era in the United States, the whole body of corporate law has developed not only the specific statutes governing corporate behavior but also all the legal institutions that are required to interpret and administer the laws created.

In one sense law is the same everywhere throughout the earth, the statement of a society's norms of behavior. It is the institutions of law that vary most among societies. The body of law, of course, has grown infinitely from early Paleolithic time to the present, as man has created his varied and more complex cultures, but the spatial and regional differences among bodies of societal law are best seen in the variation of the legal institutions.

The rules by which men are governed apply in theory equally to all inhabitants of a single political territory but differ from one political territory to another. The law embodies the whole story of a nation's development: what it has been, what it is, and what it is tending to become. We are here concerned with the spatial attributes of legal systems, because they are the touchstone by which to identify the means men use to organize and administer political territories on the earth's surface (Fig. 6.2).

In the Western world, legal systems are customarily divided into two groups: (1) the *civil-law systems*, as on the continent of Europe, in Quebec and Louisiana, throughout Latin America, and, by extension, in much of Africa and Asia; and (2) the *common-law systems*, developed in England and diffused overseas into its colonies, now become nation-states,

hence almost coextensive with the English-speaking world. In civil law, which was strongly and variously influenced by Roman law, large areas of private law are codified, a matter not typical of common law. Nevertheless, both systems share many values and both are products of Western civilization.

The essential difference between the two systems derives from the creation in England, earlier than on the European continent, of a national, efficient, and centralized administration of justice. After A.D. 1066 the unified law of the royal court of the Norman kings became the one law common to all the realm. The centralization of justice produced an organized class of lawyers, the English bar, which established its own tradition of teaching law, in contrast to the dominant role in legal education held by universities on the continent. A basic reason why Roman law was never fully utilized in England may well be the existence of an independently organized bar with a vested interest in the law as administered in English courts. Bench and bar came to defend the law, which they had created, from the encroachment of later English kings.

The adoption in civil-law countries of codes of law that revised and unified diverse local laws almost always followed historic periods that were sharp breaks from past traditions, as in France after 1789, or in Germany after 1870. The few great exceptions are in Scotland and in the Roman-Dutch law of South Africa, which operate with modified civil-law systems. Codification always implies something of a new start (e.g., the French Civil Code, or Code Napoléon, became effective in 1804; Austria's in 1811; the German Civil Code in 1900), the conscious creation of something that replaced all that had preceded.

A third legal system in the world is that of the Communist bloc, extending from Central Europe to North Korea and Laos and Vietnam in easternmost Asia. Law is essentially the means whereby the political leadership exercises control over society. Civil-law elements in the legal institutions of Communist countries have been reduced almost to the vanishing point. Expropriation and planning measures, backed by criminal sanctions, have made almost all law public law. Nevertheless, there remain some significant differences between the legal tradition of the Peoples' Democracies and their Soviet model, as well as the divergences discernible in the independent countries following their own "road to socialism."

A fourth legal system is that of Islam, based upon the Qur'ān and other religious sources. It reflects traditional religious views in dealing with family and inheritance law and preserves the jurisdiction of religious courts in such matters. Saudi Arabia and Yemen are examples of Arabic countries where Is-

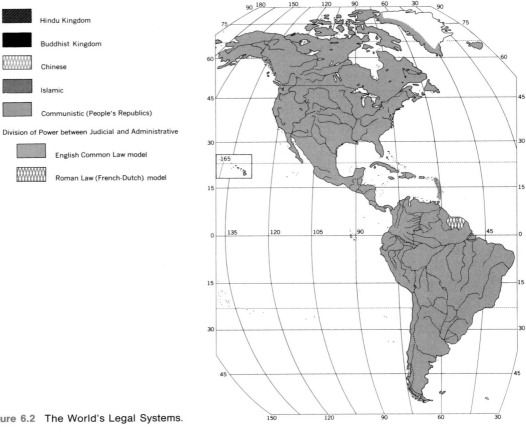

Hindu Kingdom

Buddhist Kingdom

Chinese

Islamic

Communistic (People's Republics)

Division of Power between Judicial and Administrative

English Common Law model

Roman Law (French-Dutch) model

Figure 6.2 The World's Legal Systems.

lamic law has been least influenced by Western concepts.

The traditional Chinese body of law and legal institutions formed a fifth system, derived from folk customs of Chinese and non-Chinese people of China and its borders, modified by various early legal philosophers from Confucius to Hantzu, and eventually codified into Confucian state doctrine that provided the governance of China to about 1900. The Chinese legal system strongly reflected the pragmatic social views of the ruling classes, by which peaceful relations were established and maintained throughout the empire among the great family that composed Chinese society. Civil institutional forms became highly developed in the Chinese codes, and it is from these that the rest of the world took the concept and the forms of the modern civil service. To variable degree the Chinese legal system became the framework for the evolution of the localized regional legal systems of Tongking-Annam, of Korea, and of Japan. The Communist take-over of North Korea, China, and Vietnam substituted much of Communist law for the traditional systems.

In other parts of the world, the interpenetration of common, civil, Islamic, and indigenous law has produced mixed systems. Latin American countries have a fundamental civil-law character, yet the impact of

common law has been felt through incorporating provisions of the United States Constitution into Latin American constitutions. Puerto Rico, Louisiana, and Quebec are civil-law enclaves in a common-law world; commercial law in Quebec, for example, reflects English influence. In Ceylon and Guyana, common law has widely replaced the former Roman-Dutch system. The new nations of India, Israel, and the Philippines have followed common-law methods, turning to British or American solutions to meet some of their problems. Liberia's legal system is almost entirely patterned after its nineteenth-century United States model, with the interior of the country still partly governed by customary tribal law.

Civil-law systems have spread into many parts of Asia and Africa. The main structure of the German civil code was adopted by Japan and Thailand, and strongly influenced both Greece and pre-Communist China. Austrian models were influential among the Balkan nations, prior to their absorption into the Communist orbit. Turkey copied the Swiss Civil Code (and the Swiss Code of Obligations was German-influenced in its Commercial Code and Code of Criminal Procedures) and used an Italian model for its Penal Code. The Code Napoléon has influenced most of the Islamic nations in North Africa and southwest Asia (e.g., Civil Codes of Egypt, 1949;

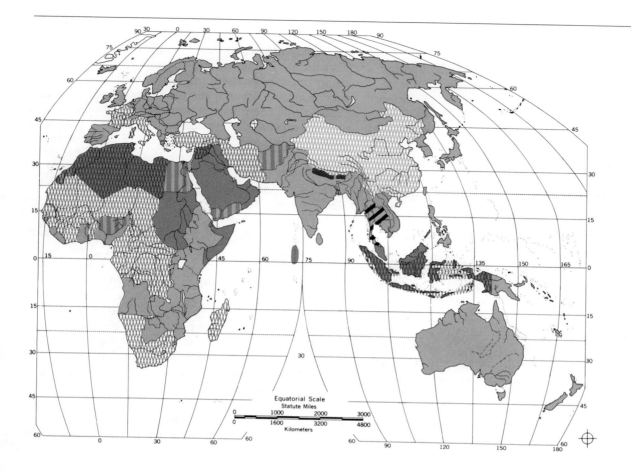

Syria, 1949; and Iraq, 1953), which synthesize Islamic law with a Western civilian-style commercial code in the law of obligations and in commercial law. The Islamic Republic of Pakistan, on the other hand, has synthesized British common law.

Scandinavian legal systems occupy a special position, with characteristics of both common-law and civil-law systems. Most statutes are not organized into comprehensive code systems, yet many uniform statutes have been enacted to be effective throughout all the Scandinavian countries, governing such matters as marriage, sales, checks, insurance, trademarks, air traffic, and social security. A model for coordination and cooperation among sovereign nation-states has been created.

RELIGIONS: THEIR ROLES AND SPATIAL PATTERNS

There have emerged only a small number of organized religious systems in which theological beliefs became codified, highly systematized, and structured. The principal ones are Buddhism, Christianity, Islam, and Judaism. In the case of each of three of these, Buddhism, Christianity, and Islam, there developed a dynamic sense of the importance and applicability of the religious beliefs to all mankind. This is commonly labeled the "missionary urge," although it is much more complex than that. These three religions deal in universal aspects of theological thought in a way that makes possible the separation of the religious elements from the basic components of a culture system. This allows the beliefs of each to be spread among all kinds of peoples regardless of the specific differences among culture systems. These three religious systems have diffused around the earth to become accepted by many different peoples possessing quite varied systems of culture. The religious beliefs become integrated into the separate bodies of culture to greater or lesser degree. As each of the religions is accepted by a new culture group, the religious symbols, interpretive systems, and specific practices undergo varying amounts of modification. In the adoption of one of these religions there takes place the development of institutional features that permit the functional aspect of the religion to play particular roles in the separate cultures. This selective perpetuation and broad diffusion of just a few religious systems is a part of the broad process of cultural convergence that has taken place in late historic times and during the modern era.

In the modern world we sometimes are impressed

more by the sectarian divergence that seems rather recently to have appeared in religious beliefs and practices than by the actually more significant processes of convergence of religious beliefs. That is, we cite the many kinds of Christianity or Buddhism or Islam as a symptom of sectarian divergence which has seemed to be an increasing trend in the modern world. This view, however, neglects the larger fact of the missionary spread of some kind of Christianity or Buddhism or Islam, which have been replacing the much greater variety of only slightly organized religions and the great but variable reservoir of local animisms.

The world map of religions (Fig. 6.3) displays this diffusionary spread of the several Old World religions, in which many specific regions show mixed patterns of religious affiliation. Similarly, the world map demonstrates that there remain large islands of simple animism and regions in which folk religions possess distinctive patterns corresponding to ethnic culture systems. China is here mapped as practicing a folk religion, rather than the frequently mapped Confucianism; Confucianism properly was a politico-ethical state cult practiced chiefly by governmental bureaucracies. Lands now controlled by Communist political ideology are mapped only in the nominal sense that varieties of religious belief

are still preserved by portions of populations. The world map illustrates clearly the worldwide sweep of Christianity in the modern period, this fact in itself being another element in the trend toward cultural convergence. Among all the living religions, the Christian Church is most actively on the missionary bent in expanding the scope of the faith and has been a strong factor in the spread of modern world culture.

Historically, the expansion of a political state has often been accompanied by the spread of the religion practiced in that state. Hinduism was carried by early Indians in their political expansion into Southeast Asia, even though Hinduism never displayed a "missionary urge." Islam spread along with Arab conquest for several centuries, and the later non-Arab political expansion utilized Islam as an entering wedge in parts of Southeast Asia in the fourteenth and fifteenth centuries. The modern accompaniment by Christianity of modern European political imperialism, therefore, is not a new feature. There has always been a close connection between political administration and a religious system—the two were joined in the appearance of the first political states. The preservation of this close association in the modern era is demonstrated by the continuance of the state religion in several modern states.

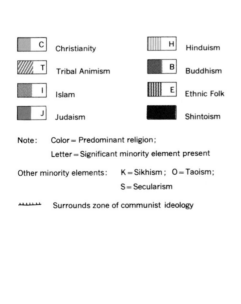

Figure 6.3
The World's Religious Systems.

Patterns of expansion and contraction of political states have, to some extent, been accompanied by the expansion and contraction of religious systems. There is not a perfect correlation in these patterns, however, for the control over a political state is more fragile than the persistence of a religion among human populations. Political rulers can be overthrown, and political states are subject to conquest by neighbors. When a religious system has been deeply integrated into a regional culture system, however, it becomes a lasting and durable core component in that culture system. There are many cases of the long-term spatial persistence of a religious system after the disappearance of the political state whose original formation or expansion carried the religion along with it. This lack of correlation between the spatial distributions of religious systems and political territories is, in fact, one of the causes for internal dissension in some political states in the modern world.

In modern times the growth of the concept of the separation of church and state has given rise to secular political systems, in the strict sense, a characteristic that then permits the spatial intermixture of religious systems to a high degree. This secularism in political administration sometimes has led both to the proliferation of sectarianism in religious systems and to the rise of secularism as a personal viewpoint, quite notably in the United States and Canada. Despite such trends, the United States, for example, remains broadly classifiable as a Christian society.

Beyond the specific role of religion as a formulative and a preservative agent in systems of culture there are many other varieties of impact of religion on regional cultures. There are large numbers of religious place-names in all parts of the earth. The calendar systems divide the year into particular segments and the patterns of ceremonial days that vary significantly from place to place. Architectural styles that go with religious systems have much wider implications than just the forms of the religious structures themselves; the degree to which religious practice is spread out on the landscape in field shrines, house shrines, sacred groves, and sacred places is a significant manifestation of religion in many parts of the earth; and among some religious systems the whole orientation of the total landscape becomes a matter of arriving at a system of mystical harmonics. To the degree to which a particular religious system perseveres in a particular doctrine, such as the vegetarianism of India or continuance of the abstinence from eating meat on Fridays among Roman Catholics of the Mediterranean Basin, religious beliefs have strong impacts upon the whole

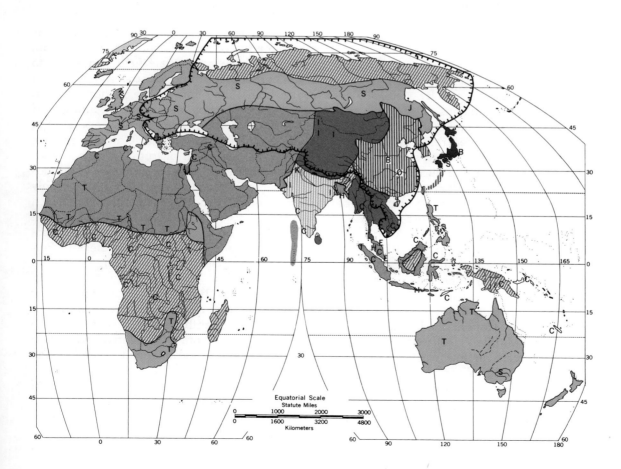

Equatorial Scale
Statute Miles
0   1000   2000   3000
0   1600   3200   4800
Kilometers

orientation of a regional economy. The taboo on the eating of pork, in Moslem lands, has led to the disappearance of the pig as an economic animal. These are but illustrations of the variable role of religion in cultural practice the world over.

LANGUAGE, WRITING, AND SYSTEMS OF EDUCATION

A language is a group achievement, an agreed-upon system for attaching meanings to a limited set of sounds and sound arrangements from the vastly greater number of possible sound combinations and orders. Most spoken languages are extremely old; some cultural groups, however, have lost their original languages and speak borrowed languages. Languages have a dynamic character; some have grown and expanded. For instance, English, which was spoken by not quite 3 million people on less than the whole of just one island in the fifteenth century, now is spoken by hundreds of millions of people the world over. Yet, other languages have remained small, have retrogressed, or have disappeared. Those languages whose internal mechanisms most readily permit changes, such as word borrowings, cultural transfer of meanings, and coinage of new words, are best able to keep abreast of cultural change and are most likely to persist.

Writing

Writing, as a device for communicating by visible linguistic symbols, is a permanent extension of spoken language and an important cultural invention (see Fig. 4.9, p. 104). There is no fixed relation between most forms of speech and the writing systems used to record them. The linkage of language with writing is a cultural achievement as a people come to adopt, and adapt, one or more forms of alphabet, numerals, and writing styles that they have received by diffusion from an often nearby source region as a result of cultural contact (see Fig. 4.10, pp. 104–105).

The major alphabets of Europe owe their present distribution to the forces of religion. Adherents of the Eastern Orthodox Catholic church learned to use the Cyrillic alphabet, as adapted from the Greek whereas the Latin alphabet of 23 letters (the J, U, and W being added during the Middle Ages) was used by the Roman Catholic church in the West. West Europeans have carried the Latin alphabet around the world. This is the way in which the alphabet and writing system used in Canada and the United States was diffused to the New World. We write an Etruscan-Roman variant of a Greek form of an alphabet invented in or near Phoenicia by a North Semitic people on the basis of their adaptation of nonalphabetic writing in still earlier cultures.

The Latin alphabet has been used by missionaries and others to create a form of writing for contemporary languages lacking developed scripts of their own (e.g., Karen in Burma) or to effect a change from a previously used form (e.g., Vietnamese). Governmental reforms have further spread the Latin alphabet (e.g., with slight variations to accommodate phonetic differences, as in Turkey, replacing Arabic in 1928; in Communist China as a supplement to Chinese writing in order to increase literacy and efficiency in mass communication). As a result, probably more people in the world are familiar with the Latin alphabet than with any other form of writing or printing. The attributes of the Latin alphabet are its simplicity and flexibility for nontonal languages, since it contains both consonants and vowels. Throughout the present-day world the greatest volume of information and entertainment is disseminated through literature printed in a Latinized writing system, and the most extensive campaigns to reduce illiteracy are overwhelmingly in those languages that employ the Latin alphabet in their forms of writing.

Educational Systems

Systems of education have long been in existence. For a millennium or more, scholars from Japan, Korea, India, and southeast Asia traveled to China—the center of learning for eastern Asia. Upon their return home, they spread the wisdom of Confucius and other sages by means of curricula based on the Confucian classics. In Western Europe prior to the sixteenth century, the Roman Catholic church, regarded as the custodian of truth and the proper agent for dispensing it, held a monopoly in education. As an international institution with a highly centralized, hierarchical administration, the Church operated its schools with a uniform curriculum based on a single language, Latin. The independent European nations that emerged from the Protestant Reformation created systems of education that were nationally, rather than internationally, focused. In each instruction was based on local vernaculars side by side with Latin schools and the universities. Prussia, in the eighteenth century, attracted wide attention with its program of universal education combined with compulsory military training. In the late nineteenth century, the Japanese government rejected classical Chinese learning. Hundreds of young men went to study in Europe and the United States, and quickly acquired the new knowledge needed to industrialize the nation. The Japanese

The Arabic alphabet was spread throughout the regions of Islamic culture. This newspaper seller in Cairo also handles magazines.

educational system was reconstructed along Western lines, new schools were introduced, and Western scholars were afforded positions of honor in Japanese universities.

It would be easy to say that there exist today as many systems of education as there are nation-states. But such a statement would be too easy and would overlook the existence of certain elements useful for purposes of general comparison. We here distinguish three categories of social philosophy (i.e., what the educating society hopes to achieve in educating its young) that serve to establish the social role of the schools. These are:

1. Belief that knowledge of the good and the true is *received* from a supernatural source.
2. Belief that knowledge of the good and the true is *discovered* in nature or in the processes of the natural universe.
3. Belief that knowledge of the good and the true *evolves* out of intellectual processes, derived from the everyday experiences of ordinary people.

The educational significance of each of these three social philosophies is that the selection of societal leaders for each is substantially different. The educational systems supported by each reflect these differences in the opportunities they afford to members of their societies.

There are a number of societies whose members believe that they have received, or are continuing to receive, knowledge of the good and the true in some extraordinary way from a source external to the temporal events of day-to-day experience. Differences occur among such societies only in the social means of "receiving" the revealed knowledge. The traditions based on *revelation* include the Jewish, Islamic, Christian, Hindu, Buddhist, and Shinto. For example, those of Jewish faith teach children, especially boys, to read the Books of Moses and other prophets as a guide for the direction of their personal affairs in daily life. Self-sufficiency and independence are the rule, whereas Roman Catholics believe that God has delegated sovereignty and the authority to interpret His will to the head of the church. Spain, for example, includes a prescribed form of religious instruction in the official system of education.

The second category of social philosophy includes those that assume "discovery" through observing objects and processes in the natural universe. The *discovery* process is a natural phenomenon whose method can be taught. Proficiency in its use can be developed and the attainment of results can only follow an expenditure of effort in promoting the search for knowledge. Education, under this social philosophy, is a matter of great importance and is usually provided at public expense to those regarded as capable of benefitting from it. Usually only a small proportion of the population is considered intellectually talented, hence worthy of being provided with intensive training.

For example, the Soviet Union operates by the method of inquiry known as dialectical materialism. Only those thought to have a special talent for discovering knowledge by this method are given a special education, accorded special privileges, and elevated to the status of a ruling elite. Public schools, free to all, train the masses for socially useful roles, but the identification of future leaders and their special education is the function of the youth organizations of the Communist Party. Another example was France, from 1789 to the reforms of 1959. Public elementary schools prepared for trades, commerce, and normal schools, whereas private primary schools led to municipal or state secondary schools whose purpose was to locate and produce future leaders with the "best minds" to direct national affairs by sending them on to great schools or universities. Highly centralized control of education insured that only those with the best records in the universities were favored in the civil service, higher administrative posts, and policy-making positions.

The third category of social philosophy focuses on knowledge that has *evolved*, or is constructed, *from human experiences.* Such knowledge is neither complete nor infallible, hence the societal institutions it creates constantly require adjustment and sometimes replacement. Since not all individuals are alike in their knowledge, experiences, and prejudices, the highest validity lies in group, rather than individual, judgments. Hence, societal authority is held to reside with the people, and societal institutions are created and controlled by society as common endeavors. Communications are facilitated and wide participation in decision-making is encouraged, so that all may benefit from the policy of diverse experiences and opinions. The decision-making role is shared by all, and no individual or special-interest group is permitted to force its will on the total group.

The present world trend in educational systems is a slow change from categories one and two toward category three. Educational programs in such diverse places as Great Britain, Japan, and the newly emerging, former colonial areas of Africa and Asia are being expanded and revised to afford a greater measure of individual opportunity and self-determination. Provision of industrial, technical, and vocational training is assisting the creation of a middle class in societies formerly divided between a higher class of educated elite and a lower class of workers. National educational systems now being revised or newly instituted will inevitably raise the general economic and cultural levels of their societies and afford opportunities for more children to rise above the economic and social level of their ancestors and parents.

THE TREND TOWARD A UNIVERSAL COMMUNITY

Most of the writings about the European colonial period have been studies of the particular, tending to emphasize differences: between white and nonwhite, dominant and subordinate, rich and poor. The differences in policies and behavior among the Europeans themselves have also been abundantly emphasized. But, although the colonial era was only a brief period in the total of human time on the earth, it made possible a vast worldwide cultural diffusion quite without precedent. Rather than creating insuperable differences, European imperialism resulted in making the earth more uniform, because it was the mechanism that spread the great cultural revolution of the West to the rest of the world.

During the eighteenth and nineteenth centuries, the Occident not only expanded its commercial exploitation and political domination but also sowed the seeds for removal of its dominant controls over the rest of the earth by transmitting small measures of its revolutionary ideas: these included enlightenment and reason, science and industry, organization of human affairs, the dignity and value of the individual person. Concepts of nationalism, self-determination, freedom, and economic development were European ideas. The meanings and implications of American independence, of the French Revolution, of British constitutional history, and of socialist theory about revolutionary action were not lost on Asians and Africans. The will of the colonial powers to maintain their empires declined as the economic value of colonies decreased, some to become economic burdens. Once colonial peoples reached critical levels of development they refused to accede to alien government no matter how efficient and enlightened.

In addition to the worldwide diffusion of aspects of Western culture, the nineteenth century saw the first steps taken toward a new kind of internationalism by mutual agreement. International treaties, voluntarily accepted, provided for new rules to be applicable worldwide. Examples of this are:

1856 Declaration of Paris on maritime law,
1864 Geneva Conventions of the International Red Cross,
1886 Berne Copyright Convention,
1899 Hague Convention on limiting methods of warfare, and providing for peaceful settlement of disputes.

In the second part of the nineteenth century, the cultural development of Western mankind was pointed to by Europeans with great confidence, as exemplifying the law of human progress. Evolution, as a process, was both physical (biological) and cultural (technological and moral). The conditions of the past had evolved into the present and would improve into the future. The idea of the inevitability of human progress was an article of faith during the European age known as "The Enlightenment."

We know much more now than we did then. The global wars and revolutions of the twentieth century have destroyed the myth of continuous progress. Vastly increased archaeological knowledge of previous societies and civilizations reveals that specific cultures have experienced both progression and retrogression and that some cultures were expanding at the same time that others were collapsing. The cultural evolution of mankind since the Upper Paleolithic has never been unilinear, but multilinear—the abstract totality is the result of a series of simultaneous positive and negative increments, fluctuating like a changing bank balance. Progress in some cultures in some times and places was offset elsewhere by stagnation, retrogression, or extinction of other cultures. Not until after A.D. 1500 were peoples able to compare the cultures of the whole earth, to use consciously the advances of one positive zone for the purposeful advancement of a negative and lagging sector.

THE IMMENSE JOURNEY

Chapter 2 of our volume began with a description of mankind's Pleistocene environment, presented the conditions of the human emergence, and outlined how mankind, as a culture-bearer, has been able to spread over the earth to become its dominant form of life. We then traced the effects of cultural divergence in creating regional cultures, to be followed by their partial combination and recombination into a smaller number of regional empires, and ended with consideration of the breakthrough into a global interlinkage initiated by Europeans. Truly, the human journey has been long in distance and immense in scope, as a cultural achievement.

For the greater part of the Pleistocene, the range of mankind remained restricted to tropical and subtropical Africa. But, by the end of the Middle Pleistocene, a fundamental redistribution of humans had taken place: a definitely *sapiens* form had succeeded in invading the northern parts of the Old World continents. Once North America was reached, expansion progressed quickly, like an irresistible tide, over the New World: only a few thousand years were required to reach Patagonia. The rise out of Africa and the worldwide spreading of *Homo sapiens* was a true humanization of the world.

As a result of worldwide expansion in the Upper Paleolithic, the human cultures of the Mesolithic were no longer merely tropical and subtropical, but panterrestrial systems of advanced life. Continuous and intensive efforts over many more millennia were required to change mankind's somewhat precarious hold on the earth. Crop growing and protoindustry were cultural inventions of great significance, enabling people slowly to fill the gaps and establish a first network of communications and trade with other groups of humans in the several regions of the world.

More than a million years were spent, mainly in Africa, in acquiring the biological and cultural capacity for a human planetary invasion. Some 30,000 years more were required for the actual occupation of the world's continents. Approximately 11,000 years—the combined Neolithic and most of historical time, until about a century and a quarter ago—were necessary to affect a consolidation of man's dominance over the earth.

The trend toward cultural convergence during the last several centuries over the earth's surface is a wholly new experience for the earth and its life-forms. It represents an era of cultural fulfillment for that unusual but ungeneralized primate who created and elaborated culture, became prolific, and made the whole planet his home. Mankind alone has grasped the concept of evolution. The way is now being opened for yet more astounding achievements, as more and more people are able fully to draw on the whole of the world's ideas, present and past, and have access to the utilization of the full range of its space, its technology, and its resources. However, since all humans have not learned the control of the intricate balancing manipulations critical to the successful operation of the earth's ecosystem there remains the danger that modern destructive technologies may have prejudiced that future.

Such has been the great accomplishment of the human adventure, for which the name "noösphere" (from the Greek *Noös*, "mind") has been suggested to distinguish the human surface of the earth, just as the terms lithosphere, hydrosphere, atmosphere, and biosphere are commonly used to distinguish the earth's other aspects. The evolution of the noösphere has created the conditions necessary to a new transformation of the earth, for such is the nature of the evolutionary process.

The second half of our text (Chapters 7 through 12) consists of a detailed examination of the conditions on the present-day humanized globe and considers the future of the spaceship Earth that will be your home for the rest of your life.

SUMMARY OF OBJECTIVES

After you have read this chapter, you should be able to:

MODERN LANDSCAPES

The varied landscapes of the world today show the results of centuries of human operations. Mankind has greatly affected both the physical and the biological ecological systems of the earth. With the exception of parts of Antarctica, it is doubtful that there are left any large areas that are truly natural. Often, even those areas we consider to be "natural landscapes" have been significantly altered, disturbed, or in some way affected by human actions. The landscapes of fully occupied and culturally mature areas are largely determined by mankind, and people are hard at work taming the "wild" landscapes of the more sparsely settled regions and the few remaining frontier fringes.

The various groups inhabiting our earth create and maintain landscapes in very different ways according to their cultural drives and technologies. It is now popular to accuse Occidental societies as the worst offenders against nature. The accusation usually implies simple mutilation or destruction. Considering large geographic regions, however, it is likely that the Chinese have made the strongest imprints on their regional landscapes, for they have been at the job of creating landscapes in their own image longer and more intensively than any other society in the world. The accompanying photographs present some examples of the different kinds of landscapes to be found in the world.

AGRICULTURAL LANDSCAPES

The most numerous of the maturely developed cultural landscapes are agricultural landscapes. Although their types are basically determined by the climatic zone in which they developed, different cultures have produced variations in the broad zonal patterns. There are numerous factors besides climate that affect the development of agricultural landscapes. Chief among them are the combinations of crops grown, field systems, methods of land survey, and the technology of the particular culture.

The wet rice fields of southern and eastern Asia produce one of the most distinctive of the agricultural landscapes (Photo 1). The cropping combinations, the planting and harvesting procedures, the small irregular fields, the terracing, and the water-control technologies are the dominant elements. Wet-field rice-cropping is very adaptable to irregular landscapes which have little naturally flat land. Intensive and ingenious human labor applied over long periods of time produces landscapes of small level fields fitted to the contours of the land. In the early stages of development, the easiest portions are brought under control and the many rough sections are left untouched. As the cultural landscape approaches maturity, however, a large part of the total area may be converted into productive fields.

Although the physical problems involved in establishing the level plots are time-consuming and require hard work the key to the control of the landscape lies in the engineering technology used to effectively control the water. The mature wet-field landscape has built into it many small and inconspicuous devices for introducing, distributing, balancing, maintaining, and removing the flow of water from the field system during the course of the year.

In the wet-field systems, ownership patterns and transportation networks are subordinate to the construction of terraced units that are in harmony with the contours of the land. The mature wet-field landscape is an integrated affair in which rice governs the cropping combination, water control is the dominant technological element, and other matters relating to the agricultural system are subordinate. The system requires continuous intensive labor organized on a community-wide basis.

A very different type of agricultural landscape has approached maturity in the central-western portions of the United States and southern Canada

Photo 1. The wet-field rice terraces of southern and eastern Asia, when developed to maturity, may use a very high proportion of the total land surface in a region. These Indonesian terraces were photographed just prior to the transplanting of seedlings from two sets of seedbeds in the left upper center. A long period of development is required to develop a whole valley into terraces.

(Photos 2 and 3). In these regions, European settlers used the rectangular method of land survey before beginning their agricultural operations or establishing their settlement patterns. The use of this survey system from the outset established some control over the ownership systems, considerable control over the transport network, and complete control over the field systems. As the cropping combinations employed relied directly on precipitation for moisture, water control had little influence on the appearance of the landscape. In this agricultural system, crops may vary from year to year at the will of the operators. The size of operating units may vary markedly according to economic manipulation. The terrain and the land patterns permit evolutionary change in operating technology, and the system is capable of response to cultural processes. Cropping patterns, settlement patterns, transport lines, cultivation lines, and other elements may affect the appearance of the landscapes, but if there is one dominant factor, it is the rectangular land survey.

Photo 2. A wheat landscape in Kansas illustrates the rectangular land survey system and the orderly checkerboard of field systems that results. The original transport network followed the survey line boundaries. This is a region of dispersed rural settlement. The larger units at the lower right are concrete wheat storage silos along a railroad line, and the smaller units are metal storage tanks.

Photo 3. (Opposite) This Wisconsin dispersed-settlement dairy farm lies within the zone of the rectangular land survey. Land use rotates among pasturage, grain silage, feed grain, and hay crops. Animal, feed, and equipment barns and auxiliary buildings are more important than human housing (lower left) in such an agricultural economic investment.

A third type of agricultural landscape is that created by the plantation economic system of the humid tropics. Products such as bananas, coconuts, rubber, palm oil, tea, and coffee are grown in massive monocropic patterns. Since large volume production of a single commodity is the chief economic goal, large operating units are the rule. Although field systems, operating technologies, transport lines, and other factors affect the visual elements of the landscape, the massed planting patterns of the mature crop plants themselves dominate the agricultural landscape to an overwhelming degree. Settlement units and processing stations often appear as small islands in a sea of plants.

Quite different are the orchard and vineyard landscapes of the earth's dry subtropical zones (Photo 4). Both deciduous and evergreen tree crops are found in some of these landscapes, although the pattern illustrated here is made up entirely of deciduous orchards and vineyards. This South African example shows an immature culture landscape. The less easily utilized sections, which are still untouched and are covered by wild-plant forms, are a prominent part of the landscape. Future pressure may result in further clearing and the cropping of a higher percentage of the total area. As the dominant cropping elements, orchards and vineyards produce partially open landscapes in which field patterns, transport lines, homesteads, and processing installations are clearly visible.

The mature agricultural landscapes in midlatitude former forest regions sometimes preserve wild trees and hedgerow shrubs, but these are most often segregated along property lines, field margins, roadways, or rocky bits of land not easily cultivated (Photo 5). Such "wild" plant elements may even become the dominant visual features of the landscape. They may also serve to mark property lines and field margins. In many parts of the earth, fully settled at some time in the past, the field patterns are odd-shaped triangles or irregular rectangles. They were formed prior to the invention of the rectangular land survey, and the legal land deeds in these areas employ the old concept of metes and bounds. (Compare Photo 5 with Photos 1, 2, and 3).

In the modern era some agricultural landscapes are beginning to take on the appearance of industrial tracts, which indeed they are with respect to their technological operations (Photos 6 and 7). The poultry and vegetable "factories", the livestock fattening yards, the urban-fringe dairies, and the commercial flower gardens have a visual appearance less like the traditional agricultural landscapes than the industrial landscapes of the cities they often surround.

Photo 4. (Opposite) Deciduous orchards and vineyards do not close off a landscape as fully as do some other cropping patterns. In this South African scene in Cape Province, there is also dispersal of settlement and of processing facilities.

Photo 5. This southern Scotland agricultural landscape is a mature landscape with almost full economic use of the cultivable land. Trees and hedgerows that have been kept are along roads, property lines, and field boundaries except for the cluster of trees around a rocky outcrop at the lower left. Field boundaries were established long ago as the region became settled.

Photo 7.   The greenhouse farms around many of the modern cities of northwestern Europe resemble factories more than traditional agricultural fields, as in this view in the Netherlands.

Photo 6.   (Opposite) As the industrialization of some aspects of agricultural production occurs, a factory landscape quite different from the traditional landscapes associated with such activities results, as in this mechanized poultry farm in Argentina.

COMMERCIAL LANDSCAPES

What may be loosely termed "commercial landscapes" belong in a different category from agricultural landscapes. Commercial landscapes vary in accordance with the economic organization, the technology, and the intrinsic features of the cultures in which they are found. Space permits only a few illustrations of the great variety of the commercial landscapes of the world. This brief essay cannot cover the whole range. It must restrict its commentary to notations about the particular items that depict the category.

At the village or small market-town level, the volume of person-to-person barter and exchange may be so small as to hardly call for a spatial unit worthy of the term *commercial landscape* (Photo 8). However, there are thousands of these miniature nodes scattered around the earth, and the great central business districts all had their origins in such patterns. In them barter and exchange is restricted in amount and kind, the institutional elements are few, and the facilities are minimal. In larger, older settlements, many transactions may still involve person-to-person exchange, but the marketplace becomes more institutionalized and has more significant spatial patterns (Photos 9 and 10). In a number of the earth's more recent settlements some of the commercial activities are of such a scale that their institutionalized elements become very strong components of the landscapes created (Photos 11 and 12).

Photo 8. The wool market in this Afghanistan scene is merely an open space in which sellers and buyers can display and examine the product.

Photo 9. In Bergen, Norway, the quays around the fish harbor serve both as a fish market and as a marketplace for flower sellers, providing formalized spatial allocation to the marketing functions.

Photo 11. (Opposite) The modern American shopping center, catering to customers using automobiles, must provide large parking lots to attract buyers. This view shows the new, fully enclosed Pekin Mall in Pekin, Illinois, opened in 1972 and photographed in the winter of 1973. The main unit encloses almost 500,000 square feet and features two department stores, a large supermarket, many small shops, plus outlying service facilities. This picture also illustrates the contemporary movement toward the rural fringes of cities, the only place that such a large establishment can find ample space. The edge of the residential section is on the upper right, and a new trunk highway cuts across the rectangular land survey pattern. The farmstead at the lower left is now out of place since it has been shorn of part of its crop land.

Photo 10. This retail marketplace in Addis Ababa, Ethiopia, shows the wide spatial spread of a particular type of institutionalized marketing procedure in which sellers rent stalls in city-owned buildings.

Photo 12. Kennedy Air Terminal, New York, exemplifies a modern commercial landscape type that has spread rapidly around the world with the development of air transportation.

The high-rise buildings of the central business districts of the earth's great cities constitute a special facet of the commercial landscape. The central city cores the world over are growing more and more similar (Photo 13). There is literally nothing left of the natural landscape, for the creation of the urban zone has normally been a full-scale re-creation.

There are, of course, many other elements of commercial landscapes. It is just that some of them, when viewed individually, are less imposing than tall buildings. For example, taken collectively in their distributions along highways near American and Canadian cities, the urban-fringe "motel rows" are becoming significant parts of the urban scene. This particular landscape element is not yet prominent in many other areas of the world. However, the automobile, road systems, and vehicular travel traits continue their present diffusion, "motel row" may well become an increasingly common element of commercial landscapes everywhere.

As the modern habit of world travel increases, the "hotel row" is becoming another important aspect of the commercial landscape. In the past, hotels have been located within or around the central business districts where they have blended in. Today, in many urban areas, there is a growing tendency for transient resident facilities to group themselves together into hotel complexes (Photo 14). The buildings themselves are visually prominent in the landscape, and the accompanying facilities—the gardens, swimming-pools, dining areas, tennis courts, and other recreational elements—are often significant additions.

Photo 13. The Plaza Colon is a key point in the central business district of Mexico City. Around it, the high-rise buildings form one of the more modern aspects of the commercial landscape.

Photo 14. Hotel row is another aspect of the commercial landscape that is beginning to dominate some areas. This view is of the beach front at Miami Beach, Florida.

INDUSTRIAL LANDSCAPES

A third category—the industrial landscape—is as old as manufacturing itself, but it is only since the industrial revolution began in the seventeenth century that this category has significantly intruded upon other cultural landscapes. In the modern pattern, there are two aspects of the industrial landscape that are spatially notable, resource-conversion and resource-extraction. An example of resource-conversion is the modern integrated steel plant. It is typical of early to mid-twentieth-century industrial maturity, and its utilitarian spatial spread produces an impressive imprint on the landscape (Photo 15). When surrounded by the other elements of manufacturing plants, creating what economic geographers call a manufacturing region, the industrial landscape assumes widespread proportions.

Photo 15. This integrated steel plant in Pennsylvania typifies the spatial alignment of a mid-twentieth century utilitarian manufacturing operation.

The second aspect of the industrial landscape is seen in the landscapes dug out or torn apart by the extraction processes necessary for feeding raw materials into the manufacturing regions (Photo 16). The ancient scars, Paleolithic through the eighteenth century, are largely grown over. Their relatively slight imprints have been softened by time and biotic processes. The scars of the nineteenth and twentieth centuries, however, are spread more widely and deeply. They will require remedial human efforts of the same proportions as those that produced them if they are not to remain as long-term blights to the landscape.

Photo 16. Many extractive industrial landscapes end-up as disfigured and disorderly wastes, but the great open pit mines, such as this copper mine at Silver City, New Mexico, create almost neat terrace systems as the pits grow deeper. There are five ore trains and numerous giant shovels in this scene, all dwarfed by their surroundings.

Increased concern about the mutilation of the landscape is slowly beginning to produce a response in industrial operations. Decorative, restorative, and camouflaging efforts have been initiated to create pleasing and even artistic landscapes around industrial establishments and to recondition landscapes in areas of extractive operations (Photo 17). In the past, mankind paid little heed to what many now consider to be the mutilation of the landscape, and old cultural habits may be hard to break. The costs of breaking the old habits involve far-reaching recalculations of cost accounting in many phases of economic production, and these costs may slow the adoption of new procedures by industry. However, there is no intrinsic reason why man cannot produce industrial landscapes that are not mutilated.

In earlier times on an earth less culturally controlled, no group paid serious attention to maintaining nature-like landscapes in areas where it carried on economic functions, since there remained so much wild land. However, mankind has long dressed and beautified selected elements of the cultural landscape. Parks, gardens, historic memorials, selected thoroughfares, and sometimes cemeteries and religious shrines have long been places where people often improved on nature. As the wild landscapes shrink to remnants, new cultural habits and new technologies must be brought to bear on the problems of creating and maintaining landscapes that are pleasing in the visual sense. Perhaps the tide is beginning to turn—finally.

Photo 17. The modern design of light industrial parks provides architectural styling and landscaping that add scenic beauty to an urban environment.

CHAPTER 7
ADVANCED RESOURCE SYSTEMS

At this point in the book it is appropriate to have a review and a preview. The objectives of Chapter 2 through 6 were the discussion of:

1. The early relationships of the human race to an earth in change;

2. The beginnings of different but basic cultural technologies;

3. The growth of different kinds of culture systems;

4. The formation of regionally centered ancient civilizations;

5. The appearance of a late and very different regional civilization; and

6. The linkage of the earth into one great operating system.

The approach used was evolutionary, and the chapters were written to be read in sequence, building understanding on the background of human development and the growth of human culture toward an appreciation of the highly variable conditions prevalent in the modern world.

Most of the balance of the book, Chapters 7 through 10, is devoted to discussion of contemporary conditions and the patterns of development in the modern world. The next four chapters may be read in any sequence, according to the interest of the reader, because they are parts of a whole, and it is the grasp of the total view of the modern earth that is important. The approach used in these chapters is analytical, and the objectives are an understanding of the nature of:

7. Resource extraction and processing, and the character of continuing industrialization (Chapter 7);

8. Spatial relations on the earth resulting from the impact of the space-adjusting and space-transforming processes of transportation, communication, and urbanization (Chapter 8);

9. The explosive growth of modern populations and the relationships among the members of the present huge total of human beings (Chapter 9); and

10. The contemporary variety in levels of culture under different systems in separate parts of the earth as a result of the operations of all the ancient and modern processes (Chapter 10).

In whatever order they are read, Chapters 7 through 10 are a necessary background to the final pair of chapters in this book. Chapters 11 and 12 consider the problems of human control over the physical, biotic, and cultural worlds as we approach the twenty-first century on a planet on which human ingenuity must be applied as never before if life is to continue in its full state of richness.

THE CONCEPT OF RESOURCES

Things that become resources vary according to the levels of culture and technology of those people seeking and using them. Processes of finding and extracting resources differ markedly under different technological procedures. The conversion of raw materials into useful products varies both according to manufacturing technologies and cultural values. Mankind has always been manufacturing something, from the first tools of stone and wood to the complex products of today. The concept of what constitutes a resource has been growing right along with the techniques of processing those resources. This chapter reviews the evolving processes and changing technologies around the subject of resources—their extraction as well as their conversion into manufactured goods useful to different levels of consumption systems.

A forest is a resource to mankind at any cultural level, from the Paleolithic to the present, but its values differ markedly according to the way in which each level of culture views it. The value of a forest to Paleolithic people lay in the directly usable natural products and in the utility of a forest as a refuge from enemies, but the forest formed a hazard in that predators and enemies could conceal themselves in it. To

Neolithic peoples the same old values and hazards remained, but the land under the forest had value as garden sites for growing crops, and the trees held little value in themselves. In feudal England wild game was legally reserved for the nobility and the king, but townsmen valued the forest for its source of firewood, charcoal, and construction timber. Rural English farmers placed the value on the land under the forest for potential cropland but saw relatively little value in the forest itself. To those escaping the bonds of feudalism, such as Robin Hood, the forest was useful as a refuge, and it provided wild game as a food supply.

The forests of seventeenth century New England were valued differently by the native American Indian population than they were by the immigrant Europeans. The Indians employed shifting cultivation in producing garden crops so that the trees had no real value, but the wild game and plant foods that could be collected from the forest were valuable. The European immigrant valued trees as commercial timber for sale to Europe, valued the land under the forest as potential cropland for agriculture, but also placed some value on the wild game and foods that could be collected from the forest. By the twentieth century New Englanders have developed new concepts as to the value of a forest. There is scenic beauty in a forest, and forests have become places to go on picnics and camping trips. Trees have high value now as a source of lumber and woodpulp for making paper. Forests now also have value as protective watersheds for city water supplies, for erosion control, and as a symbol of wild nature.

The cultural viewpoint toward elements of environment has steadily changed all through human history, and the concept of what constitutes a resource has also changed. The definition of resource, then, becomes a matter of continued rethinking about the possible utility of all materials occurring in the physical and biotic worlds. Rock, sand, gravel, soil, water, minerals, petroleum, and even air are subject to changing viewpoints. The value of air and water as resources has been fully understood only very recently in some of the more heavily populated places on the earth. Mankind began life taking the whole of the environment for granted as free and inexhaustible, but the earliest peoples understood only the very simplest elements of environment. For a long period early mankind lacked technical expertise for identifying or using, or even for comprehending the possibility of, the more complex potentials of resources. No product, natural or created, thus becomes a true resource until its utility is properly understood, its technology becomes mastered, and its output becomes utilized by people somewhere on the earth.

FORESTS: WILD HARVEST AND FARM PRODUCTION

All the temperate, humid areas of the earth once possessed natural forest growth, but today not more than 10 percent of the land surface of the earth is in productive wild forest in which reproduction is purely natural (Figs. 7.1 and 7.2). Another 10 percent of the land surface is now in spoiled or cutover condition that is quite unproductive. About 5 percent of the land surface today carries good forest growth that is chiefly planted or managed by governments or private organizations. This is forest on which growth is stimulated or controlled by cultural processes, and this figure is increasing steadily.

It is still commonly said that forests should be kept intact and left to wild biotic processes. Conservationists and urban nature lovers advance numerous reasons for this. There is, unquestionably, some emotional value for highly urbanized societies in retaining some areas of purely wild vegetation that remain under purely natural biotic processes. Human pressures on the earth, however, are becoming increasingly heavy, and such areas cannot be significantly enlarged. There is another side to this, however. All forests grow old, their great trees die, insects and diseases attack them, and there is ecological pressure upon them when slight changes take place in environments through purely natural processes. A forest left entirely alone does not go on endlessly increasing in either economic or scenic value.

The utilization of forest products ranges even more widely today than ever before, owing to technological changes and to the increasing rates of consumption by an ever larger population. Beyond wood in bulk units as lumber, the increasing need for woodpulp is matched by the demand for hundreds of varieties of gums, resins, oils, and other chemical compounds. The Paleolithic technique was to gather natural seepage of "juices" from wounds on trees, and to cut selected trees. The cutting of wild forests for their products, today, is a specialized kind of holdover technique from the Paleolithic. Attacked by modern power machinery, the remaining wild forests will not last much longer, unless preservation measures are urgently taken everywhere around the earth.

Mankind is beginning to learn how to manage forests productively. Through forest management practices, foresters are becoming able to improve on nature to produce wood products in the same way that Mesolithic mankind improved on nature by domesticating crop plants to produce higher yields. The managed forest and planted tree farm are part of progressive forest economy all over the earth today. There are many regions that still lag in these

This tree-farm nursery grows almost eight million seedlings per year. About half of them will be transplanted into commercial tree-farm forests. The seedlings shown are *Pinus ponderosa*.

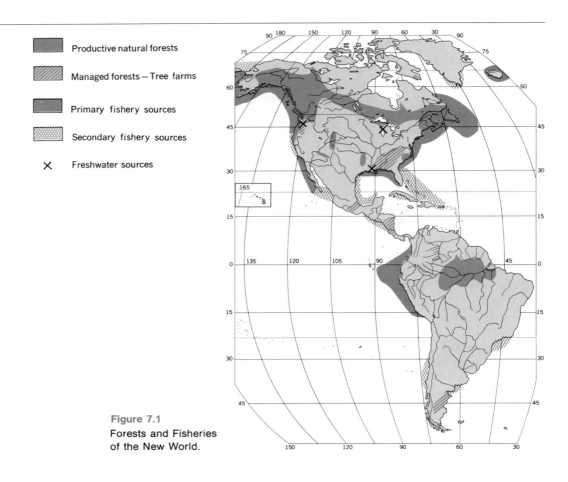

Figure 7.1
Forests and Fisheries
of the New World.

Legend:
- Productive natural forests
- Managed forests — Tree farms
- Primary fishery sources
- Secondary fishery sources
- ✕ Freshwater sources

procedures, but the techniques are diffusing steadily. The planted government forest preserve and the privately operated tree farm are becoming the effective way to grow wood products in volumes adequate to the increasing needs of large populations (see Figs. 7.1 and 7.2). Forest trees are being "domesticated" into better producers of woods, gums, saps, resins, and latexes. Whereas the Neolithic-style collecting of wild rubber must have averaged well under an ounce per acre, the selective breeding of high-yielding tree stock in the natural rubber plantation in the tropics today yields over 2000 pounds per acre. Timber tree stock is being produced today to yield more wood in shorter periods than can ever result in natural wild forests.

What is termed "sustained-yield" forestry is now reaching high technological levels of achievement. Although such patterns are not yet world wide, such areas are producing larger and larger shares of the harvested wood product. It is predictable that within the twentieth century adequate forest areas can be put into production as tree farms that not only will produce the wide range of forest products required by modern societies, but also will produce them more efficiently than nature ever did. In so doing, "wild areas" can be re-created where and as desired, although the planning for this kind of development has not yet progressed very far.

This does not mean that the remaining wild forests can be carelessly handled, but that management procedures should be applied to them that are as efficient as those now employed on tree farms. The development of modern tree farm technology does mean, however, that there are known cultural techniques that can begin to produce a renewable natural resource before all the world's wild forests have been cut. It also means that there is now available the technology for maintaining whatever patterns and designs of vegetation systems are desired, from parklands to wildernesses, providing the world's peoples will apply the proper procedures.

Not only are foresters learning how to produce more and better trees in shorter time periods, but technological advances in handling harvested timber have been going through evolutionary improvement. Where once only the main-trunk big logs were utilized, the modern wood processing plant converts a larger share of the whole tree, and a goodly share of the former wastage is utilized in the wood products, paper, and chemicals-plastics industries.

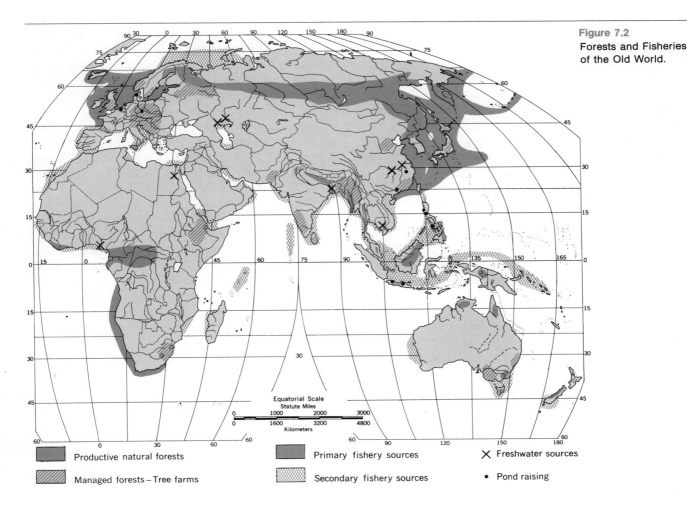

Figure 7.2
**Forests and Fisheries
of the Old World.**

Productive natural forests

Managed forests – Tree farms

Primary fishery sources

Secondary fishery sources

X Freshwater sources

• Pond raising

MINERALS AND THE MINING CYCLE

Although we use the terms Copper Age, Bronze Age, and Iron Age in historical chronologies, the true age of minerals belongs to the nineteenth and twentieth centuries.

The Distribution of Minerals

Minerals occur all over the earth, both in the solid crust and in and under the oceans. Some occur as single elements, such as free (pure) gold or copper. Most minerals, however, are compounds of elements, such as hematite (an iron ore, ferrous oxide) or bauxite (an aluminum ore, aluminum oxide). Sometimes minerals occur in high-grade simple compounds, forming a large percentage of the "rock," and these are cheaply mined and smelted. Other mineral ores are complex compounds, or occur in only low-grade deposits, and these are more difficult and costly to smelt, or they may require the process-

ing of large amounts of rock in order to secure a significant quantity of the mineral. In the case of some of the newly discovered minerals, such as uranium or petroleum, the finding techniques are still being perfected, so that the distribution of the resource is not fully known.

In many areas that have been until recently described as mineral poor, the local inhabitants were not much interested in using minerals, so that they carried on little search for them. Modern mineral prospecting in such regions often turns up considerable resources. For example, Thailand formerly was described as one of the least mineralized regions of the earth. Recent searching, however, has disclosed large deposits of several minerals. In the same pattern, several other countries, having no old history of mining and smelting, now are being found to possess considerable mineral wealth.

The whole earth, in fact, forms one very large stockpile of minerals. The poverty or wealth of any political state is related to its physical makeup, but wealth is also dependent on the cultural awareness of the populations of the state and their mastery of mining and smelting technology. Since many culture groups have, historically, either been little interested in mineral technologies or have learned them only recently, our knowledge of the total mineral resources of the earth is still far from complete. When we add to this the matter that some minerals have been discovered only in the last century, it becomes clear that the matter of mineral wealth is indeed a question of cultural awareness and of technological development.

The Mining Cycle

Mineral deposits that are mined and smelted are used up, in the sense of being dispersed, and minerals constitute a nonrenewable resource, unlike forests, which can be grown again. Most of the earth's easily mined very rich mineral deposits have been worked over repeatedly, although occasional discoveries of very high-grade deposits are still made in areas not formerly explored carefully. There is, however, a cyclic pattern to historical production of minerals in any one region. In practice this cyclic aspect is almost wholly a matter of changing technology. This can be illustrated in the following way.

The first copper miners (approximately 4000 B.C.) mined copper that they found in the naturally pure state in surface deposits, and they hacked out pieces of copper with stone tools. Not understanding the smelting of copper ores, those first miners ignored the high grade copper ores nearby, abandoned the site when they no longer could dig out pure copper, and hunted for some other site where pure copper occurred. At a later date copper miners, who had learned to smelt high-grade ores in simple compounds, came upon the site and mined some of the ore body, but they abandoned the site when the ore became low grade deep in the ground. Still later, other miners who had learned to smelt more difficult compounds or ores of lower grade came upon the site and mined the medium-grade ores, only to abandon the site when the ores of medium grade ran out. Early nineteenth century copper miners had learned to pump water out of mines and pump air into them, and could go deep into the earth for any ores over 4 percent copper. But mining ceased when the ores dropped below 4 percent, and the mining field was again abandoned. By 1920 smelting technology had improved to the point that ores of 2 percent could be mined and smelted profitably, so the miners returned to work over the old slag heaps.

Today copper miners accept as "copper ore" any large rock pile that contains at least one half of one percent copper, but at that low grade the ore deposit must be a large one to recover very much copper. And so the cyclic pattern has gone: mine to the limit

of technology and abandon the site. Return when technology has improved to the level that the ores can be worked, but abandon again when the limit has been reached. Ghost towns mark such abandoned sites, but rarely does this mean that all of the mineral sought has been removed.

Increasing volume of production has been another element in the mineral cycle. All the copper miners of the earth, in 1500 B.C., probably handled 200,000 tons of rock while taking out 100,000 tons of copper ores, from which they smelted perhaps 20,000 tons of copper. Copper miners today work ores ranging from one half of one percent to about 10.0 percent copper, but they must handle hundreds of millions of tons of rock and ore to produce the 4,000,000 tons of copper each year. Such production patterns have come to involve very complex mining operations and very costly high-quality smelters, so that modern copper mining can only be carried out on a large scale. It is totally uneconomic, today, to mine ore bodies of only a few thousand tons of ore.

Until the nineteenth century a single miner could produce sufficient high-grade ores, in any of the common minerals, to keep a small smelter in operation. At present this can take place only in a few places with a very few minerals when new finds are made. Working with low-grade ores requires large labor forces, costly mining equipment, and costly and complex smelting facilities. Most mining operations today, therefore, involve large corporations employing thousands of miners and smelter technicians, with capital investments running into many millions of dollars. Many mining operations involve recovery of ores containing small amounts of several minerals.

Thus the development of the long-term mineral cycle has involved not only the spread of mining in repetitive patterns all over the earth but also the repeated shift to larger and costlier kinds of operations, in which greater numbers of people are involved and in which the technologies have periodically increased in complexity. The solid crust and ocean volume of the earth still contain huge quantities of even the common minerals, but no longer are they so casually available. Although there is a conceivable limit to the potential mineral resources of the earth, expanding technologies should be able to cope with the problems of recovery of minerals for quite some time to come. The pessimistic view places the limit at something over a century, whereas the optimistic view puts it much further in the future. If improved recycling techniques could be devised at reasonable costs for a larger range of minerals than are so treated today, there could be a significant lengthening of the time during which present reserves would last.

AQUATIC PRODUCTS: SEA HUNTING, FISHERIES, AND FISHING REGIONS

Among the world's peoples there probably always have been fishermen, and it is likely that there always will be, because fishing can be done both for economic gain and for pleasure. Fishing has long been worldwide. The human interest in fishing has varied among the societies of the earth, and some peoples have declined to fish, although they have inhabited regions with potential resources. The dietary consumption of fishery products has been culturally implemented. All over the earth Roman Catholics have consumed fish at least once a week. Upper-caste Hindus, on the other hand, being vegetarians, never eat fish. Late in the nineteenth century Americans began reducing their consumption of ''fishy taste'' fishery products, but quick-freezing techniques are now lifting the American consumption levels. Scandinavians and Filipinos eat large amounts of fish in many forms, and the Japanese eat more kinds and larger volumes of fishery products, including aquatic plants, than any other people.

Those continental shelves along which cool ocean currents flow are zones possessing extremely rich fishery resources (see Figs. 7.1 and 7.2, which include all but the Antarctic shelf zone). These zones are cool-water regions in which minute forms of life (plankton) provide so tremendous a volume of food for many kinds of fish and shellfish that these regions support enormous fish populations. Such regions have very large bird populations that feed on the fish, but such birds are too tough and poor tasting to be useful as human food sources. On the other hand, the large populations of such sea mammals as the seals, sea otters, sea lions, walruses, and whales that have lived in these waters have been important as sources of food supplies, furs, ivory, and whalebone and whale oil.

Sea animals have been hunted along some oceanic shores since at least the late Neolithic, but their importance greatly increased after the Columbian Discoveries. European ships began roaming the world's continental fringes engaged in whaling and fur-taking, and the Japanese entered the commercial sea-hunting trade in the late nineteenth century. By the end of the nineteenth century the large populations of furbearing sea animals had been so reduced that this kind of sea hunting has now practically ceased. Whaling continued for some decades into the twentieth century, but the populations of whales have been very greatly reduced, and only Japanese and Soviet ships continue this form of sea hunting. International rules on the restriction of whaling may just have come in time to prevent the extinction of whale populations.

Many freshwater areas of the continents formerly possessed significant aquatic populations that were very useful to earlier small human-culture groups in providing food supplies. In the last few centuries, many of these local regional waters have been overfished to the point of near extinction of fish species. In some areas not overfished in earlier time, modern industrial pollution is making local waters quite unsuitable for fish life. This threat is continuing to expand its serious impact. The breeding of selected fish types in hatcheries permits the restocking of mountain streams for sport fishing. The pond raising of fish and aquatic plants, first seriously begun in south China, has now spread to several parts of the earth, and forms a possible future pattern of development in areas where fresh waters can be kept clean enough to permit such fish farming.

The modern deep-water fishery activity first began in northwestern Europe on the North Sea. Fishermen gradually worked all the eastern waters of the North Atlantic Ocean, and perhaps even before the discovery of North America they began working the shallow banks off what is now maritime eastern Canada and northeastern United States. In the nineteenth century Canadian, United States, and Japanese fishermen began working the zones of the North Pacific Ocean. More recently Peruvian waters have been fished commercially by fishermen of several nationalities, and Peru has become, at least temporarily, the leading commercial producer of fishery products. With the development of freezing equipment, large fishing craft now work most of the important oceanic zones, including such open-ocean areas as provide fish catches. The introduction of resource surveys, power equipment, refrigeration equipment, and the quick-freezing and canning facilities have greatly expanded fishery activities around the world. Improved land transport systems are moving fishery products inland from all the world's coastal zones.

The annual fishery-product catch and take is slightly more than 65 million tons per year for the entire earth. Although the yield can be increased, the resource is not vast. We must share the natural production with other top-level carnivores (other fish, birds, seals, squid, etc.), and leave a breeding stock for resource renewal. The potential sustained annual yield may be a little under 100 million tons.

What is modern about the continued employment of the ancient techniques is the use of modern materials to make them more efficient and the application of power to the operation of the boats and the gear, the new technologies in handling the take and the catch, and the processing of the yield. Early man-

The Polish vessel Pomorze is the base ship for a Soviet-bloc fishing fleet, seen here off the coast of Virginia. Tied up alongside is one of the modern trawlers that make up the fleet. Such combined fleets can stay at sea for months at a time.

kind fished fresh waters and the edges of the sea by guess and by intuition; ninth-century European fishermen began working the deep offshore waters through the accumulation of trial-and-error experience in the design of boats and fishing gear. Today, large long-distance fishing craft hunt by sonar devices through the open oceans aided by resource surveys.

Despite present modernization of commercial fishing, the revolution in world fishery techniques has barely begun. Of the perhaps 50,000 species of fish, shellfish, sponges, and related forms, relatively few are being utilized so far. Processing techniques can render many forms much more acceptable by consumers. If the modern economic pursuit of aquatic products can be restrained within the ecological balance of the natural generation of an economic good, fisheries can be a permanently productive component in any regional system.

If the competitive pursuit of the naturally productive fishery activity cannot be restricted within patterns of natural generation of fish populations, it may be necessary to alter the whole economic approach from a system of "taking" a natural resource to the technique of "farming" a resource within nationally controlled oceanic waters. The practice of "farming the oceans" is now discussed rather often, but only a few steps have yet been taken in this direction. The presently operating fish-pond assembly linked with a cannery and a quick-freeze plant promises a kind of controlled economic production that could revolu-

tionize, finally, the whole fishery industry. Controlled-water, shallow-sea "farming" of many kinds of aquatic products is not yet in extensive practice, but the technology is ready for widespread use and is catching on.

WATER SUPPLY: THE REALIZATION OF A RESOURCE

For the world as a whole, the following generalization still expresses the subjective human attitudes toward water. In the humid lands water is cheap, there is lots of it at all times and often too much; the seasonal flood is more serious than the unseasonal drought; water is the ever-present free economic good; no special systems of control over it are required other than flood prevention; and no care need be taken to keep it clean. In the dry lands water is dear; there is never enough of it; concerted efforts are needed to ensure its continued adequacy; untimely flood is feared but not fully guarded against; water is a free economic good (but special systems of control over it are required) administration of its use is specially arranged; and great care is taken to maintain the purity of supplies.

Out of a somewhat sudden rash of water problems, however, in the humid lands mankind is becoming

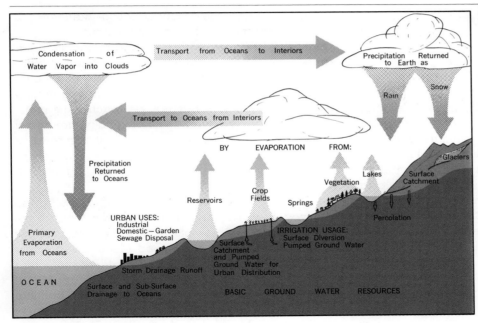

Figure 7.3
The Hydrologic Cycle in Cultural Usage.

aware of past misconceptions and the dangers of past carelessnesses. This is nowhere more dramatically illustrated than in the drastic water shortages for New York City during 1964–65, a great city in a humid region of free and plentiful water, as compared with the continued relative adequacy of water supplies for the city of Los Angeles, a growing great city in a dry region of valuable and scanty water. What is more, the waters of portions of the humid northeastern United States were terribly polluted through the casual dumping of wastes, so that many free-flowing streams almost resembled sewers in the early 1960s, whereas wastes are almost never dumped into the scarce water bodies of the dry lands. The American situation provides more startling contrasts than many other portions of the world, but it is clear that a wholesale revolution in thinking must take place about the control of fresh water in the cultural landscape. Fresh water is a free economic good that may be taken casually and without much care in only a few portions of the earth today. In most places fresh water is a valuable resource to the large world population, but it must be handled with the care appropriate to its value. What may be termed the technological revolution in modern living systems demands a revolution in the thinking about fresh water, in the storage and handling of fresh-water supplies, and in the whole attitude toward groundwater and water in lakes and free-flowing streams. If we now, in all parts of the earth, face some decades of highly irregular weather patterns, as predicted by some climatologists, the problem becomes urgent since both dry and humid lands may then experience droughts of far greater severity than those experienced in the past.

The Costs of Water Technology

Water has traditionally been thought of as a free economic good, in the sense that water is a natural part of a physical environment. The growing population of the earth and the growing complexity of cultural practices involving water have made the procurement, transport, treatment, and distribution of water increasingly expensive in all parts of the earth (Fig. 7.3). The concept of the watershed and its disciplined control have become truly effective in only a few regions within the last century. The current spread of this concept around the world today is part of a cultural process of the comprehension of a significant environmental resource. Water supply in most regions has remained a local matter, seldom highly organized, with little thought for other users within the same drainage basin. It is inevitable that cultural practice must improve tremendously in many humid regions, and around the arid margins, as the population of the earth increases. The traditional view of water as a free economic good, without liability, must be replaced by the view of water as a valuable environmental resource to be employed carefully for the good of all.

The primary developments in the whole field of water supply have become technologically mature within the last few decades. Since 1900 we have learned how to handle water in huge volumes in almost all of the ways needed. Even the desalination of nonfresh water is technologically under control, although it may yet cost more than we like to pay for water. It is human thinking and human practice that have not recently kept pace with human technological ability.

AGRICULTURE: GLOBAL INTERCHANGE, SPECIALIZATION, AND INDUSTRIALIZATION

One of the more significant results of the Columbian Discoveries was the hemispheric and continental transfer of crop plants, domestic animals, and crop-growing technologies. The basic grain, fiber, root, and fruit crops of the Old World were moved about and transferred to the New World and to Australasia. New World crop plants became basic staples in parts of the Old World; domestic animals as draft power and meat producers were scattered over the earth; and agricultural technologies were widely transplanted. The interplay of these several elements of agricultural practice was a continuing operation, as new and relatively empty lands were put under crop. Old cropping systems were renovated, and trade movements in basic commodities began to expand into a worldwide pattern. Labor and power shortages slowly began to be compensated by increasing mechanization, and crop growing gradually shifted from subsistence and local barter to distant market-oriented commercial patterns.

The accumulation of knowledge and technologi-

cal skill implicit in the occupational specialties of soil scientist, agronomist, horticulturist, plant hybridizer, geneticist, irrigation engineer, land reclamation engineer, and agricultural engineer represent qualitative improvements derived from the late nineteenth and twentieth centuries only. The continuing revolutionary development has not yet reached its maximum even in the most technologically advanced agricultural regions. More serious, the spread of contemporary knowledge has not yet reached the far countries of the earth; thus many are lagging far behind.

In the evolutionary development of agriculture in the modern world, the most advanced patterns have improved on the three old basic systems to produce Livestock Ranching, Developed Mechanized Farming, and the Modern Corporate Plantation (see Fig. 4.3, page 89). When any of these appear at their maximum scale of development in political states under Communist control, they take the form of Collectivized Managerial Agriculture. These modern types of agricultural production have reached high levels through the application of power tools, fertilization, disease-pest controls, and the breeding of hybrid crop plants and new varieties of animal stocks. Developed Mechanized Farming and Livestock Ranching have become well distributed in the

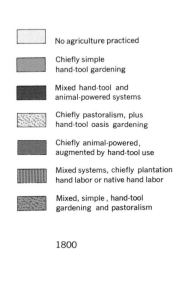
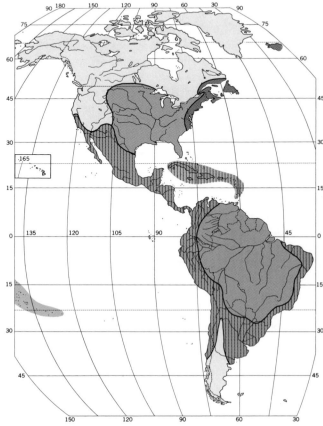

Figure 7.4
Systems of Agricultural Technology, A.D. 1800.

higher midlatitudes, and the Modern Corporate Plantation is well established in the humid subtropics and tropics. In some of the higher midlatitude regions there is coming to be less and less distinction between Developed Mechanized Farming and the Modern Corporate Plantation in management and operational characteristics.

As these modernizing developments have taken place, groups of crop plants and/or animals have been assembled from around the world into combinations that are more productive than the old regional groups ever were. This has been a slow but steady process, and one that has changed the combinations within any one region from century to century (Fig. 7.4 and 7.5). For example, in the late 1970s economic geographers refer to the American Corn and Soybean Belt as a distinctive agricultural region. Involved in the annual production cycle are corn (maize) from Middle America; soybeans from North China; wheat, barley, and oats from southwest Asia; leguminous and grassy forage plants from several regions; a number of vegetable plants from both the Old World and the New; the pig and cattle from the Old World; and the chicken from Southeast Asia. This assemblage is integrated in a very efficient manner to use land alternately and to produce particular outputs of crops or animals according to market demand. This is a highly mechanized agricultural system, employing rotation systems, fertilizers, new hybrids, and watchful attention to the market, to form one of the earth's most productive agricultural regions in this century.

REGIONALISMS IN RAW MATERIALS, POWER, AND FOOD

Economic geography is concerned with the regional distribution of production of primary commodities and their movement to processing and consuming centers. Out of the earth's environmental regionalism, certain patterns emerge as natural conditions that mankind has accepted. Although, technologically, bananas could be grown in Antarctica today, it seems not worthwhile to pay the price to do so when there are so many tropical localities in which bananas will grow easily and cheaply.

In the realm of primary production there are several axioms worth stating that affect the spatial distribution of production. In the area of mineral commodities, the normal sequence has been to begin with the richest sources processable by the easiest

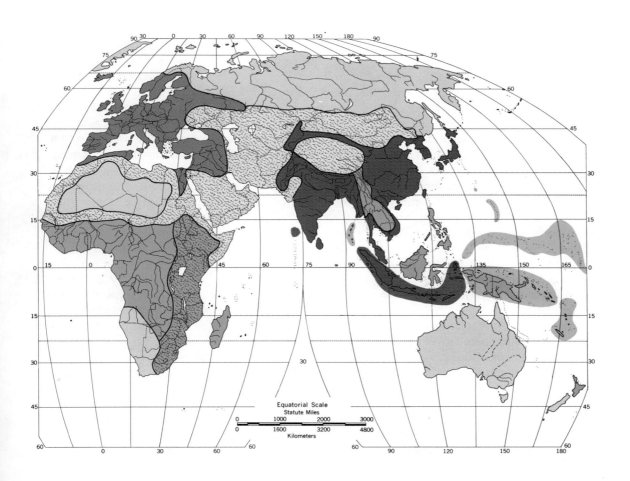

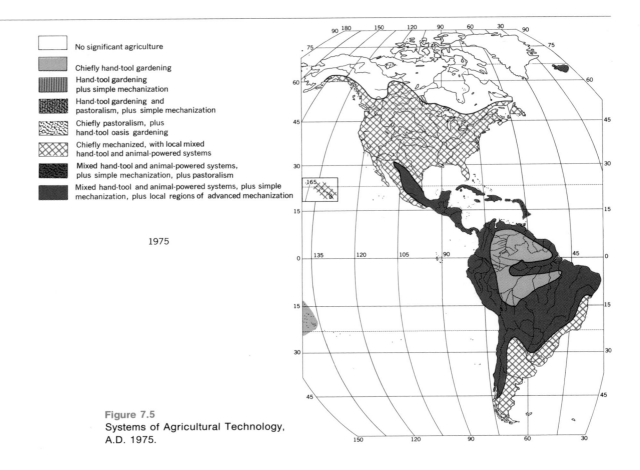

No significant agriculture

Chiefly hand-tool gardening

Hand-tool gardening
plus simple mechanization

Hand-tool gardening and
pastoralism, plus simple mechanization

Chiefly pastoralism, plus
hand-tool oasis gardening

Chiefly mechanized, with local mixed
hand-tool and animal-powered systems

Mixed hand-tool and animal-powered systems,
plus simple mechanization, plus pastoralism

Mixed hand-tool and animal-powered systems, plus simple
mechanization, plus local regions of advanced mechanization

1975

Figure 7.5
Systems of Agricultural Technology,
A.D. 1975.

and cheapest means, and to proceed to utilize less rich sources only when a new technology has been developed. In the zone of power production, theoretically, mankind has always sought to shift labor from his own muscles to some other source of power. Here people have often been controlled by cultural habit.

In the production of agricultural commodities, it is an axiom that the largest volume and the highest quality of crop yields today are not in the original ecological home of the crop plant, but somewhere out toward its ecological margins. On such a margin cultivators have improved the plant and bred varieties that will tolerate conditions not tolerable in the homeland. Here the farmers often have to play nursemaid to the crops to secure that maximum yield of high quality. It is also a notable axiom that culture groups capitalize on the natural regional monopolies in minerals, technological skills, and crop/animal products. However, there is a corresponding axiom that other people in other regions will break a natural monopoly at the first opportunity, be it mineral resource, technology, crop product, or manufactured product.

Acceptance of certain axiomatic rules of procedure, the persistence of doing things in culturally determined ways, and the technologies devised to improve on nature have all affected the spatial pat-

terns of raw material supply. Also affected were the sources and locations of power supplies, the centering in production of food resources, and the technologies of resource extraction from the earth. The breaking of natural monopolies in recent centuries has led to the almost worldwide diffusion of the resource-extracting techniques. Rates and paths of diffusion condition the spatial distribution of many of these techniques.

The post-Columbian transfers of plants and animals operated very rapidly during the sixteenth through the early nineteenth centuries, but many items have remained as monopolies until the twentieth century. Modern industrial chemistry has increasingly been making inroads on such monopolies. Some old monopoly items finally achieved a world trade distribution, only to be replaced by synthetics as products of industrial chemistry. Chinese tung oil and pig bristles, for example, have been replaced by synthetic lacquers and nylon bristles, and though there still is some trade in camphor, most of the volume used today is a product of industrial chemistry. Industrial chemistry has not yet replaced all plant raw materials, and it is likely that many of them will continue in production in their traditional homes, at least for local uses.

By linking the whole earth into one operating eco-

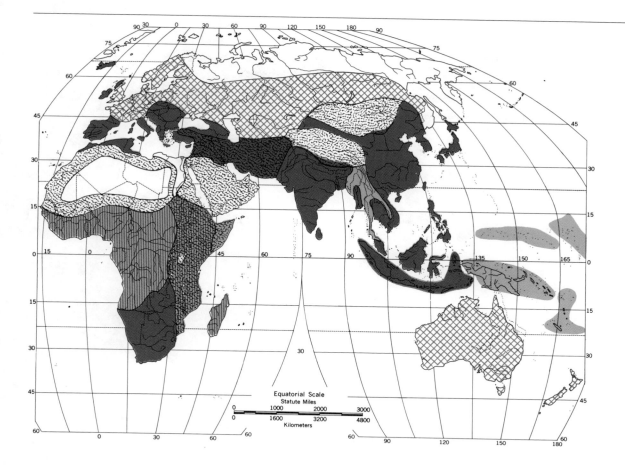

nomic system as a result of geographical explorations from the sixteenth to eighteenth centuries, and by engaging aggressively in the early patterns of world trade, the northwestern countries of the European culture realm gained an initial and competitive economic advantage over the rest of the societies of the earth. Accompanied by aggressive political attitudes taken abroad, and by colonial settlements in widely scattered regions, northwestern Europeans established early a significant control over patterns of economic and political development. As industrial technology and economic capitalism matured in European countries during the nineteenth century, the United States (and Canada to some degree) shared in the economic advantages. The maturing of the European colonial empires permitted a very wide pattern of development of natural resources around the earth, a pattern that continued without interruption until the end of World War II.

Power Supply

Here the term power (units of energy per second, expressed as the standard kilowatt [kw]) is interpreted very broadly as any means of performing work. At the highest level of technology, mankind has come a long way from the time of human muscle as the only applicable power source. In many parts of the earth, culture groups and societies have yet to apply much of the available power technology. The sources of power are many, but the problems of utilizing some of these sources are cultural in that technological procedures are involved. Power-creating technologies now range rather widely, but most of those in use are still rather elementary, and the problems of widespread application are involved (Fig. 7.6).

The human approach to the development of power began with the human body as the source of power. Through the domestication of the larger animals people gained another power source, and animal power still is a much-used source in many parts of the world, but it is destined to decline in importance. Through the windmill and the sail, the winds were tapped for power for a period of time. From the first crude "gunpowder" mix of combustible materials compounded in China through the various "dynamite" compounds to the contemporary endothermic and thermonuclear mixtures, the work potential of the "explosive" variety and the jet thrust has steadily increased, both in ability and in efficiency, and perhaps we are on the verge of a revolution in power development.

In distributional terms of total power potential,

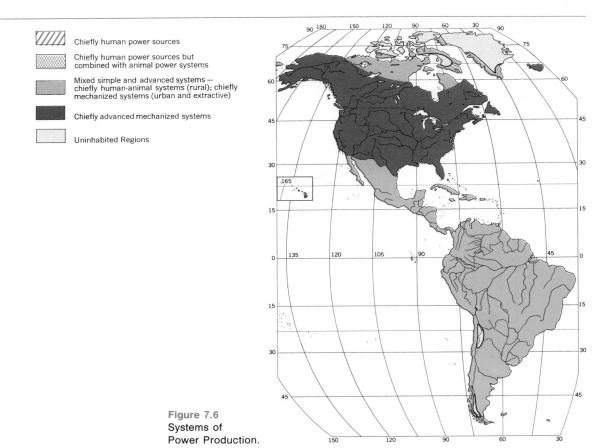

Figure 7.6
Systems of
Power Production.

power is available to every sector of the earth through some medium, but the cultural technologies are unequally distributed. In more conventional terms the northern hemisphere has far more than half the total fossil fuels, hydroelectric potential, and combustible plant materials, but it also has long had the larger share of the earth's population.

In economic and technological terms, the development of advanced power systems is neither cheap nor easily accomplished. The growth of power systems requires time, capital investment, and mature engineering skills. So far, capital funds and engineering skills have been applied chiefly in the countries already economically advanced enough to carry out such developments. There is therefore still a strong differential among the countries of the earth in power development, and it is necessary to find ways of helping underdeveloped countries create large-volume power systems while avoiding the charges of colonial exploitation.

Food Products

The zonality of world food agriculture in the modern period is compounded of three different factors.

1. There is a regional centering of production on climatic zones within which domesticated crop

plants remain controlled by regimes of insolation, temperature, and precipitation.

2. There is the transfer of crop plants about the earth from ecological homelands into similar regions elsewhere. This has produced a modern repetitive zonation of agricultural regionalisms around the earth.

3. There has been the tendency by crop growers to push crops into marginal ecological environments by breeding new varieties of old crops that possess new tolerances or by finding crop plants that naturally possess high ranges of tolerance to specific environmental conditions. Thus, in time some of the wheats have been transformed from a crop of the subtropical margin to a high-latitude crop, and wheat growing has pushed northward in both Eurasia and in Canada.

By the 1970s, under the impact of the three trends, there have been produced both duplication and heterogeneous mixture in world agriculture. The mixing of crop systems has been so subject to both human preference and human willfulness that the matters of transfer have not always been entirely logical in terms of environmental ecology.

Despite such historic changes in the traditional regional patterns of crop plants, there remains a broad zonality in world agriculture. It is less conti-

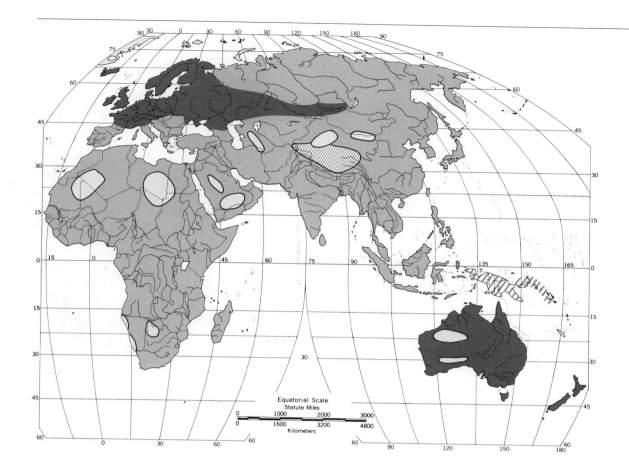

nental or hemispheric in design than zonation in accordance with the world's climates. The earth's tropics still yield agricultural products not produced elsewhere. The humid subtropics still produce particular commodities not found in the dry subtropics. The high-latitude regions yield crop assemblages that can only be duplicated in high-mountain country in the low-latitude areas of the earth.

The food agriculture of the earth today presents a broad but complex pattern of duplication and regional specialization (Figs. 7.7 and 7.8). Crop surpluses are available to move into the routes of trade to all parts of the world, but many food items are regionally available in many parts of the earth with only short-haul transportation required. There is every indication that the duplication will increase with the result that many items will be available to people everywhere. On the other hand, regional agricultural specialization is followed by specialized consumer patterns.

THE FUTURE
OF RESOURCES

Discussions of natural resources and their futures usually make several beginning assumptions:

1. *All resources will remain under the control of the political states in which those resources are now located.* This monopoly possession makes the future particularly bleak for states that are relatively poor in minerals, energy sources, or first-class agricultural lands;

2. *We now know the total resource reserves of the whole earth.* This neglects the matter of regions not carefully surveyed and the improvement in techniques of discovery and recovery;

3. *All judgments should be made within the limits of present conversion technologies.* This discounts the probability of future improvement in technology and the capabilities of scientific research; and

4. *Future resource demands can only increase greatly with continuing industrialization and under very marked population increases.* That both increasing industrialization and increased populations will be present in the future world is very clear, but the rates of increase assumed often are unrealistically high in the interests of the bleak picture.

The present statement is written from the viewpoint that the first three usual assumptions are false. We assume that in a single worldwide operating econ-

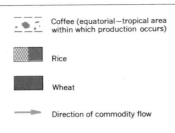

Coffee (equatorial—tropical area within which production occurs)

Rice

Wheat

Direction of commodity flow

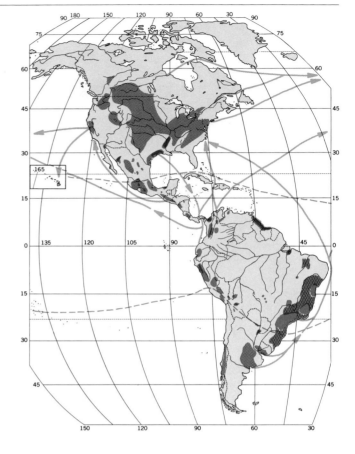

Figure 7.7
Food Shipments around the World.

Surplus production in total foodstuffs

Deficit production in total foodstuffs

Figure 7.8
The Adequacy of World Food Production.

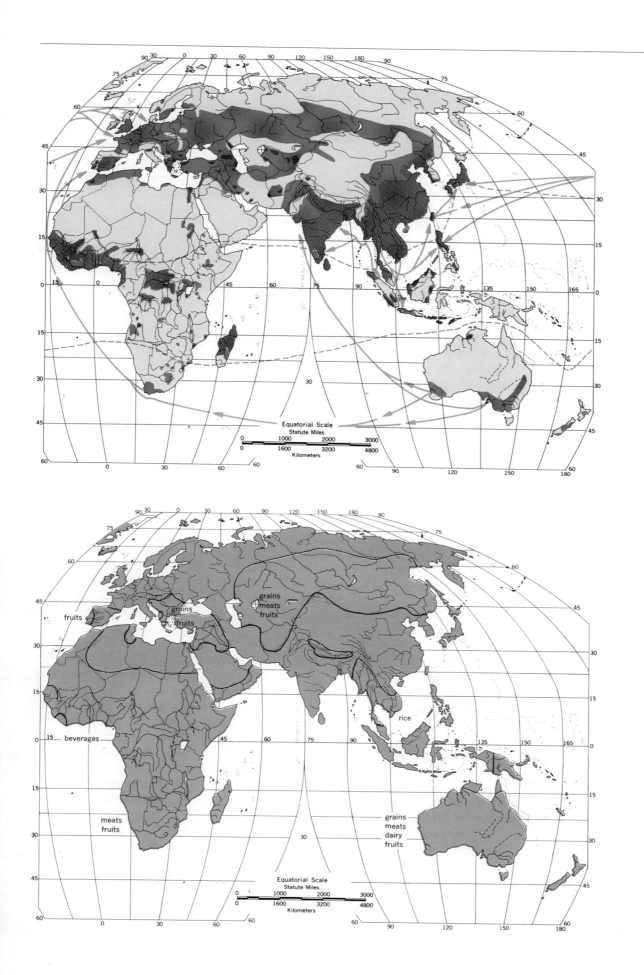

Equatorial Scale
Statute Miles
0   1000   2000   3000
0   1600   3200   4800
Kilometers

grains
meats
fruits

fruits

grains
fruits

rice

beverages

meats
fruits

grains
meats
dairy
fruits

Equatorial Scale
Statute Miles
0   1000   2000   3000
0   1600   3200   4800
Kilometers

omy ways will be found by which to share the earth's resources among those societies that can utilize them. This is necessary in order to focus on the basic question of whether the total resources of the earth have a near-future predictable point of exhaustion for a population that we assume will total seven to nine billion during the twenty-first century. We also take into account two kinds of resources often omitted from the discussion, namely, lands amenable to agricultural production and supplies of fresh water sufficient for all cultural purposes.

In the ultimate sense, the resources of the earth do have a limit, but mankind is not restricted to the use of traditional forms of resources known and used in the past. The whole crust of the earth constitutes a body of resources that will not be totally used up in the near future. As long as there are supplies of the known elements in the crust of the earth or in the oceans, then resources exist. And as long as solar energy remains available for development, the way is open for human intelligence to find a technology capable of extracting the materials and energy for use in any way that an aggressive and creative society may define.

For all the thousands of different mineral compounds and materials that are utilized as resources in some manner in the present world of complex technology, the really critical items come down to a relatively small number. Most of our resource materials are replaceable by substitute compounds in some degree. The matter of "irreplaceableness" is related to customary and habituated technology, by habituated economics, and to the fact that it has not been necessary in the past. Future needs may well require considerable change in habituated cultural practices that so far have not even been tried.

The list of basic resources (Table 7.1) is as short as it is because it is a list of fundamental items, the units of which can be combined in many separate ways in different mixtures. The list is short, also, because some of the items are chemical elements in themselves that can be combined in various ways. Probably no two lists of this sort would set down the same absolute number of critical items needed by present and future economic technologies.

On this list it may well be that the petroleums-natural gases may be the items in shortest supply, with the nearest date of exhaustion. Such an exhaustion will most certainly require significant changes in modern scientific industrial technology, but there will also have to be significant rearrangement of patterns of sharing the remaining volumes of such world resources prior to the date of true exhaustion. The mercuries, golds, and silvers have been utilized well past their level of easy and cheap retrieval, and it may well be that significant change in the patterns of usage will need to be made in the

Table 7.1 Materials Critical to Modern Industry*

Coals	Sulfurs
Celluloses	Sodiums (chiefly chloride)
Petroleums—natural gases	Glass sands (chiefly quartzes)
Radioactives	Kaolins
	Gypsums
Irons	
Manganeses	Coppers
Tungstens	Tins
Nickels	Leads
	Zincs
Nitrates	Aluminums
Phosphoruses	Mercuries
Potashes	Golds
Calciums	Silvers

*Stated as generic compounds, with specific combinations occurring in numerous constituent forms, chiefly as ores.

relatively near future, but the final exhaustion date for all of them is fairly far off. For the other items on the list, the problems are more those of improving the discovery-recovery techniques and usage technologies than true exhaustion points.

The reach of scientific theory, of course, poses problems, and some of these will be dealt with in a later chapter. Here, however, questions do arise out of previous statements. To whom do the earth's resources belong? To national political states as arranged by traditional rules? Then to whom do the resources of the world's oceans belong? What society is entitled to draw a supply of oxygen from the crustal section of the earth under the central Pacific Ocean? Which society is entitled to draw large supplies of heat from the deep interior of the earth? In the late 1970s, these, and other questions remain for discussion at international conferences. How long will such problems remain at this level when certain of the traditional resources reach acutely short supply? As long as resources are considered traditional items found *on* or *immediately below* the surface of the crust of the earth, then spatial political economics can wrestle with the problem to some end, including the use of traditional war in conquest of such needed spaces. As the issue of resources, however, turns to the issues of items to be taken from *deep within* or *under the crust*, then the traditional viewpoint becomes useless. From the standpoint of total earth resources there arises the ultimate need for a One-World kind of economic compact phrased in such a way that the traditional concept of monopolistic economic exploitation does not stand in the way of further cultural development of mankind as a species using the earth.

For almost all the items conventionally discussed as natural resources, the true shortages of the earth are not really dangerous to the continuation of human existence so long as the numbers of mankind do not become infinitely large. Much more critical, in fact, is the limit on the two items often omitted from the discussion. How much more good potential agricultural land is there that can be put into food-and-raw-material production to support a rapidly increasing total population? It is not land, alone, that must be considered, however. It is a case of land and the water to go with it. There is plentiful land in the dry parts of the earth where water is truly scarce. Desalination technology can now take care of sizable populations, for domestic purposes, in places where water is in short supply, but it is a process requiring considerable expenditure of energy. Sufficient fresh water for the endless expansion of agriculture is more than a simple technological problem. The two items of land and water, taken together, form a much tighter limit on cultural development of an unlimited world population than do the supplies of more traditional resources.

In summary, it is clear that major political, economic, and technological changes will be necessary in the relatively near future to ensure the continued supply of "resources" to all mankind, whatever the demand may be. The actual supplies of traditional resources do have eventual limits, but the point of absolute exhaustion is relatively far off. Far more critical than the supply is the possible number of consumers who may demand products of resources, and the question of how the division shall be made among the earth's future peoples. In other words, the earth still has an abundance of resources, and questions about the future relate far more to habituated human practices, including the failure to control numbers, than to the exhaustion of the resources themselves.

RESOURCE CONVERSION AS MANUFACTURING

The word *manufacture* signifies operations carried on by hand to improve the quality of a raw material. The meaning of the term has been extended so that, today, it includes all of the many and complex operations carried on in turning raw materials into end products. It includes the computer-controlled industrial plants in which machines of some kind perform all the sundry operations. However, very few materials are directly converted today into end products by a single operation. It is necessary, therefore, to distinguish stages in industrial activity and to distinguish separate levels of use for many of the materials that enter into the whole of industrial activity.

Economic geographers and economists employ several grouping systems in their discussions of in-

dustrialization. With regard to the materials themselves, it is possible to distinguish *primary raw materials*, *secondary agents*, and *end products*. The operations involved can be discussed under the headings of *processing*, *manufacture*, and *assembly*; two pairs of terms often employed for the operations are *primary manufacturing* and *secondary manufacturing*, and *light industry* and *heavy industry*. Another pair of terms often employed for the products are *producer goods* and *consumer goods*.

All the above terms refer to conventionalized stages in the industrial process. For example, iron ore as a *primary raw material* is *processed* to make metallic iron, an *agent* that is combined with *secondary agents*, the ferro alloys, to make steel in the stage of *primary manufacturing*. Some steel is *manufactured* into machine tools, *producer goods*, which through *secondary manufacture* are employed to shape beams and sheets of steel into units for the chassis and body components of an automobile. *End products* of *heavy industry* such as glass windows, copper tubing, and rubber tires are combined in an *assembly* operation to put together the automobile. Shaped plastic items and stamped insignia, as *end products* of *light industry* are also added during the *assembly* and the finished vehicle is both an *end product* in itself and an item of *consumer goods*.

Out of the Industrial Revolution has come both the application of power (derived from inanimate sources of energy) to manufacturing and the separation of manufacturing operations into many different levels with regard to the handling of raw materials. The wastes and impurities of many primary raw materials become the processed raw materials for an entirely different sequence. Many of our "products" are, in reality, only agents in some further processing of a different raw material at some particular stage in manufacturing. No single technician, in the twentieth century, working by the Paleolithic system could ever produce an automobile, a refrigerator, or a television set from its ultimate raw materials. It is not that the technician is unskilled, but that the things needed would come from hundreds of primary raw materials. Those raw materials are processed through hundreds of operations requiring intricate equipment, and subjected to hundreds of manufacturing processes by complex machinery before a sufficient number of "end products" can finally be assembled into a "consumer product."

Although the initial impact of the Industrial Revolution on the rest of the world began with the increased output of certain raw materials such as coal and iron ore, and some consumer goods such as cotton textiles, it was the series of technologies that were critical. During the nineteenth century there began the increased processing of those raw materials in many places around the earth as Europeans began to establish factories in distant lands. As manufacturing skills matured in Europe, there came the flow of finished products into the markets of the earth to begin the changing of consumer patterns everywhere. The final variety of impact of the Industrial Revolution lay in the adoption of the new technologies by local populations who, themselves, began the building of new-style factories so as to increase the productivity of their own manufacturing activities. Hand in hand with the resource extraction and resource conversion patterns went the developments in transportation. Critical to the whole complex of resource development is the matter of space adjustment and space intensification. Since the role of transportation and communication forms much of the subject matter of the next chapter, their role is merely acknowledged here.

Although this chapter is concerned with the processes that created industrialization in the modern world, we must recognize that those processes were significant to the modification of landscapes all over the world. Mining landscapes, tree farms, urban industrial complexes, and industrial ports have created new kinds of forms and patterns on the surface of the earth. Ultimately it is these changing and differing landscapes to which the geographer returns for the characterization of different parts of the earth.

THE INDUSTRIAL COMPLEX

Economic processes arrange capital accumulations, labor forces, tools and machines, raw materials, transportation mechanisms, and management organization into productive systems to turn out products people use in their daily lives. Some of the very simple systems still employed around the earth were devised during the Paleolithic and the Neolithic periods, after which time little change took place in economic systems. The advanced modern systems resulted from the creative efforts of peoples of the European culture realm during the so-called Industrial Revolution. As a result of these developments modern industrialization is able to intensify the systems of resource conversion into a productive efficiency superior to that of all previous human time. Here our concern is with the technological mechanisms functioning in the twentieth century among advanced societies. These technological systems are rapidly spreading to the rest of the earth, although not so rapidly as people in many regions wish them to. The differential in regional development of industrialization produces the contrasts now described by the terms "developed" and "underdeveloped," about which we will have more to say at the end of this chapter.

The Levels of Manufacturing

Manufacturing began in campsites, working sites, and such places as men chose to carry on their individual operations on stone, bone, wood, fibers, and other materials. The general term often applied to such simple manufacturing operations is *primitive household industry*. As simple technologies evolved and man learned to harness simple sources of power, such as the waterwheel, the windmill, a draft animal, or a group of human laborers, newer and different kinds of raw materials could be processed in greater volumes in what has often been termed *simple-powered household industry*. As technological processes and labor skills evolved still further, without development of inputs of energy, people often worked in groups to achieve a given end. Specialized division of labor in such activities was common. The volume output was greater than when people worked alone, and the site of operations often became a specialized building or working site. *Community workshop industry* is the term often applied to such activities. All three modes were common in the more advanced regions of the world—the early civilizations—by the start of the Christian Era, and they generally prevailed until the time of the Industrial Revolution. In 1600 some regions around the world could have been described as still carrying on only the simplest of operations equivalent to primitive household industry. Other regions found the first two patterns operative, and a few could be described as possessing all three.

What the early phase of the Industrial Revolution did, in one sense, was to add (not substitute) a fourth category to the possible systems of conversion of raw materials into consumer products. There are three aspects to this development, all three involving elements of advanced technology. There is, first, the employment of huge amounts of energy to the end that labor becomes the manipulator of power rather than its source. Second, there is the involved technology in the processing of raw materials by which many kinds of physical, chemical, and manipulative changes are made in the raw materials. Third, there is the development of the physical plant, the "factory building," which became a specialized instrument for specific purposes.

By the late nineteenth century the evolution of industrial procedures had produced what could be designated *energy-powered-factory industry*. In mid-twentieth century there was added a fifth level of industrial operations, one that may be termed *corporate, powered production-line industry* in which machines take over more and more of the operations, with skilled labor now tending to become machine managers. Our modern kind of industrial operation differs from that of earlier times in its

Left is a typical old-style factory. This one manufactured shoes in Brockton, Massachusetts. The belt-driven machines were powered by steam. Below is a modern operation. It is tended by the single person shown in the lower left.

complexity and in that people use power and technology to do what formerly was done by human energy.

Not all work in an industrial society is done by complex, powered, factory industry. Still at work, in his garage, workshop, or the backyard is the individual, sometimes carrying out his operation by the simplest of tools and human labor. But if he succeeds in making a good new product, his next procedure is to apply simple power to some stage of the operation. Men still gather in community workshops to carry on industrial processes in which the largest single component may be the human labor input. The above simpler organizational forms of production are both still useful and complementary to factory industry. However, the complex products of industrial society require maintenance, and, as industrialism matures, an increasing number of workers are to be found in the servicing activities, which must range from factory maintenance to the small operators who engage in repair and maintenance of consumer products.

Economic and Administrative Elements

The mainstay of a productive society is its capital resource, as formalized, organized, and employed. In traditional preindustrial societies living by cultivation, land was the chief economic good and the primary capital resource; among a pastoral people capital lay in the grazing range and in the stock of animals. Organization is only partially an economic element in modern industrialized society, but diversification and intensification are significant here too. The one-owner, one-operator economic undertaking still exists, but modern industrial society has added many more organizational forms.

Societal controls over productivity have had marked economic significance since well before the formalization of the political state, but the governmental controls operative today are unlike most of those of earlier societies in their impact upon productivity. The use of taxation, tariffs, prohibitory rules, and stimulus rules are not new by any means, but their function has been adjusted so as to permit, augment, and foster industrial productivity in a way never before achieved. Laws on machine safety for the protection of labor, and on quality guarantees and pricing levels are new kinds of controls by government not present earlier.

In the modern era some of the traditional economic institutions have been transformed into systems compatible with industrialization. In the non-Communist world, on the one hand, the evolutionary development of classical capitalism has led to widespread private ownership of stocks and bonds which, coupled with the ownership in the hands of life-insurance companies and pension funds, enables the focus of society-wide capital resources on the financing of manufacturing through private enterprise.

The modernization of the old craft guild as the modern government-protected powerful labor union has accompanied the growth of industrialism. The former kingly monopoly over particular sectors of an economy has been creatively transformed into national, state, or municipal ownership for several kinds of public utilities, operating corporations, or manufacturing facilities that are integral aspects of modern industrialism. The increasing pattern of industries controlled or operated by government is an important facet of contemporary capitalism in the non-Communist world.

On the other hand, within the Communist world the recent evolutionary development of economic institutions has focused on those elements that are contributory to the growth of industrialism, although these institutions are still undergoing change in attempts to increase their efficiency. The centrally controlled, collectivized state farm, the territorial commune operating as a separate economic entity, the state-factory system, the government retailing establishment, and the government-controlled, organized labor corps all are institutional elements worked into state planning systems aimed at promoting industrialization. The Communist system has evolved into a state-controlled capitalistic form of industrialism.

POWER PRODUCTION AND MANUFACTURING

Just as the introduction of the draft animal altered the basic nature of gardening and brought on the evolution of agriculture, the introduction of large-scale power from inanimate sources altered the construct of older systems of manufacturing and brought on the evolution of corporate, urban, powered, factory industry. The first tentative starts in power application predate the real onset of the Industrial Revolution, but it is the integration of engineering principles with inventive experimental technology and the use of fossil fuel that characterize the modern steam-engine power system. From that point onward, progress in the development and application of power has been relatively continuous, to the end that in the developed areas twentieth-century mankind can command almost unlimited volumes of power. It has been the use of power that has made possible many of the modern kinds of manufacturing operations and the ranges of activities that constitute the high level of living.

In the gross measure of total energy applicable from all sources in any given society there is a rough key to the capacity for manufacturing. In terms of mature power usage, the world leaders today are the United States, the Soviet Union, the United Kingdom, West Germany, France, Canada, and Japan. The smaller countries of northwestern Europe rank equivalently in the scale of maturity but lower in terms of total volume (see Fig. 7.6). China and India rank high in total production but not in the scale of maturity of power development, although both their levels rose significantly during the 1960s. A number of other countries, such as Australia, South Africa, Brazil, and Argentina rank relatively high in total use of power.

Speaking figuratively, each citizen of the United States in 1976 had at his ultimate disposal and use the energy that could be provided by about 25 full-grown human slaves working at normal human capacity (calculated as 0.3 horsepower). In simple figurative contrast, the total supplemental energy available to an inhabitant of highland New Guinea could be provided by one teen-age boy per person. These are only crude measures whose accuracy is not extremely high, but they do indicate the degree of revolution that has occurred in the application of power. The eventual industrial potential for power development is much greater.

The most direct concern with the production of power is its utilization in work accomplishment, in manufacturing, transport, the provision of water supplies, lighting, and heating and air conditioning. From the standpoint of raising the per capita income of members of less-developed societies, and that of modernizing old systems of manufacture and transport, the application of power is most critical. Although many of the secondary steps in the evolution of modern manufacturing have not yet begun to accrue in many of the developing countries of the earth, it is significant that the spread of power production systems is well under way.

THE CYCLIC EVOLUTION OF INDUSTRY

In the history of manufacturing is there any orderly pattern to the appearance and practice of resource-converting activities? Taking the long view, it would possibly appear that there was only a very general order of beginnings. The need to process food supplies came first, but it immediately brought on the need for tools, utensils, and agents. The biological need for clothing must be counter-balanced by the human desire for adornment, and these "needs" brought about a demand for "textiles" and for cosmetics and jewelry. The need for protection from predators brought the need for weapons and shelter, adding constructional materials and tools for housing. There is not much that we can apply in transferring our attention to the modern situation, which began with the Industrial Revolution. The latter era is one in which wholesale change, invention sequences, and expansion occurred so rapidly and widely that its cyclic nature is somewhat different from those of older patterns. However as we view the Industrial Revolution, there are some points to be made about the rise of aspects of industrialization, its regional birthplaces, and about its diffusionary transfer to the underdeveloped region.

Both simple-powered household industry (such as grain milling) and community workshop industry (such as the early wool textiles) grew to the point that the critical problem became finding a power source to run the machines. This led to the rise of coal mining as a resource-extracting procedure, coal replacing charcoal and other simple power sources. With a power source available, the issue became the means of converting that source into power, leading to the steam engine. Initial development of power found its application in the machine works that produced the engines and the mechanical devices, and in the pumping of air into mines and water out of mines.

Within the same rough time dimensions, however, some of this power began to be applied to the newer spinning and weaving machines that had been invented for the textile industries, to the making of better iron and steel products for the making of more machines, to the grain-milling industries, and to the extraction of products from the power source itself, giving rise to the most elemental of the chemical industries. Some of the products of the latter segment we would now call agents, since they became used in further industrial processes. Thus the first resource-converting procedures led to greater productivity in old manufacturing lines. Additionally, they gave rise to new kinds of industrialization by creating whole new segments of activity. From this point onward the complex maze explodes as one new development led to application in three more, one of which produced a feedback to two older items, which in turn yielded several brand-new advances.

One approach to the issues of modern industrialization suggests that power development and the processing of raw materials go hand in hand, forming initial steps in the conversion of extracted resources. The preliminary processing of minerals (concentration and other such first-stage processes), the processing of forest products, and the processing and milling of agricultural commodities are all related aspects of light industry that provide the raw materials and agents for the manufacturing industries. Light industry, power development, and agricultural processing are often the first aspects of industrialization to develop in a region that is beginning the transition from older systems to the modern powered factory. Hybrid patterns of power development and processing technologies will be present, using some older technologies and labor skills. Involved may be textile manufacturing, the making of semifinished goods, and the final manufacture of consumer products not requiring large amounts of power, highly involved technologies, or the application of complex engineering mechanics or industrial chemistry. Among the consumer products are some of the electrical manufactures, the rubber-goods industries, and the food manufactures. Assembly of finished products from imported manufactured components is a recent significant aspect of light manufacturing in a region undergoing industrialization.

As urbanization occurs or increases, and as technological education occurs, many kinds of other beginnings may follow the development of power production and light manufacturing. The logical developmental sequence suggests various heavy industries, such as iron and steel, heavy metal fabrication, the manufacture of machine tools, and certain of the chemical industries, accompanied by the continued additive of greater power production. Owing to a lack of integration, engineering inefficiencies and the wastage of by-products occur at this stage.

Figure 7.9
World Manufacturing
Regions, 1975.

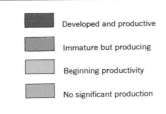

- Developed and productive
- Immature but producing
- Beginning productivity
- No significant production

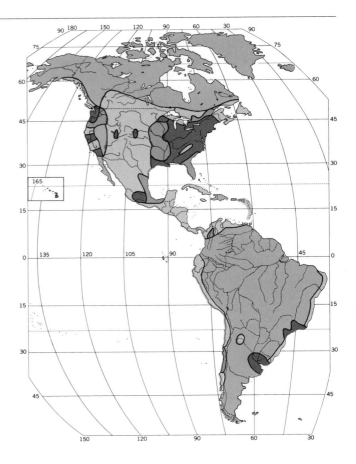

There follows the expansion of the range of manu-factures, the regional localization and concentration of industries according to environmental factors, and the industrial integration of the product and by-product, raw material, and industrial agent. Secondary types of manufacturing then gradually fit into niches in the overall industrial picture with a variety, a range, and a duplication of product. This begins to approach a kind of maturity in industrialization, at least as now defined in the most advanced industrial regions.

There is no specified sequence for all the countries of the world. It is clear, however, that the conversion of a resource into power, the feedback of that power into more resource extraction, and the branching out of resource conversion into many different lines is vital to the process of industrialization. There is a crude cyclic aspect to these simple and elementary stages of industrialization leading to mature development. Coupled with the whole dynamic of industrialization are two other processes, of space adjustment and space intensification, to be considered in the next chapter.

INDUSTRIAL LOCATION: FACTORS AND THE COMPLEX

Many contemporary economic geographers and economists study the processes involved in determining the location of industrial activity in advanced societies. Such problems are discussed here only in broad terms; the recognition of the problem itself involves a significant element in the intensification of industrial processes of resource conversion.

Wherever the processing of bulky or heavy raw materials results in a weight loss (as in cement manufacturing) or the raw material is perishable (as in vegetable canning or freezing), the manufacturing plants are located close to the source of the raw materials. Lighter and easier-to-transport raw materials are brought close to sites of hard-to-transport raw materials when more than one material, agent, or component of industrial activity is involved. Many light and easily transported finished products are often produced in a few specified centers in which

labor skills are concentrated. Such products are also often manufactured in areas where labor costs are cheaper than in highly developed manufacturing regions. This is why many American companies recently have been manufacturing in Asia and in Mexico, but labor costs tend to rise over time. This has meant that both easily transported and valuable products are traded far and wide away from their centers of production.

On the other hand, many of the finished products costly to transport tend to be manufactured close to the sites of consumption, provided the raw materials are locally available. In this kind of situation certain lines of manufacturing tend to spread widely over the earth, once the concepts and the technologies are understood and dispersed. The above are merely samples of the "rules of manufacturing", and there are numerous other rules and subrules governing the matters of location of industries that are becoming quite well understood as location theory is amplified and exemplified in studies of industrial location.

Central to the whole issue of industrial location, however, has been the recognition and comprehension of resources, the invention and evolution of technologies, the understanding of interrelations between raw materials and cultural processes, and

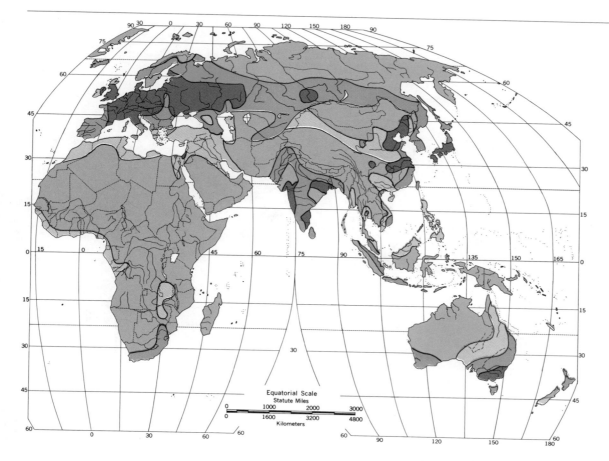

the historic progression of patterns of fabrication of the things we call manufactured goods.

THE MANUFACTURING REGION

In the twentieth century the term *manufacturing region* has been given that special definition which applies to any region satisfying the five following criteria:

1. The presence of large volumes of raw material;
2. The application of large amounts of power;
3. The employment of large amounts of labor having varied skills;
4. The location of large numbers of factories carrying on many operations; and
5. The presence of large numbers of customers within reasonable transportation range (Fig. 7.9).

Such a region constitutes an area carrying on what has been distinguished as corporate, powered production-line industry. The manufacturing regions, therefore, are those areas in which twentieth-century cultural technology is highly developed for a culturally adapted consumer population. This clearly limits the boundaries of manufacturing regions over the earth.

The modern manufacturing region is a product of the nineteenth and twentieth centuries (Fig. 7.10). As first mapped and discussed decades ago, there were but few such regions scattered over the earth. The contemporary discussion of world manufacturing identifies a larger number of mature regions and distinguishes a number of regions that are increasing their level and tempo of activity. The modern manufacturing region is one of significant urbanism, of highly developed cultural habit and custom, of strong food deficiency, and has well-developed transport ties to the rest of the earth. It is characterized by complexity and by integration of its economic activity, and it is a region in which there are concentrated not only diversified labor skills but also inventive technological skills and the concern for research into resource-converting techniques. A region of this kind possesses complex business organization that assists capital accumulation, transfer, and application to manufacturing ends. It is a region in which there are large volumes of varied types of nonhuman power available to do its work. The modern manufacturing region is possessed of advanced cultural habits that are in tune with the technological processes of production—in a word, the region's population has been brainwashed to consume the variety of products that can be turned out.

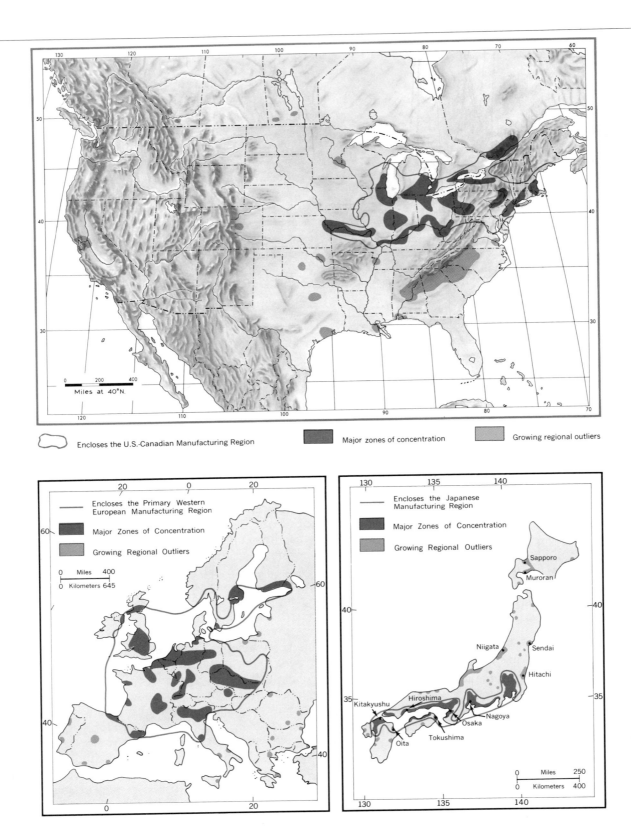

Enclfoses the U.S.-Canadian Manufacturing Region   Major zones of concentration   Growing regional outliers

Enclfoses the Primary Western European Manufacturing Region

Major Zones of Concentration

Growing Regional Outliers

0   Miles   400
0   Kilometers 645

Enclfoses the Japanese Manufacturing Region

Major Zones of Concentration

Growing Regional Outliers

Sapporo
Muroran
Niigata   Sendai
Hitachi
Kitakyushu   Hiroshima
Nagoya
Osaka
Oita   Tokushima

0   Miles   250
0   Kilometers   400

Figure 7.10
Selected Manufacturing Regions.
Top: North America
Bottom left: Northwestern Europe
Bottom right: Japan

In addition to the mature manufacturing region, as indicated, there are lower levels of immature, developing, and underdeveloped manufacturing regions, districts, or centers. Such lesser manufacturing regions are often smaller, or irregular in outline, and they are characterized by lesser measures of output, range of manufacture, and labor supply. They may differ in kind by being specialized in products, and they may lack some of the environmental advantages of the larger and more developed regions. From the viewpoint taken in this volume, such immature manufacturing regions have not yet culturally evolved the depth, range, and efficiency in resource conversion that obtains in the mature manufacturing region.

Beyond the manufacturing regions that the industrial geographer is inclined to place on his world map are numerous other regions in which manufacturing of a sort does take place. In these areas, however, it is likely that power is restricted either in volume or in kind, that technology is culturally older and not in step with that of the advanced sectors of the world, and that the whole cultural complex reflects the culture patterns of an older era or of a particular society that has not yet adopted the systems that characterize the most highly advanced sectors of the earth.

In the twentieth-century world, through corporate and state organization, many of the separate elements in the whole of industrialization are scattered over the earth. The larger corporations operate activities of separate kinds in many different countries. Venezuelan iron ore may flow to the United States to be processed and manufactured into the steel parts of motor trucks that may finally be assembled in Malaysia, whereas the copper wiring in those trucks may have been manufactured in the United States from copper mined in Chile, the aluminum parts may have been manufactured in the United States from aluminum processed in Canada from bauxite mined in Jamaica, and so on. Assembled in Malaysia also may be a Japanese and a British truck, for each of which the integration of raw materials and products was similar, and all three trucks may be fitted with tires manufactured in Malaysia from locally grown rubber. In this case three types of worldwide corporate organization had integrated manufacturing processes involving raw materials from over the earth.

The striking feature of twentieth-century manufacturing is that twentieth-century resource conversion does tie the whole earth together into one manufacturing system, the separate sectors of which operate at different levels in many different "manufacturing regions" of the earth. It is this efficiency of resource conversion that makes the modern world so different from the Paleolithic world.

THE CONTINUING GROWTH AND SPREAD OF THE INDUSTRIAL REVOLUTION

The Industrial Revolution cannot be identified as something that happened in the past. The inventive developments began in the fifteenth century, but they began simply, without formal prior plan, and by experimentation not well grounded in physical theory. The revolutionary changes have gathered strength and speed, they have broadened the ranges of their impact, and are in good stride at present.

If the eighteenth century saw somewhat of a lull in the patterns of progress of the Industrial Revolution, the twentieth century has seen a striking impetus in those patterns, based on the value and emphasis given to technological inventiveness. The lone inventor has given way to the research institute whose whole time is spent on the investigation of new means, new procedures, and new products. The concept has diffused around the world, and no longer do all the new ideas turn up in western Europe or the United States. Research not only has thought of new ways of doing traditional work but has also thought up whole new zones of industrialization never before conceived. The late-twentieth century sees more effort expended on inventive research and the possible expansion of industrialization than had occurred, possibly, in the whole of human time prior to the sixteenth century. The Industrial Revolution is thus by no means finished. It may be that merely a firm base for it has been established, with the larger share of progress to come in the future. This view concerns the systematic evolutionary growth of the Industrial Revolution, but it constitutes only one aspect.

The second aspect is the geographical diffusion of the Industrial Revolution from Northwest Europe to many other parts of the earth. That a formerly agricultural and handicraft manufacturing society in a far country now puts a high priority on the development of an iron and steel industry is an integral part of the continued growth of the Industrial Revolution on the planet Earth.

The Industrial Revolution is often held to pertain only to manufacturing and to industry. It has been a concern of this chapter to suggest that the new cultural concept of the relationship between mankind and the physical environment involves recognition of two aspects relative to the conquest of the material world, namely, the extraction of resources and their conversion into usable products. The basic element in that cultural concept is that the earth exists for the use of mankind as its now dominant life-form. It is therefore up to all peoples, everywhere, to utilize the earth to the greatest advantage possible through the most efficient technological procedures.

However, mankind has developed the modern extractive, producing, and consuming technologies with scant regard for the long-term ecological stability of the humanized earth. These technologies have ridden rough shod over the other life-forms and have paid scant heed to the maintenance of balance in the earthly ecosystem. The solution lies not in abolition of modern technologies but in devising resource recycling, waste disposal, and pollution preventive systems of technology to accompany the already developed production-consumption technologies. The total complement then could be directed toward the maintenance of the ecological balance of the ecosystem at a level compatible with the continuance of life on the humanized earth.

Conquest and control of the material world has currently outstripped control and arrangement of the social and cultural world. Efficiency on this level, too, is within the reach of mankind if mankind as a whole chooses to make the effort. After a long period of divergence in human living systems, the peoples of the earth have not yet chosen the hard choices that go with an earth operating in the truly long-term time scale. The processes of integration that go with the relatively recently instituted pattern of convergence (following the Columbian Discoveries) are still quite incomplete, but they are very necessary to the continuance of a humanized earth. The hard choices involve all the peoples of the earth cooperating in all ways at all levels of human activity. We shall have more to say on this subject in the final chapter of this book.

THE HAVES AND THE HAVE-NOTS: DEVELOPMENT AND UNDERDEVELOPMENT

The living conditions, qualities, and resources of some regions of the earth can be described in high terms, whereas some other regions can only be described in less fortunate terms. In part, these differences derive from the natural characters of the diverse and separate environments that make up our earth. In part, they also derive, however, from the unequal perception of the physical world by members of the human societies inhabiting the separate areas. Not all societies have put forth the painstaking creative efforts to make the most of their homelands. An environment is passive, in one sense, and what a human society does in that region is up to the society rather than the environment. Societies operate in territories with the tools and processes of cul-

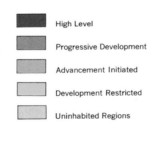

High Level

Progressive Development

Advancement Initiated

Development Restricted

Uninhabited Regions

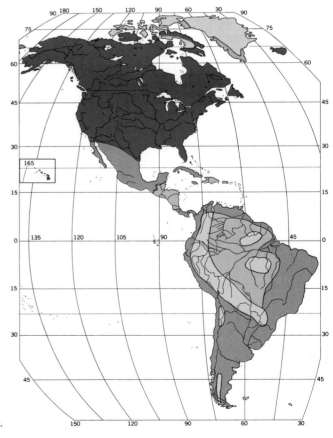

Figure 7.11

Comparative Cultural Development of Earth Resources, Includes consideration of technology, resources, organization, and perception of environmental possibilities.

ture. And those societies that have been imaginative and creative in devising efficient ways of utilizing the qualities of their environments have established total living systems superior in many ways to the living systems achieved by societies that put forth little creative effort.

Those highly successful societies have been the developers, and in comparative terms their homelands have been the "developed" regions, as compared to the "underdeveloped" regions inhabited by less creative human societies (Fig. 7.11). This is not a question of the superiority of one system of art, music, or philosophy over another, but the practical matter of developing material levels of living out of environmental resources, which is the core of the argument in "developed" against "underdeveloped." We have repeatedly pointed out this differential, since it could be found present even at the simpler levels of human culture in late Paleolithic time.

Societies that have committed themselves to the maintenance of their economic traditions and the classical systems of their pasts have lagged in what modern judgment describes under the term economic development. Both China and Japan were classic examples of the commitment to economic tradition in the early nineteenth century. Japan broke with that tradition in the late 1860s, whereas

China did not. Despite being very short of most natural resources, Japan's national determination to remodel her economic system brought steady results, and she now ranks among the industrially developed nations of the world. China has abundant natural resources, but the full commitment to the remodeling of her economic structure came about only after 1949. That the present political system is Communist does not alter the situation of a determination to modernize and industrialize the economic system. Both Japan and China have carried forward their modernization programs on their own, without massive assistance from other already developed societies.

The long history of the human occupation of the earth clearly indicates that forcefully inventive societies, those that put together superior culture systems in their own homelands, have controlled and/or exploited the territories of societies that were less creative or less energetic in developing their own territories. This has been done by infiltration of settlers, outright political conquest, or by the aggressive exchange-trade activities through which some degree of political, sociocultural, or economic dominance has resulted. In some cases complete cultural dominance has been the end product. In all such cases the conquered, displaced, and dominated have charged exploitive aggression against the society

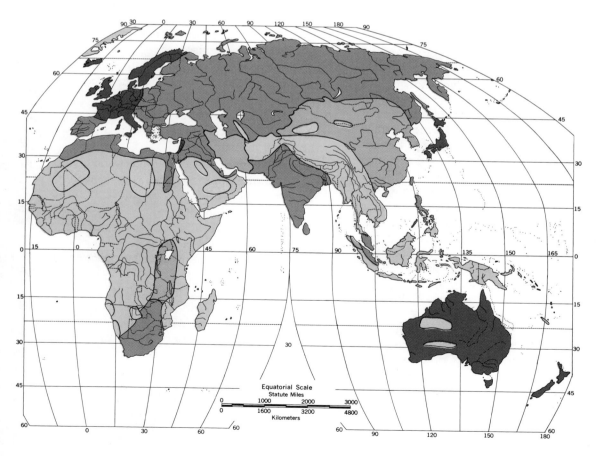

using its superior tactics. Our contemporary languages use the several equivalents of the English words "aggression," "imperialism," "exploitation," and "colonialism" for the results of such activities. In the modern era the members of the European culture realm (The West, The Occident) are being charged as sinful and malicious exploiters of the rest of the world. This very pattern has often occurred in the past in smaller "known-world" situations, but today the whole earth is involved since the economic systems of all societies now operate worldwide.

Political nationalism, as a modern cultural device whose purpose is the control of a specific territory on the surface of the earth, has reached a higher level than ever before, since the emerging international cultural "code of ethics" frowns on direct control of territory by any outside power. It is now being held by some that "economic imperialism" is replacing the old "territorial imperialism" as the "developed" countries of The West seek to continue their economic advantage over the "underdeveloped" regions, countries, and societies of the earth. The question may be asked whether it is possible that, in the twentieth century, these trends can be reversed. Is there effective cultural technology now available by which to assure parallel and equal economic development (on an earth spatially quite unequal in resources) among the whole number of politically operative contemporary societies?

Unless the long-term processes of cultural development are to be stopped, stabilized, or controlled, some leading culture group will continue to carry its creative advances beyond the levels of the least creative group, to maintain the advantage of the leading over the lagging. How, on an earth made up of hundreds of culture groups, aligned under (as of late 1977) 170 political states (each of which claims total autonomy), are the culture processes to be stabilized and controlled so that the energetic societies do not develop more rapidly than the others? Rates of progressive culture change have almost never been the same in all parts of the earth, and it does not seem possible that equivalent development can be made to take place in the near future with the present cultural technologies. The leading societies (usually stated as the most highly "civilized") have always been the most highly "developed" culture systems, and the lagging have always been the "underdeveloped."

The future leaders, having developed their own systems creatively, may not always be the same societies as the present most highly developed ones, for some of the presently lagging societies may move up in the scale. It does not seem possible that all human societies can become totally equivalent on a highly diverse earth without some master control mechanism not now operative or predictable.

SUMMARY OF OBJECTIVES

After you have read this chapter, you should be able to:

MANNERS OF CIRCULATION

In humanity's long period on the earth, a huge amount of human energy has been expended on the dull chores of trudging great distances on foot, carrying heavy things from one place to another, piling things up, and removing things from piles. Energy has also been used in the equally dull chore of breaking large things down into small things light enough to be carried. At first there were only hands and feet with which to perform these tasks, although people probably learned quite early to carry things on their

Photo 1. Human muscle has long provided the energy for doing work and human energy is still the primary source of power in many less developed parts of the earth, as shown in this canal-building project in Indonesia.

heads or shoulders. There is no real evidence to suggest when people grew so tired of dull, heavy labor they invented devices of one sort or another to make the job easier, but simple ways were developed. The shoulder pole, the head-strap, the A-frame, and the hip-sling are among the early aids used in moving things (see Fig. 4.4). In the less developed parts of the earth one still sees lines of men, women, and children laden with heavy loads. Carrying articles on the head is still common in central Africa, India, and Indonesia (Photo 1). The basket supported by a head strap is still widely used in the Tibetan borderlands; the A-frame is used in Western China and in Korea; and the shoulder pole is employed in eastern Asia from Japan to Indonesia. It is uncertain whether simple devices for carrying the lame, the sick, and exalted persons precede the generic form of the sedan chair. The sedan chair, itself formerly widely used in the Old World, was still common in rural China through the 1930s, but wheeled transport has largely replaced it in recent years. All of the above devices, except the sedan chair, were to be found in the pre-Columbian New World.

After its domestication and the development of some large and strong breeds, the dog aided man in transport by pulling the sled and the travois (see Fig. 3.11). Unfortunately, the dog was never really very useful in general transport, except in the area of the arctic fringe, where the dog became, and remains today, an important transport agent. Even here, however, the modern snowmobile may soon replace the dog sled (Photo 3). More useful than the dog for doing dull chores were the large animals that came later in the domestication sequence, such as the several kinds of cattle, the water buffalo, the donkey, the camel, and even later the horse and the elephant. The large animals could do several things for mankind, but the most important was the transport function. The first example of this was the pack train (Photo 2). The long lines of pack animals that travelled the land trade routes of the Old World for many centuries established some of the land routes still used today for commodity transport and passenger travel. In the New World, the only region with a large pack animal was the South American Andes Mountains where the llama performed rather indifferently as a carrier of light loads. In other areas of the New World, the dog was still used for the travois and the sled.

Photo 2. Pack animals provided an early energy source that has continued to allow transport in difficult regions, as shown by this small camel caravan crossing the Aïr ou Azbine in the south central Sahara.

Photo 3. The Alaskan dogsled is an example of a special regional use of nonhuman energy which developed in the Arctic fringe. The use of the dog in pulling the sled and the travois was one of the first applications of animal energy to transportation.

Photo 5. The traffic on Soochow Road, Peking, China, consists chiefly of pedestrians, bicycles, pedicabs, and two-wheeled carts.

Photo 6. (Above) The old and the new share a rural road in Hopei Province not far from Peking. The two-horse, two-wheeled cart has long been in use in the North China lowlands, but the truck wheels are a modern improvement. The motorized pedicab is Chinese-built and is used for light loads and passengers.

Photo 7. In the Philippines, the water buffalo provides the plowman with both transportation and energy applicable to agriculture. Here the plowman rides out to the fields carrying his plow, ox yoke, leaf-matting raincoat, and fish-creel lunch basket

Photo 8. The modern American wheat combine both reaps and threshes, permitting a reduction in human energy in the extensive grain farming operations in the midwestern United States.

The streams of laborers moving to and fro in dull working operations have been greatly reduced by the moving belt transport mechanism. Although developed in the United States, the device is diffusing rapidly around the world (Photo 9).

In the United States today, the application of mechanical power to the transport function has resulted in the mass movement of huge volumes of freight by rail. Passenger movement, however, has largely abandoned the railroad (Photo 10). Although railroads in Japan also move freight, the high-speed movement by train of large numbers of people is an important aspect of Japanese circulation (Photo 11). There is more to the railroad transport system than the tracks connecting distant points. There is also a requirement for eleborate terminals to assemble, sort, and disassemble trainload lots of freight or passengers (Photo 12).

In many parts of the world, the development of water transport reached an early peak and then stabilized its systems. The inland water transport system of China typifies this kind of development (Photo 13). In the United States, on the other hand, our early water-transport mechanisms never reached a high level of development before they collapsed under the pressures of the railroad technology developed in the late nineteenth century. Recently, a rejuvenation of inland water transport has occurred due to the development of new watercraft technologies which form strong contrasts with systems still in use in other areas of the world (Photo 14 compared with Photo 13). The great port city, with its docks and highly developed facilities for serving the great bulk carriers that travel the oceans between the few ports able to handle them, represents a degree of power applied to transport for both people and commodities never available in earlier times (Photos 15 and 16).

Pipelines, strictly speaking, are not modern inventions. They have been in use for many hundreds of years. However, the applications of modern technologies and the use of mechanical energy enable mankind to use pipelines to move tremendous volumes of both liquid and solid commodities (these are first turned into a watery mixture termed a *slurry*) (Photo 17). Such dissimilar commodities as oxygen, natural gas, milk, petroleum compounds, water, sewage, iron ore, and coal are moved long distances by pipeline, and the pipeline technology is still expanding.

The earliest human transport was very flexible. Individuals could disperse at will to almost any point desired. The development of formal roads, rail lines, canals, ports, big ships, and massive equipment has at times introduced a degree of inflexibility into patterns of movement. Although the contemporary truck rig traveling United States highways has definite limitations, one of the factors that has helped build up the trucking industry is the spatial flexibility it permits in comparison to the railroad (Photo 19). The small airplane provides a similar sort of flexibility when it is contrasted with the massive power and carrying capacity of large planes. The larger planes, however, have become vital in linking together the far parts of the earth and are critical to the high-speed movement of passengers and special types of commodities (Photo 18).

Photo 9. This powered rubber-belt sand conveyor moves large volumes of bulky material more rapidly and at far lower cost than the human transport system seen in Photo 1. It is a part of the American industrial system that decreases the drudgery of labor.

Photo 10. This Union Pacific freight train is powered by two General Electric U50 freight diesels, each of which delivers 5000 horsepower. Power of this type provides far more effective transportation in the United States, Canada, and some other maturely industrialized countries than do the older forms of human and animal-powered mechanisms still used in less developed parts of the world.

Photo 11. (Above) The high-speed electric trains between Osaka and Tokyo travel at 120 miles per hour and provide efficient passenger service to millions of Japanese. Japan has had more success with high-speed surface transport than any other country.

Photo 12. The Union Pacific Railroad employs automatic classification systems at many points. This is Bailey Yard at North Platte, Nebraska, where freight cars are sorted for proper assembly into train units. At the lower left, cars are being pushed inward over a high point from which they will roll down to the multiple track units according to destination. At the far right is the diesel engine repair shop.

The earliest devices that aided human movement on water—the raft, the coracle, and the canoe—were very simple and could move only a few people or small loads (see Fig. 3.4). Early development of such craft occurred in both the Old World and the New. Some similarities in the rafts and the navigating devices of the two areas seem to suggest the possibility of ancient contact between them, but the evidence available is not yet conclusive. In any case, there was much greater variety in the growth of water transport systems in the Old World. There mankind eventually developed several different traditions of boatbuilding that yielded larger boats and ships which could carry significant volumes. Loading and unloading them, however, was still restricted to single parcels of sizes and weights that men could carry (see Fig. 4.5).

The invention of the wheel in southwestern Asia began the evolution of the wheeled vehicle. Movement by cart was far easier than movement by sled, and the oxcart gradually became the standard transport vehicle for both passenger and commodity movement in the Old World (Photo 4, and see Figs. 3.11 and 4.2). From the sixteenth to the nineteenth century the horse replaced the ox in some areas of the world, but in many parts of southern and eastern Asia the water buffalo and the ox still remain the integral elements in rural transport and the two-wheeled cart is still the primary vehicle in use. In Europe, the four-wheeled wagon was developed for commodity movement, and the four-wheeled coach became the vehicle for passenger transport. In many areas, however, the development of the wheeled vehicle was restricted by the very nature of the land and the trail-road system already in existence. In China, Korea, Japan, and Vietnam the first "roads" were not built for wheeled vehicles but for foot travel. They were constructed as narrow, stone-paved paths with steps in rough country and were therefore not accessible to wheeled transport.

The United States moved from the system of poor and often muddy roads for ox and horse travel into the era of the railroad and the paved-road for the automobile in a rapid transition unlike the more gradual evolutionary processes of many of the less developed societies. In some parts of the world, simple wheeled devices, such as the bicycle and the motorcycle used in much of eastern Asia, are still basic to the circulation systems (Photos 5 and 6).

Photo 13. On the central-upper Yangtze River, a freight junk commences its downstream journey heavily laden with about a hundred tons of cargo. With its sail mast slung alongside and powered by a crew of oarsmen, this device represents a traditional water-transport operation that is cheaper than simple land transport.

Photo 14. (Above) The modern Mississippi River barge tow, with its detachable powerboat, moves thousands of tons of bulk cargo at very low per-ton cost using a very small crew. Inland water freight movement has recently been restored to a significant position in the United States.

Photo 15. This supertanker, "Mobil Magnolia," is a double-bottomed ship of 275,899 deadweight tons, 1,114 feet long, built in Japan in 1973. She will transport slightly more than 2 million barrels per trip. The deck surface is almost equivalent to two football fields placed end to end.

Photo 16. The ship "American Lark", almost fully loaded with container units of freight, represents the most up-to-date system of sea transportation. Operators in the two cabs under the overhead booms are lowering containers into place on top of the deckload. There are not many ports as yet equipped to load or unload ships of this type. Each container unit is a truck-trailer-van.

Within local environments circulation problems involving movement often relate to agricultural labor. Labor forces must move to, around, and away from farmlands. The Philippine villager keeps his work animal and his tools near his village residence, and movement out to his farm plot involves transport to the site (Photo 7). On contemporary United States farms, this kind of movement is at a minimum due to the development of mechanized equipment reducing both the volume of human labor and the frequency of movement over the large farms (Photo 8).

Although many persons dislike the patterns of trunk highways now in use in some parts of the world the automotive era has brought a degree of flexibility and an ability for mass movement to the circulation patterns of our great urban settlements unknown in the large cities of past eras. Metropolitan Los Angeles is probably the foremost example. Even though it is the butt of much sarcastic humor, few cities in the world have a similar degree of flexibility in human circulation. The Los Angeles freeway (trunk highway) system permits individual movement in virtually any direction (Photo 19).

In the opening paragraph of this essay, we referred to the dulling effect of having to walk everywhere. The moving sidewalk, escalator, and elevator systems created by modern engineering now provide almost effortless human circulation in many different circumstances (Photo 20). Although such applications are still relatively limited in distribution, they are one more part of the evolutionary chain of mechanical procedures devised to decrease the necessary expenditure of human energy, which has probably been one of the foremost objectives of mankind throughout history.

Photo 17. The men standing on the pipe at the far right provide a scale for this pipeline that will carry irrigation water 150 miles from the Sea of Galilee to the Negev Desert in southern Israel.

Photo 18. These modern airplanes, with their impressive capacities for carrying passengers or cargo, dwarf their human operators and ground equipment. Able to span the globe in relatively short flight times, they serve to bring all parts of the world very close together.

Photo 19. The modern automotive trunk highways, depicted by this four-level "stack" at the junction of several major trunk roads in Los Angeles, permit the easy, low-cost, rapid, and long-distance movement of large numbers of people. The contrast to the patterns of local transport depicted in photos 4 and 5 is startling.

Photo 20. The movement of large numbers of individuals in quick and personally effortless patterns, as shown in this view of a stairway-escalator, is an integral part of American mechanization.

CHAPTER 8
THE TRANSFORMATION OF SPACE

Fundamental to utilization of the whole earth by mankind are the human processes that transform space:

1. *Space-adjusting techniques* either shorten the effective distance in the circulation of people, goods, and ideas through travel, trade, or communication, respectively, or they integrate the content of different regions through administrative organization for political, economic, or social purposes; and

2. *Space-intensification techniques* result in increased productivity, increased population density, urbanization, and improved levels of living, by achieving new and higher orders of utilization of the earth's regions.

The word "space" obviously has different meanings for different people. The concern of geography with space relates to the surface of the earth. Space has the qualities of *extent* (length, area, or volume), *location* (qualities of site and situation) *density* (quantity contained per unit area), and *succession* (variations in quality owing to changes that occur during periods of time).

SPATIAL INTERCHANGE AND SPACE-ADJUSTING TECHNIQUES

If everything existed in the same place, there could be no differentness. Only space makes possible a dissimilarity or nonconformity from place to place over the earth's surface, and makes areal differentiation, or regional variation, a reality.

Each place on the earth's surface has a network of spatial interrelations (location, distance, direction) with all other places on the earth's surface. A thing that appears (is born, created, or built) and then disappears (dies or is destroyed) acquires and loses properties of spatial interrelations even though it may never occupy more than one place on the earth's surface. If a thing moves, it occupies one place at one time and another place at another time. This kind of change involves spatial relations just as much as it does relations in time.

The concept of spatial interchange began with the activities of early humans as they shifted their livelihood from gathering to collecting. Collecting requires not only abundance but also movement, the transfer of goods from the place where collected to another place (e.g., campsite) where processed and consumed. The separation, in space, of place of production from place of consumption is basic to the concept of spatial interchange. Various technologies, not present among food-gatherers, are necessary to achieve a collecting way of life: a technology of movement of things (transportation) and a technology of processing, as in food preservation and storage (resource conversion).

The space-adjusting techniques include all those culturally invented ways by which people change the locations of things from where they acquire or produce them to other locations where they desire them to be. For the moving of materials, exchange and trade evolved. For the movements of people, travel evolved. For the movement of ideas, communications evolved. People have repeatedly adjusted spatial interaction by inventing newer technologies to speed up movement and carry greater volumes of materials, people, and ideas across intervening

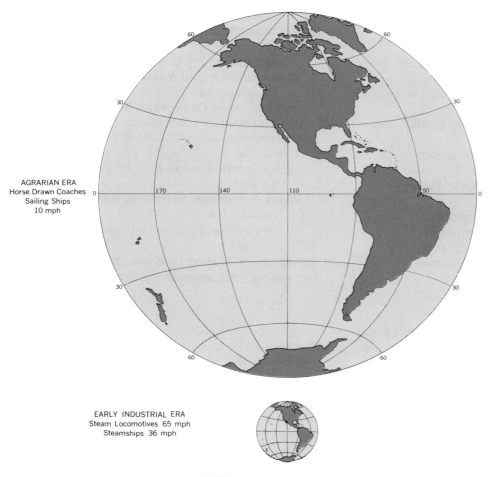

AGRARIAN ERA
Horse Drawn Coaches
Sailing Ships
10 mph

EARLY INDUSTRIAL ERA
Steam Locomotives 65 mph
Steamships 36 mph

TODAY
Jet Planes
500-700 mph

Figure 8.1
The Shrinkage of Effective Distance.

spaces. The cumulative effects of transport and communications have shortened the effective distance among places on the earth's surface (Fig. 8.1).

The space-adjusting techniques also include all those culturally invented ways by which people change the ownership of things in the process of moving, extracting, or converting resources. As raw materials move from a producer to a manufacturer, they increase in value because of the change of location. As a product passes from a manufacturer to a distributor, it increases in value because of its change of location through wholesale trade. As a product passes from a distributor to a consumer, it again increases in value because of its change in location through retail trade. Each step effects both an increase in value and a change in location.

The space-adjusting techniques further include all those culturally invented ways by which groups of people acquire access to and manipulate control over territorial spaces and the resources these contain. People have organized the highly diversified space accessible to them on the earth's surface into a set of compartments, partitioned for different purposes. The long-term trend has been to create uniform sets of conditions within each compartment, applicable to the people who come under the jurisdiction of the particular political territory. Because of cultural diversities (allegiances, sets of beliefs, communal spirit, concepts of identity), human attempts to establish political regional uniformities have created further differences in the space on the earth's surface. These culturally imposed differences further complicate the operations of spatial interchange.

MOVING MATERIALS, PEOPLE, AND IDEAS

In the previous chapter we learned that both the identity of resources and their relative scarcity or profusion are culturally determined. We also learned that it is the culture of a given people that determines the uses to which all things shall be put. It depends on ignorance or skill whether the resources shall be used at all. The various ways in which they are able to be used in turn depend on technologies for resource conversion. In this chapter we focus on space-adjusting and space-intensification techniques, for these help determine in what circumstances and in what amounts the various members of a community can have access to, and thus share in the use of, the world's resources.

The transportation "revolution" came upon mankind only relatively recently. Until well into the nineteenth century, regular travel over long distances could not exceed ten miles per hour. A mail coach,

with good weather and relays of horses and drivers, could cover no more than 225 miles per day. The fastest of the "Yankee clipper" sailing ships set a record of 436 miles in one twenty-four hour period. Now the world has increasingly shifted its transportation energy from live organisms to fossil fuels and is beginning further to change to nuclear energy.

A burst of creativity in the European culture realm brought the steam-propelled ships and trains, along with the electric streetcar and locomotives. Then came the automobile and its big brothers, the truck and the bus, followed by the diesel trucks, trains, and ships. Also there came the airplane, first propeller driven but now jet powered. Finally, there is the orbiting satellite. With great effort, the speed record on land has been pushed to just over 600 miles per hour, but anyone with the price of a ticket on an airline can travel faster than that.

The communications "revolution" has been even more dramatic than that in transportation. Only 20 years separated the invention of the telephone (1876) from that of the wireless telegraph (1895). Only 17 years elapsed between the first radio broadcast (1906) and the first transmission of television pictures (1923). Since 1963, intercontinental telephone and television contacts have been maintained through satellites. A computer has the capacity to transmit the entire contents of the 24-volume *Encyclopaedia Britannica* in three minutes.

In the present world it is easiest to transmit ideas (words, pictures, sounds) electronically. Complex ideas for instruction, persuasion, and decision-making can be sent quickly by people who travel by air. Valuable or vitally needed perishable goods can move by air freight. But most goods, as freight, are transported on land or sea at slower speeds. As a general rule, the greater the bulk and weight and the smaller the value of the goods, the less costly and slower will be the means of transportation employed.

SHORTENING EFFECTIVE DISTANCE

The study of transportation enables one to analyze the interconnections between regions or places. Involved are the location of routes; the volume, value, and directions of flow of traffic; and the influence one region has on the economic activities of all other regions with which it is interconnected. Without means of transportation for the exchange of regional surpluses, there could be no economies other than at the subsistence level. What improved transportation and communication do accomplish is to shorten the effective time required to move products from one place to another, or to shorten the time necessary to

This giant disk, located at the Goldstone Tracking Station in California, is but one element in our modern satellite tracking system.

deliver messages and news to distant places. This means that delivery can be made within shorter time periods or that delivery can be made to more distant points within a given time period.

The following discussions of transportation focus primarily on the growth of modern systems in the countries of the European culture realm in which the developments occurred. They are part of the broad set of processes that have carried The West ahead of other parts of the earth. Many of the more recent developments will seem commonplace to younger Americans who have been enculturated into a living system employing high-speed equipment and large-volume mechanisms. Many elderly Americans, however, can recall the tremendous changes brought into their lives as various of the developments came into use. Some of these improvements spread very rapidly around the earth, and others are still making their appearances in some areas. In many underdeveloped parts of the earth, high transport costs still reflect old ways, and products can be moved only slowly and in small unit packages.

In Chapter 4 explanatory reference was made to the five elements involved in any kind of transportation: route surface, motive power, mechanism of carriage, packaging of materials, and terminal facilities

at route ends. Only in the nineteenth century did human inventiveness revolutionize transportation by attacking problems around some one of the elements of the complex. During the twentieth century there has been coordination of the procedures for all five elements in many parts of the world; however, the revolution in transportation is still not complete. The current attention being given the "parking lot" problem for the automobile, the "air terminal" problem for the airplane, and the "containerizing" solution to the packaging of materials for sea transport are examples of contemporary lack of full coordination in elements of transportation.

Land Transport

The first of the means by which a person achieved the ability to travel faster than any other animal was the railroad. For decades, during the nineteenth century, the railroad was supreme, but in the twentieth century automotive transportation has been overwhelmingly successful, contributing to the mid-twentieth century decline of railroads.

The first cross-country railway began operations in England in 1825, using a locomotive engine rather than horses, at a speed of 10 to 15 miles per hour. Railway lines mushroomed in England, and by 1840 there were 76 companies and 2235 miles of track. The entire English system of internal transportation began to change. Railroad traffic developed large volumes owing to the reduced costs and increased speeds for all classes of society. Land adjacent to railroads increased in value both for commercial gardening and for industrial purposes. An expanded railroad network thus made possible the urban-industrial transformation of nineteenth-century England, one element of which was the increase in speed of movement.

The diffusion of the steam-powered railroad from England to other European countries was extremely rapid: Spain and the United States (the Baltimore and Ohio line), 1830; Belgium and Germany, 1836; France and Russia, 1837; Italy, 1840; Switzerland, 1852. The United States, in part because of its sheer size and vast distances, almost at once seized the lead, and by 1850 it had 9283 miles of track to England's 6521, Germany's 3726, and France's 2484.

Different modes of land transport developed at different times, giving rise in regions of early urban-industrial development to problems of adjustment of the older forms to accommodate the new. In the United States railroads first appeared in 1830, pipelines in 1865, automobiles in 1893, intercity trucking in 1918, and commercial airlines in 1926. The development of motor vehicles and improved highways seriously affected railroad passenger and freight traffic. The airlines' intercity passenger mileage came to exceed that of the railroads, owing to the diversion of much long-distance passenger traffic.

Many former colonial territories owe their transportation network to European-introduced technology. Route patterns, however, were often dictated by military strategy and governmental needs, and many colonial transportation systems were never economically self-sustaining. Yet the inherited systems have benefited the new nations in their subsequent economic development programs. The new modernizing nations do not have to suffer through problems of sequential readjustments but are free to choose among, or to coordinate, the several existing transport forms: railroads, pipelines, and motor vehicles, as well as water and air transport. Whatever their problems of capital formation and investment, the new nations are better able to plan wholly integrated transport systems taking advantage of all existing forms of transport with a minimum of costly duplication of services.

The automobile, truck, and bus, using internal combustion engines and burning either gasoline or diesel fuels, have become the most ubiquitous transportation vehicles in the world today. Almost one half of the total of about 260 million vehicles, how-

ever, are in use in the United States. The increasing usage is indicated by the current annual production rate of about 16 million vehicles per year, about half of which is in the United States. The gasoline engine in all forms, however, is such a great contributor to air pollution that it is unlikely to survive the century of its dominance.

Pipelines are a specialized transport system, based on the principle of continuous flow. Many long-distance pipelines are privately owned and operated, and they serve principally to move natural gas and liquids such as crude oil, gasoline, and other petroleum products. Many city-owned pipelines now move water, sewage, and natural gas. Moving pulverized coal, various ores, and other solids in suspension in water is already being done in widely scattered areas. Pipelines can now be built to whatever diameter is required. Limits on pipeline size are those of market demand, because it is most efficient to operate a pipeline at near capacity. Natural gas lines now span the North American continent, and this form of transportation is not really competitive with any other method (Fig. 8.2). A pipeline can move a ton of oil 1000 miles for about $2.50. This cost advantage is so marked that about 20 percent of all land freight (measured in ton miles) in North America is now moved by pipeline. Pipelines are being built in some other parts of the earth, but

Figure 8.2 Long-distance Pipelines and Electrical Transmission Lines in the United States and Canada, as of 1972.

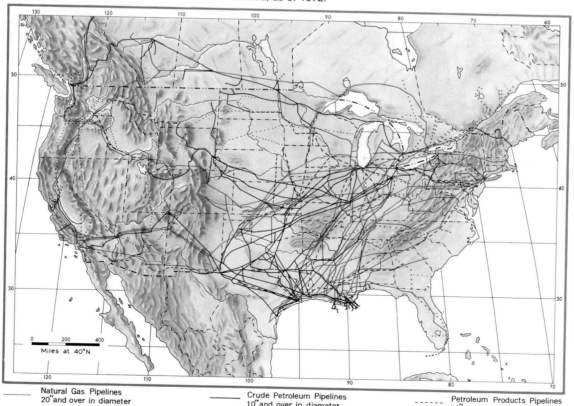

Natural Gas Pipelines
20" and over in diameter

Crude Petroleum Pipelines
10" and over in diameter

Petroleum Products Pipelines
10" and over in diameter

Electrical Transmission Lines

there remain national political restrictions on pipeline transportation in many countries of the earth.

Transmission of electricity, however generated, is less a form of transportation than a substitute for it, because the transportation of fuels is minimized or eliminated. Electrical transmission is, however, a fabulous space-adjusting technique, making possible the distribution of energy from the point of production to many points of consumption.

Water Transportation

This topic heading is ambiguous. Water, itself, can be moved, as in irrigation canals, flumes, aqueducts, urban water pipes, or sewage system pipelines. Or, goods and people can be transported in vehicles in or on water. However, the effect of space adjustment (a change in location) is the same in any case.

Transportation forms that use water in which to move are customarily divided into foreign (ocean shipping between countries), coastwise or intercoastal (ocean shipping between domestic ports), and inland waterways (lakes, rivers, navigation canals). For the United States, coastwise transportation monopolized the trade between ports in the colonial period. All intercoastal (Atlantic-Pacific) shipping was around South America, until construction of the Panama Railroad in 1854 made possible a shorter ocean-rail-ocean route. The opening of the Panama Canal in 1914 stimulated an increase in intercoastal traffic until World War II. Since the mid-1940s, however, neither coastwise nor intercoastal shipping has regained its earlier importance, owing to rail and truck competition.

Inland water transport in the United States was revived during the early twentieth century to create cheap freight transport, and to keep railroad rates down (Fig. 8.3 and see Photo Cluster No. 4, photo 14). In the United States, 358 billion ton-miles of freight were hauled in 1973. Waterways are improved and maintained by the federal government, and considered "public highways" open to use by anyone. Tolls are collected only at the Panama Canal and the St. Lawrence Seaway. The disadvantages of inland water transport are its slowness (8-mile-per-hour average speed), its seasonal character in the north owing to winter-frozen waterways, floods and droughts affecting the rivers, and the costs of freight transfers at terminal points.

Ocean transport has always included both economic and military objectives, and one must distinguish naval from merchant vessels. The latter include passenger or passenger-cargo ships, freighters, tankers, and ore carriers. Passenger liners formerly were the largest, but today the supertankers outrank all others (see Photo Cluster No. 4, photo 15). The most numerous are the freighters, with a world

Figure 8.3 Inland waterways and Commercial Seaports of the United States and Canada, as of 1973.

• Name indicates seaports handling over 20 million tons of freight in 1973.

total of 11,500 in 1974, compared to 5121 tankers.

Until World War II tramp steamers dominated the world's merchant fleet. These did not follow fixed routes or time schedules and sought cargoes wherever they could find them. In the tramp steamer trade, shipping and trading were related parts of the same business with decisions on buying and selling being made in distant ports. The twentieth-century decline of tramp shipping has been the result of several factors. The improvement of worldwide communications enabled the functions of shipping and merchandising to be separated, with decisions on buying and selling being made in a headquarters office and sailing instructions being given accordingly. The passenger-cargo liner trade expanded after 1900, many routes connecting numerous world ports. Operating on regular schedules, and travelling at higher speeds than the tramps, the liners tied the earth together in a transportation network. The rise of air traffic, however, reduced the passenger liners to the role of special-purpose vacation cruise ships, and almost no more of the huge floating hotels will be built. The big tankers and ore carriers have taken on the former role of the tramps for specialized cargo. A large firm today often operates its own fleet of ships for moving large volumes of specialized cargo (petroleum products, ores, cement, bananas, and so on). The older tramps had speeds of 10 to 11 knots, the new larger tramps about 18 knots, the modern cargo freighter 18 to 24 knots, and the supertankers about 14 to 20 knots.

Ocean transport involves not only the carrying of people and goods, but also the connections at particular ports. Loading and unloading (terminal operations) incur 50 to 75 percent of the total costs of ocean shipping. In the United States there are 237 improved commercial seaports, but the traffic, whether foreign, coastwise, or intercoastal, is concentrated at 31 principal ports (see Fig. 8.3). There is an obvious pattern of relationships between port locations and the spatial locations and developments of cities, the growth patterns of industrial regions, and the spatial networks of transportation in many parts of the earth.

Air Transport

The first successful flight by a heavier-than-air powered machine, at Kitty Hawk, North Carolina, on December 17, 1903, was the culmination of experiments made by men of many nations during the previous centuries. The airplane was not the creation of any single culture. Now, three-fourths of a century later, the worldwide character of air transport is self-evident, since the world is laced by a network of air routes (Fig. 8.4).

Commercial aviation literally "took off" after World War II. The amazing expansion resulted from the combination of a host of factors. Primary factors were the availability of skilled flight crews; the expansion of route mileage by the main trunklines; and the establishment of numerous feeder or local lines, including helicopter "commuter" services. Secondary aspects were the launching of airfreight service; the expansion of nonscheduled services and all-cargo airlines; and the institution of "air-coach," "tourist," and "economy class" services by regularly scheduled airlines. Quick movement across long distances has been the hallmark of air transport, with which other forms of transportation cannot compete. Not only has traffic been diverted from surface transport carriers, but also much new traffic has been generated (e.g., business commuting, vacation trips to far places, shipment of perishables such as cut flowers). A large passenger jet also carries a truckload of freight; a jet freighter can carry the equivalent of one American 60-ton railway freight car.

The development of the airplane into a major means of transportation gave rise to problems far beyond the capacity of individual governments to solve. The need for safety and regularity in air transport involved building airfields, setting up navigation aids, and establishing weather reporting systems. Standardization of operational practices for international services was of fundamental importance. Establishment of standards—for rules of the air, for air traffic control, for personnel licensing, for air safety—all required more than national action.

Aircraft may not move indiscriminately through air space. In the interests of safety, routes along airways are prescribed. Levels and lanes are designated, along which aircraft are guided from airport to airport. The present airway system of the United States, maintained and operated by the Federal Aviation Agency, is an elaborate set of navigational aids. There are radio signal stations, beacon lights, weather-reporting services, auxiliary landing fields, air-route traffic-control centers, traffic-control towers, radar and instrumental approach systems, and route levels.

The principal advantage of air transport is speed, which is greatest on long distance nonstop flights for which the proportion of time spent in land transport to and from the airport is minimized. Other advantages are frequency of schedules and ability to move in any direction. A flight once a week to a remote Alaskan outpost may be considered "frequent," compared to the surface transport alternatives. A further advantage is that, given an airport of adequate size, an airplane may travel to a remote area not otherwise provided with transportation facilities. The airway requires not so large an investment in route "right of way" as do roads, railroads, or canals. Alaska, the Amazon Basin, Australia, Borneo, and

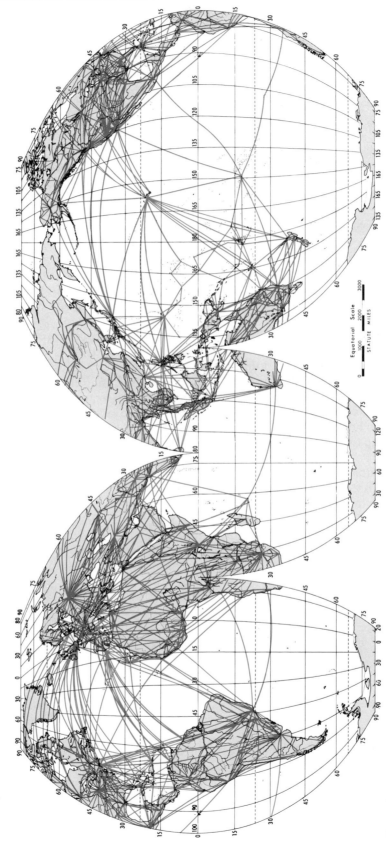

Figure 8.4 Linkage of the Modern World via Air Transportation.
Lines indicate scheduled commercial air routes as of January 1976.

Equatorial Scale

STATUTE MILES

New Guinea are all served by one or more local airlines that link their sparsely settled areas to the worldwide network of commercial air services. Again, there are clear correlations between the locations and numbers of air terminals and the extent of regional development in all parts of the world.

EVOLUTION OF MODERN WORLDWIDE TRADE

The exchange of goods and services among peoples separated across great distances on the earth's surface has been a continuing process since prehistoric times. The most spectacular advances in expansion of trade, however, have come about during the past century, resulting from the impetus of the Industrial Revolution. Every increment to resource-extracting techniques made possible an increase in raw-materials production. Similarly, every addition to techniques for resource conversion made possible a great increase in the kinds and numbers of new products available for human use. The "feedback" effect was fully operable, in that the improvements in land and water transport introduced an era of flourishing international trade.

Channels for the Distribution of Goods

In contrast to railroads, canals, and paved roads, the air and the oceans are free highways over which planes and ships can roam. But most planes and vessels, except private planes and small boats and tramp steamers, follow rather well-defined routes whose locations and directions are linked to their terminals, the seaports and airports they connect. Most of the world's oceangoing trade and air traffic is funneled through a relatively few terminals, capable of handling and servicing the large modern transportation vehicles and having sizable productive hinterlands.

The world's leading sea-lanes are shown in Figure 8.5. It clearly shows that trade moves in channels, shaped by the transport routes and by the communications network. The pattern of ocean transport routes of the eighteenth century would have looked greatly different, as sailing ships followed the earth's wind systems and the world's regions of production and consumption required fewer and different commodities than they do now.

International trade has become vastly complex. The merchant trader, the merchant banker, the freight forwarder, the import house, the national corporation are commercial institutions that have developed at each end of the channels of trade. Organized markets to serve growing populations called

for greater and greater volumes of standardized goods. Where standardization and grading could not be accurately achieved, as in wools, furs, and diamonds, there evolved international auctions with competitive bidding taking place after inspection.

Increasing Volumes and Increasing Distances

The modern expansion of trade is so intimately linked with technological developments in production and transportation that it is difficult to separate cause from effect. The harnessing of inanimate energy to industrial machinery makes possible a vast increase in the volume and variety of manufactured goods. Such commodities are no longer consumed only locally but, in varying ways, receive worldwide distribution. The increases in size and speed of railroads, trucks, ships, and planes enable a more effective distribution of goods in ever larger quantity and variety, whose sale makes possible the construction and operation of new manufacturing plants, including the cost of the machinery and purchase of inanimate energy to run them.

Mass transport now moves raw materials in huge volumes, often over great distances. Partly, this has been made possible by technological progress. The first cargo ship over 3000 tons was launched in 1899, a far cry from today's 480,000-ton supertankers that are over 1200 feet in length and 200 feet in beam. Increases in train capacities, truck haulage limits, aircraft capacities, and pipeline flow-rates all reflect the same sorts of progressive change. There is no technological limitation on the ability to move things from one place to another across the earth's surface. The limitations on space adjustment are cultural rather than physical. There must be sufficient reasons for undertaking projects of space adjustment, plans laid, engineering and administrative decisions made and orders given, financial feasibility determined, and perhaps political agreements entered into among states or nations. The logistic problems of supplying the military forces in World War II, or in the sending of the "wheat fleet" to India, the airlifting of disaster-relief supplies to Peru and Bangladesh, and the building of the Alaska pipeline are indicative of mid-twentieth-century capabilities.

Regionalisms of Markets

The ultimate basis of trade is that it serves to adjust the spatial differences between places of production and places of consumption. Most trade is bilateral, between individuals or corporations who buy and sell, import and export. Geographers have tended to emphasize both the routes and movement of trans-

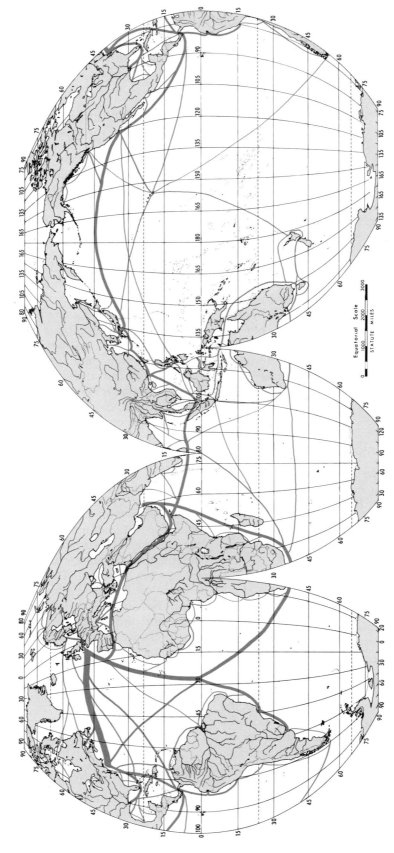

The African route-volume will decline as the Suez Canal is reopened to traffic, and its route-volume proportionately increase.

Figure 8.5 The Major Commercial Shipping Lanes of the World.

port vehicles and the volumes and flow of goods in trade. But equally important is an understanding of the multilateral and regional systems of methods of payment. When foreign exchange can be freely bought and sold in order to undertake transactions, a network of trade can extend widely over the world. Countries can then fully specialize in their best or most profitable products, if not both, and use the foreign exchange earned by their exports to buy imports from many other parts of the world.

The spread of the industrialization process around the world is drastically changing the relative proportion of commodities entering world trade. Industrial products are increasing, and the percentage of raw materials is decreasing. Over the past several decades there has been rapid economic growth in such nations as Argentina, Brazil, Mexico, Australia, India, South Africa, Yugoslavia, and Finland. All of these countries are now processing much larger quantities of their own locally extracted raw materials than ever, thereby reducing the volume of raw materials that formerly entered the export trade.

The countries of Asia, Africa, and Latin America that depend on a single commodity for most of their export earnings (Fig. 8.6) have steadily been losing ground in world trade. Except for petroleum since 1973, their main foreign exchange earnings—tropical agricultural products and minerals—have not increased in value so fast as prices on the manufactured goods that these countries purchase from abroad. In spite of increased local productivity, many countries are exporting a greater volume of goods but earning less income than they formerly did. The "trade gap" between export earnings and import needs is increasing.

The most successful examples of regional organizations for the promotion of trade come from Western Europe. In 1947 the Organization for European Economic Cooperation (OEEC) was established to distribute United States aid efficiently under the European Recovery Program (the Marshall Plan). Beginning as a program for economic recovery, the idea of cooperative unity persisted and, after American aid ended in 1951, the Organization continued as a coordinating agency for its 18 member states. Renamed in 1961 the Organization for Economic Cooperation and Development (OECD), it was expanded to include Canada, Japan, Turkey, and the United States.

The European Coal and Steel Community was founded in 1951 to pool industrial production and take a first step toward economic cooperation. The first common market for coal and iron ore, scrap, and steel was opened in 1953 among the six participating nations: Italy, France, West Germany, Netherlands, Belgium, Luxembourg. These six nations in 1957 formed a union setting up a European Atomic Energy Community (Euratom) and a European Economic Community (EEC), better known as the Common Market. A customs union has gradually removed all obstacles to the free movement of goods, people, firms, services, and capital among the member countries. Greece, Turkey, Malta, and 18 independent African countries have been made associate members. With the addition of Great Britain and eight other European nations in 1972, the EEC is now well on its way to becoming a political federation with a greater population and economic productivity than either the Soviet Union or the United States. The three Communities merged into one Commission in 1967. Europe's assets for integrating its regional economic organization were many: a developed industrial base, good transport and communication facilities, and a common purpose.

Other regional economic development and trade associations have been formed, on the European model, but not possessing all of Europe's assets. In Asia the earliest formed (1951) was the Colombo Plan, a loose cooperative system that mixes economic assistance from six donor nations (Australia, Canada, Japan, New Zealand, Great Britain, and the United States) with mutual help from 18 largely recipient nations. In southwest Asia, Turkey, Iran, and Pakistan have formed a loose union called Regional Cooperation for Development (RCD), of which a joint shipping line is in operation.

The most effective organization for economic cooperation in Latin America is the Central American Common Market (CACM), formed in 1960, to include Costa Rica, El Salvador, Guatemala, Honduras, and Nicaragua. The ten-member Latin American Free Trade Association (Mexico plus all the countries of South America except Bolivia, Guyana, Surinam, and French Guiana), created in 1960, has been less effective because of political instability and competitive economies. In 1962 some 29 newly independent nations of Africa formed the *Organisation Commune Africaine et Malgache* (OCAM), linking former French and Belgian colonies into a common organization for economic cooperation and development.

Widening the Resource Sustenance

In consequence of the Industrial Revolution and worldwide trade, a way of life has been evolving in which it is possible for an entire community to enjoy comforts, and even luxuries, such as previously were accessible only to a small minority. A material fullness of life has been realized by each society that has reached an advanced technological level. Also the greater the number of industries and occupations and the more abundant the consumer goods,

Figure 8.6
Countries Chiefly Exporting a Single Commodity.

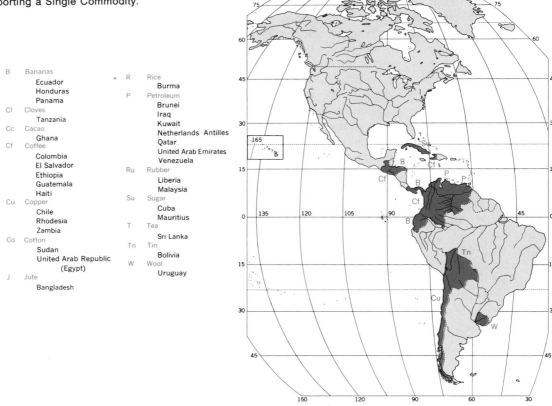

B	Bananas	R	Rice
	Ecuador		Burma
	Honduras	P	Petroleum
	Panama		Brunei
Cl	Cloves		Iraq
	Tanzania		Kuwait
Cc	Cacao		Netherlands Antilles
	Ghana		Qatar
Cf	Coffee		United Arab Emirates
	Colombia		Venezuela
	El Salvador	Ru	Rubber
	Ethiopia		Liberia
	Guatemala		Malaysia
	Haiti	Su	Sugar
Cu	Copper		Cuba
	Chile		Mauritius
	Rhodesia	T	Tea
	Zambia		Sri Lanka
Co	Cotton	Tn	Tin
	Sudan		Bolivia
	United Arab Republic	W	Wool
	(Egypt)		Uruguay
J	Jute		
	Bangladesh		

the greater is the choice in employment or business and the choice to buy whatever one pleases. Such conditions are now more available than ever to more people in those countries that are achieving a fair degree of industrialization. It is this level of living, of course, that the underdeveloped countries aspire to, and it is this level of living that is behind the urge toward industrialization on the part of the underdeveloped countries all over the world.

To emphasize how widespread are the present-day sources of production and manufacture of the commodities we use and consume in our daily living, let us examine the first hours of a "typical" American's morning.

Our solid American citizen awakens in a bed built in North Carolina from a variety of woods, including mahogany veneer made in South Korea from logs shipped from the Philippines. He throws back covers made from synthetic fibers processed from West Virginia coal, or of cotton from Texas, or linen from Ireland, or wool from Australia. These materials may have been spun and woven in Pennsylvania, or Hong Kong, West Germany, or Canada. He slips into his moccasins, made in Massachussets of leather from India and Argentina, and goes to the bathroom, whose fixtures were manufactured in Indiana and Los Angeles. He takes off his pajamas, made in

France, and washes with soap manufactured in Cincinnati. He then shaves with a stainless steel blade made in England of steel from Sweden.

Returning to the bedroom, he removes his clothes from a chair made in Denmark of teak from Thailand and proceeds to dress. He puts on his garments, all made in the United States, of wool from New Zealand, cotton from Egypt and the Sudan, buttons from Fiji, and puts around his neck a silk tie from Italy. Before going out for breakfast he glances through the window, of glass made in Illinois, and if it is raining puts on overshoes made in Taiwan of rubber from Malaysia, and takes an umbrella, made in Japan. Upon his head he puts a hat made in New York of felt from Uruguay or Iran.

On his way to breakfast, he stops to buy a newspaper, printed on paper from Canada, paying for it with coins made of copper from Arizona, Chile, or Zaire. At the restaurant he uses a plate made of a plastic from Ohio; his tableware is of stainless steel from Sweden and Japan, his glass from Belgium. He begins breakfast with pineapple from Malaysia or bananas from Honduras or perhaps a papaya from Mexico. With this he has cocoa from Ghana or coffee, blended from products of Brazil and Colombia, with cream and sugar. The milk was produced in, and delivered by tankertruck from, Wisconsin,

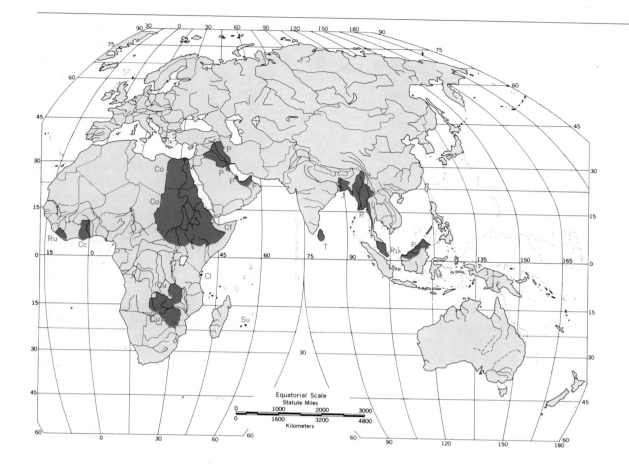

whereas the sugar was grown in Hawaii or the Philippines. The coffeepot was made in Japan of iron from Australia, chromium from the Philippines, and nickel from New Caledonia. After his fruit and first coffee, he goes on to waffles, made of wheat from North Dakota, processed in Minnesota. Over these he pours maple syrup from Canada. As a side dish he may have an egg from Delaware or Maryland, or a strip of bacon produced in Iowa, Denmark, or the Netherlands.

When our friend has finished eating, he settles back to smoke a cigarette from Virginia or a cigar from Spain or the Philippines. While smoking he reads the news of the day, printed in his town from wire-service reports of events in nearly 100 countries around the world. As he absorbs the accounts of foreign troubles, he might, if he is a good conservative citizen, express the thought that this world would be a better place in which to live if all peoples "stayed in their own backyard" and had as little as possible to do with each other, and that he, for one, was a firm advocate of the policy of "Buy American."

The Private Corporation as Space Adjuster

The twentieth century has witnessed the culmination of a new era in international trade. Numerous private corporations, as organizing and administrative devices, have burst the bounds of individual countries to become exceptionally large interterritorial operational units, thus practicing worldwide systems of space adjustment. These multinational corporations use the worldwide networks of travel and communication and can administer complex affairs with appropriate speed at reasonable costs. Within minutes, ships, trucks, and planes can receive orders that send them to distant places, expanding trade in some directions, perhaps curtailing it in others. The multinational corporation is able to allocate a "mix" of its resources—capital, administrative capacity, and technology—in order to achieve its goals. In so doing, existing spatial interchanges may take on new patterns; for example, a new assembly plant in a new country may substitute the import of raw materials for former shipments of finished products. Indeed, the movement of men and money may result in the full stoppage of physical movement of commodities, as a corporation branch plant in a foreign country becomes productive for that foreign market and imports cease.

The leadership in world business has been captured by the diversified multinational corporations. Although most of these are United States-based, there are also many with headquarters elsewhere around the world. For example, there were 161 industrial companies outside of the United States whose sales in 1970 each exceeded $500 million. The leader, by far, is Royal Dutch/Shell, a joint British-Netherlands corporation that produces gas and chemicals and operates almost around the world.

There is no doubt that the multinational corporation is an efficient form of organization for a one-world economic system, and that aspects of this pattern have great value for future developments. The large multinational corporations, however, hold tremendous power through their worldwide operations and their strong financial resources. Their economic strength is often greater than that of some of the political states in which they operate. From the viewpoint of political nationalism, such large private corporations clearly threaten the political autonomy and may threaten the economic viability of some states, since control over them is impossible at present by any one political state. Currently the multinational corporation is being blamed for many problems encountered in the underdeveloped world, and such corporations are viewed by many as prime examples of the evils of capitalism. There is a serious zone of conflict here between world political and world economic organization, for which there is no immediate solution of a satisfactory nature. We will have more to say about this kind of problem in our last chapter.

THE COMMUNICATIONS EXPLOSION

People today have the capacity to send and receive in countless ways both intended and unintended messages. The flow of communications sets the direction and pace of social change. Message sending has been a universal problem for all cultures, but the problem has varied with different cultural levels. All had the drum, smoke-signal, knotted-string, carved-stick, wampum-belt variety of techniques at the start, with the courier as a critical component. Those cultures that developed writing very early went ahead of the nonliterate cultures. Literacy improved the accuracy of communication, and regionally—as in Rome, India, China, the Maya Empire—the courier systems were highly organized, as regional administration depended on a regular volume and flow of messages.

Increasing the Efficiency of Cultural Processes

The "communications revolution" has been almost concurrent with the revolution in transportation. Regular communication over long distances could not exceed the speed of a messenger on horseback (e.g., the "pony express") until less than two centuries ago. It now seems appalling to contemplate that for most of human time on earth the process of communication was undifferentiated from other social processes. The flow of information was in accordance with traditional social hierarchies. Status relationships determined who received, evaluated, interpreted, and responded to communications. Customarily only the members of small elites were literate, and not until the late nineteenth century did the majority of the people of any nation possess such a capability. The concept of "mass media," now so commonplace, depends on "the mass" being able to read, understand, afford, and be receptive to new forms of the technologies of our "electrical age."

A modern communications system is capable of transmitting a massive flow of uniform messages to a wide audience. Today much of the function of informal, person-to-person communication in an advanced society consists of screening out and evaluating specialized information from the mass flow. Opinion-makers exist, in government, business, the professions, and other endeavors, who invest time and energy in "keeping up" on special subjects and being "fully informed." Others who may need their information are dependent on them for their detailed knowledge, as reported in lectures, articles, books, memoranda, or by television. The problem for most people in the world is to gain access to the flow of communications: through knowledge of one or more of the world's great languages, through education to literacy, through regular contact with one or more of the mass media.

A government-supported educational program is one of the cultural institutions now widely utilized by modern societies as a way to increase the spread of their cultural processes in an efficient manner to as many persons as possible at an early age. Educational systems and their agencies are formal means by which societies seek to fulfill their highest aspirations, by transmitting and transforming their cultures and by maintaining and upgrading their social systems. Educational systems are space-adjusting techniques, since they seek to create cultural uniformity and continuity over all the territory occupied by the society operating the educational system.

Eliminating Many Needs for Movement

All forms of communications are tools to be used for human purposes. One overriding purpose has been to bridge space and time. Ideas, today, can leap around the world in a matter of seconds. They are also stored (e.g., books in libraries, action on film, speech on tape) so that the present generation can learn about and profit from the experiences of its predecessors. The processes of communication include more than the transmittal of sounds as speech or the transfer of symbols as writing or printing. In addition to language, man has used gestures, facial expressions, dances, and pictures to convey meanings, often in effective combinations, as in the theater or motion pictures. Our concern with communications lie in the effectiveness of such means of space-adjusting techniques, for through communications many needs are eliminated for the movement of people or of goods.

Our present-day forms of mass communication—telegraph or teletype, telephone, radio, teletype-setting, wirephotos, phonograph, motion pictures, television—are all dependent on electricity to accomplish transmission across distances. The "electrical age," which has rendered our modern era so different from any previous period of human existence, began just a century ago. Although telegraph and telephone systems were diffused so rapidly around the world that they now operate in almost every political state on the globe, the electrical age has not yet permeated the rural hinterlands of many regions. Programs of bringing electric lighting to rural countrysides form one of the socially significant aspects of modernization going on in many underdeveloped parts of the earth.

Developments in wireless telegraphy ran ahead of the extended use of the telephone, but the telephone has become a necessity of life, both in the home and in business, in the United States and in many other countries (Fig. 8.7). The most recent development is the communications satellites, which will solve problems of intercontinental distance and heavy traffic, and make possible a truly worldwide linkage of telephone systems. This is a major step in space adjustment, permitting person-to-person long distance communication.

World War I put the newly developed "wireless set" to use in military communications. The postwar boom of the 1920s turned the wireless into the mass-produced radio set and put the radio into private homes as a means of mass communication for news distribution, education, and entertainment. The phonograph, based on Thomas A. Edison's device of 1877, was rendered practicable by electronic amplification, and by evolutionary change has become the record player of today. The transistor has replaced the vacuum tube and in so doing has ushered in an even newer "electronic age" of rapid mass communications based on the electronic computer. The speed of operations of cultural processes has taken a revolutionary jump forward, unknown to any previous era, not only in the incredibly rapid

actions of computerized programs and machines but also in the worldwide spread of inexpensive, portable radio receivers—the transistorized radios of pocket size that have reached nearly all the world's peoples. The facsimile, a device for photographic transmission, and the laser light for long-distance transmission, together with other new inventions are making possible the transfer of vast amounts of data at fantastic speeds.

Motion pictures and television, first in black and white, then in color, followed by tape-recording of sounds and kinescopes of both pictures and sounds, are all part of the twentieth-century technology that has revolutionized the worldwide communications to reach mass audiences. For many new nations television is a "status symbol," and there are more television receivers than telephones (Fig. 8.8).

Channels for International Communication

Most people receive the bulk of their ideas about other parts of the world from the same media that bring them local news. These domestic media rely principally on national and international wire services. The "big five" of these are: Reuters (serving 110 nations), Agence France-Presse (outlets in 104 countries), Tass (for Communist countries), and Associated Press (7600 outlets in 80 countries), and United Press International (6500 subscribers in 111 countries). In addition, there are some 178 national or specialized news services.

A second major source of information from abroad is radio and television. In 1970 there were more than 13,000 radio transmitters and 500 million radio receivers throughout the world, with the greatest rates of increase in Africa and Asia as inexpensive battery-powered transistor radios have become widely distributed. Television is spreading widely also; the number of television receivers outside the United States now vastly exceeds this country's 85 million. The airwaves are truly international, since it is easily possible to tune in on programs originating in neighboring nations. Most nations of the world now have some form of international broadcasting service.

The most recent trends in international communications consist of regional, continental, and intercontinental linkages. Eurovision, for example, links the television networks of western European countries with those of Tunisia, Yugoslavia, and Israel. Intervision includes the Soviet Union and six eastern European countries and exchanges some programs with Eurovision. The Asian Broadcasting Union and the African Radio Television Union became operative in 1964. Worldwide live television, via satellite relays, has become common in the past decade. The world is certainly a more communications-linked community when an important event, on tele-

NUMBER OF PERSONS
PER TELEPHONE IN 1971

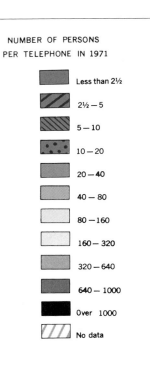

	Less than 2½
	2½ — 5
	5 — 10
	10 — 20
	20 — 40
	40 — 80
	80 — 160
	160 — 320
	320 — 640
	640 — 1000
	Over 1000
	No data

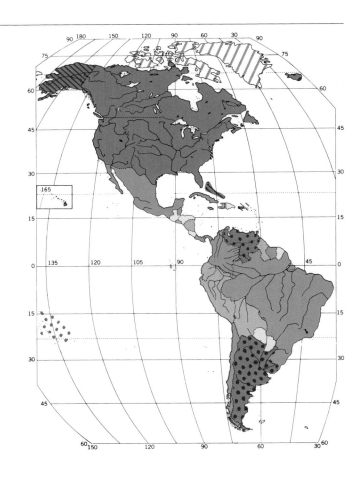

Figure 8.7
Distribution of
the World's Telephones, as of 1971.

NUMBER OF TELEVISION STATIONS
PER COUNTRY IN 1970

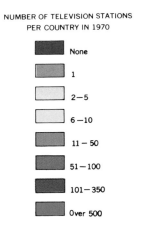

	None
	1
	2 — 5
	6 — 10
	11 — 50
	51 — 100
	101 — 350
	Over 500

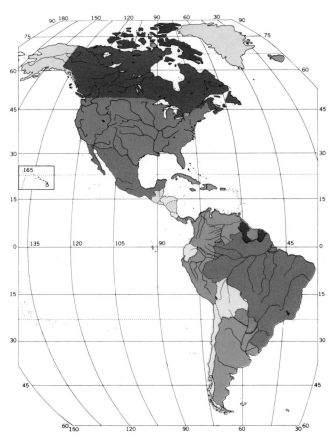

Figure 8.8
Distribution of
the World's Television Stations,
as of 1971.

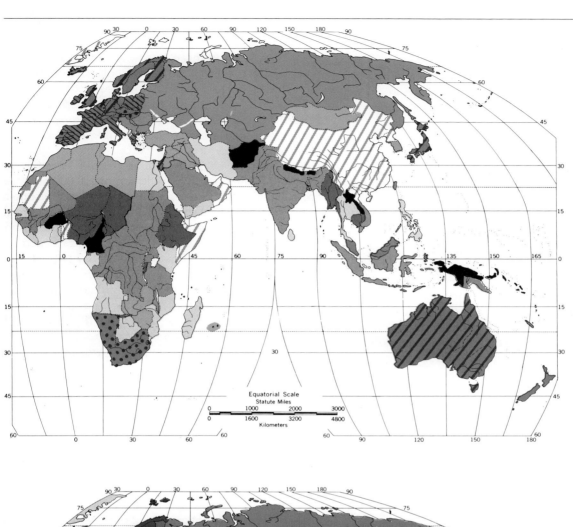

Equatorial Scale
Statute Miles
0   1000   2000   3000
0   1600   3200   4800
Kilometers

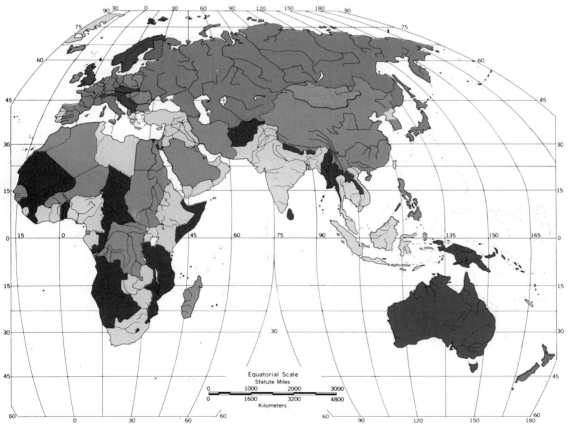

Equatorial Scale
Statute Miles
0   1000   2000   3000
0   1600   3200   4800
Kilometers

243

vision, can be witnessed simultaneously by 1 billion viewers.

Books and magazines travel widely, either in their original form or in translation. More than a third of all translated books are from English to other languages, followed, in decreasing ranks, by Russian, French, and German titles. Many popular magazines have many overseas subscribers; the *National Geographic Magazine* each month sends over a half-million copies abroad. Others, such as *Time* and *Reader's Digest*, print special overseas or international editions. The United Nations and its agencies publish more than 50 periodicals, many in several languages.

Most countries rely heavily on foreign-produced films for their many theaters. Films are second only to radio as the most widely available mass medium. Japan is the world leader in film production, followed by India and Italy, with the United States a poor fourth. There are many other forms of international communications: trade fairs, sporting events, cultural exchanges, travel, international education, private businesses, government intelligence, and diplomatic reporting services. The means and the opportunities for learning about the rest of the world are exceedingly diverse, yet all are interlinked and serve to correct or to reinforce one another.

THE CONCEPT OF SPACE INTENSIFICATION

The space-intensification techniques include all those processes that permit an increase in the use of space beyond that made possible by nature on the earth's surface. For example, the arts of civil engineering, architecture, and city and regional planning are major contributors of techniques for concentrating human activities in relatively small spaces. A modern city is thus a great engineering work, which, by adding levels and further usable area over the same land surface, permits an astonishing intensification of space utilization. Urbanism, thus, is a key element in the processes of space intensification.

Intensified use of space involves great concentration. Centripetal action and centralization become dominant in order in achieve high densities. Space-intensification techniques focus on particular sites and expand outward from such points to others. In intensifying available earth space, emphasis is placed on a new kind of relationship in the human ties with the earth: the view is vertical rather than extensive. The objectives of space intensification are to improve upon or "do better" than nature, either by increasing the rates of productivity for raw materials (fish, furs, forests, farms) beyond those previously achieved or by creating more usable surface area than previously existed.

Cities as Concentrations

In 1800 only about three percent of the world's people lived in the 750 urban areas of more than 5000 inhabitants; there were but 45 cities of over 100,000 population. Asia had the largest total population in urban agglomerations; Edo (Tokyo) was probably the largest single concentration the world had yet seen. In 1800 London was Europe's most populous city, yet with fewer than 1 million people. Twenty-one European cities with a population of 100,000 or more contained only about one thirty-fifth of Europe's people. By contrast, in 1900 there existed 11 great cities of over 1 million inhabitants each, London and Paris being joined by Berlin, Calcutta, Chicago, Moscow, New York, Philadelphia, St. Petersburg, Tokyo, and Vienna. By 1970 there were 174 cities of 1 million or more inhabitants that included well over 440 million people, or nearly one-eighth of the world's total. The 26 largest cities in 1970 each had over 4 million people in their urban agglomerations; their total of over 180 million people was five percent of the world's population.

Cities are also concentrations of human activities. In modern cities one expects to find political parties or factions, chambers of commerce, credit associations, labor unions, factories, churches, schools, newspapers, welfare agencies, humane societies, fraternal and civic associations, museums, art galleries, zoos, auditoriums, parks, playgrounds, business districts, residential areas (including slums), sanitation plants, and public transportation. The impress of high population densities creates a tremendous increase in demands for special services: streets, public water supplies, public sewage systems, garbage and trash disposal, police protection, fire protection, civic centers, schools, libraries, parks and playgrounds, transportation systems. Complicated administrative systems arise to handle the interrelated, complex problems of engineering, law, finance, education, and social welfare. Cities that become huge in size create problems that become huge in size, such as the residence arrangements, financing public services, the technological disposal of the many forms of urban wastes, and the means and manner of renewal of first-built portions of the urban areas. The operating complexity of most modern cities is such that all its systems as devised thus far seem to be very costly and at the same time, not work too well.

Evolution of the Industrial City

Not until about 100 years ago did the proportion of the human population concentrated in cities begin to increase significantly. The earliest cities were the products of societies that relied principally on animate sources of energy: the muscles of people and animals; the burning of wood, straw, and dung; and the forces of wind and falling water. Although possessing metallurgy and such devices as the plow and wheel, the earliest cities were preindustrial, supported by social surpluses in food production that could be collected, stored, and redistributed.

The stimulus for change from the preindustrial to the industrial city derives from commercial forces (not necessarily private secular forces only): the merchants, financiers, and landlords. The trading cities that changed most under a moneymaking economy were those in which private investment had the greatest freedom to accumulate private profits. Merchant adventurers expanded production, widened markets, furthered technological change, and brought raw materials and finished products from afar. In the emphasis on profitmaking speculations, more and more of the economic life lay outside the control of the traditional preindustrial towns and cities, which became ever less self-sufficient.

The production of goods that required safe and speedy distribution was the modern incentive for technological exploits that gave successive rise to stagecoaches, canal systems, docks and to warehouses, railroad lines, and streetcars. Mass transportation moved not only goods but also workmen; thus for the first time in history, walking distances between homes and places of work no longer set the limit to a city's dimensions. The expansion of public transportation made possible a vast horizontal extension of the nineteenth-century city.

Space was money; that is, land had a market price. Urban congestion, represented by the close-packed erection of tenements four and five stories high, resulted not from·efforts to provide residences for people but to create sources of sufficient rents to correspond with the income demanded by high-cost land. After it was discovered that there were even higher profits to be obtained from industry than from trade and rents, the nineteenth century saw the rise of industrial cities, generated by the mines, the factories, and the railroads. Coal and iron were the basic raw materials; the steam engine supplied the steady output of inanimate energy. Industrial cities were built so rapidly that their basic improvements over preindustrial cities were few: usually incomplete and inadequate systems of water pipes, gas mains, and sewers. Once the concentration of factories began, the increase in urban populations was overwhelming. Manchester in England had about 6000 people in 1685; about 40,000 in 1760; 303,382 by 1851.

The steam engine changed the scale of activity. Industrial factories became both large in size, to effect economies, and concentrated in form, to draw power from the central source. More labor was needed in the large factories, and more people thus lived in residences close by. New population centers settled around industrial sites located on or near coal areas, along the railroad lines, in the transport junctions, and at the ports. Factories, railroads, and residential slums were the prime constituents of the industrial cities. Factories occupied the best sites, near water—for transport, boiler supply, cooling agent, or solvent, and also for the discharge of wastes. Factories and railroads became formidable concentrations; the city's residential area consisted of the jumbled sectors not used for the railroads, switching yards, dump heaps, and factories. Utilities and government, police and fire protection, hospitals and schools, plumbing and sanitation, were often belated additions. Smoke, fumes, dirt, and noise were omnipresent and affected masses of people, who had to accept these conditions as inevitable concomitants of city life.

The breaking of the bounds of city growth was an achievement of trade and transport. But in the realization of increased size the industrial city fed on its own products. Public hygiene or sanitation was a nineteenth-century achievement. New standards of light, air, and cleanliness were made possible by the molding of large glazed drains and the casting of iron pipes for use in bringing fresh water from distant sources and disposing of sewage. Early in the present century improved street lighting became common in most urban residential areas. Provision of such services became a city responsibility, and for the first time the sanitary improvements known only to the palaces and the elite of preindustrial cities were extended to entire populations.

Expansion to Metropolis and Megalopolis

There are many other reasons for city growth in addition to favorable situation (coastal or river port, railroad center). There is always the role of accident, of personal decision, of the local invention of a new technique, or of deliberate action by planning. Especially important is the action of a nation-state. It may select the biggest city as its capital (e.g., London) or, more commonly, the capital city becomes the largest city because it is the capital (Madrid, Berlin, Tokyo, Djakarta, Buenos Aires). Not all of the world's great cities are political capitals, however; many are economic centers (Milan, Sydney, New York). And not all political capitals are excessively large; some are wholly created according to plan (Canberra, Brasilia), others are nineteenth-century restorations (Athens, Rome) or European creations for colonial rule (Rangoon, Lima). Contemporary central place theory, concerned with the location and spatial ordering of cities and towns, has developed a significant body of knowledge concerning urban location, growth, and hierarchical status, which, in the modern world, can do much to explain the worldwide developing patterns of urbanization.

The impulses that have generated modern city growth were not those from within—the merchants, artisans, and city officials. Rather, the modern city is the result of four interrelated sets of external forces:

1. The centrally administered political nation-state;
2. The vastly improved communications that permitted the organized deployment of great numbers of people and materials;
3. The rational, capital-using and ultimately mechanized material economy; and
4. The sheer increases in human populations, since not all the burgeoning population could continue to reside in rural countrysides.

Some cities grew to become regional centers, each with its satellites and subsidiaries, and towns arose on the outskirts as former crossroad hamlets suddenly acquired vitality when industrial promoters chose their sites for new factories or an enterprising builder perhaps backed by a transit company developed the nucleus of a residential suburb. The new communities were functional offspring of the major urban centers, as colonies under the control or influence of a mother city. Urban areas were no longer single cities and became instead a metropolitan sprawl. The word *metropolis* literally means "mother city."

Urban growth concentrated large population densities in major towns. Improvements in transportation facilities often only spread congestion into new areas. As commercial and other functions increasingly occupied the downtown business district, the numbers of residents declined in the city's central wards, only to be replaced by even higher daytime populations at work in the city's soaring skyscrapers that marked every metropolis by 1915. Factories grew in size and were relocated on more spacious sites outside central business districts. These factories were often located beyond the city limits where taxes and land costs were lower, as soon as high-voltage electric power and surface transportation became available. Factory suburbs and dormitory communities appeared as functional subsidiary neighbors to the great core cities, which assumed the character of regional capitals.

Thus a further image emerges, a wholly new concept of regional "cities" covering hundreds of square miles. Urban growth has already created several vast agglomerations of varying texture (*megalopolises*), in which farms and pastures so intermingle with intense developments, factories, and shopping centers that to distinguish the individual patches as city, suburb, or country is a meaningless exercise; there will be many more such in the future.

Vertical Intensification of Space Use

The symbol of verticality in space use is the multistoried skyscraper. An immediately conjured image is that of New York's Empire State Building, whose 102 stories reach 1250 feet and which reigned supreme from 1931 to 1971 as the world's tallest building. The title then passed to the 110-story twin-towered World Trade Center near the tip of lower Manhattan (see photo, p. 123), but in 1973 Chicago claimed the title with the 1450-foot Sears Roebuck & Co. Tower. The Sears Tower is comparable to a small city in that its air-conditioning equipment could cool 6000 homes, and its electrical system could serve 147,000 people. Such skyscrapers are now becoming common around the earth, and some other city may claim the title in the near future.

Many materials have been used in high-rise construction. Brick, flat limestone, steel, glass, aluminum, all have their advocates, and the steel skeleton closed in by curtain walls is now worldwide in use in large cities. But of all these perhaps nothing is as symbolic as concrete. More architects have used concrete with greater freedom and imagination since 1950 than in the previous 100 years, thanks to major developments in structural design and new methods of fabrication and construction.

The means toward verticality in space use include the elevator (or lift) and the escalator for rapid passenger and freight transport. With safety devices added, elevators came into widespread business use during the 1870s, at the same time that typewriters made office workers productive. After electricity (substituted for hydraulic power in 1889) simplified their operation, the use of elevators was extended to hotels and apartment houses. Elevators, which enabled top floors to command premium rents, were but one element in high-rise construction. Taller buildings required cast-iron and steel-frame construction, sheathed in fireproof materials.

The larger buildings were made practicable by the transportation facilities that brought crowds of people to the downtown shopping districts and office centers. They stimulated improvements in interior heating and lighting, in sanitary facilities, and especially in telephone communications to improve business efficiency. These features were advances of the 1880s. The first moving stairway, or escalator, was installed in 1892. Now, of course, the standard practice is to combine the use of power ramps and escalators (the capacity to move 8000 persons per hour is almost 30 times that of a single elevator) for continuous traffic flow at the lower levels with "local" and "express" elevators to the upper floors of tall buildings. The self-monitoring electronic devices installed in elevators since 1953, to work automatically without attendants, can be programmed to adjust elevator

The Sears Roebuck
Tower soars far above
the other tall buildings in
this early evening view of
Chicago.

locations in anticipation of expected changes in traffic flows. Through such control systems vertical transportation no longer has any need for human supervision.

Vertical intensification of space use results in the greatest concentration of people in urban areas that the world has ever experienced (Fig. 8.9). Unfortunately the customary census enumerations record where people reside (sleep) rather than where they work. Thus the central core (the Loop) of Chicago in 1950 reported 11,000 residents, whereas 275,000 per-

sons were employed there in the daytime. Residential densities in cities are considered low if they fall below 5000 persons per square mile, yet this figure is higher than the greatest recorded rural densities for Java, a most congested island by rural standards. In the zone immediately surrounding Chicago's Loop and extending two to three miles from the center, residential densities are over 30,000 per square mile. Beyond, at distances from four to six miles, densities are about 15,000, and in the eight-to-ten-mile zone they average about 12,000 per square mile.

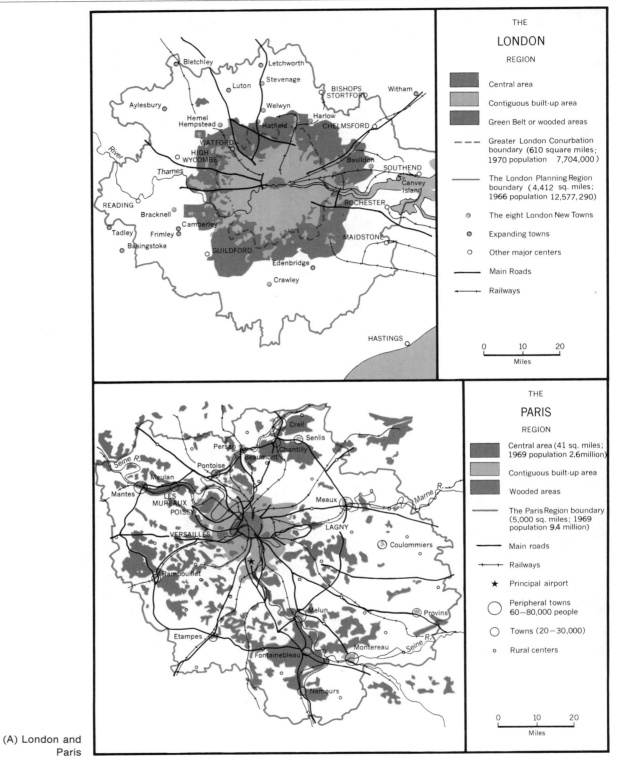

THE
LONDON
REGION

- Central area
- Contiguous built-up area
- Green Belt or wooded areas
- - - - Greater London Conurbation boundary (610 square miles; 1970 population 7,704,000)
- —— The London Planning Region boundary (4,412 sq. miles; 1966 population 12,577,290)
- ● The eight London New Towns
- ◉ Expanding towns
- ○ Other major centers
- —— Main Roads
- ┼┼┼ Railways

0   10   20
Miles

THE
PARIS
REGION

- Central area (41 sq. miles; 1969 population 2.6 million)
- Contiguous built-up area
- Wooded areas
- —— The Paris Region boundary (5,000 sq. miles; 1969 population 9.4 million)
- —— Main roads
- ┼┼┼ Railways
- ★ Principal airport
- ◯ Peripheral towns 60–80,000 people
- ○ Towns (20–30,000)
- ○ Rural centers

0   10   20
Miles

(A) London and Paris

Figure 8.9 Some Great World Metropolises, as of 1970.

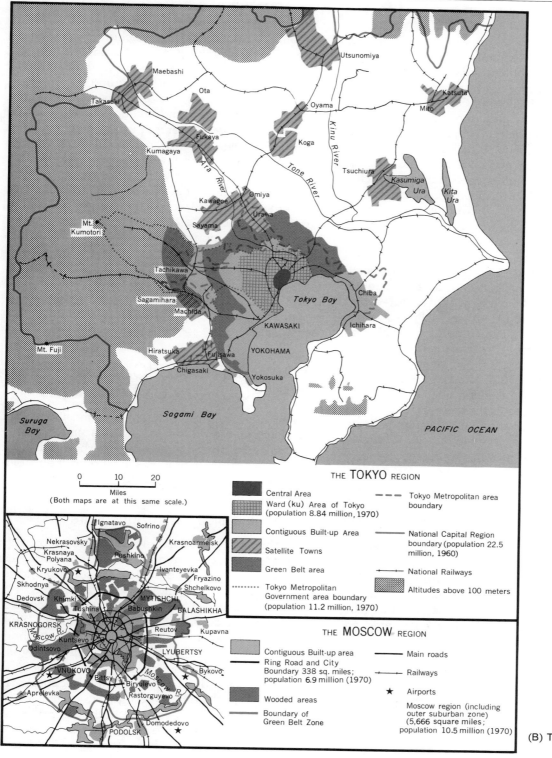

THE TOKYO REGION

- Central Area
- Ward (ku) Area of Tokyo (population 8.84 million, 1970)
- Contiguous Built-up Area
- Satellite Towns
- Green Belt area
- Tokyo Metropolitan Government area boundary (population 11.2 million, 1970)
- Tokyo Metropolitan area boundary
- National Capital Region boundary (population 22.5 million, 1960)
- National Railways
- Altitudes above 100 meters

0   10   20
Miles
(Both maps are at this same scale.)

THE MOSCOW REGION

- Contiguous Built-up area
- Ring Road and City Boundary 338 sq. miles; population 6.9 million (1970)
- Wooded areas
- Boundary of Green Belt Zone
- Main roads
- Railways
- Airports
- Moscow region (including outer suburban zone) (5,666 square miles; population 10.5 million (1970)

(B) Tokyo and Moscow

Urbanization: A Worldwide Process

Human populations have been forming into cities of various kinds for almost six millennia, but at no time before the nineteenth century did the urban areas in any country contain more than a small fraction of that country's total population. Only recently has it become possible and necessary to have the *majority* of the people of any country living in urban concentrations. From about 1850 to 1920, this was characteristic only of Great Britain: today all industrial nations are highly urbanized and nearly 40 percent of all the world's peoples live in urban places. Furthermore, the process of urbanization is rapidly accelerating. At the present rate of growth, more than half the world's people will live in urban areas with populations of 100,000 or more by the year 2000. With the appearance of urbanized societies, the twentieth century is experiencing a new and fundamental change in man's social evolution. That social revolution has not yet produced urban culture systems yielding satisfactory living patterns for much of mankind, and urbanism—the growth of cities—presents a serious cultural challenge. The challenge must be faced since urbanism is not likely to fade away if human life continues on the earth.

Urbanization is the rise in the *proportion* of a total population that is concentrated in urban settlements. As a change in a ratio, urbanization should not be confused with absolute growth. Urbanization results from the growth of both the numbers and sizes of cities, but is not the same as the growth of cities. Urbanization, an increase in the urban/rural proportion, has a beginning and an end, whereas cities may continue to grow in size as a result of general population increase with no change whatsoever in a stabilized, high-level urban/rural balance.

Now proceeding apace over the world is the transition from urban-based civilizations to urbanized societies. The urbanization of society is the most comprehensive, profound, and unprecedented social change affecting the present world. What was described as the process of cultural convergence is taking place in the context of gigantic regional cities in which most human beings soon will be living (see Fig. 8.9). The creation of predominantly urbanized societies is spreading outward, principally from the United States and Europe under the guise of economic or technological development, to engulf former city-based civilizations in a wholly new, higher-level arrangement and organization of mankind.

The 55 countries already urbanized are distributed widely over the globe. Thirteen, including Australia, New Zealand, Uruguay, Sweden, Belgium, West Germany, and Israel have over 80 percent of their populations living in urban areas. Eleven, namely Argentina, Chile, Venezuela, Canada, United States, United Kingdom, France, Netherlands, Denmark, East Germany, and Japan are in the 70–80 percent category. Twelve, including the Soviet Union, Czechoslovakia, Greece, Spain, Iraq, Taiwan, Mexico, Cuba, and Colombia are in the 60 percent group. Eighteen countries, including Puerto Rico, Panama, Peru, Brazil, Ireland, Switzerland, Italy, Austria, Hungary, Poland, Bulgaria, and Algeria each contain over 50 percent urban dwellers.

Around the world six extensive urbanized regions are clearly evident: northwest Europe, northeastern North America, the Pacific coast of Anglo-America, around the mouth of the Rio de la Plata, central to southern Japan, and southeast Australia. The urbanization process is unfolding in other smaller zones as local regional developments, with big cities appearing all over the world. Although the mass of world population still consists of rural, village, or town dwellers, such a clear trend toward city life has set in that one may well speculate what the pattern of clustering of world population will be, say, for the year 2500.

SPACE INTENSIFICATION IN RESOURCE EXTRACTION

Modern high-level productivity, an intensification of use of existing earth space, has become possible only because of the existence of urban-industrial economies and their products, including their abilities to harness inanimate energy.

Space Intensification in Agriculture

The technology of traditional agriculture was always subject to, and dominated by, the sociocultural aims of the society using a particular agricultural system. In the North American Midwest, after about 1880, two conditions caused a relaxation of these traditional cultural controls over agricultural practice. These were the marked and continuing rural labor shortages and the failure to integrate rapidly newly immigrant populations into the sociocultural systems of Canada and the United States. The labor shortages set the farmers of the Midwest onto the path of labor-saving technological revolution in the operations of agriculture. This created a new agricultural system, that of mechanized agriculture (see Figs. 4.3, 7.4, and 7.5; and see the discussion of the Transformation of the Midlatitude Forests and Grasslands on pp. 148–149). The looseness of the sociocultural situation permitted changing economic technology to become so dominant over long-term sociocultural aims that sociocultural patterns began to adjust to changing technology rather than to control it. It is perfectly true, of course, that a changing social outlook among the long-term residents of Canada and the

In many parts of southern China, as shown above, farm machinery remains very simple and large amounts of human energy are still required. In contrast, as shown below, the grain belt of the United States and Canada is highly mechanized. This allows farmers to make use of much larger areas of land.

United States was aiding and abetting the changing technology of agriculture. The resulting mechanized agricultural system not only operated with less labor per farm, but began to increase productivity per acre, setting in motion a pattern of space intensification.

The North American agricultural revolution began by adding horses to the small traditional teams to increase their power. In the twentieth century the shift was made to mechanical power, thus increasing the rate of change. From the incubator on the North American continent, these twentieth-century technological changes have been, and are, diffusing around the world. First affected were Argentina, Australia, and the Soviet Union. Japan, and varied sectors of the humid tropics and subtropics, each in its own time and system, then began to follow the lead of the nineteenth- and twentieth-century agricultural revolution in which mechanized and scientific technology are changing sociocultural patterns and economic relationships throughout the realm of agriculture. This is another era of the marked impingement of technology on systems of social culture, and the social unrest and economic turbulence of many parts of the world are the manifestations of the strength of modern scientific technology.

About one-fourth of the world's total cereal production is rice, and more than half of it is grown in the tropical and equatorial regions, mainly in Asia. Traditional yields are deplorably low by modern production standards. Average yields for tropical countries seldom exceed 1800 pounds per acre. The average Indian yield is about 1350 pounds, that of Thailand about 1450, and that of the Philippines about 1100 per acre. By contrast, temperate-zone countries, such as Japan, Australia, and those of the Mediterranean, have rice yields that range from 3600 to 5400 pounds per acre, partly as a result of better water control, land preparation, weed and pest control, and other practices. The quest for higher rice yields is proceeding in just those tropical regions where the food problem is most acute.

Most rice varieties can be classified loosely into either *indica* or *japonica* types. Indica types are typically those of the tropics: tall and weakstemmed, with a profusion of long, drooping, pale-green leaves, and late-maturing seeds. The grain usually remains dormant for a period following maturity, an important trait in the tropics where high humidity can cause grain to germinate before harvest. The tropical varieties of rice evolved through selection by many generations of farmers who required plants that would thrive under conditions of low fertility, poor water and weed control, and a rudimentary agricultural system. Improvements in cultivation practices, using traditional varieties, produce disappointing yields. Nitrogen application, for example, leads to taller and leafier indica rice plants but to little or no additional grain.

By means of recent modern scientific breeding programs a new and more desirable short-statured type of rice plant has been achieved combining the virtues of the traditional indicas with additional desired characteristics of high yield, nitrogen respon-

A large part of the productive land in Indonesia has been terraced to permit wet-field rice production, as shown in this scene on the island of Bali.

siveness, lodging resistance, and short growth duration. Yields, wherever tested, ranging from 5000 to 9000 pounds per acre, are from three to seven times the national averages for Southeast Asian countries. For the first time the contemporary Asian farmer has a food plant able to respond to high levels of nitrogen and attain rice yields that are astounding for the tropics. Consider the possible impact on the Asian food situation: if 15 percent of the rice lands of India could yield 1800 pounds per acre more than at present, then India's annual 10 million ton foodgrain deficit would be eliminated. In the Philippines an extra 900 pounds per acre from 10 percent of the rice lands would render that rice-importing country self-sufficient. Self-sufficiency was achieved, temporarily, in 1971–1973. A "Green Revolution" is now in process in Southeast and South Asia, a clear example of space intensification.

Plant geneticists have made tremendous strides by selecting characteristics from one species and passing the genes that carry those characteristics to a closely related species. Hybrid corn, nematode-resistant soybeans, and rust-resistant and short-straw wheat are among the more important results. Frequently the gene-passing must be accomplished through a third, or intermediate, species. Rust resistance, for instance, was passed from a wild grass to wheat through emmer, a feed grain. Even though recent advances have been impressive, there remain many serious problems yet unsolved. Some of the recent advances have even provoked serious problems of a new sort.

Progress in animal genetics, however, has been slower, primarily because animals are immensely more complicated organisms than plants. One of the research objectives, for example, is to enable animals breeders to transfer desirable characteristics, such as disease resistance, from wild to domestic animals.

INTENSIFICATION IN RESOURCE CONVERSION

The assembly line and mass production were techniques that enabled production to increase markedly, and twentieth-century industrial firms to expand at an unprecedented rate. Unlike the "entrepreneurial," or "one-man," rule of the early stages of the Industrial Revolution, one man as "boss" could no longer maintain a constant vigil on daily events, formulate managerial decisions, and initiate action. As a consequence, the first half of the twentieth century was marked by the growth of organizations whose managerial structures were pyramidal or hierarchical. Responsibility and authority were delegated to subordinates and enterprises were divided (dispersed) for purposes of management and control. Such an organization most effectively limits the flow of information between levels in the hierarchy, transmitting only a small fraction of the information that is created or becomes known at each level to the next higher (or lower) level.

The early 1940s witnessed the emergence of the electronic computer, with its near-infinite capacity for storing data and its near-instantaneous capacity for manipulating and retrieving them. A computer can handle all tasks that are "programmable," that is, those that can be precisely defined in structure and procedures. And the computer can do so more quickly, more accurately, more completely, and more economically than can human beings. Electronic data transmission and computation are capable of analyzing minute-by-minute activities of production and summarizing the analysis to management with a minimum of error and time lag. Centralized electronic decision-making is widely practiced today. The airline and the hotel reservation systems, signature-validating systems, police-record files, air-defense command, the satellite-tracking systems are just some of the uses. Once again it has become possible for one person (the manager), or a small group of persons (the executive board), to possess all the information necessary for production control.

The centralization of information and decision-making made possible by electronic data transmission and computation systems is reversing the trend toward specialization that prevailed in the earlier part of this century. Former individual or semiautonomous units are being made into larger ones; and what were single-purpose companies are able to expand rapidly into diversified activities. This explains the remarkably similar patterns of expansive growth among the world's largest business corporations during the mid-twentieth century. (See "Economic and Administrative Elements," pp. 203–204, and "The Private Corporation as Space-Adjuster" pp. 239–240.)

Electronic computer-based production systems are effecting a "second" Industrial Revolution, based on the techniques of systems analysis, which is still in the making. The computer can be applied to fundamentally new tasks as components to control such complex activities as air traffic patterns and petroleum refining. These latter are the truly new space-intensification techniques, because they have never existed before or have never been capable of being performed before. By speeding intercommunication among a large number of activities, by rapid and error-free calculations, and by rapid compilation and reporting of data, an almost unlimited range of choice is possible in designing control systems wherein the computer monitors and adjusts produc-

This plant in Arizona produces paper board for use in packaging. Many of the operations are controlled by a computer which is monitored by a supervisor at a central location.

tion to achieve explicitly stated objectives.

A fully effective computer-using production system integrates the design of the product, the design of the process by which the product is made, and the design of the control system that monitors and adjusts production to achieve set objectives. Fully automated factories are now in existence (machine tooling of engine blocks, petroleum refineries) in a few cases. Although they represent a radical departure from conventional techniques, they provide sufficient outstanding examples of what can be accomplished, as the approach becomes more widely adopted.

Automation is a new way of analyzing and organizing work, not a mere extension of mechanization. New machines, such as electronic computers, numerical tool controls, massive transfer machines, and the instrument panel of an oil refinery, have been made possible by the new concept. Automation links production, information-handling processes, or both into an integrated system by applying the principle called *feedback* in the design of machines that can control their own operations and be free from the limitations of their human operators. The idea of automation is not new; what is new is its widespread application. The fundamental importance of automation is the ability to create automatic information and control systems that both multiply productivity and efficiently distribute commodities to ensure guaranteed minima of life and that also free men to undertake new tasks not previously possible.

New Sources of Water

The need for developing unexploited sources of potable water is well understood. Analyses of United States water reserves from conventional sources (i.e., underground aquifers, flowing streams, and lakes) indicate that the nation's water use now exceeds, or soon will exceed, locally available water supplies. Conservation and improved use of existing sources, although helpful, will be inadequate to meet the projected increase in need. The practical alternatives in seeking new sources of water are the following:

1. Develop untapped conventional sources, including interbasin transfer.
2. Collect, process, and reclaim sewage or effluent waters.
3. Exploit saline water (brackish or seawater) by construction of desalting facilities.

The first of these alternatives extends the conventional means of resource extraction and space adjustment. The latter two alternatives, through their application of modern technology, are examples of space intensification in resource conversion.

Water renovation, to meet drinking-water standards, requires full treatment, an absorption process involving activated carbon plus chlorination. To approach a full recycle system would also require inclusion of a desalting process. If treatment of effluents sufficient to permit their discharge are considered as sewage costs, and only the final steps of renovation are charged as water-supply expense, production costs would approximate 50 to 60 cents per thousand gallons. Presumably much of this renovated water would be used to recharge existing underground aquifers, particularly in areas where ground waters are approaching the upper limit of salinity tolerable in potable water supplies. The high-purity level of the renovated water will reduce the salinity level of the total mixed waters and make additional, now marginal, sources immediately available as usable supply without the expense of developing entirely new sources.

Of all the earth's water, 97 percent is in the ocean. Desalination is a proven and feasible technology, and shore-based commercial plants have been producing fresh water from the sea since the 1940s. The conversion of seawater to high-quality fresh water is an extraordinary achievement in the application of space intensification to resource conversion. The high quality of product, the full-year seasonal reliability, and the potentially close physical relationship of source to point of use are factors favoring desalination processes.

SPACE INTENSIFICATION IN SPACE ADJUSTMENT

New technologies of miniaturization and microminiaturization involving radical new techniques that pack components in ever-tinier packages but with multiplied powers, are being applied to communications, transport, industrial processes, and to living spaces.

Satellite communication represents a vast extension of humanity, because communication is possible almost without regard to distance. That is, unlike the cost of messages sent by telegraph cable, telephone cable, and microwave relay, the cost of satellite communication is nearly independent of distance. Extensive use is being made of the *Relay*, *Telstar*, *Syncom*, and *Early Bird* communications satellites to bring events and programs of foreign origin, as they are happening, to television screens in the United States. Future communications satellites will be larger, will cover nearly all areas of the world, and will be capable of broadcasting directly into the home.

Electronic films are being used increasingly to reduce further the space requirements of television equipment, computer logic and memory circuits, two-way communications systems, missile and spacecraft controls, and pocket radios. As distinguished from photographic film, electronic films are

delicate tattoos of electronically active material which have been condensed from hot vapors onto cold, hard insulating surfaces such as glass. Uses of such films may eventually lead to a television camera only half an inch square, a hand-held battery-operated computer, a superconductive computer memory capable of storing 250,000 bits of information on a glass slide five inches square, videotape that can store pictures optically for later readout by an electron beam.

Computers promise a millionfold increase in man's capacity to handle information. Already familiar is their use in solving business problems and in increasing the efficiency of industrial operations. But computers are finding other applications as well—in medicine, the library, education, and in credit and finance. Tape-recorded and processed cardiograms, scoring of personality tests, and automated simultaneous testing of blood samples are being accomplished at various medical centers. Evaluation of information, its programming onto tape, and its instantaneous retrievability by computer are the bases for the building of centers of scientific information exchange from which those to whom it is available will have virtually push-button access to up-to-date information. The ultimate impact of the computer lies in the integrated data-processing system. A person need only carry a bank identification card for insertion into a communication device between store and the computer of the bank. This represents the direction in which the advanced urban culture systems are moving, even though it will be some decades before such devices will be worldwide and in use by all people. Mankind may need to accept such patterns unless the total population of the earth can be held to a level compatible with simpler systems.

A SUMMING UP: INCREASING INTERDEPENDENCE

The space-adjusting techniques of transportation, trade, and communication have all been used on an increasing scale of frequency, distance, and volume to create interlocking worldwide networks that serve one another. No one of the techniques could have been developed to its present order of magnitude without the existence and support of the other techniques. The end result is that the world's people now have available more frequent, faster, and more voluminous means of, communications, travel, and transport than ever existed. For example, just over a century ago, Hawaii was a native kingdom linked by slow sailing ships to other ports of the world—a far cry from the present, multi-cultural American state linked by telephone, radio, and communication satellite to the rest of the world. Now, just one among 21 airlines—international corporations selling transportation—connects Hawaii to 127 cities and 87 countries on six continents.

Modern means of communication have unified the earth, but they have not created uniformity. The discovery that the earth is "global" does not change the fact that peoples are, first of all, parts of their local culture systems; they have their own interdependent relationships of person to person, and humans to nature, their own traditions, preferred institutions, and ways of living.

Certain values have been idealized by all the great world religions as the highest of which mankind is capable: loving kindness, gentleness, nonviolence, brotherly helpfulness to all others. These values express the essential cooperativeness or sense of mutual aid that is fundamental to the life process. Awareness of the contemporary hazard of mutual destruction is more intense and widespread than ever before. Nuclear fission and the atom bomb did not cause international hostility; it is the other way around. International (interculture group) hostility has long existed among mankind. But nuclear fission and warheads have made international hostility a hazardous and outmoded concept for furthering human existence. More people than ever now understand that war is incompatible with the continuance of an urban-industrial, technological level, and way, of life.

By the end of the nineteenth century the European nation-states were bound by a vast network of international organizations: unions, tribunals, commissions, bureaus, congresses, each of which had limited functions, but which, in sum, expressed a felt need for international peace and order. The twentieth century has continued the trend toward interdependence and greater integration, expanding its regional focus from Europe to the entire world. At the same time, European-derived political nationalism has spread to the ends of the earth in recent decades. Aspects of these two trends are in contradiction and competition, but there are signs of hope; the trends toward unity are being furthered by different, interlocking, and concurrent movements. We need here only mention the political and economic integration of western Europe and North America that ensured the recovery of Europe's economy after two devastating wars, the worldwide agreement that led to the formation in 1944 of the International Civil Aviation Organization (ICAO) to coordinate the facilities and service for air transport, and the establishment of the United Nations in 1945 as a world organization for the maintenance of global and collective security.

The essential thing about our world is the vast amount of cooperation that exists, even between countries that are opposed to each other in political ideology. Much more publicity is given to conflicts and disasters than to the achievements in cooperation. But why should more attention be focused upon events rather than processes, when our evolving world depends on cooperation and not on conflict? The network of cooperation among nations both in technical and political fields continues to be more widely spread and more closely woven than ever.

There is beginning to emerge an enlarged view of social responsibility, involving concern for, and loyalty to, mankind rather than to nation-states, and a conception of the latter as subsystems of a larger whole. Differences among cultures and individuals are to be valued as subsystems, both as ends and means. A genuine system of world integrative relationships need not be predicated on the elimination of conflict, but calls for the use of increased skills in conflict management without overt violence, emotions of violence, or undue tension.

The space-intensification techniques include all those ways in which humans utilize space to its greatest advantage for themselves. In part, this has been achieved by the "feedback" process in which the products of industrial technology are applied to intensify resource extraction, resource conversion, and space adjustment. But it also has meant the supplanting, in part, of wild and cultivated landscapes by an artificial or human-made landscape. The creation of urban concentrations has led the way to a new era of human densities on the earth; urban areas are demonstrative of how to support the most people in the least space. Cities, thus, are the most efficient areal technique invented by mankind for maximizing production and consumption.

The evidence of the twentieth century indicates that mankind has expressed a preference for living in, or in close contact with, urban areas with their potential amenities for prolonging and enhancing life. The urbanization process and its industrial-technological by-products of improved shelter, food, health, and transportation are avidly followed and sought by most of the world's people. The urban way is in process of worldwide diffusion without much need of persuasion or compulsion. However, in the late twentieth century mankind has not solved the problems of operating large cities successfully, and there are many who decry further urban growth and seek refuge from it. Huge urban settlements are still young institutional developments, and there is very great need for the invention of socioeconomic, technological, and organizational systems that will make them practical. The earth now has too many people to permit a return to dispersed and village-living systems. And man still has much to learn about the ecological implications of his new urban systems. It is the cultural and biotic environments, rather than the purely physical one, that now present the technological and evolutionary challenges for the whole of the future of mankind.

SUMMARY OF OBJECTIVES

After you have read this chapter, you should be able to:

Page

1. Distinguish, with several examples, between a space-adjusting technique and a space-intensifying technique.   227

2. Describe a familiar area (e.g., your home and neighborhood) in terms of the four qualities of geographic space.   227

3. Explain what is really meant by the statement "Our planet has shrunk as a result of modern 'revolution' in transport and communication."   228

4. Explain why nation-states are both techniques *for* spatial adjustment and, at the same time, a *hindrance* to spatial interchange.   229

5. Discuss why so many different forms of transportation are considered necessary.   229

6. Elaborate on the idea that new nations have an advantage over older industrialized nations in establishing an integrated transportation system.   231

7. Describe the pattern of the major routes of international trade.   235

8. Analyze your activities for one day and itemize all of the things you use or consume that are extracted, converted, or produced outside of the United States.   238

9. Explain in what ways a multinational corporation can engage in space adjustment.   239

10. Describe your dependence upon products of "the electrical age" and the "mass communications revolution." How are these space-adjusting processes?   240

11. Analyze Figures 8.7 and 8.8. and explain the similarities and the differences in the patterns of distribution of the world's telephones and television stations.   241

12. Explain why cities, as we now know them, began as products of the Industrial Revolution.   244

13. Describe how metropolitan areas form, with some becoming megalopolises.   245

14. Present reasons why intense verticality in space use is also dependent upon "the electrical age."   246

15. Differentiate between urbanism and urbanization.   250

16. Explain how such countries as Japan, Taiwan, and Kuwait can be considered urbanized societies, since they were never colonies of European nations.   250

17. Understand how a State Agricultural Experiment Station and an Agricultural Extension Service are agents for space intensification.   252

18. Describe how the use of computers and automation can intensify both resource conversion and space adjustment.   253

19. Write an essay outlining how the urban-industrial way of life is synonymous with dense concentrations of people linked by intricate systems of communication, transportation, and trade, and the degree to which this way of life has diffused worldwide to create a trend toward worldwide integration and interdependence.   255

CHAPTER 9
THE POPULATION SPIRAL

---KEY CONCEPTS---

The Evolution of Population Produced Four
    Billion People
    The Human Capacity to Reproduce Is Very
      Great
    Primitive Society's Vital Statistics Indicate
      Slow Early Gains
    Thomas Robert Malthus Stated a Principle of
      Population
The Prevalence of People Is Very Pronounced
    The Distribution of People Over the Earth Is
      Uneven
    Increased Human Longevity Is Not the Same
      Everywhere
    There Are six Different Phases of Population

    Change
The Population Explosion Belongs to the
    Twentieth Century
World Population Might Triple in Your Lifetime
Population Problems Are Culturally Created
    People as Consumers Vary Among Different
      Cultures
    The Long-Run Balance to Resources Must
      Involve Population Control
    Cities Present Problems of Human
      Concentration
    The Vital Transition Is Part of the Multiple
      Transition
    Modernization Is a Worldwide Phenomenon

The human population of the earth totalled slightly more than 4 billion in 1976. The numbers of humans exceed the sum total of all the other 500-plus species of primates. Among the mammals, mankind is at present the most abundant single species. This figure, and the numerical dominance by mankind, are the direct result of the cumulative achievements of the development of culture by the human species. The world has never before experienced, among the higher animals, such a population-growth pattern as has taken place among human beings in recent centuries. For example, in 1977 seven countries had populations of more than 100 million each. These were the People's Republic of China, 850 million; India, 623 million; The U.S.S.R., 259 million; the United States, 217 million; Indonesia, 137 million; Japan, 114 million; and Brazil, 112 million. Another group of slightly smaller countries are Bangladesh, Pakistan, Nigeria, and Mexico. Each of the latter

countries today possesses more people than did the whole earth only 6,000 years ago when the first advances of civilization were beginning to shape new ways of living.

THE EVOLUTION OF POPULATION

The astounding feature is that such an abundance of human beings could ever have occurred at all. The other animals that exist in huge numbers are those—unlike humans—whose body size is small and whose reproductive cycles are short. There are only about sixty species of mammals of our size or larger. The human being is at a great disadvantage in the reproductive cycle as compared with most other life-forms. So much time is spent in reaching sexual maturity, as a result of culturally induced delay between the onset of puberty and actual childbearing, that the average human generation covers a span of about 25 years.

The human specialization lies not only in the long postponement of the onset of fertility, but also in the marked elongation of the fertile period (about 28 to 30 years). It is a notable cultural achievement that the average human life span is now more than twice that of other primates. That more and more individuals have an increased opportunity for successful completion of their full reproductive cycles is a consequence of biological adaptation to the acquisition of culture. As a result, however, the numbers of mankind can now be described only in terms of overabundance in too many parts of the earth.

Human Capacity to Reproduce

The number of children actually born to each couple is generally far less than the maximum number possible. The decisions to produce fewer children than the maximum possible are personal, but invariably culturally influenced. If these cultural impediments to productivity were cast aside, some very disastrous patterns of population increase would result.

Around the beginning of the nineteenth century, English and American women were having an average of 7 or 8 children, and families of 15 to 20 were not unknown. For example, John Wesley was the 15th child in a family of 19 children; Benjamin Franklin was the sixth son in a family of ten children; and the author, Washington Irving was the eleventh and last son in his family! Today, there are many places in the world (such as among settled Arabs in southwest Asia) where each woman has, on the average, about eight children. Under conditions in which women live through their full span of reproductive years and exercise their full reproductive capacities, they would bear on the average well over eight children. Since almost half of all infants are girls, each woman could contribute, on the average, four potential mothers to the next generation. If we assume that there are four generations in a century, a wholly unimpeded exercise of human productive capacity could give a 256-fold increase in the span of 100 years:

One mother could produce

4 girl babies of second generation, who could produce

16 girl babies of third generation, who could produce

64 girl babies of fourth generation, who could produce

256 girl babies by the end of the century after their great-great-grandmother was born.

Continuation of the fertility pattern of colonial America would have yielded an approximation of this trend. Obviously, the pattern has not prevailed, or else the population of the present United States (without any further influx of immigrants since Independence) would have exceeded by a thousandfold the present total population of the world.

It is not difficult to see that a 256-fold increase in one century or a 16,384-fold increase over two centuries could give rise to overwhelming problems. These figures are cited only to indicate what is biologically possible, that the potential for population growth is explosive. Fortunately, no large population or nation has ever produced at maximum capacity. The actual number of live births (fertility) has always been far below the maximum reproductive capacity (fecundity).

What is, and has been, the actual situation of population increase? For our answer we look first at the human fossil record and contemporary primitive societies to secure some indications of human living conditions prior to the development of food production. Second, we shall look at the actual record of population growth over the world to the present time. Our concern is, first, to arrive at an understanding of why a difference exists in human reproduction between a biologically maximum capacity and a culturally inhibited productivity of lesser quantity, then we wish to determine the order of magnitude of this difference. If mankind is biologically capable of reproducing 256-fold in a century, but is not doing so, then at what rates has it increased?

Primitive Society and Its Vital Statistics

The difficulties in estimating possible populations, their growth rates, and their areal densities, for the earlier Paleolithic are enormous. During the earliest portion of the human time span, mankind can only have engaged in simple and unspecialized food-gathering of a sort not much more efficient than the systems of other omnivorous primates. In time, with the learning of basic skills, early Paleolithic peoples must have developed some primary systems that were not purely ecological, as with the primates, but depended on cultural abilities to convert simple resources into more usable materials and foodstuffs.

About the best that can be done, still, in such patterns of estimation is to observe the sizes of populations that can be supported by contemporary gatherers and collectors, determine the areas of territories over which they range, and calculate their population densities, assuming that conditions in the past were somewhat similar. Australian aborigines, for example, inhabited a continent that provided meager resources for the support of people having only very simple equipment. Populations had to be dispersed in relatively small groups. The search for sustenance required long seasonal migrations during which aged or infirm persons sometimes had to be abandoned. The inhabitants of the whole island-continent of Australia at the time of its discovery by Western man probably numbered between one-quarter and one-third million. The crude population density of aboriginal Australia was thus of the order of magnitude of one person per ten square miles. Locally, where circumstances were highly favorable, a density of as many as one person per two square miles can be supported by gathering and collecting, in a semi-sedentary system.

The life cycles of the earliest humans, and the length of an average generation, were very short. Only rare individuals survived to pass through a full reproductive cycle into old age. Human deaths from predators, diseases, and famines were so common that the extremely high mortality could only have been offset by high fertility. Thus, both birthrates and death rates were high. Since mankind survived, births obviously were fractionally greater than deaths.

Cultural traits and circumstances, quite early, must have placed many "roadblocks" in the way of maximizing births. As mating systems became marriage systems, ceremonial customs and taboos must have come into operation to create a considerable gap between fecundity and fertility. Marriage customs can postpone a first child until long after a female reaches puberty. Many ancient customs block the biological sex function, whereas others lead to abortion. The historic continuance of infanticide suggests an ancient custom widely practiced in circumstances in which the young could not be cared for. The conclusion must be reached that the growth of the human population during most of the Paleolithic was so slow as to be almost imperceptible.

In the mid-to-late Paleolithic, with the occupation of the temperate lands that yielded good returns to hunters and specialized collectors, there must have begun to occur some relatively important increases in human numbers. As the Paleolithic drew to a close, and the plant-and-animal domestications began to occur, human numbers again must have increased relatively rapidly. We cannot effectively estimate increases during either of these periods. Starting, however, from a biologically valid estimate of perhaps 125,000 individuals during the earliest Paleolithic, estimates suggest a total approaching 10 million people by about 9000 B.C. To accomplish this, starting 2 million years ago, all that is necessary is a growth rate of about two-tenths of one percent per century. For a regional tribe of roughly 1000 people, toward the end of the era, that rate would produce a

Table 9.1 World Population Growth over the Past 11,000 Years*

Time	World Population Total
9000 B.C.	Between 5 and 10 million
Time of Christ	Between 200 and 300 million
A.D. 1650	500 million
A.D. 1850	1,000 million (1 billion)
A.D. 1925	2,000 million (2 billion)
A.D. 1962	3,000 million (3 billion)
A.D. 1975	4,000 million (4 billion)

*Compare with Table 9.3, p. 267.

net gain of only one person every fifty years, a rate not noticeable during a lifetime.

Reconstruction of population growth patterns since 9000 B.C., indicates an enormous increase, with certain periods having more rapid rates of increase (Table 9.1). The early Neolithic period of crop/animal production stands out, as does the period just after the onset of civilization, when the culture hearths began their formulation of culture systems. Most of the sheer growth of numbers, however, has occurred in the last few centuries (Fig. 9.1). Whereas hundreds of millennia were required to accumulate 10 million people, the increase to a total of 1 billion came rather quickly (9000 B.C.–A.D. 1850). A doubling of that total, however, came in the amazingly short period of 75 years (1925). Only 37 more years were required to reach 3 billion, and a scant 13 years to reach 4 billion (1962), which meant a doubling in only the 50 years since 1925.

For all the long time span prior to the beginnings of cultivation of food, population growth among the world's gatherers and collectors could only have averaged about 0.02 per thousand per year, or a two percent increase in a thousand years. By contrast, the rate of present world population growth is nearly 20 per thousand per year. Hence there has occurred a thousandfold increase in mankind's rate of growth in numbers during his habitation of the earth. Of the total number of people that were ever born, nearly four percent are now living, but, however, living longer than their ancestors ever did. In creating a humanized earth, mankind has experienced a fantastic acceleration of its growth rate.

Thomas Robert Malthus and His Principle of Population

The person credited with first drawing attention to the problems created by the trend toward a rapid increase in human population is the Reverend Thomas Robert Malthus. In A.D. 1798 there was published in London a small book entitled *An Essay on the Principle of Population as it Affects the Future*

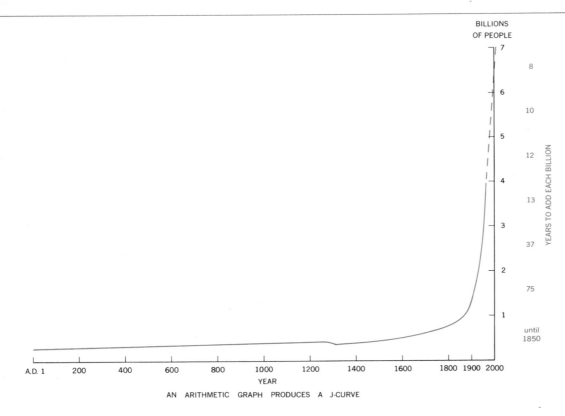

BILLIONS
OF PEOPLE

YEARS TO ADD EACH BILLION

8

10

12

13

37

75

until
1850

AN ARITHMETIC GRAPH PRODUCES A J-CURVE

Figure 9.1
World Population
Growth.

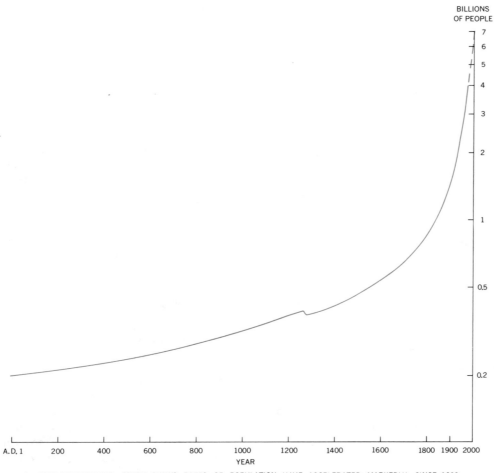

BILLIONS
OF PEOPLE

A SEMI-LOGARITHMIC GRAPH SHOWS RATES OF POPULATION HAVE ACCELERATED MARKEDLY SINCE 1600

Improvement of Society. In the longer and more carefully documented second edition, in 1803, Malthus discarded anonymity when setting forth the basic statement of his principles.

Malthus said that, once unchecked, populations tend to grow in a geometric progression—1, 2, 4, 8, 16, 32—doubling in successive generations about every 25 years. However, he reasoned, man's food supplies, so necessary for his continued existence, could not be increased in more than arithmetic progression—1, 2, 3, 4, 5, 6. A doubling of food production over a 25-year period by improved farming methods or by opening new lands to cultivation carries no assurance that another doubling can occur in the next 25 years. During each period the increase in food production could not be expected to be equal, Malthus contended, as greater energy and ingenuity are required to grow more food on marginally less valuable land not previously in use. Hence, Malthus concluded, population growth inevitably outruns food supplies.

Malthus's *Principle* stated that:

1. Population is necessarily limited by the means of subsistence.
2. Population invariably increases where the means of subsistence increases, unless prevented by some very powerful and obvious checks.
3. These checks, and the checks which repress the superior power of population and keep its effects on a level with the means of subsistence, all are resolvable into moral restraint, vice, and misery.

Malthus's "moral restraint" included continence and delayed marriage. These were the preventive checks that he felt man could exert to curtail his own numbers and reap the rewards of better living conditions. But he was not hopeful that the mass of the population would exercise sufficient restraint, even under the threat of poverty. He foresaw that the positive checks on population growth were starvation, disease, and war.

Malthus was not so much wrong as he was premature. He recognized that migration and improved techniques of production would temporarily postpone the difficulties engendered by population increase. But he could not have been expected to foresee the tremendous burst of productivity in the modern West during the nineteenth and twentieth centuries, which brought progressive release from the positive checks on population growth. Malthus also could not foresee that Europe during the same period would exercise a preventive check by drastically reducing its birthrates through the spread of birthcontrol practices. It was not poverty and disease in Europe but the improved living conditions and rising aspirations that motivated the trend toward

birth regulation. Not necessity among the suffering poor induced restraint, but the exercise of human choice (combining foresight with prudence) among the middle and upper classes started the trend toward reduced fertility. The modern trend, which resulted in a slowing of Europe's rate of increase, came late, however, after several centuries during which Europeans and their descendants increased faster than the peoples of the rest of the world.

THE MODERN PREVALENCE OF PEOPLE

The Distribution of People Over the Earth

No one can ever know exactly how many people there are in the world at any one time. Births and deaths in many countries are recorded incompletely, or not at all. Censuses are not even taken for large parts of the earth, and some censuses that are taken are quite inexact. The estimates in current use are collected from each country and are evaluated by the Statistical Office and the Population Division of the United Nations. The midyear 1977 estimate was that the world total population stood at 4083 million. Breaking this down by continents results in some marked contrasts (Table 9.2).

Table 9.2 Population by Continents, 1976

Area	Number	Percent
Asia	2,287,000,000	56.9
Europe (including U.S.S.R.)	733,000,000	18.2
Africa	413,000,000	10.3
Latin America	326,000,000	8.1
North America	238,000,000	5.9
Oceania (including Australia)	22,000,000	0.6
TOTAL	4,019,000,000	100.0

The world's population is, and always has been, unevenly distributed over the earth's surface (see back endpaper map). Some areas are empty or nearly so, whereas other areas literally swarm with humanity. It is truly astounding that about two-thirds of the people now present on the earth live on about seven percent of the land surface, whereas more than half the land surface supports only five percent of the people. The four regions of greatest population concentration—eastern Asia, south Asia, western and central Europe, and the northeastern United States—are the positive results of mankind's historic adjustments to, and development of, world resources. The eight great nearly empty areas—Greenland, northern Canada and Alaska, Siberia, central Asia,

Thomas Malthus, shown left, predicted that the human race would one day have problems caused by overpopulation. Above, the surviving members of a family in Bangladesh, one of the most heavily populated parts of the world, seem to offer proof of the truth of Malthus's prediction.

the Sahara, the Amazon Basin, central and northern Australia, and Antarctica—are the negative results of the same historic adjustment.

Asia contains two great areas of population concentration (see back endpaper map)—eastern and southern. Within each general area there are clearly observable variations of intensity. In eastern Asia the zones of densest population are (1) the three southern islands of Japan; (2) the western zone of Korea extending north into the Manchurian lowlands; (3) several regions within China—the North China plain south to include the Yangtze River lake basins and delta, the Szechwan basin, the lower valley of the Hsi River, and the southeast coast; and (4) the offshore islands of Taiwan and Hainan. Southeast Asia, in contrast, has only small clusters of dense population concentrations, the two largest being coastal Vietnam and the island of Java. The great populations of Bangladesh, India, and Pakistan are prominent (1) from the Ganges River delta westward to the Indus River, and (2) along both coasts of peninsular India. Southwestern Sri Lanka is also part of this high population cluster of south Asia. These east and south Asian clusters have been persistent and dominant: as early as 2000 years ago they contained at least one-half and possibly three-fourths of all mankind.

The striking feature of Europe is the concentration of population that focuses on the Low Countries at the southern end of the North Sea. The heavily populated coastal zone extends from the mouth of the Elbe River in Germany to the mouth of the Seine River in northern France. Two arms extend inland: One goes south up the Rhine River to Switzerland, down the Rhone River through southern France and thence to northern Italy. The other arm extends east along the southern margin of the North European Plain into the Ukraine. In addition there are several separate clusters: in England and northern Ireland, around Barcelona along the east coast of Spain, and around Oporto in northwest Portugal. However, north of 60° N. latitude (north of Stockholm) Europe is nearly empty. Another sparsely settled zone occurs in the depression north of the Caspian Sea. In general there is not the sharp transition in Europe that there is in Asia between the heavily populated areas and the empty ones. The correlation between peoples and plains, so prominent in Asia, cannot be so clearly illustrated in Europe, where the hill zone from France to Poland contains a denser population than most of the plains of European U.S.S.R.

The population pattern of Africa is particularly striking. The densest zone is in the lower valley of the Nile River, a phenomenon similar to the dense zones of Asia and quite unlike anything else in Africa. The clusters are widely scattered: along the Atlas coast in the northwest, the southern Gold Coast and Nigeria, the eastern highlands between Lakes Tanganyika and Victoria, and around the coastal ports of South Africa.

The fourth world region of greatest population concentration is in eastern North America. The area between Boston, Montreal, Chicago, and Washington, D.C., contains the densest population clusters, but the concentration of population is less clear-cut than in either Asia or Europe. A spotted or clustered pattern is even more noticeable than in Europe. Whereas the eastern United States is somewhat uniform, dense population in the western half is definitely clustered, as in Southern California, the San Francisco Bay region, or around Puget Sound in the northwest. The empty lands are in the great north (Alaska and Canada), extending southward through the western mountains and deserts into northern Mexico.

In South America the population is arranged in clusters about the fringes of the continent, concentrated in the highlands, in the temperate south, or along the coastlines, with each cluster isolated by empty, or nearly empty, land. Clockwise from the eastern tip of Brazil are the clusters around the cities of Recife, Salvador, Rio de Janeiro, São Paulo and Santos, Montevideo and Buenos Aires (in this largest cluster Montevideo contains almost half the population of Uruguay and Buenos Aires more than one-quarter of Argentina's population), and in the north around Caracas and in the uplands of Colombia, with Bogotá as the principal center. Strikingly apparent are the huge empty interior between the coastal zone and the highlands and the sparsely inhabited zone south of the 40th parallel.

Australia is the most nearly empty of all the continents except Antarctica. There are five nuclear cities: Perth in the southwest, and Adelaide, Melbourne, Sydney, and Brisbane in the southeast, but there are no extensive "filled-up" areas anywhere. The contrast between the nearly empty expanses of Australia and densely settled Java to the north is as great as between Mongolia and North China, or between the highlands of Tibet and the Ganges plain of north India.

Not only is the greater part of the world's population concentrated in four major regions, but these, except for Java, are all located in the northern hemisphere. The densely populated regions are further located in the midlatitudes of the northern continents of Eurasia and North America. Only tropical south India and Southeast Asia are the major exceptions. These midlatitude regions are the sites of the historically most productive agricultural lands, with at least a 100-day growing season, a favorable precipitation-evaporation ratio (or supplemented by irrigation), and relatively extensive lands of low relief, suitable for cultivation.

The four major world regions of greatest population densities are not alike. Peoples of the dominantly agrarian cultures extending from Egypt across south Asia to China are concentrated within the alluvial valleys of great rivers where irrigation agriculture has been most highly developed. The most populous zones of Europe and North America lie in areas where access to coal and iron in combination has encouraged the growth of modern manufacturing and of cities based on industry and transportation. The northwest European belt of dense settlement is an example. It stretches from central England eastward from French Flanders through the Low Countries and the lower Rhine into Bohemia and Polish Galicia. Petroleum and natural gas, in more recent decades, have served other industrial concentrations, as along the Gulf Coast of Louisiana and Texas or in Southern California.

The world's sparsely settled areas are of three main types: high-altitude and high-latitude, arid and semiarid, and tropical or equatorial. The sparsely settled parts of the world are dotted with small spots of denser settlement, depending almost entirely on local resources. Where water is available in arid regions, as on the alluvial piedmont of Sinkiang (westernmost China) or in exotic rivers such as the Amu Darya or the Nile, oases or strips of dense settlement are present. Where commercially exploitable industrial minerals exist, particularly petroleum (Arabia or south Sumatra) or metals such as copper (northern Chile, or Arizona, Nevada, and Utah in the United States), isolated dense settlements in otherwise empty regions also occur.

Some general conclusions can be drawn from an examination of the map of world population distribution (see back endpaper):

1. People are to be found in numbers only where there is a supply of usable fresh water (but cultures differ widely in the efficiency of technologies for securing water and rendering it usable).

2. People are sparse in high-latitude mountain uplands and subarctic lands with inadequate growing seasons.

3. People in numbers in agrarian-based civilizations are a function of the amount of level land in relation to the amount of water available.

4. People in numbers in industrial-based civilizations are a function of the amount and quality of industrially usable minerals, particularly coal and iron, available to an area (either locally mined or imported via world trade).

Increased Human Longevity

Very great differences in rates of growth of population have occurred among different regions of the earth. During three centuries of the modern era, 1650–1950, the world's population underwent a five-fold increase, from about 500 million to over 2.5 billion. In this period it was Europe and the overseas areas settled by Europeans which experienced the most rapid population growth (Table 9.3).

Many technological, social, and economic changes had their origin in Europe. These have combined to accelerate the rates of population growth in many countries. The principal effect has been a drastic drop in death rates, resulting in an unprecedented increase in the average longevity of life. For example, the average life expectancy of a Roman at the beginning of the Christian Era was no more than 30 years. For people in western Europe and North America, by 1650–1700, it was about 33 years. As recently as 1880, in New England, the life expectancy had climbed only to 48 years. Today, life expectancy in western Europe, Canada and the United States, the Soviet Union, and Japan ranges as high as 74, compared with a world average of 59. All insurance companies have their estimates for different parts of the earth, and they will "bet" on these

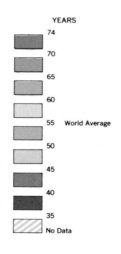

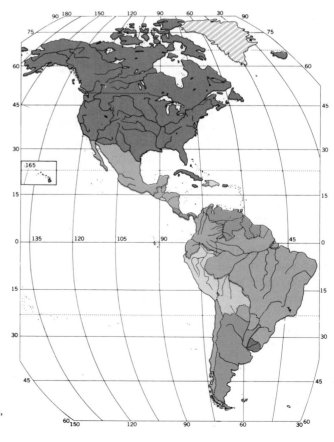

Figure 9.2
Life Expectancy at Birth of the World's People, by Country, as of Midyear 1976.

Table 9.3 Population Growth by Continents, A.D. 1650–1950

Region	Population (in millions)		Increase
	A.D. 1650	A.D. 1950	
Europe (including U.S.S.R.)	103	590	6 times
North America	1	168	168 times
Latin America	12	168	14 times
Oceania (including Australia)	2	13	6½ times
Asia	327	1450	4 times
Africa	100	200	2 times
TOTAL	545	2,584	5 times

figures by selling policies based on them (Fig. 9.2). During the present century there has been an increase of more than 23 years to the average life expectancy in the United States, a rather remarkable fact.

Over the span of human existence there have been two doublings of the expectation of the length of life from birth. From under 20 years in the earliest prehistoric period it almost doubled to 35–40 years for people in the preindustrial societies, and it has almost doubled again in the expectancy found in today's advanced societies. Perhaps no other achievement of modern civilization has been so glorified as having added so much to human happi-

ness. At the same time this very achievement now threatens the existence of future human happiness for all mankind.

Three factors contributed to the decline in death rates and these were first evident in Europe and areas of European settlement:

1. The rise in levels of living owing to
   a. technological advances, and
   b. peace and tranquility as a result of the establishment of relatively powerful and stable central governments.
2. The achievements in public health (environmental sanitation) and improved personal hygiene

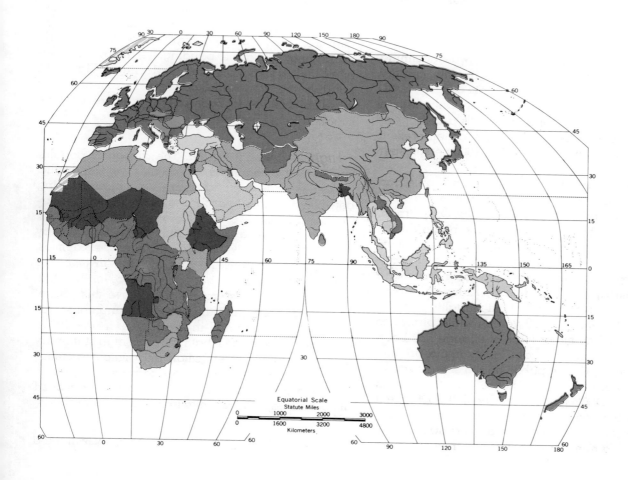

(e.g., purifying food and water, thus reducing parasitic and infectious diseases).

3. The growing contributions of modern public health (e.g., recent progress in insecticides, chemotherapy, and measures in sanitation).

What happened was an upset of the previous near equilibrium between high birthrates and equally high death rates that typified the previous millennia of human evolution. The European realization of the significance of the old Chinese invention of vaccination for smallpox was but the first of a number of dramatic discoveries that led to reduction in the numbers of deaths among infants and children. The increased application of new findings in biology and medicine, combined with improved agricultural production, better transportation, and the emergence of an industrial society, created a set of circumstances that drastically lowered death rates and thereby greatly increased the efficiency of human reproduction. Increases in longevity and increases in population have been the result. Data from France provide an example of the effects of increasing longevity:

Era	Births	Alive at Age 1	Alive at Age 20	Alive at Age 60
Eighteenth century	1000	767	502	214
Present-day	1000	984	956	760

While mortality in western Europe was declining sharply toward the end of the eighteenth century and in the early part of the nineteenth century, fertility maintained its usual high levels. Only with a considerable time lag (of as much as a century or more) did birthrates begin to decline to their present levels between 12 and 20 births per 1000 persons per year.

Until 30 years ago most of the world's people had a life expectancy at birth no greater than did western Europeans during the Middle Ages. Only Japan, among nations of non-European stock, had appreciably increased longevity through decrease of its death rate. But now the situation has been dramatically changed. Death rates have been, and are being, drastically curtailed (e.g., Algeria, Mexico, Costa Rica, Malaysia have all cut mortality rates by more than 50 percent in 20 years), while birthrates continue high. The major force of change has been scientific and technological. For example, in late 1946 and early 1947 the public health services of Sri Lanka (Ceylon) had the houses of the island sprayed with DDT. The death rate fell 40 percent within a single year. As a direct result of ever more people living longer lives, population growth is now greater among non-Europeans than among those of European stock. Most of the rest of the world is entering its phase of "death control" that European peoples have already experienced.

Differing Patterns of Population Change

A true picture of the world population situation is that different areas are experiencing different rates of growth. A map of world population distribution (see back endpaper map) is like a snapshot or still picture taken of the surface of a kettle of boiling soup, whereas in reality some elements are stationary, or nearly so, and others are in rapid motion.

In mid-1977, Latin America, at 2.7 percent, has the most rapid rate of annual increase in population growth. The annual increase is: the birth rate minus the death rate, combined with the plus or minus factor of the net immigration or net emigration. Then comes Africa at 2.6 percent, Asia at 2.0 percent, Oceania at 1.3 percent, followed by the U.S.S.R. at 0.9 percent, North America at 0.6 percent, and Europe at 0.4 percent. The highest annual growth rate is that for Kuwait and Libya at 3.9 percent. Honduras and Mexico, each 3.5 percent. By contrast, the United States has an annual growth rate of only 0.8 percent, which is still higher, however, than any of eighteen countries in Europe.

The examination of the rates of annual increase for countries for which statistical information is available, reveals six phases of population change (Fig. 9.3). These are: (1) high stationary; (2) early expanding; (3) late expanding; (4) recovering stability; (5) low stationary; and (6) declining. Figure 9.4 portrays the worldwide patterns, by countries, of these six categories. Each of the stages reveals the situation described below.

High Stationary Phase. The growth of population in this category is nil, or only very sporadic. Birthrates and death rates are both high, ranging between 35 and 60 per thousand each year. Agrarian populations live near, or at, subsistence levels. Tendencies toward population increase are offset by frequent and rarely controlled famines, epidemics, floods, and droughts. For most of human time, the entire population of the earth, wherever located, must have been just fractionally above this phase. Until the modern period, many small and conservative societies remained in this phase. In the 1970s, there are no countries that typify this high stationary phase, although a few small and simple societies may still be at this position. Modern health services and measures for combating famine have removed almost all population units from this category.

Early Expanding Phase. This category is marked

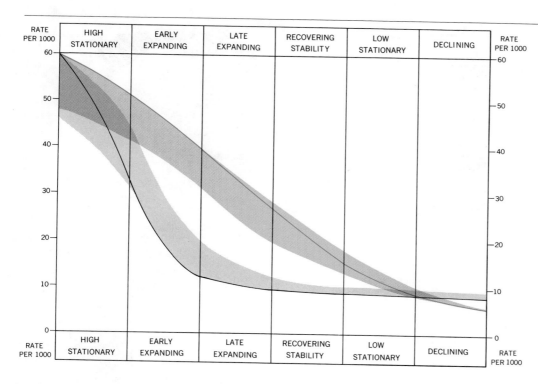

Figure 9.3

Six Phases of Population Change. These result from differences in rates of change in birth rates and death rates.

by a continuation of high birthrates but with a lowering of the death rates. Birthrates of 40–52 per thousand against death rates of 10–35 per thousand per year are typical. The differences between these figures provides large net increases in population. Countries experiencing this category of change are predominantly agrarian, but have benefited from technological advances. These have led to increased agricultural production, the beginnings of industrialization, and the improvement of transportation and communications systems. Public health services have begun to check disease and epidemics. School systems are re-reducing illiteracy. Relatively, internal stability and peace are present.

Countries now in this early expanding category include most former colonial territories of Asia and Africa. In addition, this category prevails in Southwest Asia, Mexico, Central America, northern South America, and Brazil. Modest rates of increase have meant huge additions in population to those countries whose populations were large before entering this stage, such as Pakistan.

Late Expanding Phase. Birthrates have declined in this category, but death rates have declined more rapidly, so that a net population increase continues to occur. Countries in this category typically have birthrates ranging from 30–35 per thousand per year, with death rates less than 20 per thousand, or birthrates in the 20s and death rates near or below 10 per thousand. Relatively advanced agriculture and modern industry have brought urban patterns of living accompanied by efficient modern sanitation and public health services. Countries in this late expanding category include Cuba, Chile, Egypt, Sri Lanka, Thailand, Singapore, Indonesia, China, Taiwan, and South Africa. These are mostly countries that are experiencing conflicts between political or religious ideologies desirous of expanding population and the humanitarian or economic needs for population stability. India and Egypt, for example, can exercise a choice between raising the level of living for their present populations through increased production, or diluting the present relatively low level of living by further increases in total populations.

Recovering Stability Phase. This category is the result of birthrates falling toward previously lowered death rates. Countries in this category typically have birthrates of 20–10 per thousand per year and death rates below 10 per thousand. The difference, however, is still sufficient for a population to double in between 46 and 99 years. Some 29 countries now are in this recovering stability phase, and include Canada, the United States, Argentina, Uruguay, France, Spain, the U.S.S.R., Japan, and Australia.

Low Stationary Phase. The low stationary phase is marked by birthrates and death rates that are both low and nearly equal, so that the population is kept relatively stable in numbers, with growth rates of 0.7 percent or less annually. During 1976 some 11 countries of Europe were in this phase, including those with the world's highest levels of living. Outside Eu-

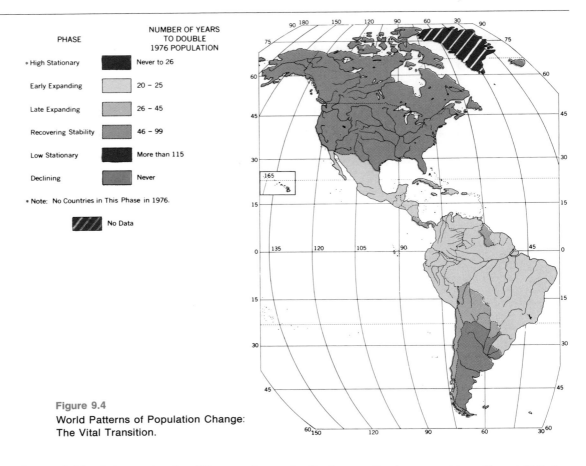

Figure 9.4
World Patterns of Population Change:
The Vital Transition.

rope, only Grenada and Martinique, in the West Indies, have attained status in this category, although Bulgaria and Switzerland, each with less than one fourth of their populations under the age of 15 years, seem the next most likely candidates. The countries now stablest in population are West Germany, Austria, and the United Kingdom, which have growth rates of less than 0.25 percent annually.

Declining Phase. This category is marked by an actual decline in total population through an excess of deaths over births. In some cases the death rate is not excessively high, but the birthrate has dropped to an extremely low level. In other cases, death rates formerly were excessively high, as in the depopulation of certain islands through epidemics of disease, so that remaining groups within the reproductive age span are small. For example, the Marianas Islands declined from over 50,000 people to less than 1800 between 1521 and 1764, following European contact and introduction of new diseases. There are but few areas in this category at present, with only the city of Berlin and East Germany illustrating it.

The foregoing six categories are represented in the mosaic of world population growth at the present time (compare Fig. 9.4 with the back endpaper map). Each country is experiencing its own phase of population change. Such change can also be viewed

dynamically for particular countries. Great Britain, for example, was in the high stationary category before the advent of the agricultural and industrial revolutions. With advances in medicine, sanitation, and rising levels of living sustained by overseas emigration and colonialism, Great Britain moved successively through the early expanding, late expanding, and recovering stability phases to the present low stationary category.

Each of the six categories of population change is gradational or transitional, one with another. A particular country or region may be a "borderline" case and in successive short periods alternate between one category or another; or a country, after many years in one category, may switch to another category as a result of an abrupt change in either its birth or its death rate. Two examples will suffice to demonstrate such changes. France, for the ten years between 1936 and 1946, when its death rate exceeded the birthrate, was the outstanding example of a country in the declining phase category. After World War II France became very population conscious and experienced a relatively marked excess of births over deaths. It moved from the declining to the late expanding category but has now returned to recovering stability status. During the 1930s the United States was in the low stationary phase, but for 20 years after World War II it was in the late expand-

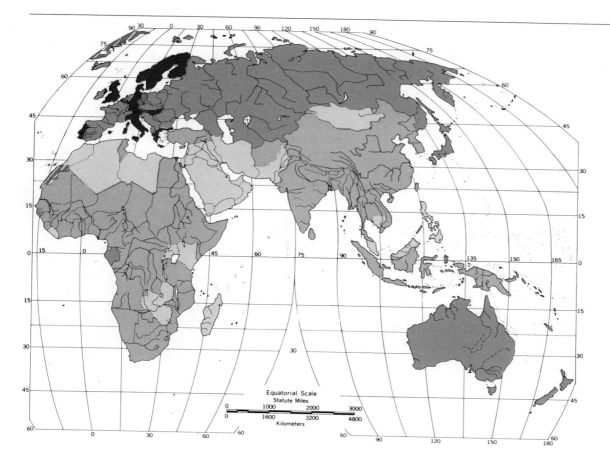

ing category. However, since 1954 the annual death rate has nearly stabilized, fluctuating upward only in those years marked by influenza epidemics, whereas the birthrate has declined continuously since its peak high in 1957. Only during the 11-year period 1954–1964 did registered births exceed four million each year. The 3,136,965 born in 1973 was the smallest number since the World War II years of the mid-1940s. Those relatively fewer babies born during the depression years of the 1930s were the child-bearing generation of the 1960s and early 1970s, and the trend toward smaller families diminished the number further. The result is that the United States once again is in the recovering stability category. But since it is the children of the post-World War II baby boom that are now marrying, it will require a continued low birthrate to remain there, or to shift to the low stationary phase. Which way will we go?

The Contemporary Population Explosion

Why should we be concerned about the effects of the contemporary population explosion? All unforeseen, the creation of culture has led to a fundamental change in the balance of births and deaths that control human numbers on earth. In the slow transition from primitive societies to the modern urban-industrial civilization, people applied death control

before they learned to apply a simultaneous control to births. This changed the balance between births and deaths that for so long kept populations well within the limit of local resources. Ours is the age of the intense peopling of the earth, a wholly new phenomenon. Compared to most of human time on the earth, our present rate of growth is something like the speed of an express train compared to the pace of a tortoise.

The inherent quality of the earth's basic resources has not changed. The earth continues to intercept solar energy at the same rate as in past time periods. Neither has there been a change in the inherent nature of the human animal. Men had the same basic structure, arrangement of muscles, and brain 10,000 years ago as today. Rather, the change in human numbers has resulted from the changes in methods that people have invented and learned to apply in their discovery and treatment of the earth's resources.

A decisive factor in the people-resources ratio is the coupling of increased longevity to the marked rise in the per capita consumption of food, materials, and energy. People in remote corners of South America, Asia, and Africa are learning to use European-American-derived means of production, transportation, and communication and are adjusting their institutions and ways of life to patterns of modernization that include an increase in their rates of consumption. The Western nation-states, in the cause of exporting their ways of life to other peoples, are engaged in a great competitive enterprise to raise the "levels of living" (read "rates of consumption") for most of the rest of the world. Can everyone become as profligate as a modern American?

The Population Ahead

The present rate of increase in world population is estimated at 1.8 percent per year. This may seem rather insignificant, yet it is truly an astounding growth, as may easily be demonstrated (Table 9.4). From a base of four billion people (1975 world population), a growth rate of 1.8 percent per year will result in a tripling of the world's people in 62 years and such a rate would produce in 675 years (A.D. 2650) a population density of one person for each square foot of land surface presently on the globe. In 1655 years (A.D. 3630), in theoretical continuation of the present growth rate, the mass of humanity would weigh as much as the planet Earth itself! Obviously that extreme cannot happen, but world population increased by 64 million in 1977, alone. This rate of growth amounts to a new United States every four years, or a new Chicago every four weeks! Only fifteen countries on the earth have populations as large as the current world annual increase.

The present era with its growth rate of 1.8 percent per year is most abnormal, and the conclusion is clear: the present growth rate cannot possibly last much longer without exhausting simple living space, not to speak of exhausting the resources of the earth. A few nations only have faced up to the problems of restricting their rates of growth to bring about better balance between human numbers and their respective territories. Very great differences in rates of population growth continue to occur among the different regions of the earth. Africa, Latin America, and Asia, areas that already contain three-fourths of the world's people, are now increasing at faster rates than the rest of the world (Table 9.5).

In the Western Hemisphere a reshift in population relationships has recently taken place. Before Spanish conquest, aboriginal Meso-America (principally Andean South America and highland Mexico-Guatemala) contained a larger Indian population than did North America. Despite the subsequent widespread decimation of the North American Indian

Table 9.4 Projected World Population Growth (at 1.8 percent annual increase)*

Year	World Population Total	Number of Years to Add One Billion People
1975	4000 million (4 billion)	13
1988	5000 million (5 billion)	10
1998	6000 million (6 billion)	9
2007	7000 million (7 billion)	7
2014	8000 million (8 billion)	6.75
2021	9000 million (9 billion)	6.25
2027	10,000 million (10 billion)	5.5
2032	11,000 million (11 billion)	5
2037	12,000 million (12 billion)	

*Compare with Table 9.1, p. 261.

Table 9.5 Projected Regional Population Growth, by Continents, 1975–2000

Region	Population (in millions)		Increase	Percentage of Total World Population	
	A.D. 1975	A.D. 2000		A.D. 1975	A.D. 2000
Africa	401	815	2.0	10.1	13.1
Latin America	324	606	1.9	8.2	9.8
Asia	2255	3612	1.6	56.8	58.1
Oceania (including Australia)	21	33	1.6	0.5	0.5
Europe (including U.S.S.R.)	728	854	1.2	18.4	13.8
North America	237	294	1.2	6.0	4.7
TOTAL	3967	6214			

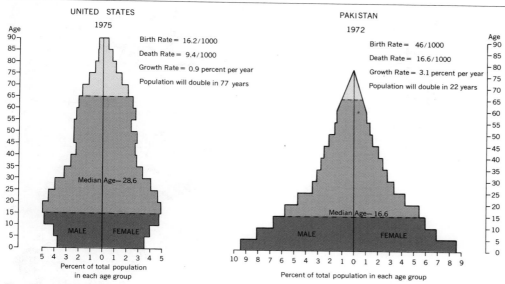

Figure 9.5 Population Pyramids: Comparison of The United States and Pakistan. Data arranged by sex and age group, in percent of total population.

population, the earlier and more rapid economic development of Anglo-America, coupled with heavier European immigration, created a larger population than that of Latin America. But since 1960 the number of persons in Latin America once again has exceeded the North American population. By the end of the twentieth century, Latin America's population will be more than twice that of Anglo-America; indeed, during the last half of the twentieth century Latin America's absolute increase of people will equal that of the whole world population as of 1650.

All the differentials in growth of regional populations in the Western Hemisphere are but details (however startling) compared with the overriding significance of what is happening in Asia (Fig. 9.5). The projected increase in Asia's population during the last 40 years of the twentieth century is as great as the population of the whole world in 1925!

POPULATION PROBLEMS: A CULTURAL INTERPRETATION

In a very real sense, all problems of rapid population growth are of mankind's own creation. Each major upward step in numbers has followed some major discovery or invention: agriculture, the initiation of urban life and trade, the harnessing of inanimate power, or the establishment of public health services. In recent centuries cultural evolution reached a state that enabled Europeans to precipitate a technological, industrial, scientific, and democratic revolution that made possible the application of scientific med-

icine and profoundly altered the long-established near-equilibrium between birth and death rates all over the world. Many of the problems of modern living (urban congestion, suburban sprawl, longer-distance journeys to work or recreation, automobile parking problems, higher costs of land, refuse disposal, and pollution abatement) are directly related to an increase in numbers of people competing for space, facilities, or services.

In most other parts of the inhabited world the problems are more serious: hunger and poverty are spreading, lowering the existing levels of living and further reducing the chances for individual prosperity. The gulf in subsistence levels that exists today between countries is widening, and within countries the same thing is happening between social groups. Life is better than ever for the middle and upper classes in Western industrial countries, but the majority of the world's people still live close to the subsistence level and squalor reminiscent of medieval Europe. The odds are stacked against the so-called developing countries of Africa, Asia, and Latin America, which are unable to add to their per capita wealth and productivity because too large a share of their people are not yet economically productive (under 15 years of age, as shown on Fig. 9.6) and their rates of population growth are greater than their rates of economic development. Thus funds expended for immediate needs (food, clothing, shelter) are not available for capital investment (education, equipment, research) to increase future productivity.

Mankind, however, has not only been a culture builder, but also a problem solver. Becoming aware of the true consequences of too rapid population

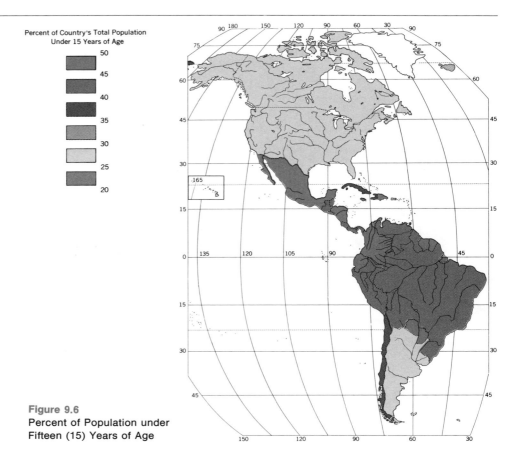

Percent of Country's Total Population
Under 15 Years of Age

50
45
40
35
30
25
20

Figure 9.6
Percent of Population under
Fifteen (15) Years of Age

growth, some people everywhere are becoming active in programs aimed at population controls. The United Nations, finally, is tackling the problem, and so are agencies in numerous individual governments. Currently India, Pakistan, Taiwan, Japan, and Korea sponsor active programs of birth control. Malaysia, Ceylon, Hongkong, Barbados, and Puerto Rico have programs sanctioned by their governments. Singapore, Tunisia, Turkey, and the United Arab Republic (Egypt) have formulated population policies. In its own way, the People's Republic of China now has embarked upon a program aimed at reducing the growth rate. Many other countries are in the beginning stages of formulating population programs, and some of these will begin to be effective in the near future.

The purpose of all such programs is to ensure gaining control over future human increases, so that means other than the traditional Malthusian checks can provide solutions to future problems. If enough countries in Africa, Latin America, and Asia can mount effective programs that reduce rates of total increase, it may be possible to postpone the date of reaching the total population listed in Table 9.5. There are signs that the programs of limitation are beginning to slow down rates of increase. Effective population control, however, is not yet a universal program, and the problem of growth remains.

Civilization has increased the average span of a human life and cut down on the inroads of death by epidemic disease. We have not yet quite conquered drought when it strikes in continental proportions, and famine continues as a natural check. We have not yet eliminated war, but there no longer are the large-scale slaughters that once went with conquest. We are learning that unchecked population increase can only spell misery for much of the earth. How we achieve an effective population policy for the earth as a whole, and just what limits must be set, remains a problem for peoples of the world in the next few decades.

The question not only concerns sufficient food and an adequate dietary standard for billions of additional hungry mouths but also involves opportunities for education and employment and the development of social standards of living compatible with human dignity. There has been perhaps too great a tendency to seek a quantitative solution to human problems, whereas, in truth, the total answer can be given only in qualitative terms.

People as Consumers

The size and growth rate of a population are major contributors to the extent of that population's impact on its environment, whether local or global. Each

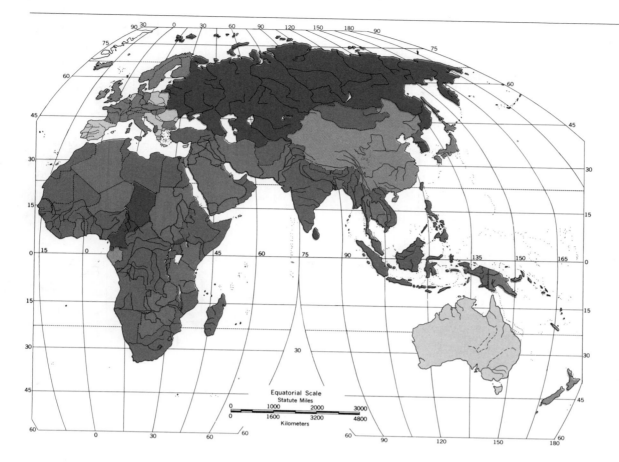

individual utilizes renewable and nonrenewable resources, which can be expressed statistically as per capita consumption (Table 9.6).

For example, total annual energy consumption in the United States increased by 160 percent between 1940 and 1975. The combined consumption of aluminum and steel increased by 140 percent during the same period, whereas population increased by only 60 percent. Per capita consumption is racing far ahead of population growth. The effect is an increase in the negative impact on the environment.

Consider the problem of providing nonrenewable resources, such as minerals and fossil fuels, to a growing population even if per capita consumption were fixed rather than increasing. As the richest supplies of resources and those nearest to centers of use are consumed, we must process lower-grade ores and drill deeper and extend the network of supply. All such extensions result in an increase in the use of energy. Per capita costs and the impact on environment escalate dramatically. Also, the cost of maintaining environmental quality at a given level increases disproportionately as population rises. Reducing contaminants per unit volume to lower and lower levels is an illustration of the principle of diminishing returns. Municipal sewage is a good example. Removing 90 percent of the suspended solids, 80 to 90 percent of the biochemical and chemical oxygen demand, and 60 percent of the resistant organic material costs about 11 cents per 1000 gallons in a large sewage treatment plant. The achievement of a disinfected and potable water supply, reusable for groundwater recharge, costs four times as much.

In a developed society, because people live longer, their high rates of lifetime consumption demand larger volumes of resources. The developed countries make the heaviest negative impact upon the global environment. For this reason it is population growth in the advanced countries that must be regarded as the most serious in the world today. To maintain itself, the Netherlands, for example, appropriates the raw materials and the products of many lands outside its borders. It imports 100 percent of its corn, rice, cotton, iron ore, antimony, bauxite, chromium, copper, gold, lead, magnesite, manganese, mercury, molybdenum, nickel, silver, tin, tungsten, vanadium, zinc, phosphate and potash for fertilizer, asbestos, and diamonds. It annually consumes the equivalent of 47 million metric tons of coal and is the second largest per capita importer of protein in the world.

Existing practices of resource consumption by the developed nations are compromising the aspirations of most of mankind for their share of the world's goods. The 6 percent of the world's people residing in the United States account for about 30 percent of

Table 9.6 Selected World Food Consumption Patterns

| Country | Data Years | Percent Cereals Produced | Kilo-calories per Day per Capita | Percent Animal Sources[a] | Protein from all Sources | Grams per Day per Capita | | | | | | | Annual Consumption per Capita | |
						Cereal Grains	Roots	Sugars	Nuts, Seeds, Pulses	All Types, Meats[b]	Milk Products	Other Fats and Oils	Tea in Grams	Coffee in Pounds
Australia	66–68	400	3280	45	92	233	149	141	?	289	629	38	2350	1.44
Bolivia	64–66	30	1900	14	46	239	421	50	10	61	59	13	?	?
Brazil	67–69	70	2020	15	65	242	522c	109	87c	84	201	15	?	?
Burma	64–66	110	2210	6	44	419	6	25	20	17	48	18	?	?
Canada	66–68	1400	3180	46c	96c	186	215	137c	13	247	639	58	989	4.04
China	64–66	95	2170	9	57	387	247	10	39	47	9	7	950	?
Czechoslovakia	64–66	70	3180	27	83	348	295	103	8	167	356	44	?	1.70
Egypt	68–69	55	2500	7	80	565	28	44	24	31	135	19	844	0.19
Finland	66–68	90	3050	45	96c	229	270	110	5	114	930c	54	?	10.19c
West Germany	66–68	90	3290	42	82	192	303	95	12	209	562	74	144	3.18
India	66–68	90	2070	6	48	370	29	49	65	4	116	10	358	?
Indonesia	67–69	94	1790	3	40	302	276	21	42	11	4	9	100	1.00
South Korea	66–68	90	2520	4	67	571c	173	12	19	18	3	2	?	?
Mexico	64–66	105	2580	11	66	379	35	109	77	55	157	23	?	?
Mongolia	64–66	100	2380	37	93	374	50	46	?	302c	237	8	?	?
Portugal	66–68	80	2900	18	79	337	280	57	45	84	165	48	?	1.72
Sudan	64–66	35	2160	15	59	282	421	31	22	55	286	18	850	?
Turkey	64–66	99	3250	10	78	474	113	41	35	39	219	29	590	0.08
U.S.S.R.	66–68	90	3280	29	92	428	378	106	19	106	426	33	295	0.14
United Kingdom	66–68	45	3190	42	90	203	278	135	17	214	593	62c	4023c	1.68
United States	66–68	215	3330c	40	96c	178	145	132	22	301	667	62c	327	7.58

Sources:
[a] Includes fishery elements.
[b] Excludes fishery elements.
[c] Highest consumption per capita.

the annual consumption of the earth's nonrenewable resources (e.g., 37 percent of the energy, 33 percent of the synthetic rubber, 28 percent of the tin, 25 percent of the steel). Developing the underdeveloped countries means being able to afford the technology to mine lower-grade deposits and pay for shipping costs, something that the underdeveloped countries, as their populations and needs grow and their means remain meager, will not be able to do. Poor countries are the economic dependencies of the advanced countries, the principal consumers of the world's nonrenewable resources and foodstuffs. We will have more to say about "the rich and the poor" in the next chapter.

The waging of war contributes heavily to the profligate use of manpower, industrial productivity, energy, raw materials, and transport facilities. During the decade of the 1960s the world spent the equivalent of $1870 billion for military purposes. Six countries—the United States, the U.S.S.R., China, France, Great Britain, and West Germany—accounted for more than four-fifths of the total expenditure. Those countries having more than 2 percent of their populations in the armed forces are Cuba, Honduras, Luxembourg, Israel, Jordan, Laos, Vietnam, Taiwan, and North and South Korea. An annual expenditure of more than 200 billion dollars (about $60 for every person on earth and about 7½ percent of the world's total gross national product) supports the activities of 22 million persons in the armed forces of all countries. This exceeds what the world spends each year for public education and public health combined. It is 22 times the amount given annually in foreign aid. It is also as much as the income produced in a year by the 1.8 billion people in the poorer half of the world's population. And yet developing countries continue to buy arms at a rate outstripping the growth in their gross national products.

The Long-Run Balance

There is a basis for assuming an average long-term human need for just over 2000 calories per day per person for biologic existence, yielding a total annual per capita requirement of about 750,000 calories. The assumed 5,000,000 inhabitants of the mid-Paleolithic era would have consumed about 3,750,000,000,000 calories per year (more briefly expressed as 3.75×10^{12}). The present world total population need, for food alone, is a bit less than twice *per day* what the Paleolithic total was *per year;* the present annual food requirement is not far from 2.5×10^{15} calories (Fig. 9.7). It is obvious that such an amount must be grown or manufactured, since gathering and collecting could never yield this total from the wild resources of the earth. It is probable that the earth under proper ecological care can go on producing traditional plant and animal foods for a somewhat larger population than that on the earth at present. A shift to an almost vegetarian diet could perhaps enable support for a population of about five billion people on a continuing basis.

We can make no sound assumption regarding the nonfood energy requirement for mid-Paleolithic man, but there are population groups resident on the earth today that probably require not much more nonfood energy than did Paleolithic man. Most of the earth's present population, however, lives at levels far above the Paleolithic. There are grounds for an assumed average annual consumption of energy per capita at about 2.5×10^7 (25,000,000) calories, yielding an annual world requirement of about 8.25×10^{16}. The wide discrepancies in present world consumption levels range from about 8.25×10^5 to about 8.1×10^7 (that is, from 825,000 calories per person per year to about 81,000,000). The latter figure is near the average often employed to characterize per capita energy consumption per year in the United States during the late 1960s.

It is obvious that some of the earth's present inhabitants require only a little energy beyond that contained in the food they consume, whereas other inhabitants consume or utilize nearly 100 times as much energy as needed for biologic continuance. The latter consumption is involved in what is customarily termed a high level of living. Such a level utilizes huge amounts of coal, hydroelectric power, petroleum, natural gas, wood products, and other energy sources to process the industrial commodities (the natural growth items themselves consuming energy) purchased, used, and "consumed."

Assuming that the United States level of energy consumption is lifted only a reasonably small amount during the next two or three decades and that the rest of the world makes outstanding progress toward modern industrial consumption patterns, what will the requirements then be? There are too many unknowns and assumptions in this question for the answer to be soundly realistic, but the calculation may be made for the sake of argument. Assuming the average world consumption at about one-third present American rates, the annual energy needs for a 5 billion population could approximate 1.4×10^{17} calories. The known recoverable fossil fuels in the standard resource table approximates 2.7×10^{19} calories, yielding an expectable exhaustion date of about 200 years, give or take a little for presently unexpected future finds and for recovery at somewhat less than currently estimated rates. But this obviously would mean power-conversion equipment that could substitute fuel sources, and it would mean a total world sharing of the fossil fuel resources, of which the United States has an overly large share in its coal and oil shale deposits.

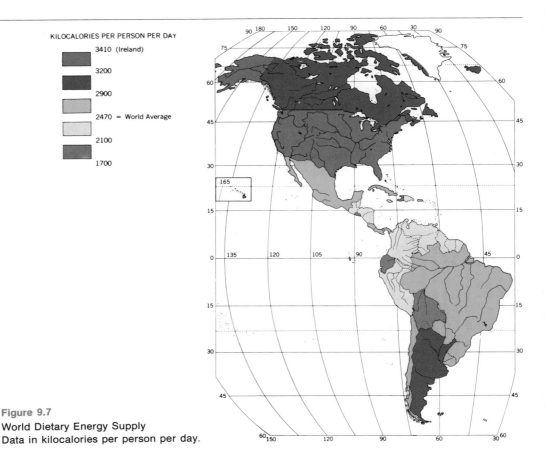

KILOCALORIES PER PERSON PER DAY

3410 (Ireland)

3200

2900

2470 = World Average

2100

1700

Figure 9.7
World Dietary Energy Supply
Data in kilocalories per person per day.

Any further increase in world consumption would shorten the period prior to exhaustion. Our calculation does not include water power, nor does it include the uranium group of power minerals, and it makes no allowance for the use of solar energy in direct applications. Neither does it involve the possible increase in efficiency of resource conversion, which could stretch out the use of known resources. Future improvement in energy production can be very marked. There already is considerable sharing of total energy consumption in the shipment of manufactured goods from one region to another, but the foregoing assumptions on energy sharing would involve a far greater application, probably in ways not now taking place.

Reviewing the two phases of the problem suggested above—food production and nonfood energy supplies—yields the conclusion that population holds the key to the future. For the earth's present population a permanent food supply can be assured, and energy reserves are sufficient for at least several centuries. For a population of about five billion the food potential may be close to a limit, and energy reserves demand conservation measures. For a population much beyond five billion, the technological and ecological problems loom very large indeed. The present four billion world population will grow larger before universal human restriction in popula-

tion growth comes into effect. Resolution of the problem is less a matter of industrial technology and energy resources than a matter of human population control.

PROBLEMS OF URBAN CONCENTRATION

The Background to Urbanism

As the human populations of some regions increased to significantly larger numbers during the Neolithic, people began accumulating at particular settlement points. Several kinds of settlements have been used (see Fig. 3.9, page 65). A hamlet became a sizable village through the increase of its native-born members and through the attraction of latecomers. The well-situated village sometimes grew larger through attracting increasing numbers of people who found it a convenient place to carry on different kinds of activities. These activities seldom were carried on at sites different than the points chosen for residence. Some villages grew so large and carried on such varied activities that we have distinguished them as towns. A few towns, at least, must have existed as early as the late Neolithic. Some towns continued to

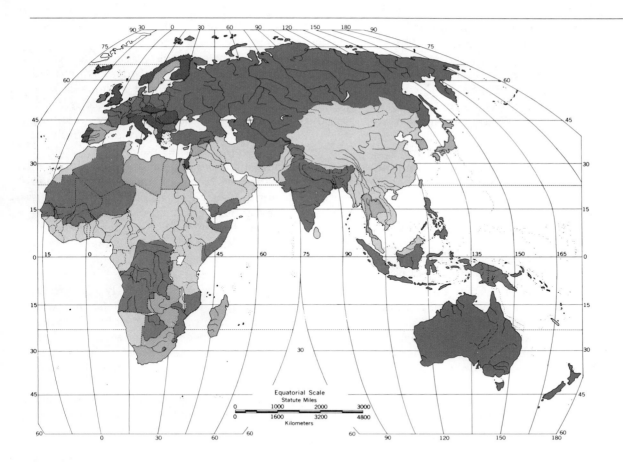

grow in size and range of activity. Once a village became a sizable town, the accumulation of people and all kinds of activities in unordered arrangements began to create problems of different kinds. These problems continued to be troublesome to the inhabitants until some problem attracted the attention of a group of people. Solutions to problems were probably sought by changing some arrangements within the settlement, but such solutions must have been pragmatic efforts to ease problems without removing them entirely.

As towns became more numerous and grew larger in size, the numbers of problems around town life must have grown steadily greater. The problems of large settlements finally produced some thought and at least a little planning and, as the onset of civilization occurred, there was invented a new form of settlement that we distinguish as the city (see the section Writing, Urbanism, and the Political State, pp. 99–104). The city was usually founded by a ruler who combined elements of religious and political leadership. The city was a formalized settlement symbolic of the power of the ruler, and of the existence of the state and of the society of which it was the capital. As a formal settlement, usually newly founded, the city took on certain elements of form and shape that were related to pursuit of certain kinds of activities and to formalized lines of access to

and through the settlement. The ruler's palace, the chief temple, certain other civic buildings, and a city wall with formal gates and defensive works were first laid out and established. In the first cities certain kinds of economic activities were apparently located outside the city at the time of founding, and certain other functions were arranged for inside the city.

Depending on the particular cultural inclinations of the society that elaborated the formal city as a capital and chief settlement, the planning of several aspects of urban life were arranged in different ways. In some societies, India for example, the residential quarters of particular classes became designated parts of the formal layout of the city. In other societies there was no apparent zoning of any economic activities. Some societies apparently never founded new cities at the beginning of a rulership. The new ruler simply took over an already existing settlement and then made some rearrangements in the layout of its political and religious structures and institutions. Other societies left the activities and residences of private citizenry entirely to unofficial disposition.

Cities have always presented problems to their builders, and the present era has no urban problems that are unique in basic terms. Although the cities of the modern world are larger and have many more of the primordial problems, they suffer them to an acuter degree than did the cities of the distant past. Where to live, in what kind of quarters, and with what kinds of neighbors; where to work and how to get there and back; where to shop, and how to get delivery of the goods; how to dispose of unwanted trash; and what to do with free time and where to spend that time; these and related problems have always bothered the city dweller. Seldom throughout history could everyone choose the ideal, or individual, solution to each and every issue. The gathering of people together into large groups immediately induces differential choices by individuals to many of the issues of daily life.

The palace of the ruler, the fine quarters of the rich and influential, the hovel of the poor in an alley, and the hut of the latecomer in a shacktown suburb outside the city wall near an approach road, all these were present in the earliest city. Once the city was built, occupied, and grew somewhat larger in population, there began to occur the congestion that called for rearrangement of the details of streets and buildings within the city wall. Urban renewal is not a product of the twentieth century only. Congested alleys became slums; markets grew overcrowded; suburbs blocked access to the city; and new temples expropriated land from prior residents, who seldom could better their situations when being forced to move.

The disposal of trash in many early cities was taken care of simply by throwing it into the streets (for which archaeologists are very thankful). The disposal of sewage and garbage in many early cities was left entirely to individuals, and it often went into the streets, too. The Indus cities, however, are notable for their clean streets and drains, and the Chinese early developed systems of removal of all sewage, garbage, and street sweepings to the agricultural plots and fields for use as fertilizers. European cities were both slow and late in organizing sanitary systems. Water supply remained a private matter within the city, although urban planning early took on the issue of controlling a source for urban water by providing wells, the enclosure of a water source within the wall, or both. Internal transportation within the early city remained the problem of private individuals, and different culture systems developed different means for the movement of both people and goods. Congestion within the city, however, is an age-old problem.

In the ancient world in which the city was born, the settlement where the ruler resided and where government and the religious institutions operated was "the city." No other settlement had the same high status in society, though there must have been lesser places that functioned as marketplaces for traders. As long as the territorial extent of the state remained small, the single city remained dominant. The Greek city-states exemplify this pattern.

In those states that expanded to such territorial dimensions that all administration was not managed from the capital, there developed systems of subordinate cities. These in almost all cases were small-scale, small-size replicas of "the city," and they functioned politically, religiously, and culturally as extensions of the capital. They served to exhibit the local presence of the symbolic strength and unity of the society. Although some of these became quite large, and carried on trade and handicraft manufacturing, they could be considered only the symbolic and territorial reflections of the capital, which retained the rank and status as the primate city in cultural terms.

Inevitably, in larger territorial states, settlements did arise that were concerned entirely with trade, with the processing of primary products, and with handicraft manufacturing. Without status as extensions of the capital, however, such settlements remained rude kinds of "marketplaces" in the perceptions of the inhabitants of the state, and their place-name systems label them so.

Within that traditional world in which "the city" served as the ranking symbol of a culture system, the percentage of the total population living in the city was quite variable. In the small area such as the Greek city-state, perhaps as much as 75 percent of the population lived in urban places in some cases. In the larger territorial states, however, this percentage seems seldom to have been as high as ten percent, and often it was apparently no more than five percent. In these states, the capital, and the subordinate regional cities, were attractive places to visit, and rural folk made pilgrimages to them. The economic restrictions of such cities, however, prohibited rural-to-urban migrations for purposes of permanent residence. Only a very few of the ancient cities in large states became large settlements in modern terms (Table 9.7). These conditions continued into medieval times, and the generalizations hold true to perhaps the tenth century in the Old World, and until the sixteenth century in the New World.

Transitional Urbanism

As the linkages of parts of the Old World became greater through the centuries after the appearance of civilization and the formation of the territorial political states, the human demands for goods of other lands promoted the increasing importance of trade throughout the Old World. That many such marketplaces had no symbolic status as cities probably mattered not at all to the merchants of those times. The earliest of these patterns cannot be traced with any degree of certainty. By the time we reach the centuries immediately prior to the beginning of the Christian Era, however, the record becomes somewhat more specific. Goods from the Roman Empire were traded along the east coast of India, and Arabic traders turned up annually in southern China. Goods from regions of The Baltic Sea were available in the eastern Mediterranean, and Chinese silks moved overland to be worn by the rich of Rome. The marketplaces of the Old World were thriving, and the settlements in which such trade was carried on often were growing large and more important. A good many of these places wherein traders met were capital cities or subordinate regional centers, that combined both symbolic and economic functions.

During the centuries of the Christian Era, prior to the beginning of the Columbian Discoveries, it appears that the definition of "city" underwent a gradual change. No longer was the purely symbolic politicoreligious aspect the sole characteristic of the city, although many such cities clung to their status. There were such diverse developments as the growth of the "free city" in western Europe, the activity of the handicraft trading towns of the Hanseatic League, the rise of the Italian trade entrepôts, the appearance of the port-principality in Southeast Asia, the recognition of the university towns in Italy, and the formation of the march-site frontier settlement in many parts of the Old World. Villages concerned with agrarian or mining primary production were growing into towns, and villages were shifting from agrarian bases into handicraft manufacturing patterns in many regions of the Old World in which populations were markedly increasing. Politicoreligious symbolism was being infiltrated by new cultural elements, and there was a gradual broadening of the concept of what constituted an urban settlement. The "city" was coming to be any large urban community carrying on any of a wide range of functions. In this process, secular economic considerations were complementing traditional symbolic functions but, in many places, were gradually replacing them.

After the Columbian Discoveries these broadening trends continued to expand against the traditional definition of what constituted a city. Port towns assumed new importance, and these often came to hold sizable populations. As the Industrial Revolution began to gather strength there began to occur new kinds of towns as manufacturing centers appeared and gathered large dormitory communities of labor forces. Port towns sometimes combined these functions, to become large cities. Port towns in other parts of the earth began their periods of rising growth as the centers for a new international flow of trade. Such diversely located and formerly small places as Rangoon in Burma, Calcutta and Bombay in India, Boston and New York in the New World, and Cape Town in South Africa were started on long-term trends of growth, to be joined slightly later by a host of other port cities that never had held status in traditional terms. In the face of the rise of these new, secular settlements, the old politicoreligious symbolism of the traditional capital city was replaced by a modern, secular economic concern. Urbanism was taking on both new forms and a new meaning.

Modern Urbanism

The city continues to be, around the earth, the creative and dynamic symbol of the living system to which rural, village, and small-town residents aspire, and to which settlements they are attracted as expressing the preferred way of life within their culture realms. Young people in the United States and Canada are attracted to the city today because, in modern language, "that is where the action is." The small town, the village, and the rural farmstead are the "dead place where nothing ever happens." Despite its faults, the city remains the attraction for many human beings, and there are only a few who turn their backs on the city and its way of life. Such people often reject most aspects of modern culture, and not just the city as a place of residence. There are a good many who now are choosing the smaller urban place, but in doing so they help to bring many more aspects of the big city with them.

A great many people in all parts of the world are of two minds about the contemporary large city. They seek its values while disliking many of its problems and difficulties. They continue to live in the city, however, because they recognize that almost no aspect of their present ways of life can be maintained outside the city. There is a tendency among many people, in all parts of the earth, to think in retrospect about life in the city—life was better in the smaller cities of the past. Perhaps it was, in certain respects, but life in the older cities was also poorer in many ways.

The large city of the twentieth century has become so large and so complex, so varied and so much in flux, that critics of city life (who rarely live in strictly rural environments) compile overwhelming lists of things to dislike. The city has been described as too big, too crowded, too full of immigrants, too noisy, too smelly, too grimy, too commercialized, too heartless, too impersonal, too pushy, too ostentatious, too gaudy, too wild, and too uncontrolled. For many of the contemporary critics of urbanism, the comparison becomes that of the wilderness, the green agrarian

Table 9.7 Historical Trends in the Sizes of Selected Great Cities. All Figures Are in '000's

City and Country	Date Founded	Any B.C. Figure	1st Cent. A.D.	7th Century	10th Century	14th Century	16th Century	1750	1800	1850	1900	1920	1930	1940	1950	1960	1970–1974
Accra Ghana	1886	—	—	—	—	—	—	—	—	3	39	61	95	125	175	390	850
Addis Ababa Ethiopia	1887	—	—	—	—	—	—	—	—	—	52	80	185	120	390	430	910
Alexandria Egypt	Ancient	400	400	180	10	30	35	6	6	210	315	540	610	750	1,035	1,515	2,060
Amsterdam Netherlands	A.D. 13th Century	—	—	—	—	4	14	220	200	290	510	640	750	785	890	910	1,140
Athens Greece	Ancient	155			30	30	17	10	12	31	180	540	895	1,115	1,345	1,810	2,425
Baghdad Iraq	Ancient			700	1,000	100	10	15	80	60	160	200	360	500	540	815	1,250
Baltimore United States	1634	—	—	—	—	—	—	6	26	300	510	680	820	905	1,165	1,420	1,730
Bangkok Thailand	1782	—	—	—	—	—	—	2	50	300	540	560	835	850	890	1,130	2,800
Barcelona Spain	Ancient	—	5	10	—	40	30	70	110	240	550	700	990	1,080	1,280	1,560	1,850
Berlin Germany[a]	1237	—	—	—	—		12	115	170	1,045	2,000	2,425	4,530	4,500	3,335	3,270	3,340
Birmingham England	750	—	—	—	2	2	15	50	70	480	520	1,695	1,900	2,070	2,585	2,775	2,980
Bogota Colombia	Ancient						20	25	30	110	110	240	285	365	675	1,240	2,500
Bombay India	Ancient						10	80	175	720	780	1,295	1,350	1,660	3,335	4,090	5,100
Boston United States	1630	—	—	—		—	—	16	35	560	1,075	2,210	2,330	2,505	2,230	2,700	2,600
Brussels Belgium	600	—	—	—	2	18	32	55	66	120	340	520	620	910	1,110	1,350	1,700
Bucharest Rumania	Ancient					2	25	25	34	325	560	1,070	1,200	1,260	1,325	1,425	1,645

This page presents a continuation of a population table (figures in thousands) with no column headers printed on this page. Cities are listed in rows; the final column gives the founding date.

City																	Founded
Budapest, Hungary	2,060	1,805	1,620	1,695	1,430	1,320	1,100	325	54	35	20	15	2				Ancient
Buenos Aires, Argentina	9,400	7,000	5,215	3,500	2,750	2,400	950	215	40	14	—	—	—	—		—	1580
Cairo, Egypt	5,600	3,745	2,505	1,525	1,140	1,100	595	255	265	300	200	500	150				Ancient
Calcutta, India	7,350	6,140	5,155	3,400	2,055	1,390	1,085	795	200	120	—	—	—	—		—	1596
Canton, China	2,500	1,900	1,600	1,100	1,100	800	740	950	800	350	100	300	120	220	50	5	Ancient
Capetown, South Africa	1,100	805	560	300	260	220	135	20	20	2	—	—	—	—	—	—	1652
Caracas, Venezuela	2,145	1,230	695	300	170	92	82	50	30	26	—	—	—	—	—	—	1567
Casablanca, Morocco	1,460	1,000	645	400	250	235	89	20	—	—	2	—	—	—	—	—	1515
Chicago, United States	6,985	5,960	4,935	4,530	4,425	3,565	1,715	405	—	—	—	—	—	—	—	—	1803
Chungking, China	3,500	2,300	1,400	950	620	610	400	350	250	100	50	50	20	10	5	5	Ancient
Cologne, Germany	1,790	1,375	965	1,065	975	700	435	100	41	44	37	57	21	15		—	A.D. 50
Copenhagen, Denmark	1,480	1,260	1,170	1,000	795	730	460	375	100	79	30	5	—	—	—	—	1167
Dakar, Senegal	670	530	350	200	195	175	165	110	3	2	—	—	—	—	—	—	1617
Damascus, Syria	3,100	2,310	1,735	650	800	375	205	160	90	80	57	80	200	250	—	40	Ancient
Delhi, India	4,445	3,540	2,665	2,135	2,015	1,395	295	150	160	500	130	100	10	—	—	—	Ancient
Detroit, United States	4,250	3,760	3,015	2,700	2,215	1,050	380	35	5	2	—	—	—	—	—	—	1701
Dublin, Ireland	600	700	675	575	335	310	275	255	165	125	26	20	4				Ancient
Florence, Italy	470	435	365	335	320	255	210	110	79	74	65	61					100 B.C.

Table 9.7 (Continued)

City and Country	Date Founded	Any B.C. Figure	1st Cent. A.D.	7th Century	10th Century	14th Century	16th Century	1750	1800	1850	1900	1920	1930	1940	1950	1960	1970–1974
Genoa Italy	Ancient				15	85	62	72	90	110	270	525	575	645	680	765	930
Glasgow Scotland	A.D. 550	–	–			2	2	43	85	635	1,070	1,395	1,685	1,720	1,910	1,960	2,010
Hamburg Germany	A.D. 825	–	–	–		10	23	90	130	350	895	1,360	1,710	1,810	1,950	2,170	2,405
Hangchow China	Ancient	10	20	200	250	400	350	350	500	700	350	475	505	540	640	835	1,100
Havana Cuba	1515	–	–	–	–	–	3	35	60	230	550	625	775	1,000	1,080	1,550	1,700
Hong Kong	1843	–	–	–	–	–	–	–	–	5	190	750	800	1,500	1,560	3,075	4,105
Hsian China[b]	Ancient	100	100	1,400	300	150	160	175	225	260	300	200	200	190	600	1,500	1,900
Ibadan Nigeria	Ancient								25	785	900	1,000	700	800	985	1,465	2,600
Istanbul Turkey	Ancient		25	500	450	150	650	700	570	800	1,125	650	700	800	1,000	1,950	2,250
Jakarta Indonesia	Ancient						52	100	92	71	115	300	525	1,000	1,750	2,850	4,500
Johannesburg South Africa	1886	–	–	–	–	–	–	–	–	–	100	540	600	650	865	1,155	1,400
Karachi Pakistan	Ancient	–	–	20	100	3	3	20	20	20	120	240	300	430	1,085	1,885	3,425
Kharkov USSR	1656	–	–	–	–	–	–	–	10	50	200	390	800	910	770	960	1,225
Kiev USSR	Ancient			10	45	15	15	28	20	56	300	460	530	920	850	1,155	1,560
Kinshasa Congo Dem. Rep.	1881	–	–	–	–	–	–	–	–	–	15	24	50	100	210	415	1,300
Lagos Nigeria	1472	–	–	–	–	–	2	2	2	4	100	120	180	190	270	420	2,100

284

City	Founded	500	400	300	225	175	110	80	30	22	21	13					
La Paz, Bolivia	Ancient																
Leipzig, Germany	1015	605	590	620	690	700	625	530	210	32	30	15	5	–			
Leningrad, USSR	1703	3,850	3,390	2,950	3,650	2,100	720	1,140	765	300	140	–	–	–			
Lima, Peru	1535	3,315	1,520	950	600	210	225	120	83	54	40	1	–	–			
Lisbon, Portugal	Ancient	1,500	1,335	1,160	940	785	640	365	275	240	215	80	35	15	15		
Liverpool, England	1207	1,825	1,740	1,665	1,365	1,335	1,200	685	650	75	25	1	1	–			
London, England	Ancient	12,945	12,465	11,700	11,500	8,125	7,740	6,480	1,950	865	675	150	40	25	30	40	
Los Angeles, United States	1769	9,675	6,490	4,010	2,600	2,100	1,000	195	2	1	–	–	–				
Lyon, France	43 B.C.	1,225	830	730	710	695	665	490	330	110	115	120	35	20	10	5	
Madras, India	Ancient	2,600	2,090	1,700	765	690	525	505	720	820	400	200	100	100	50	10	
Madrid, Spain	Ancient	2,990	2,260	1,590	1,080	940	740	540	410	170	125	6	4				
Manchester, England	A.D. 919	2,540	2,525	2,510	2,430	2,415	2,305	2,150	590	80	27	10	2	–			
Manila, Philippines	ca. 1500	4,100	2,705	1,475	900	600	285	205	155	80	30	3	–				
Melbourne, Australia	1835	2,200	1,850	1,350	1,100	950	775	485	15	–	–	–					
Mexico City, Mexico	Ancient	9,000	4,825	2,950	1,675	1,175	675	345	330	130	110	300	20				
Milan, Italy	600 B.C.	1,750	1,490	1,250	1,150	920	650	490	275	135	125	110	100	30	25	80	
Montreal, Canada	1642	2,440	2,040	1,440	1,165	1,200	740	325	90	16	7	–	–				
Moscow, USSR	1147	6,750	6,100	5,520	4,980	2,630	1,120	1,100	600	240	160	30	22				

Table 9.7 (Continued)

City and Country	Date Founded	Any B.C. Figure	1st Cent. A.D.	7th Century	10th Century	14th Century	16th Century	1750	1800	1850	1900	1920	1930	1940	1950	1960	1970–1974
Munich Germany	A.D. 750	–	–	–		10	15	30	40	200	500	680	760	945	1,110	1,290	1,500
Nagoya Japan	Ancient			5	10	15	50	70	100	110	260	610	770	945	1,030	1,590	2,255
Naples Italy	600 B.C.			30	30	60	275	125	430	450	560	850	760	825	990	1,160	1,300
New York United States	1626	–	–	–	–	–	–	13	65	700	4,240	8,050	10,230	10,930	12,340	14,115	16,075
Osaka–Kobe[c] Japan	Ancient			40	20	70	275	400	450	500	1,040	2,535	2,850	3,500	3,820	4,125	4,740
Paris France	Ancient	6	10	25	20	275	260	560	550	1,325	3,330	4,965	5,885	6,050	6,300	7,285	8,175
Peking China	Ancient		5	40	60	400	670	950	1,100	1,400	1,100	900	1,265	1,250	2,500	5,500	8,000
Philadelphia United States	1681	–	–	–	–	–	–	20	70	790	1,420	2,300	2,665	2,695	2,940	3,655	4,355
Prague Czechoslovakia	Ancient				5	40	70	60	80	120	385	745	840	975	930	1,000	1,050
Quito Ecuador	Ancient				20	25	30	30	30	70	45	75	100	130	210	315	500
Rangoon Burma	ca. 550	–	–	2	3	5	5	10	20	100	230	340	375	500	695	885	1,715
Rio De Janeiro Brazil	1555	–	–	–	–	–	1	30	45	515	750	1,325	1,675	2,150	3,050	4,700	7,215
Rome Italy	Ancient	320	650	50	35	30	55	155	160	250	485	750	890	1,270	1,605	2,020	2,920
Santiago Chile	1541	–	–	–		–	2	10	21	100	205	500	720	950	1,275	1,905	2,600
Sao Paulo Brazil	1554	–	–	–	–	–	1	5	5	22	265	600	900	1,425	2,450	4,375	8,405
Seoul Korea	Ancient					100	150	170		180	255	300	380	935	1,525	2,400	4,660

City	Country	(population figures, most recent → oldest, in thousands)																Founded
Shanghai	China	6,000	7,200	6,000	3,750	2,200	1,700	480	250	100	60	25	10	—	—	—	—	1050
Shenyang	China	3,750	2,500	1,700	850	500	290	250	180		200	50	20	—	3	—	1	200 B.C.
Singapore		2,115	1,635	1,020	600	490	195	200	65	3	3	0	0	2	—			Ancient
Sydney	Australia	2,720	2,150	1,650	1,350	1,040	475	385	185	2	—	—	—	—	—			1788
Taipei	Taiwan	2,150	1,330	750	325	220	135	120	50	10	5	—	—	—	—			1708
Tashkent	USSR	1,385	965	660	580	350	320	310	70	25		15	5					Ancient
Tehran	Iran	3,250	1,840	1,075	625	340	325	280	120	30	20		5					Ancient
Tientsin	China	4,500	3,500	2,200	1,000	900	800	700	325		80	25	5	—	—	—		A.D. 986
Tokyo	Japan[d]	14,840	13,545	8,180	8,560	6,065	5,705	1,740	780	1,000	500		—	—	—	—		1457
Varanasi	India	660	475	350	275	265	250	215	175	185	180	100	25	85	75	55		Ancient
Venice	Italy	385	370	310	280	260	170	155	140	145	160	150	100	35	25	4		A.D. 68
Vienna	Austria	1,890	1,875	1,860	1,760	1,870	1,860	1,660	1,000	230	170	45	20	20	5			Ancient
Warsaw	Poland	2,665	1,845	1,300	1,300	1,140	1,060	725	310	75	30	10	5	—	—	—		965
Wuhan	China[e]	4,250	2,500	1,200	900	1,000	1,300	450	300	175	165	130	120	110	100	70	45	Ancient

Sources: Tertius Chandler and Gerald Fox, *3000 Years of Urban Growth*, Academic Press, New York, 1974; Josiah Cox Russell, *Medieval Regions and Their Cities*, Indiana University Press, Bloomington, 1972; Adna Ferrin Weber, *The Growth of Cities in the Nineteenth Century*, Cornell University Press, Ithaca, 1963 (reprint of the original Macmillan publication of Vol. XI, *Columbia University Studies in History, Economics and Public Law*, 1899); United Nations Dept. of Economic and Social Affairs, Population Studies, No. 44, 1969, *Growth of the World's Urban and Rural Population; Encyclopaedia Britannica*; numerous other minor sources for scattered historic figures.

The selection was made to include cities historically important at different dates rather than to tabulate a current list of the world's largest cities.

A dash in a column indicates the city had not yet been founded, or that it had a population of less than 1,000. A blank in a column indicates no accessible figure.

All figures must be considered relative estimates whose accuracy may be open to dispute. For totals under 100,000 the actual figures are presented, but figures over 100,000 have been rounded upward or downward to eliminate odd terminal digits.

For 1850 and later years most figures are for metropolitan areas, but other figures are available for different spatial units.

[a] Figures for Berlin, Germany, are for the whole city, both East and West.
[b] Figure for Hsian (Sian, Xian), China is for the 8th century.
[c] Figures prior to 1900 are for Osaka only.
[d] Figures for Tokyo include Yokohama after 1900.
[e] Wuhan is the current name for the urban entity including the three historic cities of Hankow, Hanyang, and Wuchang.

countrysides, the parklands, the self-contained village, and the small town with the detached neighborhoods—places they often come from but to which they do not return.

Cities do contain cultural values, and these vary with the level of the demand of the urban resident, but in daily life they often are overshadowed by the difficulties with which residents must cope. Chiefly, however, the city, in the minds of many, does not present its advantages and cultural values in easily accessible terms to all members of an urban community. No really large city can possibly do so—four million people cannot attend the same musical concert, nor can they visit the same art gallery during a brief showing.

The modern city has become overwhelmed by its very size and population. Its formal plan, its political management, its peace-maintenance systems, its economic structures, its zonal arrangements for the carrying on of particular functions, and its systems of social structuring are basically those of the traditional city in most cases, and these do not serve well under the heavy pressures of huge populations. Despite innovative changes in the details of how some of these features work, the management of cities remains inefficient and ineffective in face of the demands placed upon it by its too-large numbers of residents (see pp. 244–246).

Modern urban residents, it should be noted, demand far more of their "city management" than ever before in history. Now expected from "city management" are a great many kinds of things, at rates of cost that are low, frequencies that are great, and degrees of efficiency that have come to be the measure of modern life. The urban resident now expects the provision of light, water, heat, power, garbage and trash disposal, sewage removal and treatment, and the egalitarian maintenance of law and order. To that list one must add the paving and sweeping of streets; transportation systems; communications systems; easily accessible marketing facilities; control over disease and the purity of food; plentiful, clean, and well-managed parks and playgrounds; recreational facilities; educational facilities that provide adequately for all; emergency care facilities; and entertainment facilities. Above all, somehow, urban residents hope that the higher cultural values of the city will be made available to each resident. That almost none of these were institutional aspects of the ancient city where urbanism began is simply not accepted: In the modern era "the city" should provide the higher way of life for all. In the face of these kinds of demands, it is not at all certain, however, that the institutional structures of the modern city can cope with the problems increasingly loaded upon it.

One of the elements in the "problem" of the modern city is its continued growth in population, with the cumulative escalation of the set of demands. There are those who suggest that the city has a limit beyond which it cannot expand endlessly. This viewpoint suggests that cities, as we know them, cannot be expected to cope with the whole of the population of a society. Questions then arise as to who should live in cities, and how that determination should be reached. Communist China has taken steps in this matter, with a continuing campaign of returning urban residents to the rural countrysides. Few other societies are, as yet, prepared to face the problem in that way. Who, and how many, shall live where? Perhaps here is the crux of the problem of urbanism, a dilemma to which the continued human development of culture has now brought all mankind.

The Dilemma of Modern Urbanism

Urban populations are increasing rapidly on all the inhabited continents (see Table 9.7). Cities everywhere are burgeoning with people, but urbanization in recent years is taking place more rapidly in the underdeveloped nations than in the highly advanced ones (Tables 9.8 and 9.9). The traditional patterns of association in the villages are transferred to the city, which thereby resembles a number of contiguous villages. These dense cities of rural culture are a new phenomenon in the world.

It is in the cities of the developing countries that problems inherent in the gap between men and available jobs are most evident. In almost every developing country urban population has grown faster than rural. Within the last thirty years waves of migrants from rural areas and small towns have poured into the metropolitan areas until the cities are bursting at the seams. People have fled from the land as opportunities failed to keep pace with the numbers seeking them, or they have been pulled toward the cities because of the prospects of jobs, education, or the excitement of a new life. Slum and squatter colonies mushroom within the inner city and on marginal lands on the periphery. These new urban colonies do not compare favorably with the "proper" urban settlement areas because their inhabitants are poor and cannot achieve the levels of living that go with standards of urban living that have been thought to denote urban life styles.

Most migrants are young and frequently bring their families. They come in waves, moving into areas where relatives or families from the same village or town have settled. Sometimes, as in Lima, Peru, their "invasions" have been resisted with bloodshed and killings. Mostly, the slum dwellers and squatters are seriously underemployed. Unfamiliar with the city and untrained, they are frequently forced to take marginal jobs.

Table 9.8 Representative Long-Term Trends in Urbanization

Percentage of Total Population Resident in Cities Over 10,000

Country	1800	1850	1890	1920	1950	1970	1975
World average	4	6	8	19	33	39	40
Austria	4	6	16				51
Belgium	13	21	25		60		87
Brazil	2	5	11	16	36	56	59
Canada	3	13	30		63	76	76
China	6	8	8	9	11	17	18
Egypt			16		30	42	48
France	19	14	26		54		70
India[a]	8	8	10	11	17	18	22
Indonesia	2	4		7		18	19
Italy	9		21		42		53
Japan	18	18	21	26	37	72	72
Mexico			13	31	43	59	62
Netherlands	29	29	33	47	55	78	78
Pakistan[b]	3	3	4	5	10		27
Portugal	13	13	13		32		32
Sweden	10	10		19	30	45	80
United Kingdom	21	39	62	65	80	79	77
USSR	4	9	12	15	39	56	60
United States	4	12	35	51	64	73	74
Venezuela	4	6	14		54	76	76

Sources: Growth of the World's Urban and Rural Population, 1920–2000, United Nations Dept. of Economic and Social Affairs, Population Studies No. 44, New York, 1969; Population Reference Bureau, *World Population Data Sheets* for recent years; various scattered sources.

[a] Through 1920 includes Pakistan and Bangladesh.
[b] Estimates for what in 1976 is Pakistan.
Blanks indicate no available equivalent data.

The outsider's image of the slum dwellers everywhere is that they are the poor, the disorganized, the alienated, and the crime-ridden. Yet these same slums and squatter colonies often have a vigorous developmental quality. There are networks of organizations, ranging from societies to protect the slum dwellers from those who would drive them away, to religious organizations and athletic associations. In the undeveloped countries many of these organizations have their roots in the village customs of mutual cooperation. Programs such as mapping campaigns to invade the lands they decide to occupy, establishing a network of paths and roads, and struggling with the authorities to have urban services extended to them, are exercises in popular participation that create identity and often political power.

The changing nature of the basic economic aspects of modern technological culture systems has altered the whole basic structure of life in societies that begin to modernize. In some developing countries the majority of the population is no longer required on the land. In some others the surplus population can no longer be accommodated on the land, but urban secondary activities cannot yet accept them. The altered patterns of economic systems today require large numbers of people as labor forces in secondary and tertiary activities. Such labor forces cannot live in scattered rural locations. The evolutionary changes in secondary and tertiary activities have pulled people in from rural locations, and the developmental changes in primary production systems are increasingly pushing them off the garden plots and farms. The more modernization becomes an effective force in many societies, the greater the pull-and-push forces move people out of the rural countrysides. This dual process will continue to increase in the future all over the earth, as modernization takes increasing hold.

The excessive growth of human populations in all parts of the earth since about 1850 constitutes a pressure factor that complicates enormously the adjustments to the changing nature of modern life. That excessive growth has flooded, and will continue to flood, the cities all over the earth. This often is construed as a failure of the city as an institution. This is less a failure of the urban institution than it is a failure of the human race to initiate systems of controls over population growth. Population control systems should have been initiated along with the advances in modern public health and sanitation systems and with the maturing of agricultural systems and manufacturing systems, almost a century

Table 9.9 Estimated Rates of Recent Urbanization

Percentage of Total Population Resident in Cities Over 10,000[a]

	1940	1941	1946	1947	1948	1950	1952	1954	1955	1956	1958	1960	1961	1962	1964	1965	1966	1968	1970	1972	1974	1976
Algeria					24	24		23				31					39	42	45	49	52	52
Australia			69					79					82				83			86		86
Brazil	31					36				36		46		47			52	54	56	57	59	59
Bulgaria			24			28				33	35	38		40	45		47	49	52	55	58	59
Canada		54					62					69	70			74			76			76
Chile	52					55		60				68							76			76
Egypt				30		30										39	41	42	42	43	44	48
Finland				25		32	33	34		35	37	38		39	43		45	48	52	55		58
Iraq						37	37	38		39	40	44		47	49		52	55	58	60	63	64
Mexico	35					43	43	44		44	45	50		53	54		55	57	59	60	62	63
Netherlands			54	54		55					75	80		80	79		78		78		77	77
New Zealand			60			61						63			63		68		80			81
Peru	35					31			34		44	44					50	51	52	54		55
South Korea					20	20			24			28					33		41			48
Taiwan						54		56		57		58		58					62			63
Thailand	8			10		11				12		18										20
USSR						39	42	44		45	47	49		51	52		54	55	56	58	60	60
Venezuela		30				54	56	56				63	67	69	70		72	74	76		77	77
Zaire						6						8		9	10	11	12	13	15	20		26

Sources: *Growth of the World's Urban and Rural Population, 1920–2000,* United Nations Department of Economic and Social Affairs, Population Studies No. 44, 1969; Population Reference Bureau, *World Population Data* sheets for recent years; various scattered sources for individual countries.

[a]Years for which percentages are given are those for which reasonable and comparable data were available.

Slums and overcrowded areas are as old as cities themselves, but the modern city often has more of both than did the ancient city. Also, the contrasts in living conditions are greater in the modern city. The view on the left shows the stark constrast between modern high-rise building in Bombay, India and the traditional squatter's huts. Below, the Brooklyn section of Port au Prince, Haiti, reflects the poverty of many families who have moved into the city.

ago. Had this been achieved in approximately balanced fashion, the earth's population might well have been contained at a level, perhaps, of 2.5–3.0 billion. And had that been the case, much of the "urban problem" would never have arisen above traditional levels. Such foresight in population control did not occur, for mankind seldom has collective foresight. It is only now, when confronted by explosive population growth that mankind as a whole has come face to face with perhaps the most serious cultural problem in all human history. Unable to be accommodated on the land in traditional terms, rural people who now are surplus have nowhere else to go but to the towns and cities. Since neither institution was developed to service endless populations, some other kind of institutional structure is now required. Moving surplus people to the city is no effective answer. The problem then can be shifted from the failings of the city to the question: Can mankind cope with the problem of too many people?

THE MULTIPLE TRANSITIONS

The "population problem" is an outgrowth of the vital transition (often referred to incorrectly as the demographic transition) that was described (pp. 268–271) as the six successive phases of population change owing to the reduction in death rates later followed by reduction in birthrates. The theory of the vital transition may be stated thus: On reaching a sufficient threshold of socioeconomic development, a traditional population will pass from a near-equilibrium, in which high levels of mortality tend to cancel out high levels of fertility, to a modern near-equilibrium, in which low fertility almost equals low mortality. But since the decline in births lags behind the decline in deaths, a rapid acceleration in population growth is ensured. Population increase occurs during the vital transition.

The vital transition is imbedded in, and inseparable from, a concurrent series of other transitions, as well: in technological development, in socioeconomic development, from low to high levels of production and consumption, in occupation, in education, and in increasing mobility. Each of the several transitions creates feedback effects on all of the others; the whole is an interacting system of processes that are concurrent, circular, and cumulative (Fig. 9.8). The systems model presented in Figure 9.8 emphasizes advancements in health, productivity, and longevity that accompany increasing levels of living. Successful development is associated with increasing levels of consumption and of capital formation as well as with decreasing levels of, first,

mortality and then of fertility. Successful development results in, and from, a balance between mortality and fertility at the lowest mortality rate attainable with the resources available. Improved health and greater longevity increase the returns from human resources. Decreasing fertility decreases the burdens of dependency. The result is a maximum improvement in levels of living.

Some of the other transitions are only crudely charted in Figure 9.8. The occupational transition comprises that series of changes whereby premodern societies evolve from heavy concentration in primary (raw materials) production to an almost totally different modern mix of primary, secondary (processing), and tertiary (services) occupations. The educational transition is a series of steps from a totally illiterate society to one in which a majority of the young adults have attended college.

The mobility transition also moves through several phases, such as:

1. from traditional communities of sedentary farmers firmly fixed in localized territories, through the

2. shaking loose of migrants from the countryside into a variety of paths such as into towns and cities, to rural frontiers, and to foreign lands, followed by the

3. slowing of the flight from the land, and the rise of migration between towns and cities, culminating in the

4. high-level aggregate circulatory movement of large shares of the population. In this stage, the numbers of people traveling, the frequencies, the distances spanned, and the reasons for movement (tourism or recreation) all increase markedly as personal affluence has been achieved.

Development of the phases of the several transitions does not operate equally in all societies, since diffusion of culture, and the intervention of external factors, may accelerate, telescope, or modify the rates of progress. It appears to have taken Great Britain and Sweden almost two centuries to move through their mortality-fertility-mobility transitions, whereas it required less than 80 years in the cases of Japan and the U.S.S.R. In the collective sense these several transitions are today considered as comprising what is termed *modernization*. Similar processes of cultural change, however, have characterized each of the past eras of massive cultural change, as for example, the achievement of "civilized" levels by different societies, once civilization had appeared in some regions.

The model presented in Figure 9.8, and the sequential developments suggested in the preceding paragraphs, assume continuing optimum rates and patterns of change. The model is prescriptive only for

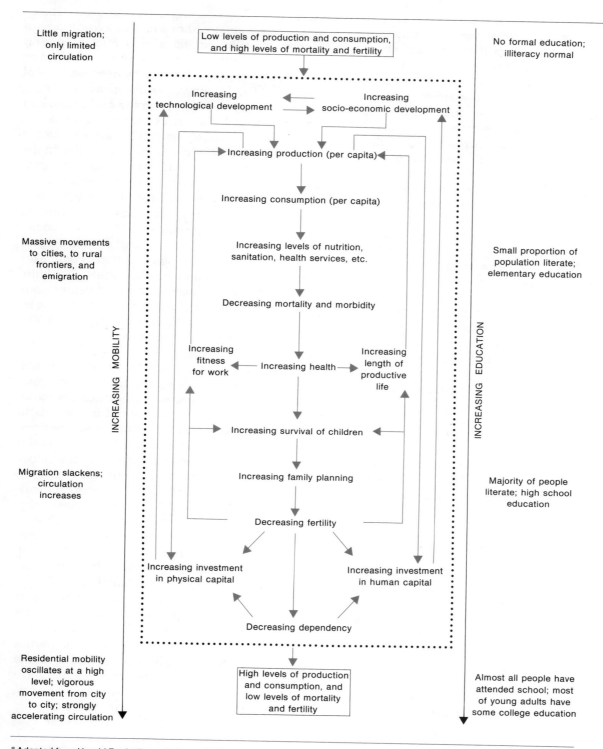

Little migration; only limited circulation

Low levels of production and consumption, and high levels of mortality and fertility

No formal education; illiteracy normal

INCREASING MOBILITY

Massive movements to cities, to rural frontiers, and emigration

Increasing technological development

Increasing socio-economic development

Increasing production (per capita)

Increasing consumption (per capita)

Increasing levels of nutrition, sanitation, health services, etc.

Decreasing mortality and morbidity

Increasing fitness for work

Increasing health

Increasing length of productive life

Increasing survival of children

Increasing family planning

Decreasing fertility

Increasing investment in physical capital

Increasing investment in human capital

Decreasing dependency

Small proportion of population literate; elementary education

INCREASING EDUCATION

Migration slackens; circulation increases

Majority of people literate; high school education

Residential mobility oscillates at a high level; vigorous movement from city to city; strongly accelerating circulation

High levels of production and consumption, and low levels of mortality and fertility

Almost all people have attended school; most of young adults have some college education

*Adapted from Harold Frederiksen, *Science,* 14 November 1969, pp. 837–47, and Wilbur Zelinsky, *The Geographical Review,* vol. 61, no. 2, April 1971, pp. 219–49.

Figure 9.8 System of Multiple Transitions

an ideal case. There is great danger in thinking that every separate societal case will operate smoothly, with all sequential stages fitting neatly together, whatever the area, the time period, and the external factors, to produce the fully "developed" society. This is obviously not the case. The model can serve, however, as an analytical device in the examination of the varying aspects of progressive change in different societies. Reference to the separate transitions, and their comparative degrees of change, may serve the diagnostic purpose of determining where and how change in a given society has *not* developed properly.

Further, the model depicted in Figure 9.8 is concerned with the achievement of full "development." It does not deal in those aspects of change that relate to the older society in decline, in which a society has passed its zenith, and lost dynamic characteristics and adaptive ability through increasing inefficiencies of its institutional structures. Studies of cultural development have been most often concerned with the ways, means, and efforts of societies to reach higher and more mature levels of life. Human knowledge and skill in the perpetuation of those high levels is not yet very pronounced, and the implications for societal decline, as recorded in past history, have not been translated into effective preventive processes. As authors, we are not yet able to suggest a model dealing with prevention of societal decline, and we turn back, therefore, to a further consideration of modernization as a contemporary culture process/complex.

Modernization

Modernization is both a process and a worldwide phenomenon. It may have begun in western Europe, and its early stages of diffusion over the earth were often labeled "Westernization." Then involved was the complete restructuring of other societies on the European model. This has proved both impossible and ineffective. Modernization, on the other hand, involves the West at the same time that it involves societies elsewhere on the earth. Modernization encompasses wholesale processes for the transformation of all societies, and for the development of new and different patterns of spatial interrelations among societies all over the earth.

That change has become rapid and enormous in extent everyone realizes, but consider for a moment the degree of change. Over the last century we have increased the possible speed of travel and the ability to control diseases by a factor of 100, potentially usable energy resources by 1000 times, the power of weapons and speeds of data handling on the order of a million, and the potential speeds of communication by a factor of ten million times. Within the past 25 years, many cultures have moved into an age of jet planes, missiles, satellites, and nuclear power, computers and automation, a service and leisure economy, mass higher education, widespread use of television, superhighways, agribusiness, oral contraceptives, to say nothing of marked and increasing environmental pollution and crises of the cities.

The kinds of changes mentioned above make it imperative that the spatial organization of life on the earth be transformed. Such change, however, requires also the transformation of all relations between mankind and the environment of the earth, both in matters external to any society and internal within that society. At the same time that some societies are taking on these new cultural elements, there remains the practice of other culture complexes characteristic of the late Neolithic levels of culture. The conflicts between these two totally different kinds of mankind-environment relationships produces, within many societies, cultural problems for ways of life that result in cultural trauma for the society as a whole.

Modernization is a most complex cultural process, that involves the wholesale patterns of culture complexes. Modernization requires integration of material and social culture complexes. It involves both innovative and adaptive change under severe pressure of time; it also requires patterns of accommodation among quite different aspects of culture. Modernization, for each society, is an internal process that must reach all members on at least approximately compatible terms if it is to be relatively successful. Persons not sharing in the broad processes of change become problem misfits in a society and thereby inhibit the whole pattern of development.

Modernization, through its increasing acceptance, has become the screen through which many of the concerns of the underdeveloped world have been viewed and then acted upon; but its use also distorts our understanding of change. One of these is the implicit directionality of the concept: to be modern is somehow better than being traditional and is, in any event, inevitable. Modernization, as access to power and material goods and better health and literacy, contributes benefits that are thought to have great positive appeal. Does modernization, however, also necessitate acceptance of social values and behavior according to the models of urban-industrialized cultures?

There is no single, simple answer to the societal approach to modernization that will yield the most successful results. Specific programs must include certain basic economic developments, such as power production, effective transportation systems, efficient systems of communication, the improved means of supplying commodities to a population, and the development of at least some aspects of scientific agriculture. Beyond these basic changes in systems of material economy, however, there must be the development of institutional structures that achieve societal goals. No two societies, probably, can work with the same institutional systems in moral, social, and political transformation, and each must seek its own way in the renovation of its traditional culture system.

Postrevolutionary China and Cuba both have suppressed the material incentive as a primary social and economic sanction. Economic development is directed as much toward social-psychological goals as toward increasing production, in which people come to see work as a duty instead of a means to individual self-betterment. In Tanzania, for example, the goal is *ujamaa vijijini* (familyhood in the villages) a uniquely African style of socialism based on a primary subsistence economy, a pattern of nucleated village settlements, and a traditional form of egalitarianism. Since 1967 this goal, which places a higher priority on social and economic welfare of people than on rapid increases in the gross national product, has been the national mandate set for the country's 15 million people, whose educational, economic, and political structures are being reshaped toward the realization of a particular pattern of life. Leadership in Malaysia, a political state made up of Malays, Chinese, and a smaller number of Indians-Pakistanis, must seek integration of social and cultural institutions among its divergent peoples in order to create a sense of common citizenry and nationhood. These problems are more difficult for Malaysia than are the elements of economic development, as such, and more critical to success.

The Example of Japan. At the end of World War II, Japan was a nation ego-deflated and downhearted. Shorn of an overseas empire, it was almost devoid of internal supplies of raw materials. No one under age 33 can remember that Japan, for now it is the world's third-ranking economic superpower (next after the United States and the Soviet Union). Industrial plants supply the markets of the world with shoes, ships, and steel, cameras, cloth, and cars, TV sets, tape recorders, and transistor radios. Japan produces one-sixth of the world's steel and half of its ships. Fully 30 percent of Japan's exports go to the United States, its major customer. In return, Japan is the largest overseas market for United States exports. How and why this achievement has been possible is worth considering, for it is the best example on earth that modernization does not equate with either Caucasians or the West.

The Japanese, under government guidance and financial support via low-interest loans and low corporate taxes, adopted policies of population stabil-

ity, heavy industrialization, and meager military buildup that have channeled their vigor into peaceful pursuit of economic development. Japanese society values harmony, order, security, and dedication to work. It is also the largest homogeneous society on earth, containing only tiny racial and linguistic minorities. Reconstruction after wartime destruction used the latest technology focused on automation. Productivity soared, with remarkably large annual increases during the decade of the 1960s. The result was that Japan became the greatest force for postwar stability in Asia.

Japan, however, is still very vulnerable, since it remains almost totally dependent on overseas sources for its energy and raw materials, and strongly dependent on foreign markets for its manufactures. A three-week supply may be all the industrial stockpile on hand; maintaining the flow of imports is vital. To search these out, the Japanese are scouring the world for new sources and contracting for existing supplies. Oil exploration is going on simultaneously in the Persian Gulf, the Gulf of Siam, Colombia, Australia, and New Guinea. Copper is being obtained from Chile, Zambia, Brazil, and the Zaïre Republic (Congo); nickel and iron from Australia; coal and lumber from Canada and the United States. Japan derives 78 percent of its energy from oil and only one percent from natural gas. Of that oil nearly half comes from Arab countries; and half is used in Japan's industries. Thus Japan has considerably less scope in saving energy than the United States.

Scarcity of arable land, a relatively intensive and small-scale agriculture that is now achieving high productivity, changes in dietary habits and food consumption patterns, and liberalization of agricultural imports combine to make Japan one of the major world markets for agricultural products. Japan purchases about one-tenth of the world's agricultural exports. For example, most of Japan's corn, wheat, grain sorghum, soybeans, and hides and skins come from the United States.

Other problems have also become apparent. The country is running into labor shortages, and so wages are rising. Skimping on internal investment in housing, roads, and pollution control has unbalanced social conditions. In the 1970s the fantastic economic growth rate is slowing, the average work week is dropping to 40 hours, and priorities are shifting to meeting the social needs for schools, hospitals, sewer systems, parks, housing, and pollution abatement. In their transition toward "postindustrial" society the Japanese are placing great emphasis on producing services for their people (e.g., health, education, recreation) and less on producing industrial goods. Industrial growth at any cost based on low wages and the sacrifices of personal amenities is no longer the principal national goal.

At each phase of the multiple transitions the definition of what constitutes progress keeps changing. Now concerned about rising inflation, increased costs of fuel and materials, residential crowding, inadequate social amenities, and massive environmental pollution, the Japanese are revising their priorities upon understanding that the social costs of modernization have not been charged wholly to the economic forces that have increased production and consumption.

SUMMARY OF OBJECTIVES

After you have read this chapter, you should be able to:

MONUMENTS TO CULTURAL SYSTEMS

Since at least the middle Neolithic, people have been erecting monuments of different kinds for a number of purposes. Certain patterns are visible in the kinds of monuments built by different culture systems. In the photographs that accompany this essay, we have presented a collection of some of these symbolic structures arranged in roughly chronological order. Monuments reflect many things about culture systems besides the architectural concepts employed by their designers and builders. Monuments may demonstrate the material technologies developed by particular cultures, but they also communicate the attitudes held by different societies, and they reflect something of the dominant value systems. Some of these values are pragmatic human values, but others are symbolisms significant only to particular cultures. The fact that monuments were constructed in the first place, the type of pragmatic-to-mystical feelings they are meant to bring about, and the amount of human and economic resources that are expended on them tell us much about the builder's views of their society.

Religious concepts and beliefs have been among the strongest human motications for building monuments throughout history. Religious attitudes have been expressed in different ways. In many cases, however, the monuments serve to symbolize particular cultural attitudes beyond the precise forms and shapes given to them. The architectural forms themselves reflect regional and cultural variation in material technologies, and systems of architecture. Monuments also reflect different systems of esthetic expression. For example, Hindu temples often depict sculptures of people and animals, whereas Islamic mosques never depict the human form but show geometric designs in stone.

Debate continues concerning the astronomical uses of some of the megalithic and cyclopean stonework found in the eastern Mediterranean Basin and around the coasts of western Europe (Photo 1 is a late example). Much of this stonework is related to burials and concepts of the hereafter, but much of it also relates to formal religious concepts.

The stones used were of enormous weight. They were cut and worked crudely into shape with primitive tools. Often, they were transported over long distances which required a huge expenditure of human energy. These factors, coupled with the rough similarities in the formal construction of stoneworks built so far apart, are evidence of a widely distributed and very strongly held set of beliefs. Although these prehistoric stoneworks are only fragmented remnants, their profusion, particularly in Brittany and southern England, contributes an important element to the creation of distinctive cultural landscapes.

Photo 1. The late megalithic stonework of Stonehenge, in southern England, dates from about 1800–1400 B.C. It had religious and possibly astronomical functions typical of many rough stone monumental constructs found in the Mediterranean Basin and coastal western Europe dating from about 4000–1400 B.C.

The Egyptian pyramids (Photo 2) were the religious monuments of a highly organized system in which the political ruler was also conceived of as God. Not only did a deceased king deserve a tomb but a god deserved something more than a common burial mound. In the politicoreligious system of Egypt's Old Kingdom, the state belonged to the God-King, and the whole of society owed a deceased God-King an eternal tomb-home requiring the highest level of craftsmanship and the largest possible expenditure of economic resources. The pyramids of the Fourth Dynasty (ca. 2600–2413 B.C.) demonstrate a control over an effective technology and a level of craftsmanship not matched by later Egyptian works. Still, the whole series of pyramid tombs contributed distinctive elements to a particular kind of cultural landscape.

Photo 2. The Cheops Pyramid (ca. 2600 B.C.) was constructed as a mausoleum for the Egyptian God-King Khufu, and the Sphinx (ca. 2550 B.C.) was sculpted as a monumental portrait of the God-King Khaf-Re. Both are sophisticated Egyptian monumental concepts.

The most common religious monuments in historic times have been temples designed for symbolic worship, shrines to those believed to be above the mortal level, sacrificial altars, monasteries, and structures built to provide space for worship by massed congregations. One or more of these monumental structures may be found in all those regions of the earth where culture systems have reached sophisticated levels. A number of distinctive technologies, systems of architecture, and methods of decoration have been expressed in structures of this type. The more notable examples have often become centers of religious pilgrimages as well as points of interest for the modern tourist. They make a distinctive contribution to the quality of the cultural landscapes around the earth.

Among the temple systems of the world there are striking differences in both architecture and function that influence their contributions to cultural landscapes. Function has definite bearing upon the form and shape of religious structures. Greek citizens visited the temples individually or in small groups, but there were no services for massed congregations (Photo 3). This feature is also true of Hindu, Buddhist, and Taoist temples (Photos 6 and 8 are Buddhist). Islamic and Christian religious structures are used at certain intervals by massed congregations. Large covered spaces must therefore be provided in humid regions, although in the dry zones the Islamic mosque often has only an open paved courtyard in which the worshippers kneel (Photos 7, 10, 11, and 13). Chinese Confucian temples and the Chinese imperial temples were not used for public worship. At Confucian temples, regional officials performed symbolic rituals on behalf of the regional societies, and only the emperor was supposed to offer ritual worship at the Temple of Heaven on behalf of all members of the Middle Kingdom (Photo 9). Mayan monuments were built for symbolic worship and sacrifice (Photo 5). Whatever their shape or function, religious structures have always been significant elements in cultural landscapes, and the great temples of the world stand forth as symbols of many aspects of the various culture systems.

In contrast to the religious monuments is the pragmatic and utilitarian kind of monumental structure such as the Great Wall of China (Photo 4). The eastern section of the Great Wall has been maintained in good condition. It serves as a permanent landscape feature symbolic of the line of historic demarcation between civilized China and the barbarian outer world of central Asia.

Photo 3. The Parthenon on the Acropolis was constructed as the chief temple to the goddess of the city of Athens (Athena) in the fifth century B.C. and represents the high point of Doric architecture. Portions of modern Athens, including the Athens Hilton Hotel, lie to the right.

Photo 4. The eastern end of the Great Wall of China has been maintained in good condition as a historic cultural symbol. The Great Wall was completed in the third century B.C. by linking together several short sections of older walls to form a military line of protection against pastoral raiders from central Asia.

Photo 5. The incompletely restored pyramid-temple of Kukulcán (the plumed-serpent god) at Chichén Itzá, northern Yucatan, covers a smaller temple built to a jaguar god in the fifth century. The present structure was built by Toltec occupants of Chichén Itzá in about the thirteenth century and stands in an immense square containing a ball court and several theaters. Each of its four stairways has 91 steps leading to an elaborate temple at the top.

Photo 6. The Sai-in at Horyuji Temple, Nara, Japan, is the main hall of the west wing of the Horyuji complex. Built in the seventh century A.D. on modified Chinese architectural principles, it is one of the world's oldest wood buildings and is now classified as a National Treasure.

Photo 7. The original Christian cathedral of St. Sophia was completed in an old quarter of Istanbul by Emperor Justinian in A.D. 537. In 1453 it was somewhat modified and converted into an Islamic mosque, the Hagia Sophia. Since 1935 the complex has served as a museum.

Photo 8. The Angkor Wat, near Siem Reap, Cambodia, was the crowning monument of the Khmer kings. Built during the twelfth century A.D., the great temple complex suffered neglect after 1432. Restoration began late in the nineteenth century and was well along when military action again began in Cambodia in the 1960s. Some damage occurred during the Communist takeover.

Photo 9. The Hall of Prayer for Good Harvests is the primary structure of the Temple of Heaven in the southern section of Peking, China. Originally built in A.D. 1420, it was struck by lightning and burned in 1889. It was rebuilt according to original design about 1900.

Photo 10. Construction of the Basilica of St. Peter in Rome, as seen from across the Piazza San Pietro, was begun in A.D. 1506. There were numerous changes in plans before the structure was basically completed in 1615.

Photo 11. The cathedral of St. Basil (the church of Vasily Blazhennoi) stands in Red Square, Moscow, with one of the Kremlin gates at the right. It is an example of Russian Orthodox religious architecture. The cathedral was built under Ivan IV, in 1555–60 to commemorate the conquest of the Khanate of Kazan.

Photo 12. One of the last of the great German castles to be built is Neuschwanstein in the Bavarian Alps, completed by Ludwig II in 1886 after a 17-year building program. The architecture and interior decoration reflect Wagnerian themes. Castle building by ruling monarchs and princes was a particular expression of several culture systems, and was indulged in most notably by Europeans, Islamic Indians, and the Japanese.

Photo 13. Mont Saint-Michel, on the north coast of France, became a minor religious center and a pilgrimage point in the eighth century A.D. The oldest existing buildings date from the tenth century. From the twelfth to the late eighteenth centuries the rocky hill was the seat of monasteries for several religious orders. Used as a prison in the mid-nineteenth century, it was later restored and is now a national monument.

It may be asserted with some basis in economic fact that some of the most imposing monumental structures were egocentrically motivated and were made possible only through the diversion of a society's economic resources by the holders of political power. The modern world may or may not be better off for the diversions that created the Egyptian pyramids, the Taj Mahal, or the great castles of Europe, but it accepts these monuments as symbols of their particular historical culture systems (Photos 2, 12, and 14).

In regions having representative governments, the structures that house administrative organs below the national level have often become monuments accepted as cultural symbols. Nowhere is this better illustrated than in the United States. The Pennsylvania statehouse, for example, has become enshrined as a symbol of the American system of representative government. Even county and municipal agencies have began to receive quartering in monumental buildings some of which have become imposing elements in urban cultural landscapes.

Photo 14. Construction on the Taj Mahal, near Agra, India, began in A.D. 1632 on orders from Mogul Emperor Shah Jahan as a mausoleum for his best-loved wife. The plans were drawn by a council of architects chosen from India and southwestern Asia.

In the past century, the United States has witnessed what might be described as a pragmatic trend in monumental construction, a trend that may indicate the increasing secularization of American life. Religious buildings no longer receive the immense economic resources that went into such monuments as Angkor Wat. Instead, resources are being put into such utilitarian constructions as Hoover Dam (Photo 15) and ostensibly utilitarian athletic stadiums (Photo 16). These modern American structures are prominent elements in our cultural landscapes and, in their own way, constitute monuments to our cultural system.

Finally, international monuments may symbolize the attempt of the collective human society to rise above individual regionalisms, political and economic nationalisms, and particularized culture systems. One might view the United Nations complex in New York City (Photo 17) as a monument holding forth to all mankind the symbol of a future world in which various cultural systems may one day be effectively integrated on our single planet Earth.

Photo 15. Hoover Dam, in southern Nevada, was completed in 1936 and may be said to represent a modern utilitarian level of monument construction. The power plants are below the 726-foot structure, and Lake Mead extends for 119 miles.

Photo 16. The open playing field gave way to the coliseum and stadium in which the seating facilities dominated the scene. Now modern engineering has made possible the total enclosing and weather-proofing of the playing field. This is a view of the facilities of the Houston Sports Association in Houston, Texas. The enclosed Astrodome is in the upper center, the baseball field is in the lower right, and automobile-parking facilities surround both.

Photo 17. The Secretariat of the United Nations complex in New York City, looms high above its surroundings. Its flags are a symbol of hope for cooperation in a troubled world.

CHAPTER 10

THE EARTH IN THE TWENTIETH CENTURY

The surface of the earth in the twentieth century presents the most interesting aspects, and some of the most complex conditions, that have ever been present on our planet. The scenery that attracts tourists around our earth is more colorful and varied than at any time in the past, and the variety in human living systems is far greater than ever before. These modern patterns are the result of three different sets of processes and activities. Physical processes have shaped the present features of land and sea into more complex spatial conditions and regional variations than ever before existed. The nonhuman biotic processes have produced greater variety, beauty, and complexity in plant and animal life than were present earlier. And the human biotic and cultural processes have produced the modern human species both in such numbers and in such variety of culture systems that the richness of human life is far greater than at any earlier time. It is true, of course, that the modern development of human cultural technologies has created impacts on our earth more intense than those produced by any of our more remote ancestors. Those impacts have also created some problems in the functioning of the most complex ecosystem the earth has ever known.

There are three challenging sets of problems facing the human species on the contemporary earth. First is the creation and application, in the near future, of sound ecological procedures that will restore and maintain conditions within the limits of balance for biotic life on the planet. Second is the creation and application of rules and principles whereby the human species, in all its variety, can continue to utilize the resources of the earth to share, maintain, and improve civilized living systems. And third is the development and application of cultural controls over the rates of population growth so that mankind can continue to live on the earth.

One-sided creation and application of exploitive cultural processes in recent centuries has probably pushed the earth's ecosystem into an unbalanced condition from which it cannot easily recover so long as these processes continue unchecked. It is *probably* true that if mankind continues to neglect the creation and application of ecologically balancing processes, the conditions could become so disturbed that much of biotic life would be endangered, including the whole of mankind. It is not true, however, that the modern earth is a a ruined environmental system beyond repair and management. Such a pessimistic view neglects consideration of the enormous natural patterns of change taking place from the late Pleistocene into the present and the workings of natural ecological processes that have long governed the progression of life on the planet earth. These processes remain active at present, entirely apart from human activity, and they will continue despite any pattern of human action. The pessimistic view also neglects the basic self-interest of mankind in ensuring the preservation of human life on a thriving and interesting biotic earth.

Many of those who accuse mankind of despoiling the natural earth by laying out cultural landscapes fail to look around themselves to observe the patterns of biotic adjustment by which plants, birds, rodents, and smaller life forms adapt to cultural environments with apparent ease, quite willing to accept opportunities to live with people in crop fields, tree farms, gardens, parks, woodlots, hedgerows, and roadsides. It is true that certain life-forms find cultural landscapes unacceptable, but it is equally true that other forms find such environments both attractive and productive. Rare is the cultural landscape in which some biotic assemblage cannot find suitable niches in which to work out an acceptable living system.

BASIC HUMAN IMPACT PROCESSES

Within recent years many people (particularly urban Americans) have come to view any action disturbing or altering the surface of the earth as inexcusably destructive. From the beginning, however, human life has always been an intrusion upon the natural world; people have struggled incessantly to live on a level higher than that of the animals, which are a part of the natural world. The whole basis for cultural advancement beyond the level of primitive life has been to improve on nature for the good of mankind, and it is now too late to turn back. Traditionally, human activities that alter the wild elements of nature and make for better human living conditions were thought to be constructive, developmental, and progressive. This human modification of environment involves the removal of wild vegetation from any surface considered usable by mankind and the introduction of crop fields, settlements, mines, road systems, and other cultural artifacts. Such activities include the elimination of wild animals considered dangerous to human life and their replacement by animal forms either domesticated, or conditioned, to live with mankind. They also include all of the recreational and aesthetic features satisfying the wide-ranging psychologies of peoples of different interests all over the earth. The human impact on the earth has been the cumulative general processes of turning wild landscapes into cultural landscapes contributing to human satisfaction.

The long-continued human process of developing cultural landscapes on the face of the earth is one which slowly replaces natural wildernesses—those "empty" areas occupied only by sets of plants and animals. The human race has been continuously involved in reducing wilderness to "productive" landscapes, and no natural biotic or physical conditions have significantly interrupted this expanding human occupancy of the whole earth. The present human population of the earth is proof that the process was practical and productive for mankind. Yet the landscapes of today's earth are largely cultural ones, and unless the human race can rapidly find a way to limit future increases in numbers, the final result can only be the extinction of all remaining wildernesses on the face of the earth.

The preservation of some wilderness areas has values and is worthwhile when so few are left, but problems are involved. Necessary to the maintenance of any wilderness areas are:

1. the rapid spread of the concept of the value of a wilderness,
2. the development of control and management technologies so far little practiced, and
3. the ultimate ability to limit human population to numbers that will not force the use of remaining wildernesses for human support.

Some of the purposefully constructive processes have had, of course, unexpected side effects that are now being recognized as destructive. All such destructive activities have a long history behind them, beginning at a time when mankind had not yet learned that cultural activities could produce results destructive in the long-term pattern. Culturally induced soil erosion is far older than crop growing, and the impacts on forests began with the fire-hunting procedures employed by earlier, simpler societies. Air and water pollution is ancient, although the scale of "production" of pollutants was relatively small in earlier time.

None of the points made above are really new in cultural geography, for geographers have long observed, mapped, counted, and criticized aspects of both constructive and destructive human processes. The importance of the issues has only recently been understood by large numbers of people the world over, and their present concerns about the problems involved are of recent origin. Unfortunately, not enough of the world's population is yet really seriously concerned, and for many people living conditions pose day-to-day problems of greater subjective importance. In the historical development of human living patterns, we have not yet taken a look at the variety of human conditions now present around the earth as a whole. The balance of this chapter is devoted to a few exemplary discussions of the nature and variety of human occupance of the earth in the twentieth century.

VARIATIONS PRODUCED THROUGH HISTORICAL DIVERGENCE

Although the earth has been linked into one operating system, and important events now become known worldwide within a few hours, there remain many differences in human living systems, regional patterns, and individual life styles. Not all these differences can be dealt with in one short chapter, but we can present samples of the patterns of variety now found on our earth as a result of the long period of development of separate systems in different regions.

Critical to an understanding of that variety and difference is an appreciation of the rate of increase among the world's peoples, with many individuals now being born into culture systems that are quite unlike. In the time it takes to read this sentence the population of the earth will have increased by eleven or twelve people. Thinking of people as vital statistics, the 1975 rate of increase was about 144 persons per minute, or roughly 8650 per hour, 208,000 per day, and approximately 76 million per year. At the end of 1975 the total population of the earth was about 4 billion, or roughly 800 million families. Fortunately, they are scattered very widely over a large part of the land surface of the earth (see back endpaper map).

About half the world's population consists of young people, and if all couples marrying each day have families of three to five children, there will be a lot less space per person in the very near future (see Fig. 9.6). Although many couples in the United States, Canada, and northwestern Europe, now plan for no more than two children, many young people in the rest of the world have not yet adopted such attitudes. Owing to the benefits of modern preventive medicine and the public health programs, infant mortality has declined. People everywhere are now living longer so that the rate of world population

increase has been going up in recent decades. Currently this rate is at its highest level in human history for the earth as a whole.

Variations Among People

A great many of the differences to be seen around the world derive from the choices made by culture groups as to how and where they will live, and how they will carry on their lives. Many of these choices derive from decisions made a long time ago by earlier members of the separate culture groups, but the patterns persist and are part of today's world.

People as Settlers. Our 800 million families are located in many different dwelling sites (Fig. 10.1). Perhaps 100 million families (500 million people) live in dispersed homes scattered in the rural landscape. There are about 3 million villages, in which about 290 million families live (1500 million people). There are perhaps 75,000 small towns, housing about 110 million families (550 million people). About 300 million families (1450 million people) live in roughly 1300 large towns and cities. These proportions change slightly every year, since some rural inhabitants migrate every year to the towns and the cities.

Rural settlers change the surface of the earth slowly and in small amounts as farms increase in area. Villages grow larger in population without strongly changing the landscapes around them. The towns and cities, on the other hand, are increasingly becoming imposing artificial environments as power-using techniques are employed to construct new kinds of buildings that become very visible in the landscape.

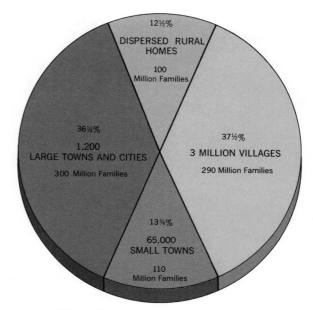

Figure 10.1 World Residence Patterns.

Populated Prior to 1800

Populated Since 1800

Empty Areas (under 2 people per square mile)

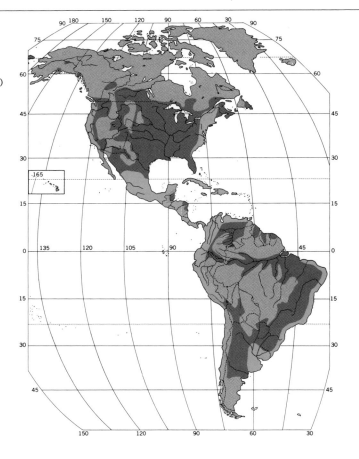

Figure 10.2A
World Settlement since 1800.

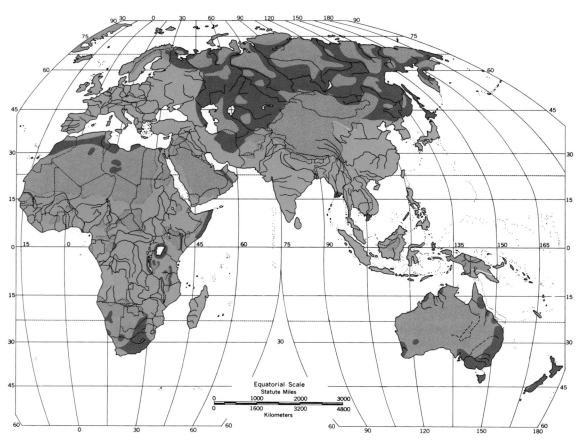

Equatorial Scale
Statute Miles
0   1000   2000   3000
0   1600   3200   4800
Kilometers

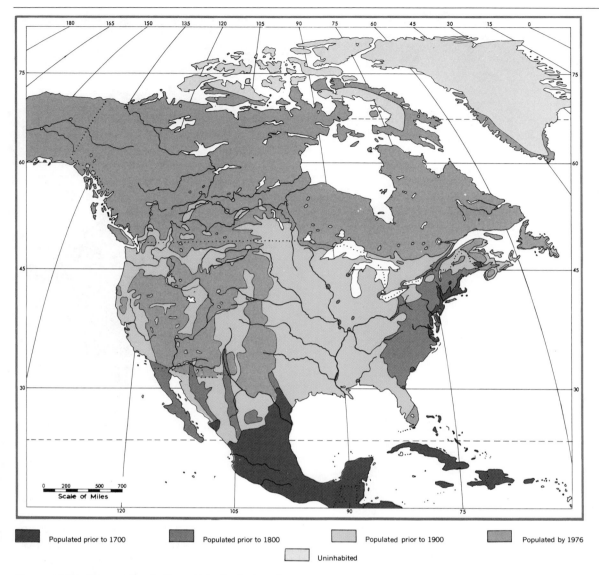

Populated prior to 1700   Populated prior to 1800   Populated prior to 1900   Populated by 1976

Uninhabited

Figure 10.2B Progressive Settlement in North America.

Although more numerous than ever, people still do not live everywhere on the earth's surface (Fig. 10.2). The chief unsettled spots are the really cold margins and the higher mountain lands, totaling roughly one-fifth of the land surface. Second are the truly dry areas of the world, scattered in fragments among the continents, empty except where some special circumstance has made occupancy worthwhile; these account for perhaps another sixth of the world. Third are the very hot and humid sections of the earth, principally in South America, Africa, and Southeast Asia, in which present-day occupancy is more notable than in the other empty sectors. The tropical one-eighth of the world's land area once thought unlivable is increasingly sustaining large populations and perhaps will be the first "empty zone" to disappear. Today, however, most of humanity (about

80 percent) occupies the areas of relatively mild conditions and is thus concentrated on half the world's land area.

People as Ethnic Groups. In the strict sense, mankind is but a single species. Among all people 46 chromosomes are normal, and all people can interbreed. However, people have always distinguished different groupings of the human family on some specific basis. Skin color has been one common criterion, and such other criteria as shape of head, color of eye, shape and size of the nose, color and texture of hair, and so on may be used as classifiers of ethnic types (see Photo Essay No. 6, both photos and text). Different groups perceive these factors differently and attach varying social significance to them. In some parts of the world such recognizable

Each area on the map is occupied by a different ethnic group.

The intent is to demonstrate diversity and not relationships.

☐ Uninhabited

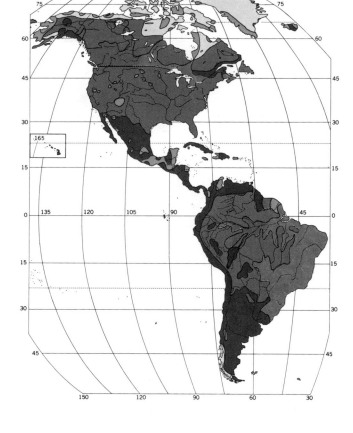

Table 10.1 World Religious Affiliations, 1975

Religion	Estimates in Millions	
Christianity		
Roman Catholic	578	
Eastern Orthodox	123	
Protestant	323	
		1,024
Islam		
Sunni	465	
Shiah	64	
		529
Hinduism		478
Buddhism		
Theravada	105	
Mahayana	163	
		268
Shintoism		75
Taoism		52
Judaism		14
Sikhism		9
Minor organized sects		21
Subtotal		2,470
Ethnic folk religions	605	
Tribal and ethnic animisms	193	
Subtotal		798
Secularized descendants from Christianity	303	
Secularized descendants from other faiths	165	
Unknown	231	
Subtotal		699
TOTAL		3,967

biological attributes have formed the bases for group organization, whereas in other parts some aspect of culture has been the source of group cohesion. Frequently, human grouping is predicated on mixed series of biological and cultural items, with the term *ethnic group* often used to express the result (Fig. 10.3). *Nationality* is commonly used when elements of political organization play a leading role. *Minority* is still another term for grouping people, usually with reference to a group that is proportionately less numerous in some political, social, or economic sense. According to purpose, the world's people can be ranked in any number of groups, from at least 3 to as many as 400.

People as Religious Believers. The basic distribution pattern of the world's major religions as established prior to the twentieth century was discussed in Chapter 6 (see Fig. 6.3, page 160). The affinity between persons of like religion, whether Christian, Buddhist. Jewish, or other is a clear force for motivation of group behavior in the modern world. A difference in religious belief, unfortunately, is also a force, in this case for dissension and conflict. Recent events in Ireland, Israel, Syria, Lebanon, India, Bangladesh, and the Philippines bear out the

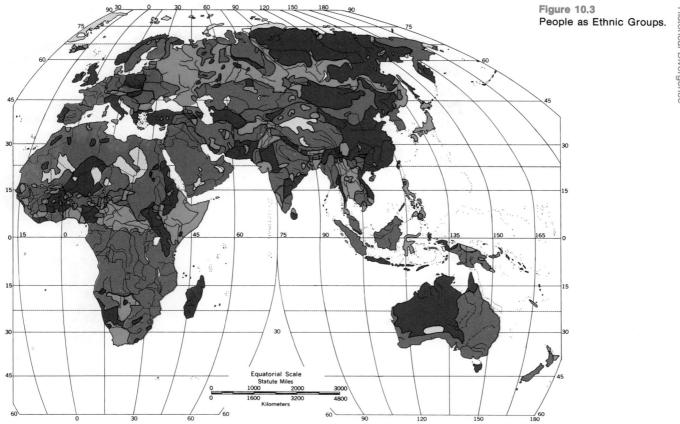

Figure 10.3
People as Ethnic Groups.

strong feelings peoples have for their religious faith, and the degree to which they will associate nonreligious matters with their religious convictions. There are few political states that have populations belonging to only one religious faith, for religious systems have carried on missionary activity that has spread most living religions widely around the earth. The tabulation of members adhering to particular religions is highly generalized and makes no effort to separate sectarian elements of particular faiths (Table 10.1).

People as Linguists. The term *language* normally presumes an ability that is restricted to human beings. As a group of people adopts specific techniques for uttering sounds, spoken language emerges.

Languages developed when people were scattered widely over the world in small groups and hence evolved differently. At one time or another some 6000 to 7000 spoken languages have been used. There are still nearly 3000 different systems of composing sounds in use in the modern world in addition to almost twice as many dialects. The language pattern of the twentieth-century world sets no bounds on regions (Fig. 10.4). Most people, including most Americans, are monolingual. Many people—

even whole cultural groups, principally minority ones—are bilingual and participate easily in several culture realms. Fewer individuals and societies (Switzerland is an unusual example) are truly multilingual and more cosmopolitan in their abilities and in their effectiveness to communicate directly with a larger proportion of the world's peoples.

The most widely used language today is "Mandarin" Chinese. It is spoken, with dialectic differences in particular areas, by perhaps 650 million people. The next most common language is English which, with its dialectic differences, is spoken by about 360 million people as their primary language. Spanish, Hindi, Russian, Arabic, Japanese, Indonesian, Portuguese, German, Bengali, French, Italian, and Urdu are other single languages, each spoken by more than 50 million people. These 14 languages include more than 60 percent of the world's people. At the other extreme are a few dozen languages, each spoken by only a few dozens of people. Some 500 languages perhaps account for no more than a million persons. India, southwestern China, the island world of Southeast Asia and Oceania, central Africa, and Soviet Asia are the regional areas of greatest contemporary variety in human speech.

Indo-European, originating in southwest Asia, is the largest family of languages with technically sim-

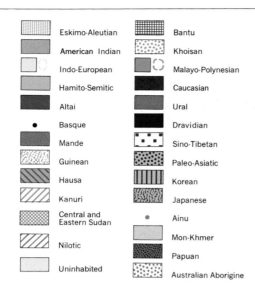

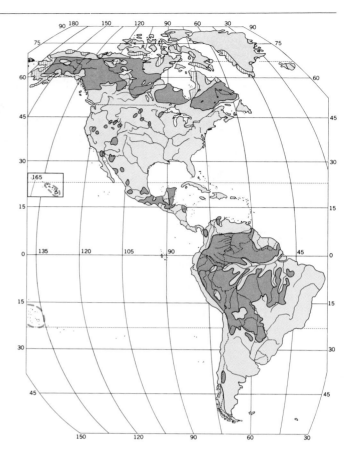

ilar speech systems. Speakers of these languages are currently distributed worldwide. Sino-Tibetan comprises another great family of languages and is concentrated in China and Southeast Asia. The Ural-Altaic family apparently originated in Siberia but is now scattered from eastern Europe to central Asia. Hamito-Semitic languages (which are chiefly north African), the Bantu language group (which is numerous in central Africa), and the American-Indian family—now partly preserved by linguistic anthropologists—constitute other large families of speech. It is now recognized that the New Guinea languages make up a large family grouping. Some languages cannot be assigned to family groupings. Basque is the standard example of a language whose origin and family relationships are unknown, but Korean, Japanese, Vietnamese, and even Greek are hard to place.

Early mankind tried and discarded many systems of writing before the alphabetic system spread and became dominant over all the world except China and those areas that adopted the Chinese system (see Figs. 4.8 and 4.9). Written language has been difficult to adapt to the human tongue, and no single international language, either spoken or written, can effectively express all the variations of human speech and thought. Language is one of the strong cements of culture; it becomes the tool of group cohesion in the hands of willful leaders. The United States in particular is most fortunate in possessing the basic elements of a single language, for it is one of the things that has helped to blend members of the country's diverse ethnic subcultures.

People as Dietary Specialists. Well over 1000 of the roughly 1800 domestic species of crop plants, and almost all of the domesticated animals, provide products eaten by someone as a food source. But all the world's peoples do not eat all of these items, and there are many patterns of cultural choice that derive from many kinds of background reasons. Regionalisms and specializations of various kinds exist at every level of culture still present on our earth today. At several points in earlier discussions we pointed out the development of many of these regional specializations as parts of developing culture systems, and here we review these variations only briefly.

The United States and Canada form a meat-consuming region in contrast to Japan, a fish-consuming region, and also form a coffee-consuming region in contrast to China, a tea consumer (see Table 9.6). The region of the United States and Canada is not the only high meat-consuming region, of course. Mexico and southern Asia are notable consumers of the hot spices, but western China (unlike the rest of China) also consumes hot spicy foods. The Mediterranean Basin is a historic consumer of olive oil, and migrants from that region have carried that preference very widely around the world. The varieties of cheeses to be seen in an American supermarket

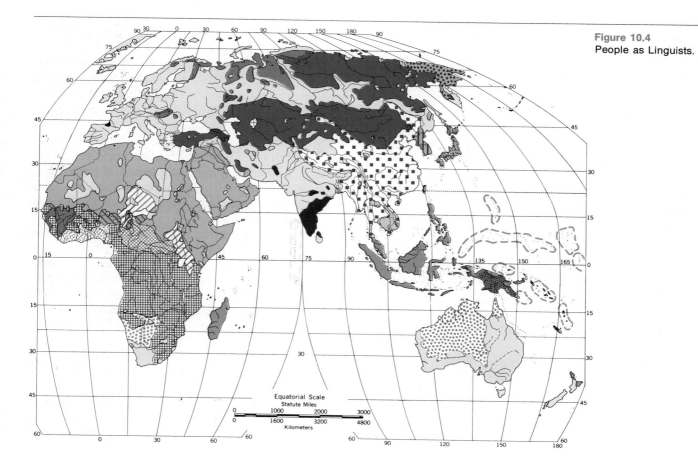

Figure 10.4
People as Linguists.

expresses a wide tolerance and acceptance of cheese as a food by Americans in the general sense, but individuals within the American consuming public may buy only certain of those cheeses. The cheese food habit is less than worldwide and expresses historical patterns of a very distinct sort.

Not all Chinese eat rice as a preferred daily staple food, for the northern Chinese have long belonged in the wheat-eating group. Americans belong to the wheat-eating regionalism but not to the rice-eating group that persists in its dietary preferences from Japan around and through southern Asia and into parts of central Africa. The Andean Highland of South America continues to be a white potato consuming region, but in the post-Columbian period the white potato has spread very widely, and there are numerous regions where people eat large volumes of the starchy food, including northwestern Europe. Maize, the corn of the American corn belt, remains a staple for many people in Middle America, and its spread today is very wide around our earth, but there are large numbers of culture groups that still regard corn as an animal food only. The persistence of preferences for asparagus, garlic, oatmeal, coconut, jack-fruit, yams, or okra; and for pork, goat meat, blood pudding, or highly pickled fish indicate historic patterns that were regional in origin but that

remain both regional and individual in our modern dietary systems.

Many causal factors may be isolated in the production and consumption regionalisms of the current world. The tastes, preferences, and changing habits of culture groups are basic to the development of world patterns of commodity production. The growth and availability of transportation, aided by the application of capital resources, facilitates the satisfaction of these preferences. Despite the effectiveness of modern transportation the issue of value versus bulk continues to loom large in the world picture of regionalism in both supply and consumption. Highly valuable items of small bulk move widely over the earth, whereas items of large bulk but low value are restricted to localized distribution. Out-of-local-season fresh fruits such as cherries may be shipped widely around the earth today, and there is a growing trade, and consumption, in such items as mangoes and papayas, whereas white potatoes are seldom shipped far in their natural bulk state.

Variations in Societal Structure

Changes in the way human societies structure themselves, derived from principles of social, political, and economic organization, have been taking

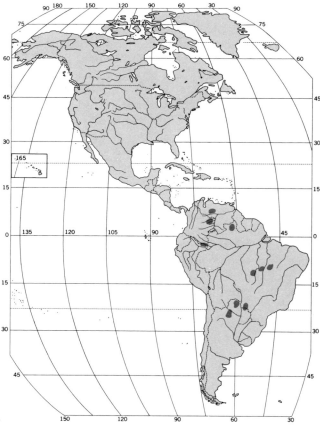

place since early human times. Our earth today shows many varieties in such patterns, and we set down only briefly some examples of structural variety.

Group Living. Scattered around the world are remnant populations that still preserve some of early man's organizational structures. In many early human groups, a simple sharing of living problems, techniques, and rewards was characteristic. Organization took various forms, depending on how the food supply was derived, how the group lived together, and what social bounds kept it unified. The issues of food and shelter and of selection of proper mates in marriage may have been basic, but social arrangement of groups for the practice of magic, for recreation, for control of territory, for the expression of intergroup friendship or enmity, and for those primordial urges toward individuality was involved in the motivations that led early people to organize their forms of living.

The simplest, and possibly the earliest, of these groupings was the band of unspecialized *gatherers*, or *collectors*, then a small mobile unit without permanent housing (Fig. 10.5). This pattern still characterizes a few of the simplest migratory economies pursued by remnant populations in odd spots

throughout the tropics. In contrast to the unspecialized collector group is the specialized collector group, which, because it concentrated on a localized resource, was sedentary and used a living system developed at a later date. The almost total self-subsistence of such groups, either fully sedentary or periodically mobile, is seldom found in the world today, but approximations still appear. In the drier portions of the world semimobile and mobile groups still preserve some of the organizational patterns of an early pastoral sort, but these groups have also become fragmented and largely absorbed into more specialized economies.

Among slightly larger units, to be measured in terms of a few hundred people, were amorphous groupings frequently and simply referred to as *communal*. Organization among such peoples was often a temporary matter, subject to change and realignment as specific needs arose. Economies were frequently combinations of simple crop growing, gathering, and hunting of game, fowl, or fish. Housing also took various forms: associations of hamlets, loose villages, settlements of separate family houses, or the articulated "longhouse" that sheltered a whole village. Once population totals rose above a few hundred, a variety of suborganizational units appeared. The expansion of the European peoples

Subsistence bands and groups

Nomadism still relatively operative

Nomadism remnantal or restrictive

Sea gypsy variant of nomadism

Figure 10.5
Systems of Group Living.
Subsistence Gatherers and Collectors.

over the earth, after A.D. 1500, encountered large numbers of groups employing all manner of living patterns. All of the simplest groupings can be termed elementary and exist today in remnant distribution only, representing the preservation of the early efforts of people to bring order into the simple problems of group living on the earth.

Multigroup Living. Eventually people conceived an organizational form both more specific and effective than the aforementioned types. Tribal structures and confederacies are more advanced forms, but the most effective one so far has been the political state, which is thought to have been a social invention of the classical Ancient East of southwest Asia. This soon became the state of the God-King, dealt with in an earlier chapter.

Historically, every small "country," kingdom, principality, and state—and every colony, empire, commonwealth, and union—has employed some modification of the early concept of the state. Greece, in its prime, abolished the God-King but kept the core city and slavery. Early Indian states elevated the role of religion and social structure, but kept the king who, with his Brahman-caste minister behind him, was almost a God-King. The late European national states made one last bid for the Divine Right of Kings.

Our American democratic state divorced religion from temporal power, made the God-King an elective administrator, elevated the rights of the individual, but kept the power of taxation and the right of conscription in time of war. The Communist state has returned to an earlier practice in respect to the power of the chief officer, by taking back into state control the power over people as persons, over resources, wealth, and production.

A nation-state has at least the six following characteristics:

1. An expanse of territory delimited by boundaries,
2. An effective administrative system that controls the area of the state,
3. A specific set of legal institutions used in administration,
4. A resident population holding citizenship in the state,
5. An economic structure that supplies much of the support to the population, and
6. A system of communications and transportation.

A nation-state has three basic functions to perform:

1. Maintain its internal cohesion through its internal administrative system,

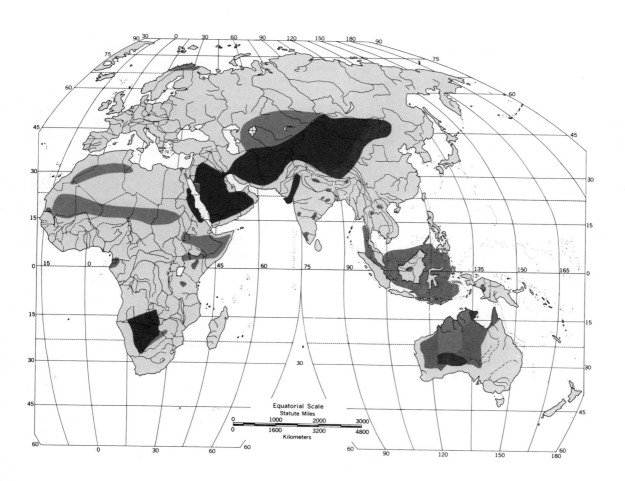

Equatorial Scale
Statute Miles
0   1000   2000   3000
0   1600   3200   4800
Kilometers

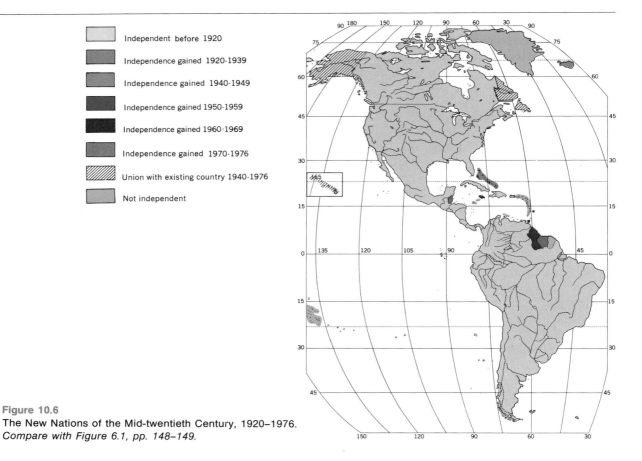

☐ Independent before 1920

▨ Independence gained 1920-1939

▨ Independence gained 1940-1949

▨ Independence gained 1950-1959

■ Independence gained 1960-1969

▨ Independence gained 1970-1976

▨ Union with existing country 1940-1976

☐ Not independent

Figure 10.6
The New Nations of the Mid-twentieth Century, 1920–1976.
Compare with Figure 6.1, pp. 148–149.

2. Finance its costs of operation, both internally and externally, and

3. Guard its territorial security against outside pressures.

Many of the long-established states have both the characteristics and perform more than the minimal number of functions. Many of the newly independent states, however, are nation-states by international tolerance and courtesy only, since they do not possess the characteristics or perform the basic functions. The proliferation of "nations" and former colonies as independent states since World War II has introduced many small, weak, ununited, and poor states into the world system. More than 70 political states have fewer than five million people each and, of these, 39 have fewer than two million each. It is not imperative that all political states be large and very populous, but the very small states have great difficulty performing all needed state functions.

Within the broad framework of today's varied expressions of the political state lie many variants in terms of their political systems. A few states still are ruled by kings. Slavery still exists in fact if not legally, in some political states. A few states have no formal system of taxation of the citizenry. Some political states are very simply structured, whereas oth-

ers, such as the United States, comprise literally thousands of elected and appointed governments within governments.

Special-Purpose Organization. Almost every person in the United States "belongs" to some special-purpose group, organized to further some specific cause or to satisfy some special social urge. But in this Americans merely exhibit a human tendency: people everywhere, and apparently almost from the beginning, have "belonged" to special-purpose organizations, some that have been purely social, serving as an outlet for human gregariousness or providing status, and others that have carved a quasi-official position in the society of which they have been a part. In the United States these range from such groups as the Antique Automobile Club of America, the Society of the Friendly Sons of St. Patrick, the American War Dads, the Daughters of the American Revolution, and Rotary International to the American Heart Association, the Institute of Radio Engineers, and the American Rocket Society.

The anthropologist, examining a simple society, notes the presence of the Men's House, the several age groupings, the grouping of the skilled warriors, and that of the skilled Old Men. In other, more advanced societies other kinds of organizational

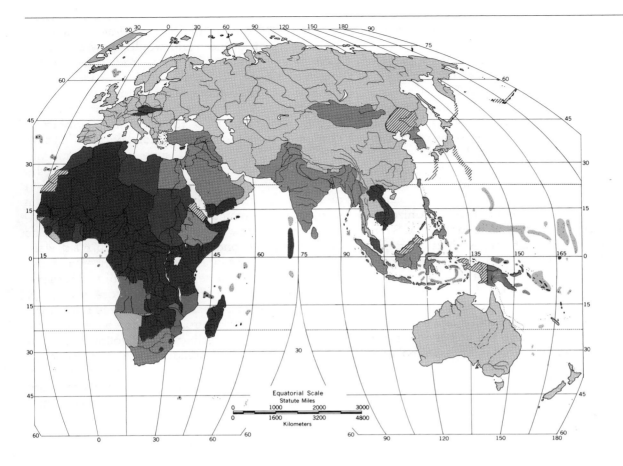

grouping appear. In many instances these special-purpose groupings are status symbols for the membership; in others they represent the effort to serve special needs of a segment, or the whole, of the society. Such special-purpose organizations have become more numerous and have often taken on an international character.

Variations in Spatial Units

In administrative terms the peoples of the earth have repeatedly changed the spatial units by which the earth is organized. There are both international and internal systems for such patterns, and we present only two samples of this on the international level. But to remind ourselves of the variety in administrative matters, the United States has a particular spatial pattern on the international political level. Internally, however, it consists not only of 50 separate states, but of court districts, federal reserve districts, military corps areas, and many more such spatial units. The United States is divided into over 3000 counties, and there are over 78,000 spatial units pertaining to local governmental administration alone. Most large political states around the earth have such multiple internal spatial patterns, although the

United States may have created far more of these than any other political state.

The Political Earth. The political map of the earth undergoes distinct change at the end of every major war, and at other intervals of political change. In the twentieth century the political map has been changing constantly in almost every part of the world, and it is likely that during the balance of the century there will be additional important changes. Not that this necessarily involves further major wars, for the evolution of the political organization of space on the earth is today in active flux. Colonies become independent, peoples formerly organized as tribes organize political states on modern lines and seek admittance to the United Nations (see front endpaper map and Fig. 10.6).

In 1945 there were 71 independent political states in operational condition. At its origin in 1945, the United Nations numbered 50 political states. But, since the end of World War II, many former colonial territories have achieved their independence. Thus, by November, 1975, the United Nations had admitted the 143rd independent state, and a few more were still seeking admittance (Fig. 10.7). For a period of time, at least, the political map of the world continues to become more complex as the few remaining colonial holdings achieve independence and struggle to operate as political states.

Freedom of the seas, control of territorial waters, and freedom of the air above states are experiencing change. Although the demarcation of political boundaries actively continues, conflicting claims to earth space are perhaps more numerous than ever. Despite preserving every boundary stone of the separate national entities, the political map of western Europe is overlaid with a series of such new political lines as those formed by NATO, the Iron Curtain domain, and the European Coal and Steel Community (see Fig. 10.7). Not all political organization tends toward great combines. Although the U.S.S.R. has blocked out the world's largest state, Andorra, Monaco, Liechtenstein, San Marino, Bahrain, Singapore, and Brunei continue to exist.

What is true in our lifetime has been true over the long term in the past. A historical atlas illustrates a bewildering series of political states. From the time of widespread tribalism, through the era of city-states, to the first empire and on toward our present political map, thousands of spatial patterns of political organization have been in force. Many present ones seek historic precedent in the formality and outline of some previous state, but many strike out anew.

The Currency Bloc Earth. A late Pleistocene economic map of the world would show only local subsistence regions and a few trade zones within which exchange was commodity-for-commodity barter. The composite economic map for the contemporary earth makes it clear that production and commodity movement is a rather highly elaborated pattern that embraces the whole earth in a one-world trade community. Despite the one-world impression that comes out of commodity flow charts, the earth is not yet united into one trading community in all senses. Transportation facilities have made that single trading community possible, but certain barriers remain to slow or deflect actual trade. Whereas the barriers to economic exchange in 3076 B.C. were physical problems of transport, distance, and the nature of landscapes separating populated regions, the barriers in 1976 were chiefly those set up by people themselves. These consist of trade blocs, tariff regions, politically controlled trade zones, and the like. Such terms as "sterling bloc," "dollar bloc," and "ruble bloc" suggest humanly instituted economic barriers to commodity flow. The map of currency systems is still a rather complex one for an earth that might otherwise function economically as one integrated unit (Fig. 10.8).

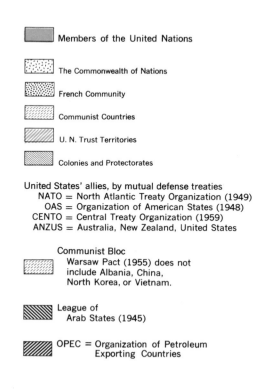

Members of the United Nations

The Commonwealth of Nations

French Community

Communist Countries

U. N. Trust Territories

Colonies and Protectorates

United States' allies, by mutual defense treaties
NATO = North Atlantic Treaty Organization (1949)
OAS = Organization of American States (1948)
CENTO = Central Treaty Organization (1959)
ANZUS = Australia, New Zealand, United States

Communist Bloc
Warsaw Pact (1955) does not include Albania, China, North Korea, or Vietnam.

League of Arab States (1945)

OPEC = Organization of Petroleum Exporting Countries

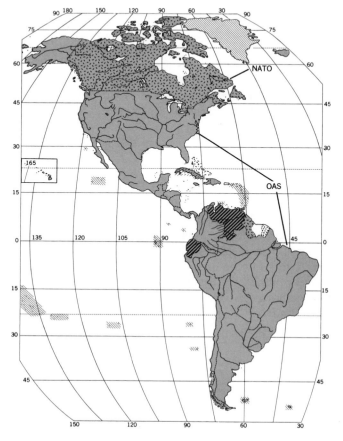

Variations in Economies

The contemporary world possesses the greatest range of economic structures that have ever existed on our earth, for the differences have been growing greater in the modern period despite the trends toward convergence in the broad cultural pattern. This range of economies forms a very long continuum. It is easy to place some of the markedly different structures at points on the continuum, but it is very hard to place each and every society at its proper relative position on the line of development from the most simple and primitive to the most complex and advanced. Growth, change, and advancement of the societies of the earth that are operative at present has been very uneven in the last five hundred years, and their culture patterns have been further complicated by the way in which the industrial and trade systems have developed. As an aid to understanding present variations in economic structures of the earth, we set down a brief resume of their historical development as a prologue to the discussion of contemporary variations.

Prologue on the Background of the Contemporary Economic World. In A.D. 1500 the conditions of the earth were on the verge of change, but those conditions were not desperate, and human life was probably not all that exciting, either. There were then perhaps 450 million people inhabiting the earth as a whole. Of these, about 80 million were Europeans, and among them were people who were far more curious about the earth than were those of any other region. The Europeans did not then live any better than peoples in some of the more advanced societies elsewhere around the earth. Perhaps the Chinese then had the most satisfactory level of living, followed by some culture groups in Peninsular India and by the townspeople of the north Italian trading cities. The *standards of living* (the level at which one hopes, or would like, to live) had not yet begun to lift markedly above the traditional *levels of living* (the actual level at which one lives) in any part of the earth. If the Chinese, Peninsular Indians, and north Italian city residents were then among the leaders of the earth, the inhabitants of Australia, southern Africa, eastern South America-Amazonia, and northern Siberia certainly lagged far behind and were near the low end of the continuum. The societies of the rest of the earth fell somewhere in between. The lagging societies in A.D. 1500 were still operating with Paleolithic technologies, thousands of years behind the

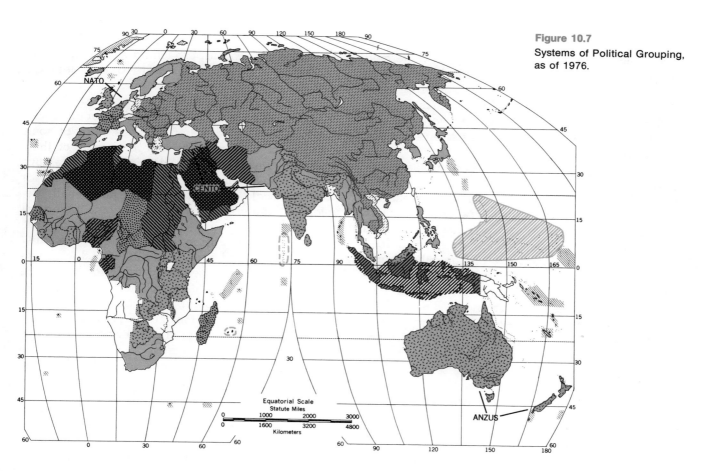

Figure 10.7

Systems of Political Grouping, as of 1976.

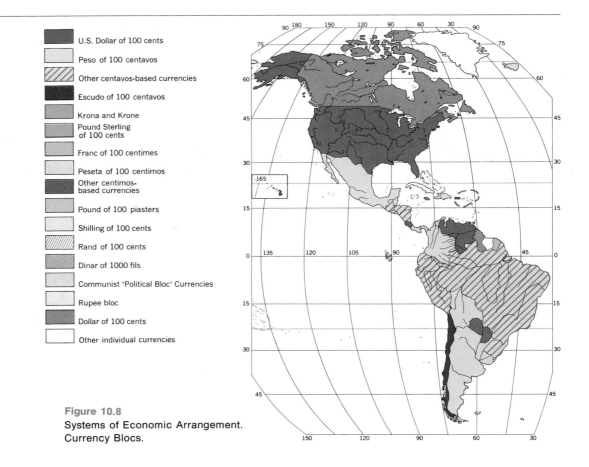

Figure 10.8
Systems of Economic Arrangement.
Currency Blocs.

leaders, who had advanced to what may be termed the proto-industrial level. For this level, agriculture was productive, handicraft manufacturing was highly organized and required many skills, and the wide range of commodities produced were well distributed throughout the respective trade regions by well-organized trading systems.

The different societies of the earth occupied homelands according to their ability to hold regions that suited their socioeconomic and politicoreligious levels of culture. The Paleolithic-type societies were small in population per group, and each group held a quite small territory that contained few resources. The proto-industrial societies, on the other hand, occupied much larger territories, or, in the case of the north Italian trading cities, dominated the economic activity of the Mediterranean Basin. These societies were linked together in various ways so that the grouped pattern amounted to a large region from which the natural and cultural resources could be drawn fairly freely. Thus, the Chinese, in A.D. 1500, controlled and operated a territorial expanse that really amounted to an "economic world" of its own. In the same way, Peninsular India comprised an economic world, and the European region formed a third such world (in which the north Italian trading cities were then the core). In A.D. 1500 Southern Mex-

ico-Guatemala was approaching such a situation, although the technologies were a little less advanced.

These leading societies, the most economically advanced of their time, were the "developed" societies of A.D. 1500. All other societies were at some lower point on the continuum, and most of them could be termed "underdeveloped" in some relative degree, with the most seriously lagging societies far down the scale toward the bottom. Each such "developed" region drew tribute from, and otherwise exploited, the marginal satellites located around it, and each exported its manufactures to those satellites. The skills and cultural attributes of the "developed" societies diffused outward and into those marginal territories. In A.D. 1500 none of the "developed" societies could exploit, or have contact with, territories far beyond the marginal satellites. The linkage systems—those of effective systems of transportation and communication to far parts of the earth—had not yet developed. None of the "developed" societies were able to make contact with, or trade with, the whole of the earth at that time.

The Columbian Discoveries emphatically altered the above situation by linking the whole earth into one system of organized contact. From the first voyages of discovery to about A.D. 1600 the various soci-

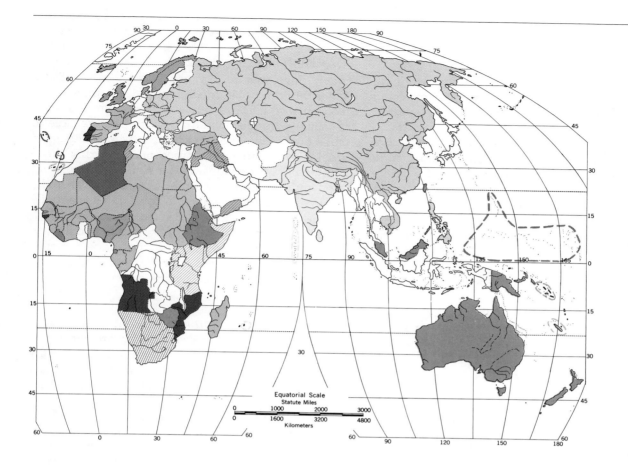

eties of the earth were in a state of shock at the occurrences. By about 1600 the European societies had begun to formulate systems of contact with the rest of the earth, and the trading companies (the East India Companies were the first) began to probe the earth for trading possibilities and to stake out claims to trading rights in territories around the earth. No other "developed" zone challenged the Europeans or entered into competition with them. This meant that the future directions of economic development were at the discretion of the European societies.

Western European societies took to the sea lanes in noisily competitive patterns of world search, trade, and exploitation of all the new products and resources. Russia, which had been a marginal satellite of the West European socioeconomic realm, quietly took the land route eastward across Eurasia, and was successful enough that the Russians in due time confronted the British on the northwest coast of North America. Among the competing West European societies, which were rapidly maturing into separate national political states, the privately organized trading companies experienced more and more difficulty in their attempts to control trade in specific territories. Home governments finally cancelled the company charters and took over control to initiate

formal colonial systems. Portuguese and Spanish patterns had involved territorial controls from the start, and most of their colonies became independent in the early nineteenth century when other West European patterns were still operating through private companies in some regions. The Russian efforts were always government controlled, but since the zone of Russian expansion was contiguous to the home territory, the control over trade zones was easily and effectively turned into the consolidated territorial control that simply expanded the Russian political state.

Modern colonialism ran its course in the overseas territories operated by West European societies. While that course was being run, there came the modern advances in agriculture, medicine and public health, transportation, communications, industrial-manufacturing, economic organization, and political organization. These features brought maturity to European societies, and to new ones in lands settled by Europeans. Capitalism, as an evolutionary advance in applied economic organization, was a device that greatly facilitated the maturing of other aspects of modernization in the already advanced European culture world. It also provided a mechanism for dealing with the rest of the earth. The old-fashioned tribute drawn from satellite margins by the rulers of "developed" societies turned into the "profits" reaped by corporations that dealt with the world as a whole.

In every colonial area the Europeans carried out some kind of developmental program introducing new cultural, technologial, and economic features. These programs varied greatly from colony to colony, but they ranged from initiating educational programs designed to train labor forces and civilian bureaucracies in new and different skills, to the building of ports and land transport systems, to the setting up of agricultural experiment stations, to the developing of public health systems, and to the establishing of radio communications and air transport facilities. The technological level of such programs lagged the general level of progress in technological development throughout the world as a whole. Many such features were introduced for the purpose of increasing production of those commodities sought by the home country as raw materials, consumer products, or articles of trade. It is often overlooked, however, that such programs were the very first steps taken to alter the traditional economies, that the incentives for change lay with the Europeans rather than the local populations, and that local populations resisted many efforts. Unplanned and uncoordinated though they were, the cumulative volume of such programs was inherited by local populations on achieving independence, and this inheritance formed the base of the economic structure of the new state.

The technological advances mentioned in a previous paragraph were dramatic happenings taking place at a faster rate of cultural change than the world had ever known, allowing no period of time for adjustment to new patterns by less advanced societies, since further advances followed so rapidly on the path of earlier ones. But even more dramatic than the technological changes were the startling rates of population increase that affected societies everywhere around the world, except for the smaller Paleolithic groups near the bottom of the continuum. Some of these are barely holding their own, a very few are making a cultural transition into the modern world, but some of them are dying out. Among the most advanced societies, the highly "developed" ones, population increases are beginning to be brought under control, and it is the large number of "underdeveloped" societies in the mid-portion of the continuum that continue to have rates of population increase far out of proportion to their advances in economic and social organization.

Political organization, in the form of the centralized national political state, is the one feature of European modernization that has spread rapidly around the whole of the earth, particularly since World War II. From the 71 independent societies operative at the time of the organization of the United Nations, the number has now expanded to over 170. Political nationalism, in outrunning economic development, has produced many political states that are not independently viable units in the modern world. These states, claiming gross inequities suffered under modern colonialism, are able to make public demands on any society they choose to label "developed" and "imperialist." Notably, those being called to account are the West European societies and some of their former colonies that rapidly progressed to the "developed" level, but it is equally notable that Soviet Russia, the most successful of all the colonial imperialists, is not called to account. In the present staging of international developments, there is the possibility of a kind of world confrontation the earth has never before known.

Agricultural Economies. The most productive types of agriculture in the world today are carried on close to the regions of greatest industrialization by people who are culturally part of advanced industrial societies. The most productive forms of agriculture are carried on in parts of the United States and southern Canada, here referred to as the North American region. Higher yields per acre can be found in some other places in the world for specific crops, but their total production is small and the costs are relatively high. On the average, the North American farmer produces food for about 40 people off the farm. Although some ecological criticisms can be levelled at certain North American farming practices, these farmers stand as the safeguard against famine in many parts of the world today. The North American farmer is being pushed into even greater productivity because agricultural economies in so many other parts of the earth no longer can regularly feed their populations.

In earlier chapters we repeatedly discussed the evolution of agricultural systems and the growth of productivity in crop production (see pp. 85–89, 148–149, 192–193, 250–253, and Figs. 3.5, 4.3, 7.4, and 7.5). North American agriculture today is a highly mechanized and commercialized operation carried on by a relatively small part of the total population. It has developed to the level at which it can turn out large volumes of food and industrial materials to provide for urban residents who are engaged in many other forms of economic activity. In many other parts of the earth, the great majority of the population is engaged in some form of gardening that feeds few people beyond those engaged in it. Some other areas find over half the population engaged in some one of the older traditional forms of agriculture, in which surpluses do occur but in comparatively small amounts. The gardening economies in recent centuries have spawned large population increases that lived reasonably well so long as good agricultural land was plentiful. The returns from that gardening, however, steadily went into population increases in the last two hundred years rather than into capital savings to be invested in upgrading the system of agriculture. Good lands have now become so scarce in many parts of the earth that populations are being forced into further intensification programs on existing lands. This added labor level has often reached the point at which yields are growing slightly greater in total but are becoming less and less per person. This pattern is perhaps most marked on the island of Java, where today several times the agricultural population cultivates only a little more land than was in cultivation in 1800, and the per-capita share of produced foods is decreasing. Even when some share of the crop pattern is produced for an export commercial market, the level of productivity is relatively low, and the returns do not provide sufficient financial surpluses to upgrade the technological levels of agriculture.

Around the earth today large numbers of people are engaged in every level of gardening and agriculture from simple shifting cultivation through industrialized agricultural production (see Fig. 4.3). The largest share of the earth's population, however, is still engaged in some form of gardening or traditional agriculture rather than in the advanced and modernized forms of agricultural production that characterize the highly advanced societies.

Industrial Economies. The Industrial Revolution that began in northwestern Europe did three things in its creative growth that altered traditional manufacturing systems. It employed inanimate energy on a large scale to replace human muscles; it utilized powered machines (rather than hand tools) to make things repetitively in large numbers; and it developed a tradition of conducting research into still newer and better ways of producing things. In these three features lies the difference between traditional manufacturing and modern industrial fabrication, since the new features permitted the production of kinds of finished goods that traditional methods never could produce. These three aspects show up as the large-plant power systems, the integrated factory systems, and the research laboratories.

Some kind of manufacturing is carried on wherever there are people. We discussed the evolution of manufacturing in earlier chapters as the evolutionary addition of new systems to continuing older systems (see pp. 65–68, 90–92, 201–204, and Figs. 7.9, 7.10, and 7.11). What the industrial revolution did was first to add a powered factory system to the complex of older systems and, more recently, to add still another system in the powered production line system that incorporates machine controls over production.

It has been conventional to divide manufacturing activity chronologically into historic pre-industrial and modern industrial eras on the bases of the first two features mentioned above. Translating the locational aspects of this division onto a distribution map divides the earth into two simple patterns, traditional handicraft manufacturing and industrial fabrication. This is too simplistic a way of viewing the present status of the growth of industrialization. Modern industrial procedures have been spreading widely around the earth in the last two centuries, including considerable diffusion into former colonial lands during the colonial era. A better way of viewing the patterns was presented in Figure 7.9, wherein the evolutionary development is recognized as taking place in the contemporary era, with different regions at different stages of growth. Even that map overgeneralizes the developmental aspect since it does not distinguish between basic heavy industry and lighter, secondary manufacturing operations as these develop within any one region (see the section on Cyclic Evolution of Industry in Chapter 7, pp. 205–206).

Many of the first undertakings in modern industrial activity consist of the processing of raw materials rather than the fabrication of finished goods. Such industrial processing is almost worldwide today, whereas the fabrication of finished products has a much more limited distribution. More recently, some of the multinational corporations, however, have been both fabricating and assembling light industrial goods ever more widely around the globe. The chief factor operative in this spreading pattern, of course, has been the competitive cost of labor, and for two or three decades this was, and still is, a significant differential. The costs of labor, however, tend rapidly to become approximate in all areas, and in the near future there will be very few areas of the earth in which "cheap labor" will be still available.

There remain, of course, strong differentials between the mature industrial society and the society that is just beginning to industrialize. The mature industrial society has achieved a very large and flexible corps of skilled labor, and it possesses a complex, interrelated, and integrated factory plant with both large and wide ranges of activity. In its operational procedures many by-products become the raw materials for other procedures. The mature industrial society possesses and regularly uses complex patterns to subcontract the manufacturing of components and to establish integrated systems for assembling them into finished products. And, very significantly, the mature industrial region practices

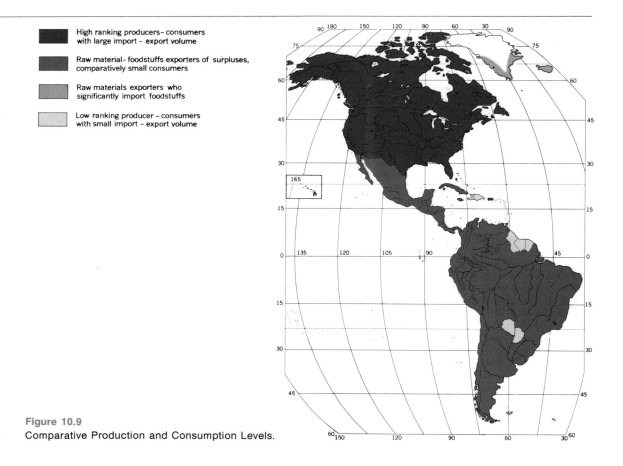

High ranking producers–consumers
with large import – export volume

Raw material–foodstuffs exporters of surpluses,
comparatively small consumers

Raw materials exporters who
significantly import foodstuffs

Low ranking producer – consumers
with small import – export volume

Figure 10.9
Comparative Production and Consumption Levels.

its tradition of industrial research and development (colloquially abbreviated to R&D). The differentials between the mature industrial region and the beginning industrial regions are strong and significant over time. Further, they tend to persist, for the research-and-development function tends to develop new technologies and new products so rapidly that the regions of lesser development are always kept in the position, so it seems, of trying to catch up.

Along with the high level of economic organization and production in the advanced society, of course, goes a high level of consumption of manufactured goods. The highly advanced society, therefore, has a very large total trade volume consisting of large amounts of imports of raw materials, fuels, food commodities, and consumer goods, as well as a large export trade in industrial agents, semimanufactured items, and finished industrial goods. A share of both the imports and the home manufactures go into domestic consumption by the population making up the industrial society. The United States is, of course, a primary example in this type of economic turnover, for the United States imports tremendous volumes of materials from all over the earth. On the other hand, the United States exports to the world as a whole at the present time. In contrast to this pattern is the case of the agricultural society carrying on only simple

levels of manufacturing that employ only local materials. Such a society has a relatively low volume of imports and only a narrow range of exports, along with a relatively low consumption rate (Fig. 10.9).

In the development of the industrial economy, of course, the matter of financial resources plays a strong role, and capitalism as an "industrial tool" has functioned very effectively on the side of the society that could muster the resources to invest in industrial undertakings. Here it is undoubtedly true that the West accumulated capital resources in the early era of international trading, but we believe there has been too much complaint from many of the societies of the earth that they cannot get ahead because they were exploited by the colonial imperialists. Capital accumulations for the beginning establishment of industry do have to be made, but the experience of Communist China (which certainly claims to have been exploited by the West) suggests that with real determination a society can accumulate its own capital resources from within.

For all the strong differentials between such mature industrial regions as northwest Europe (where modern industrialization began), or northeastern North America (the second industrial region to develop maturity), and the late-starting region that is still immature, the situation is not static. Other indus-

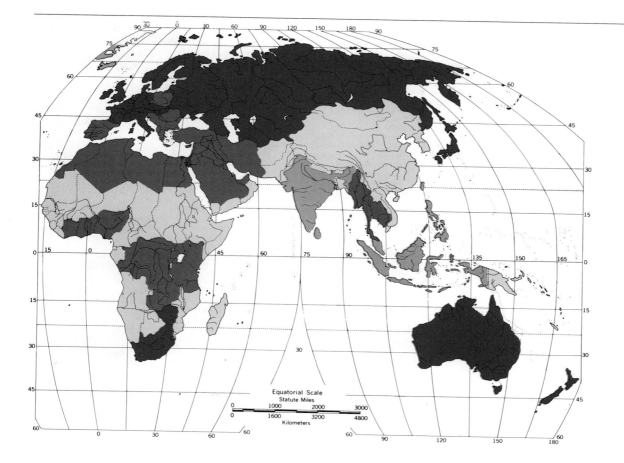

trial regions are approaching maturity (see Fig. 7.9). Japan, significantly, is approaching industrial maturity, as is a region within the Soviet Union, and one in Australia. Chinese and Indian industrial regions are coming close to mature status, also. Industrialization is not something that can be achieved in a decade, and it will be some time before many of the present nonindustrial areas of the earth achieve significant levels of maturity.

The Rich and the Poor. Any citizen of the United States or Canada who has traveled very widely around the earth is familiar with the viewpoint that all North Americans are regarded as rich by everyone everywhere else. And Marco Polo returned to Europe in the late thirteenth century with the emphatic impression that China was a rich country and that the Chinese lived better than did the Europeans at the time. There is basis for both sets of beliefs, but many outside the United States refuse to believe that any North American can be poor. The subjective differential between the "rich" and the "poor" has long been present among many of the peoples of the earth (Fig. 10.10), and see p. 336.

The persistent pursuit of a higher standard of living than the level at which one lives, at any time, has not been a constant universal objective of all ethnic

components of the human race. Being thought rich is one of the crosses that must be borne, in a sense, by members of a modern mature industrial society for the display in his culture system of imaginative initiative, hard work, and persistence in fashioning the best possible pattern of living out of the resources of a region. For over two hundred years there has been no better statement of the psychological outlook of "an American" than the phrase, "There has to be a better way to do it." The concept of "progress" that improves life has been entangled with the concept of "happiness." A quite different viewpoint has been expressed by many a member of a non-Occidental society when, after working hard at a job for a few days, he quit the job to live in relative idleness and comfort until it became necessary to work a little more to earn enough by which again to live in comfort. If both the above generalizations are stereotypes, so are many of the contemporary views concerning the rich and the poor all over the earth.

The maturing of the European culture realm, in the seventeenth to early twentieth centuries, brought out more strongly than in any other contemporary culture realm the viewpoint that the resources of the earth could be exploited to improve the level of living of human beings. Whatever its origin, members of the European culture realm worked steadily

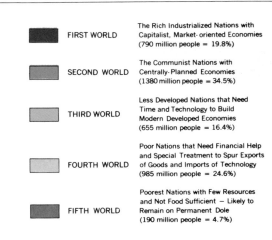

FIRST WORLD — The Rich Industrialized Nations with Capitalist, Market-oriented Economies (790 million people = 19.8%)

SECOND WORLD — The Communist Nations with Centrally-Planned Economies (1380 million people = 34.5%)

THIRD WORLD — Less Developed Nations that Need Time and Technology to Build Modern Developed Economies (655 million people = 16.4%)

FOURTH WORLD — Poor Nations that Need Financial Help and Special Treatment to Spur Exports of Goods and Imports of Technology (985 million people = 24.6%)

FIFTH WORLD — Poorest Nations with Few Resources and Not Food Sufficient — Likely to Remain on Permanent Dole (190 million people = 4.7%)

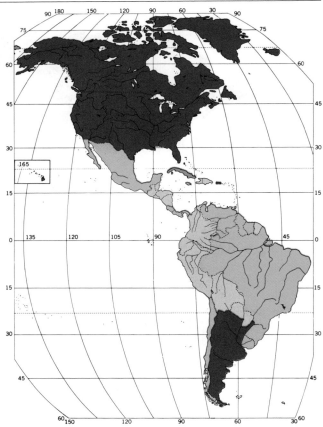

at improving the levels of living for themselves. The viewpoint, often attributed broadly to the East, that people should live in harmony with nature, did not carry with it any such concept, for one of its points was that material things were less important than the happiness that could be attained without strenuous exertion. Until late in the nineteenth century, many of the culture groups around the earth declined to accept any aspect of the European viewpoint; instead, they were content to pursue their own concepts of happiness while the Europeans continued to pursue their separate objectives. The European outlook, however, has spread all over the earth, as members of other cultures increasingly watched the European accomplishments. The concept displayed was gradually comprehended, then slowly accepted and, in the twentieth century, is acted upon ever more aggressively. The economic and social history of non-European societies recounts the difficult role of the pioneers of those societies in their efforts to gain acceptance for economic procedures taken from Occidental practice. The improvement of the levels of living, by the substitution of European standards of living for the traditional standards of the particular culture group, has been in process all over the earth in the twentieth century, and this change involves changing, also, the concept of the use of resources. Among many non-European societies this substitution is not more than a generation old, although pioneers in many societies have been pushing the motivation forward for over a century.

The viewpoint concerning the use of resources by industrial countries is largely responsible for the rising wave of protest over what is termed their "unfair exploitation" of material resources located in less-developed countries. In a roundabout way, the drive for political independence by the former colonial lands owed something to the changing concepts regarding levels of living, higher standards of living, and the use of resources. This changing set of view-

points is clearly behind the pressures currently being placed on the West and on the group of industrialized societies (which now includes Japan among its members) to share the resources of the earth with societies whose industrialization has not taken place. This pressure extends to the view that the West should also help to finance programs for improving the economies of the less-developed countries so that the members of those countries may reach and enjoy higher standards of living.

Involved in the shift of viewpoints, of course, has been the extraordinary growth of population around the earth as a whole, which in turn has raised questions of economic productivity. It is chiefly in those lands in which little initiative was formerly expressed toward improving agricultural productivity that the real pinch has come; for there true poverty is spreading as populations outgrow the productivity of their agricultural economies. The attitude is now established almost worldwide that, in times of crisis in such a society, it is the obligation of rich countries to support the poor of the less-developed countries.

The separation of the political states of the earth into the rich and the poor ignores, of course, the internal imbalance within individual political states and societies. Canadians and Americans are well aware that not all their citizens are rich, and they also know that in many countries now labeled poor

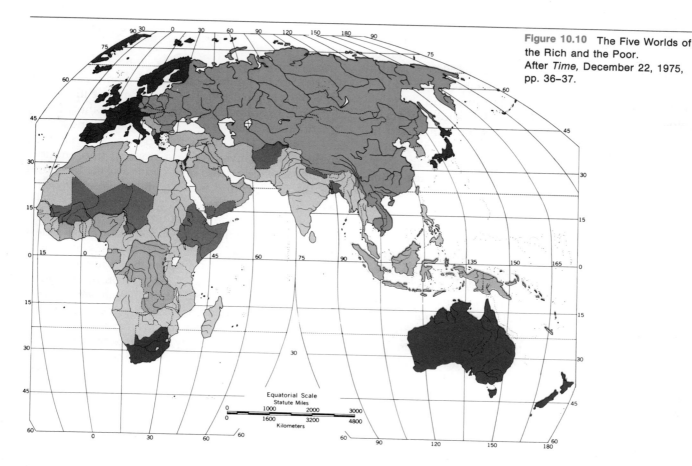

there are many rich families. The discussion on simplistic worldwide terms attempts to make political capital of the basic distinction, while ignoring totally the facts of internal socioeconomic stratification.

Just who are the Rich, the Poor, the Developed, the Less-Developed, the Industrial, the Non-Industrial, the Haves, the Have-Nots, the Leading, and the Lagging? Repeatedly, in earlier chapters, we pointed out the existence of a historical differential between what we termed the leading and the lagging, and noted that a given society did not always remain in a leading position. We also pointed out the problems of the haves and the have-nots. The contrasts between the aggressive society that has worked at creative development of culture, at improving technology, at adopting new ideas and ways of solving problems of human living, and the passive society that has followed the conservative role of keeping things as they were comfortable, has been with the human race since it emerged from the animal kingdom. Those contrasts have often provoked open conflict, and the record of wars among societies of the earth reflects the confrontations that could not be solved peacefully. We have also pointed out that, on our single planet as the home of many separate societies, the future sharing of resources and technology is bound up with the problem of the continuance of the human race.

There are many measures by which to distinguish the relative status on the continuum of development, and the use of a particular measure produces different listings of political states. By any of the measures usually employed, the United States heads the list of highest ranking states, and Bangladesh (now that it is independent) is often at the bottom of the list. In recent years, by mixed patterns of classification, it has been common to divide the countries of the earth into three Worlds, in which the ranking often is similar to this:

1. First World: United States, Canada, Switzerland, Sweden, New Zealand, Australia, United Kingdom, France, Norway, Denmark, Finland, Iceland, Belgium, West Germany, Austria, Netherlands, Spain, Portugal, Italy, Greece, Israel, Japan, South Africa, and Argentina.

2. Second World: U.S.S.R., East Germany, Czechoslovakia, Hungary, Bulgaria, Romania, Poland, Mongolia, China, Albania, North Korea, Vietnam, Laos, Campuchea, and Cuba.

3. Third World: All other political states.

Quite obviously the above is not strictly a matter of economic development, for the Second World is the Communist World, regardless of degree of development of any single state. Yugoslavia is often omitted from the list on the score that it is a Communist "irregular;" Poland, although it is also somewhat of an irregular, is always put on the list. Notably, the countries of the First World are members of the European culture realm in which the Industrial Revolution took place or to which it spread relatively early, except that Japan and Israel are now included in this group. Spain, Portugal, Greece, South Africa, and Argentina are marginal members of the First World, and are often left off that list. There are other problems with this casual type of classification, for example, the inclusion of the very small states that often simply ignored in other listings.

As the question of degree of development continues to be debated, such criteria as illiteracy, degree of labor skills, rates of capital investment, balances of trade, rates of investment in research and development, percentages of employment in agriculture, and per capita gross national product should begin to refine bases for judgment. Much of the current discussion is biased because it rests strictly on economic criteria, on political criteria, or on emotional criteria and, with such criteria, many different classifications will result. We present in the accompanying map a classification of five worlds, in which the three criteria are per capita income, illiteracy, and the balance of trade (see Fig. 10.10). This classification is by no means the highest possible refinement of the question of just how poor is poor, but this map is representative of the socioeconomic patterns that are currently being discussed.

Variations Through Time

In most cultural landscapes there are a few features that persist for long periods of time as monuments of the past (see Photo Essay No. 5). These are the exceptional elements of landscapes, by their very persistence. Most cultural landscapes alter over periods of time through periodic rebuilding and rearrangement of the works of the peoples occupying particular localities. Around the world, however, there are very different time patterns involved in the duration of particular sets of structures.

The Persistence of Local Customs. The travel ad reads: "Go back 3000 years to fairy-tale towns and folklore provinces—go by jet plane to live in luxury in modern hotels." How is this possible on the earth—our one world? The very lure of the ad, of course, is that of the far country, where things are different. To the American, having a short history, who lives in "today," with the expectancy of living in "tomorrow," this means ways of doing things that belong to the "day before yesterday." Were the ad written for the occupant of the fairy-tale town, to lure

him to North America, it would be written in quite different terms.

American travel literature exploits the "differentness" arising out of persisting regional customs in other parts of the earth. These customs are the habits of people expressed in their houses, clothes, foods, arts, and manufactures—in short, in their landscapes and cultures. The American way, in contrast, puts much stress on inventiveness, novelty, and change. Another way may stress preservation of the tried and true, the familiar and the old. Still another way may combine the two extremes.

The world is currently made up of peoples who follow many combinations of the old and the new (compare Figs. 10.11 and 10.12). In terms of human technology, a few local regions have just emerged from the Stone Age, and a few are on the verge of the Space Age. The element of time shows its clear imprint in the thousands of local regions, ethnic groups, and culture systems now active in the world. Human custom, technology, the attitude toward the world and the environment, and the composite will of the group vary from region to region and from time to time.

Revolution as a Regional Phenomenon. As applied to human life on the earth, "revolution" denotes widespread change in both areal and population organization. The term is used here in the broad sense and not just in a political context. Revolution proceeds from dissatisfaction with the way things are; it seldom begins with total dissatisfaction but normally derives from extreme unhappiness about a few matters. Dissatisfaction may spread from one group of causes to others and end by producing large changes over a wide range of items. Revolution is precipitous, occurring at a faster rate than normal change. It does not occur often in a given region, but when it does appear it may create a notable break with the past. Since few revolutions in human living are ever even reasonably total, there is normally much retention of prerevolutionary elements.

As applied to human culture, to living systems, and to the landscape in which man lives, most change is nonrevolutionary. Most regional landscapes are slow accretions of elements of different age. But change is almost continuous in some aspect of the living pattern or the landscape itself, either despite man or else effected by his hand.

By slow accretion the landscape, and the pattern of human living in it, are altered (Fig. 10.13). If by chance a long interval of gradual change does not bring maladjustment, a given region may slowly acquire a degree of stability, and a culture group a tradition of continuity. But sometimes maladjustment within the region itself, or in a nearby region, brings human dissatisfaction to a boiling point that promotes revolution. Then occurs the drastic change that introduces a break in the ways of human living in a portion of the earth and often carries with it major changes in the landscape.

The Spread of Change, and Its Nature in Regions. Changes in human living patterns result occasionally from the invention of new developments within a region, but more customarily they arise from the introduction of new features from outside, the process known as *diffusion.* The region may be a large area of the earth's surface or a very small one. If a large zone, it may possess sufficient ecological variety (in the sense of total environment) to contain subregions in which new developments may take place. New features may be passed along through trade or barter, they may be introduced by human migration from one region to another, they may be seen and acquired by the itinerant wanderer for his homeland, or they may be diffused as ideas or techniques, later to be reproduced in a new place.

As applied to regions, cultural change of the kind suggested produces new manifestations in the landscape itself. New items displace old, take their places beside them, or form a veneer over the old while never totally causing them to disappear. In the travel ad at the beginning of this section the process is evident: 3000-year-old fairy-tale towns cannot be too far from jet-age airports, and new hotels with American plumbing and air conditioning have intruded on the landscape.

THE MOSAIC OF A HEAVILY POPULATED WORLD

Composite Regionalism

For purposes of mapping, the earth may be divided, according to any of various criteria, into major and minor component areas, with lines drawn between them, regionalisms discerned, and patterns of rank established. By geographic region the geographer normally means a cohesive unit, with a certain number of criterion boundaries falling into place around it. Areas that have no such cohesive unity, since they are frequently bisected by the boundaries of particular criteria, are sometimes called *shatter belts.* Both regions and shatter belts are essentially abstract concepts, but are employed as tools of convenience.

Like the geographer, the anthropologist deals with issues of the regionalism of man on the earth. Normally he starts with a broad concept of culture as his most significant criterion. This corresponds to the geographer's use of climate or landforms. Subcul-

tures, particular aspects of culture, and regional biological anthropology next concern him, and ordinarily he adds a time series of criteria for the extinct cultures that can be studied only archaeologically. This volume attempts to combine the concepts and conclusions of both geography and anthropology; hence the factors of regionalism that we consider here are more numerous than those commonly employed by the geographer.

In this view, the earth is a large complex of regionalisms, a complicated, three-layered, translucent mosaic-mural. The fragments forming the mosaic are of three kinds: *earth space, people,* and *culture.* The units of earth space exhibit such aspects of nature as physical size and shape, landform, climate, drainage, vegetation, animal life, and soil. The units of culture exhibit the customs and ways of human living, settlements, field systems, buildings, roads, mine and erosion scars, politically organized territories, languages, religions, clothing patterns, social organisms, manufacturing, music, art, food and game patterns, agricultural and trade systems, and others. The units of people exhibit such aspects of population distribution as ethnic groups, migrations, density variations, birth and death rates, age structures, and regional totals. The separate pieces of the mosaic can be "seen" in their three layers, pieces of each being different in size, shape, density, intensity, value, gravity, and color. Viewed quickly, certain bold motifs stand out, comprising all three kinds of pieces. Seen in detail, other motifs become highlighted. Here and there motifs show up in which earth space, culture, and people seem in rather full unity; these are the units most geographers would call geographic regions. But parts of the mural have no unity, and the separate designs of the three layers do not at all agree—these are the shatter belts.

Certain great groups of pieces show up in master designs on the composite (Fig. 10.14). They form the great realms such as North America, Latin America, and western Eurasia. Each possesses a series of smaller designs, motifs, and patterns that are individual, yet particular elements transgress the smaller units and the larger masses alike. For example, the Indo-European languages spread from southern Asia through western Eurasia into two New World realms, and the food habit of the bread-wheat consumers ranges far throughout the mural.

Changing Regionalism through Time

In the long-term duration of man's habitation of the earth, environmental values have actually changed in practical terms. During the late Pleistocene much of Canada was covered by ice, whereas much of the Sahara was well-watered. As Canada became deglaciated and reoccupied biotically, its value rose in

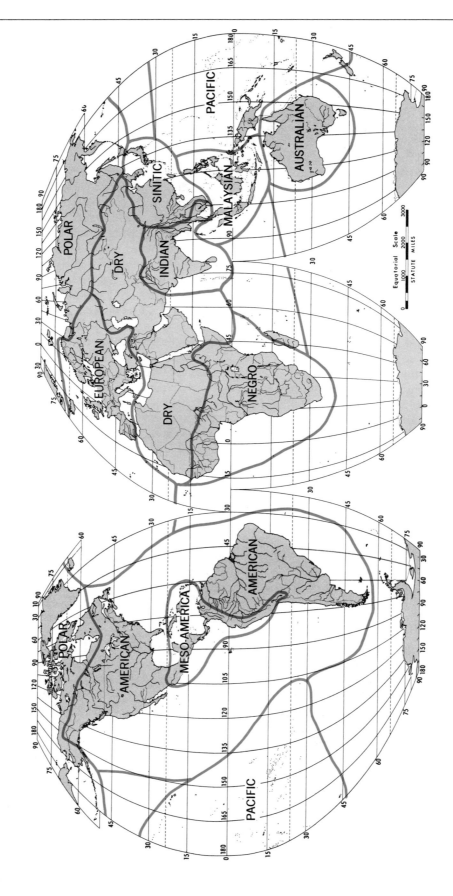

Figure 10.11 Culture Worlds, A.D. 1450.

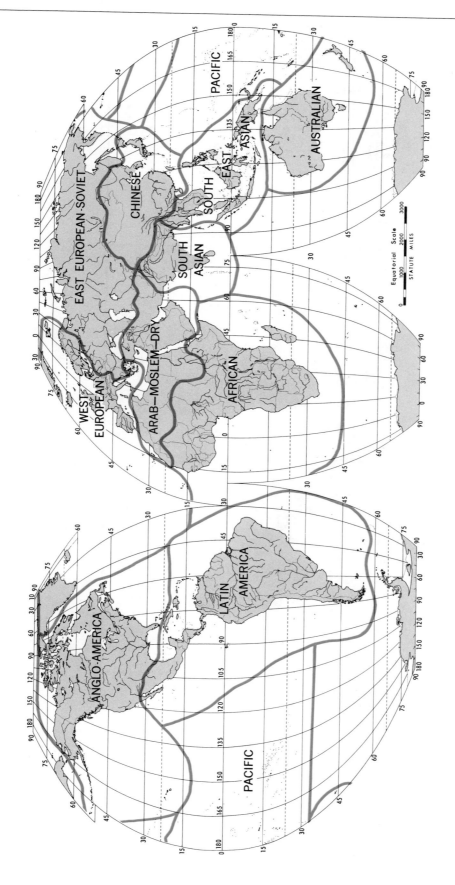

Figure 10.12 Culture Worlds, A.D. 1976.

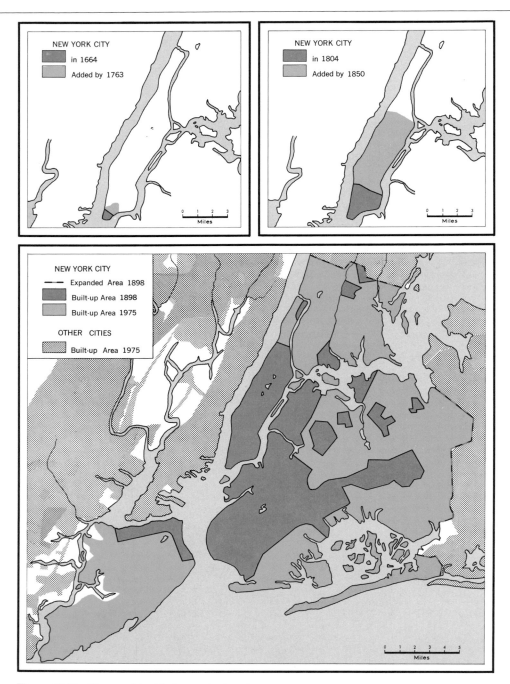

Figure 10.13 Time and Its Marks on the Landscape.
The Example of New York City and Its Environs, 1664–1976.

immediately usable terms, but an increasingly drier Sahara declined in immediately usable values. Early humans possessed so little technology that they could use only immediate values. To some modern people who possess advanced technologies, both Canada and the Sahara (despite its drier condition) have high values never suspected earlier, and the regional parts of the most valuable areas differ from those of an earlier time.

The history of the occupation of the earth by mankind, therefore, embodies a triple time variable—physical/biotic/environmental time, culture time, and human time. Each aspect of our mosaic—earth, culture, and people—has waxed and waned, increased and decreased in value, rejuvenated and declined, been new and old. The state of affairs in any single region of the earth today is a composite of these three variables. To omit the time factor for any region is to portray less than the whole.

The Significance of Regionalism

Although a great many single factors operate in the physical/biotic environment, and people function according to their ability, the pattern of life in an area of the earth is determined by group living. But since people are thinking creatures with a power of initiation beyond that of other animals, they can alter, convert, or rearrange many of the conditions of their occupancy of the earth. Traditionally they have done so by areas and regions. Although the pros and cons of a "geography of Antarctica" can be argued for a time before human explorers ever set foot thereon, our real concern is for regional environments containing people and culture.

It is this interplay of people and culture in regions that writes our story and fashions the shining bits and pieces in the rich mosaic of the inhabited earth. Although human cultures operate through human groups, group culture is almost infinitely variable, since it may alter with time and space. People, themselves, are variable in numbers, both in time and space. Physical environments also vary, but their rates of change are less sharp and less rapid than those of culture or those of people.

We must always realize that the human ability to build, alter, destroy, rebuild, and maintain is not the same the world over. Simple generalizations for the earth as a whole, drawn from any single region, almost always exaggerate, either by overstatement or understatement. Although there are regions in which humanity appears to be the dominant element in landscape change, human abilities in those regions have in fact not yet reached that level.

There are endless differences in what the geographer conventionally calls the man/land relationship. These patterns can be measured and spatially stated in two-dimensional terms. We prefer to think in the threefold terms of people, culture, and earth. As the twentieth century enters its final quarter, the three-dimensional mosaic comprises the most complicated set of regional designs that ever existed on earth. We should also recognize that in the future people will complicate the great world mural further by redesigning it endlessly. Both the demands people make of their environments and the landscapes they create in them will change continuously in the future.

SUMMARY OF OBJECTIVES

After you have read this chapter, you should be able to:

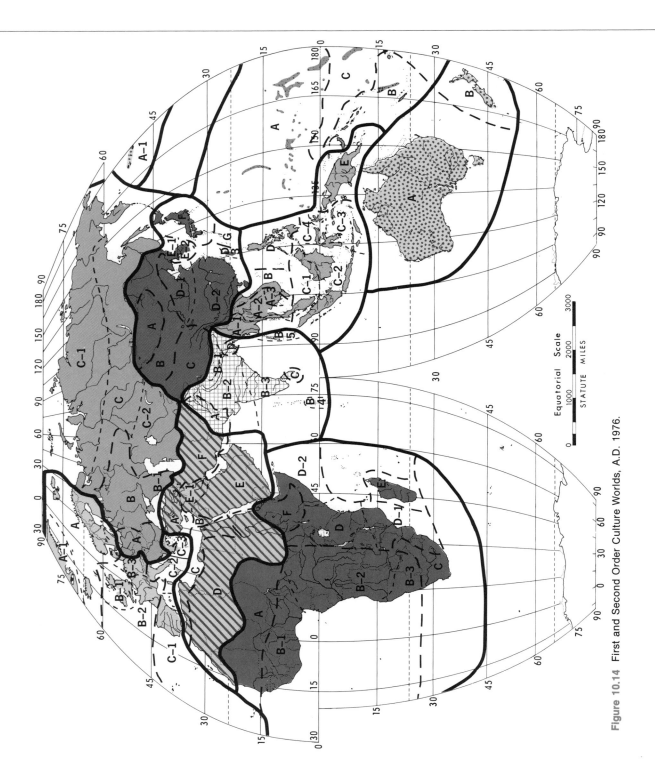

Figure 10.14 First and Second Order Culture Worlds, A.D. 1976.

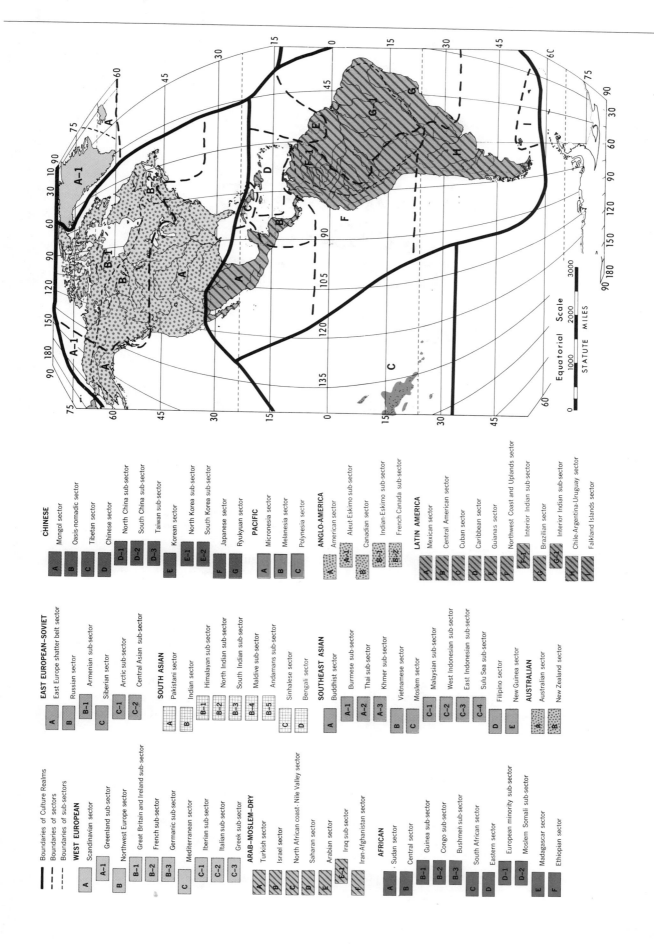

Boundaries of Culture Realms
Boundaries of sectors
Boundaries of sub-sectors

WEST EUROPEAN
A Scandinavian sector
A-1 Greenland sub-sector
B Northwest Europe sector
B-1 Great Britain and Ireland sub-sector
B-2 French sub-sector
B-3 Germanic sub-sector
C Mediterranean sector
C-1 Iberian sub-sector
C-2 Italian sub-sector
C-3 Greek sub-sector

ARAB-MOSLEM-DRY
A Turkish sector
B Israel sector
C North African coast- Nile Valley sector
D Saharan sector
E Arabian sector
E-1 Iraq sub-sector
F Iran-Afghanistan sector

AFRICAN
A Sudan sector
B Central sector
B-1 Guinea sub-sector
B-2 Congo sub-sector
B-3 Bushmen sub-sector
C South African sector
D Eastern sector
D-1 European minority sub-sector
D-2 Moslem Somali sub-sector
E Madagascar sector
F Ethiopian sector

EAST EUROPEAN–SOVIET
A East Europe shatter belt sector
B Russian sector
B-1 Armenian sub-sector
C Siberian sector
C-1 Arctic sub-sector
C-2 Central Asian sub-sector

SOUTH ASIAN
A Pakistani sector
B Indian sector
B-1 Himalayan sub-sector
B-2 North Indian sub-sector
B-3 South Indian sub-sector
B-4 Maldive sub-sector
B-5 Andamans sub-sector
C Sinhalese sector
D Bengali sector

SOUTHEAST ASIAN
A Buddhist sector
A-1 Burmese sub-sector
A-2 Thai sub-sector
A-3 Khmer sub-sector
B Vietnamese sector
C Moslem sector
C-1 Malaysian sub-sector
C-2 West Indonesian sub-sector
C-3 East Indonesian sub-sector
C-4 Sulu Sea sub-sector
D Filipino sector
E New Guinea sector

AUSTRALIAN
A Australian sector
B New Zealand sector

CHINESE
A Mongol sector
B Oasis-nomadic sector
C Tibetan sector
D Chinese sector
D-1 North China sub-sector
D-2 South China sub-sector
D-3 Taiwan sub-sector
E Korean sector
E-1 North Korea sub-sector
E-2 South Korea sub-sector
F Japanese sector
G Ryukyuan sector

PACIFIC
A Micronesia sector
B Melanesia sector
C Polynesia sector

ANGLO-AMERICA
A American sector
A-1 Aleut-Eskimo sub-sector
B Canadian sector
B-1 Indian-Eskimo sub-sector
B-2 French Canada sub-sector

LATIN AMERICA
A Mexican sector
B Central American sector
C Cuban sector
D Caribbean sector
E Guianas sector
F Northwest Coast and Uplands sector
G Interior Indian sub-sector
G-1 Brazilian sector
H Interior Indian sub-sector
I Chile-Argentina-Uruguay sector
J Falkland Islands sector

343

TYPES OF PEOPLE AND KINDS OF COSTUMES

TYPES OF PEOPLE

In looking through these pictures, almost everyone can find at least one person who resembles some acquaintance in spite of differences in "race," nationality, or the costume worn. This points out one important fact about people: human beings are much alike the world over. This is true because all human beings belong to a single species, and this species forms a closed mating community in which the basic genetic elements are common to all. Regional or continental isolation of groups never reached the level of separation into different species. Continuous intermixing among the many different groups has kept the whole population of the earth within a single species.

As the human species spread out over the earth, different regional groups went through biological changes adaptive to the environments in which they lived for long periods. This long process produced several kinds of biological differences among regional groups, such as variations in skin color, the structure of the nose, certain aspects of blood composition, and certain aspects of body build, height, and weight. It is not possible to relate all physical characteristics to environment, however, despite a common tendency to do so. Many biological features are not adaptive, although they are inheritable because different groups possess different genetic combinations. Hair color, form, and texture, the shape and size of the ear, and eye color are inheritable but are not adaptive to the environment.

Two cultural practices have tended to prolong biological differences, whatever their sources. The first is the tendency of a population group to mate within its own membership even when two or more groups live in the same general region. The second is the historic tendency for some groups to migrate into regions where they do not have easy contact with any other group. The first factor remains important, but the second no longer occurs to any marked degree in today's world. The over-all capacity for variation within the human genetic system is very great. Natural selection, in response to both environmental and cultural pressures, does produce change in human beings over long periods of time. The long-term effect of persistent disease may both reduce and increase the range of variability in a population group.

Cultural factors, such as the pressures toward social class/caste uniformity, religious grouping, or political restrictions tend to concentrate the inheritance of biological features, separating broad regional populations into subgroups over periods of time. Tolerant social attitudes, on the other hand, tend to mix genetic combinations rather rapidly. In grouping the "races" (long-term mating populations), many different bases such as political citizenship, geographical location, or the kinds of biological traits mentioned above, may be used. Although there continue to be arguments about "race," almost all such arguments involve social, economic, or political issues, and thus are quite subjective.

The only logical and simple system for grouping modern mankind is according to the historical locations of the origins of certain evolutionary trends. This yields three racial families—Caucasoid, Mongoloid, and Congoid. These originated in southwest Asia, eastern central Asia, and central Africa, respectively. Each set of family stocks has varied markedly from the earliest to the most recent examples. Beyond this simple classification, ethnic stocks may be separated into over 400 types by applying particular criteria to increasingly specific categories.

Photo 1. These Choco Indians of Panama, poling their dugout canoe on a river, are short in stature but are sturdily built variations of the American Indian Mongoloid family ethnic stock. Living in a tropical environment, they find no need for protective garments and so wear only the culturally prescribed loincloth and a little jewelry.

Photo 2. The Chimbu of New Guinea have the hair and some facial features of the Congoid racial family, but most Papuans are of mixed racial ancestry. The one garment shown is the Chimbu version of the traditional wrap-around "sarong."

Photo 3. This Zulu girl of the Congoid racial family is laden with the decorative elements of adornment that serve the functional purposes defined by tradition in her culture group.

Photo 4. (Left) The garments worn by this Navajo American Indian woman conform to the traditions of modesty prescribed by the American culture, but her necklace, bracelets, and hairstyling are those of her Indian cultural heritage.

Photo 5. The modern population of the Union of India has physical characteristics derived from considerable early mixture among uncertain components with a large element of later Caucasoid racial family inclusion. They also have several mixtures of culture-group garment styles. In this photo the wraparound tradition is represented by the *sari,* worn by the two women at the left. Many younger people with modern educations now wear European style clothing in public, as shown by the two girls at the right.

There are obvious differences in the physical appearances of the separate varieties of the modern ethnic stocks that inhabit different regions of the earth. As a result of living apart, Choco Indians tend to resemble each other more than they resemble Chimbu tribesmen (Photos 1 and 2), and Algerian Bedouins resemble each other more than they resemble Japanese (Photos 6 and 10). Differences in skin pigmentation are the most obvious visual aspects of biological diversity (Photos 2 and 3 compared with 13, 14, and 15). Differences in hair color (Photo 1 compared with 16) and hair texture (Photos 2 and 4) are almost as obvious. Other features are also quickly recognized. Many Europeans are resigned to being called "big-nose," which in some parts of the earth has come to mean "European" (Photo 8 compared with 14, 15, and 16).

Racial characteristics are biologically inheritable qualities which determine body structure, blood type, and gene combinations. Facial features are also inheritable, but superficially similar combinations occur outside of inheritance pools to provide those occasional resemblances so often noted. There are distinctive features among the facial characteristics of the basic ethnic stocks, and a comparative review of the photos will bear this out. Photos 1, 4, and 7 show the American Indian subvariety of the Mongoloid racial stocks. Photos 10, 11, and 12 represent the Asian Mongoloid stocks; Photos 2 and 3 present the Congoid ethnic variety; and Photos 13, 14, 15, and 16 illustrate the wide variation to be found among the Caucasoid racial stocks. Photo 8 presents one of the hybrid ethnic types now common throughout the modern world. The Nepalese are chiefly north Indian Caucasoid stock, but the Himalayan mountain country is a zone of mixture between the Caucasoid and Mongoloid stocks, and individuals vary accordingly. Photo 9 presents one of the archaic Mongoloid types, the Paleo-Asiatic Koryak. Photo 17 shows the kind of ethnic mixture produced in the modern era of widespread human migration. It is a group of Americans which represent elements from all the major racial stocks of the earth.

Photo 6. Bedouin women show a distinctive ethnic composition, and they wear the full-covering garment that is derived chiefly from the wraparound but also has some minimal elements of tailoring.

Photo 7. (Left) The Peruvian highland folk, living in a cool climate, have added some European elements to their garment system, but they also retain the older poncho wraparound garments.

Photo 8. (Above) The Gurkha peoples of Nepal are of mixed ethnic origin, including both Caucasoid and Mongoloid elements, and have lightly built physiques. They wear garment styles that combine simple tailoring with the traditional wraparound elements.

Photo 9. This Koryak man from eastern Siberia wears fur garments that demonstrate the earliest of the tailoring traditions. His clothes are form-fitting, giving full protection to all parts of the body.

Photo 10. The Japanese racial ancestry includes considerable mixing of racial family elements at a very early date but little added mixtures in recent centuries, yielding a relatively uniform modern population. These women wear the traditional national dress, derived from an early tailoring tradition that preserves elements of the wraparound tradition.

COSTUMES AND ADORNMENT

This cluster of photos has been arranged in a roughly evolutionary sequence of types of adornment rather than by any kind of racial criteria. Earliest mankind, originating in a mild climatic region, needed very little protective covering. However no human group has ever gone totally naked once they have devised decorative and functional types of body adornment. Eventually, people learned procedures of adornment and developed those technologies. The Choco Indians (Photo 1) illustrate three of the five basic types of body adornment. They wear the necklace, typifying decorative "jewelry"; they display distinctive hairstyling; and they wear the loincloth, typifying garment-wearing. The other two types of adornment are deformation and the use of cosmetics, neither of which is represented in these photos. The Chimbu tribesman (Photo 2) wears the same three types of adornment as are worn by the Choco. The Zulu girl's adornment (Photo 3) appears more decorative than functional, but it serves the older accepted purposes of that culture group. The clothing of the Navajo woman (Photo 4) appears more complete, but the critical items illustrated are the necklace and the hairstyling. Her garments conform to American concepts of modesty. The loincloth (Photo 1) is a simple rectangular unit of textile material held in place by a waistband. The narrow width is typical of cloth produced by a simple weaving loom. The Chimbu tribesman (Photo 2)

Photo 11. The Chinese include both very early and much later strains of the Mongoloid racial family. Modern garment styles are chiefly modifications of the traditional styles, in which loose-fitting garments were made using simple tailoring techniques; the women, above, wear traditional oiled-silk long gowns common in South China, and the men, below, wear the loose, tailored army-style jacket popularized by Communist Chinese leadership.

Photo 12

wears a "sarong," a simple wraparound garment worn in the style that was long the normal pattern in the warmer parts of the earth. The wraparound may be produced in almost any size by sewing simple loom units together. In warm and hot regions, the sarong eventually replaced the loincloth among many culture groups. In regions where additional protection from sun, wind, rain, or cold was needed, the wraparound became either large in size in order to cover more of the body or heavier in weight to provide added protection. The Indian *sari* (Photo 5, left), the Bedouin burnoose (Photo 6), and the Peruvian poncho (Photo 7) are variations of the larger or heavier form of the wraparound garment adapted to particular cultural preferences or needs. The Nepalese couple (Photo 8) wear their traditional version of the sarong along with other garments found necessary in their cool highland homeland.

The art of tailoring evolved in the arctic fringes of Asia. Pieces of fur were cut and sewn together to make form-fitting garments that provided protection to all parts of the body. The Koryak man (Photo 9) wears a fur cap, an inner jacket, trousers, a long outercoat reaching to the knees, gloves, and boots. From eastern Asia, the use of furs and tailoring diffused southward into northernmost China. The Chinese combined the use of furs and tailoring with the older wraparound garments to produce their distinctive long gown styles in which the furs for winter garments are worn furside inward. Tailoring also diffused from eastern Asia into other parts of the Asian continent. The Japanese women (Photo 10) wear particularly a stylised version of the Chinese type of long gown. The modern Chinese have worn combinations of tailored jackets and trousers with the long gowns, but Communist China has popularized a civilian variation of the loosely tailored army uniform style (Photos 11 and 12).

Photo 13. This Rumanian girl wears a distinctive national costume combining culture-group textiles and patterned weaves in a semi-tailored outfit that preserves much of the wraparound tradition.

Photo 14. This Greek family illustrates another variant ethnic strain of the Caucasoid racial family. They wear garments belonging to the modern occidental style termed ''Western dress,'' in which tailoring is a strong element.

Photo 15. Many occupations are marked by specialized clothing styles. These Boston police wear a rather standardized uniform common to police personnel all over the United States.

Photo 16. This Australian girl of British ancestry wears a modern machine-knit woolen sweater that is the industrial version of the hand-knit tradition of garment-making that originated in medieval northwestern Europe. Her pleated skirt is a modern tailored version of the older wraparound skirt.

Photo 17. These New Yorkers on Fifth Avenue demonstrate the wide range of ethnic backgrounds present in the population of the United States. They wear a wide range of informal variations of "Western dress" styles and display a wide variety of hair styles.

CHAPTER 11
HUMANKIND AS A GEOMORPHIC AND BIOTIC AGENT

THE CHANGING HUMAN VIEWS OF THE EARTH

In the time of mankind on the earth, there have been three different views of the earth. These are: (1) the Paleolithic view of the earth as nature's world; (2) the Neolithic view that mankind could "tame the wilderness;" and (3) the twentieth-century view that the earthly ecosystem must be maintained in ecological balance. This first section discusses these three separate views.

As mankind evolved above the animal level, the human race came into a physical and biotic world controlled by natural physical and biological processes. This Pleistocene world was changing under the competition of new evolutionary forms of life among both the plants and the animals. Early people accepted this world as the only one they knew. Paleolithic people developed skilled means to appropriate natural products, but they devised no means by which to increase the productivity of nature. They did develop techniques for concentrating collectible resources in particular localities, such as driving animals by fire. Paleolithic peoples picked and ate many of the fruits, dug a share of the roots, gathered a fraction of the available seeds, killed numbers of game animals, and caught sufficient fish to work out simple patterns of subsistence. These patterns remained subject to the variations in seasonal weather, long-term climatic conditions, and the processes of natural ecologic adjustment and balancing. The human view of that world was that it was nature's world, and that mankind could utilize all those things that nature produced.

When late Paleolithic peoples began to produce changes in the natural habits of plants and animals, they began to change their view of the world around them. In the crude beginnings of the first domestication processes, late Paleolithic peoples began to alter the patterns of natural ecological development, and as the processes matured, groups of people in a few regions began to direct the productivity of the earth according to their needs and purposes. This increase gave an upward lift to the human birthrate in a few areas and also gave humankind a competitive advantage over lesser forms of life. As Neolithic culture systems and material technologies diffused over parts of the earth and systems of production replaced systems of collection and appropriation, people began to see the natural world in a new and different way. Increasingly, they began to understand that human action could alter the patterns and productivity of nature. Cultivated gardens became more valuable than the wild area the gardens replaced. "Taming the wilderness" slowly evolved into a concept that meant both increased productivity and human progress.

During the Neolithic period there could have been no real human concern that wild landscapes, wilderness areas, and wild biotic populations of plants and animals were being permanently destroyed, because at that time the wild areas of the earth were so vast as to seem endless in extent. Neolithic mankind still felt threatened by the natural world, which loomed about the small clusters of human settlements. Appeasement of the gods was necessary to ward off evil, and dangerous predators, and to ensure success to the newly cultivated gardens. These efforts involved religious rituals and sacrificial offerings to win the favor of the gods toward continued human existence.

Had mankind never accomplished plant and animal domestication, the view of the world would have remained that of the early Paleolithic peoples: mankind is dependent on, and must live within, the productivity of the natural biotic world. Domestication procedures did succeed, and gave rise to the concept of a humanly managed biotic world of crop plants and farm animals, with landscapes arranged according to the respective perceptions of culture groups.

As civilization appeared, with its organized concepts of societal life, the Neolithic view of "taming the wilderness" was adopted as the civilized view of economic development of productive activity. The early populations of the newly civilized societies developed organized procedures for altering the natural flow of streams, deforesting local regions, reducing wild plant and animal species in local regions, building cities, erecting monuments, polluting water supplies, and creating waste heaps. Whatever the long-term impacts of such actions were, they were regionally localized, and none created major threats to the biotic systems of the whole earth. The positions of the leading societies were solidifed through their advances in sociocultural organizations and through specific technologies that aided productivity. Such patterns of development characterized the slowly advancing level of life in many parts of the ancient world, utilizing the basic concept that taming the wilderness was a means of improved productivity. The concept had been integrated into culture systems.

As the Columbian Discoveries linked the earth into one great operating system, technological advances began to occur that made possible huge increases in population. These brought the local and regional impacts on the earth to a cumulative level that became increasingly worldwide. With the recent technological advances of the industrial revolution, human technology for the first time has reached proportions that threaten to interrupt seriously the operations of the earth's natural ecosystem on a planetary basis. This has come about because technological change centered on productive aspects of resource utilization without considering problems of ecological balance.

Taming the wilderness has continued to be a sign of progress during the 10,000 or so years in which people have converted wild landscapes into cultivated field patterns and settlements. The basic Neolithic concept was never questioned. Developed areas came to exceed the remaining wild ones, except in the deserts, the polar fringes, and the very high mountains. Neolithic gardening and animal husbandry became institutionalized into human economic systems that have been carried forward during the historic era and that remain operative today. The continued cultivation of the agricultural landscapes remains necessary to the sustenance of the very large population of the earth today. The Neolithic view that the physical and biotic world justifiably could be altered to support the human species remains the predominant view of the world held by most peoples. In this view replacement of wild landscapes by cultural landscapes still is believed to constitute progress and development.

In the last few decades, other views of the physical and biotic world have been put forward with increasing strength and have met with increasing acceptance. These modern views contradict the traditional and institutionalized views that have prevailed for so long. Wild landscapes, wilderness areas, spectacular landscapes—sometimes even or-

Even today, the taming of the wilderness continues in a few regions of the earth. This scene shows the opening of the first road through a forested mountain area in Central Alaska.

dinary and mutilated scrubby forests—and particular species of trees, animals, and birds are now held to possess uncountable value. "Taming the wilderness" has become a grievous human sin from this new viewpoint. Any action (apparently) that threatens any area of the earth not already maturely developed is now considered wrong and destructively exploitive. Development has been redefined as spoliation. These views were not initiated in this century, for their roots go far back in time, as among the early devout Buddhists.

The new viewpoints regarding the earth have been elaborated and accepted widely in the occidental world only in the late twentieth century. Not all occidental peoples accept them, however, and few peoples of other culture worlds can afford to accept them fully. The new views and values are most frequently advanced by urban residents having little contact with, or participation in, the primary production of food supplies, minerals, and forest products. These views combine two concepts, the preservation of the remaining wild plants and animals, and the development of the ecological concept of natural balance in the earthly ecosystem. The combined conceptual view is that through the failure of preservation of the wild, humankind is progressively destroying the ecological structure of the ecosystem; and through modern technology, it is progressively upsetting the principle of ecological balance.

There is difficulty in arriving at a rational judgment concerning the "preservation of the wild" in the midst of lack of knowledge and strongly held emotional opinion. There are wild populations, of both plants and animals, on the one hand, that are naturally declining through early Holocene climatic changes that established new food chain and population-pyramid conditions adverse to their long-term continuance. Some of these species, and populations, might be able to endure under even adverse conditions in a world untouched and uninhabited by mankind. In our present world, however, these are probably doomed to eventual extinction. Other species and populations may be sufficiently adaptable that, with the assistance of mankind, they may still find acceptable conditions in the present world. Mankind, on the other hand, has unquestionably increased the severity of naturally adverse conditions for many species and populations of both plants and animals. Only careful restorative efforts can establish conditions under which such populations may continue to exist. The increasing development of wild life refuges, and bans on hunting and killing, are steps in the right direction. For those species and populations that already have become extinct, the discussion is academic, though instructive as to the importance of preservation movements and policies designed to restore and maintain conditions acceptable by endangered species.

The failure to consider problems of ecologic balance as such, however, is only one side of the modern problem. In the evolution of modern industrial technology, productive efficiency was defined as "at the lowest possible cost." Industrial research focused on technologies of extracting, processing, and manufacturing that *produced* without any corresponding focus on *reusing* or *recycling*. Wastes were piled, polluted water was dumped, worn-out manufactured materials were thrown away. Research in the broader aspects of reusing and recycling has not produced effective results so far because it was never considered a useful part of the productive process of taming the wilderness. For example, strip coal mining can be done with restorative technology integrated into extractive technology, but the concept of "lowest possible cost" long prohibited any study of landscape restorative techniques, let alone their application. Industrial chemical research has been extensive on the extraction of useful chemicals from forest timber, but almost none was conducted, prior to 1960, on how to clean up the large volume of water used by a paper mill. Advanced societies still are committed, through tradition, to past concepts of progress. It is against these traditional viewpoints that the new views of the world have been conceived.

As the serious threat to the biotic world became widely recognized, there first emerged a view of the earth as an ecosystem already so seriously damaged as to be unable to withstand *any* further development. This, essentially, was a static, negative view of the biotic world. This view became popular among some groups who still take a somewhat fanatic position against any further development. More recently, however, there have been emerging dynamic viewpoints that recognize humanity as part of the earth's ecosystem, that understand the ability of an eco-

system to balance itself, and that hold that industrial technology can assist those balancing processes.

A rational view of the whole complex of natural and humanly influenced trends in physical and biotic change is needed. There is also need to understand that the earth, far from being static, is undergoing continual evolutionary change and that this change will continue, entirely apart from any effects produced by human action. What is needed is a recognition of the human capability to develop technologies that are both preventive and restorative in ecological terms. Finally, what is needed is a recognition that continuing development of the resources of the earth is mandatory in the face of increasing human demands for food and basic consumer products. Taken together, these views, suggest that a great deal more research must be conducted in industrial ecology. People must learn and adopt ways of producing while both preventing environmental damage and restoring what damage is done. They must find and adhere to ways of fostering permanently the ecological balancing of the ecosystem.

NORMAL ASPECTS OF CHANGE

During the Pleistocene-Holocene period, the surface of the earth reached its most complex condition thus far. Physical and biotic changes that have created this complexity occurred at rates at least as rapid as those of any earlier interval in geologic time. Even though modern mechanical technology can shape small units of the earth, mankind is not yet able to make changes equivalent to those resulting from natural physical and biotic processes. We express pride in, or disgust at, what seem to be massive physical changes involved in clearing sites for new housing developments, building dams that create lakes, digging canals, and leveling hilly tracts. And we take pride in, or express sorrow at, some of the new developments in breeds of crop plants, or the extinctions of animal or bird life, that have taken place in recent time. None of these "accomplishments," however, match the changes that occur in nature through physical or biotic processes. Many of these processes go on entirely apart from any human action, and mankind inhabits a planet that continues to undergo evolutionary physical and biotic change at a rather rapid rate. It is almost impossible to state any accurate and complete measures of the rates at which these changes take place, but some useful comparative approximations can be set down.

The Geologic Norm of Physical Change

Erosion of the surface of the earth continues in areas

of sufficient relief, both above and below sea level, where flowing water can move loose materials down a slope. Deposition of transported material is part of the process of change, whether the depositional site be temporarily downstream or finally in the oceans. In exposed areas surface winds perform the same erosional/depositional functions.

Rivers remove from the land surfaces of the earth not less than 100 million tons of material in solution and suspension every day. No human effort can prevent this kind of activity and, although human action can increase it in some areas, it is a natural process. Careful management of the land can also retard such natural removal in some areas. In addition, perhaps 30 million tons of material are shifted every day from one site to another along river courses. In this phase human action actively interferes with natural processes. In building the Panama Canal only twice as much material was moved as is moved by river action in a single day. The Mississippi River, alone, may discharge into the Gulf of Mexico as much as 4 million tons per day in flood season. That culturally induced erosion is not solely responsible for the enormity of this discharge is clear from a review of the geological history of the Mississippi River drainage basin. A single major dust storm may move 100 million tons of material in four or five days. On occasion great volcanoes discharge millions of tons of ash, rocky ejecta, and lava in two- to three-day eruptions, and they inject unmeasured amounts of dust into the upper atmosphere. No human action has yet been able to concentrate the physical power equivalent to that of a single major hurricane, or that of a major earthquake along a fault structure.

Natural processes have altered every square foot of the earth's surface many times over, both gradually and in sudden bursts. Through the primary physical/chemical processes of weathering and the movements of materials by water, wind, gravity, and glacial ice, the earth is under constant readjustment. Through such secondary processes as sedimentation, volcanism, and faulting, the surface crust of the earth is constantly being rearranged. Entirely unrelated to human action, mountain building goes on at an active rate, countered by the processes of denudation. These patterns of change often seem so slow as to be inoperative, but the current rates of change are probably near the peak rates achieved during the late Pleistocene. It is against this rate of geologic change that comparison must be made with the rates of change induced by human action.

The Biologic Norm of Biotic Change

The biologic norm for biotic change cannot be stated any more accurately in quantitative terms than can the geologic norm for physical change, but unquestionably since the human ascendancy on earth, these changes have occurred at a very great rate. The earth has gone through at least four glacial stages involving at least nine climatic shifts of major proportions. The climatic shifts, in turn, have caused biotic adjustments that were continental in nature. Readjustments to the most recent ice age are still taking place in the biotic world. During the height of the Wisconsin glacial stage neither plants nor animals were living along the shores, or in the depths, of what are now the Great Lakes of northern North America. Even today, who indeed can state precisely what form of wild biotic assemblage is the "proper" one for the peninsula of upper Michigan? By the end of the Pleistocene, many plant and animal species were apparently extinct. Other species survived only as remnant populations until recently.

The period since the onset of the Pleistocene has been marked by the natural extinction of species and the emergent speciation of new forms. As species have recently adapted to new but not too drastically different environments, subspecies and varieties have multiplied. Throughout geologic time one common aspect of biotic change has been the replacement of older species that were unable to adapt to changing environments. They have been replaced by newer species that could adjust to continuing conditions of change. The extinction, or near-extinction, of great numbers of nonadaptable species is voluminously recorded in the fossil record. Conversely, the very adaptability of the angiosperms as plant forms, for example, has contributed to a high rate of biotic change. Some of the aspects of physiological change in the human species is part and parcel of Pleistocene biotic change.

NATURAL CYCLES AND CHAIN SYSTEMS

The earth's ecosystem functions by the transfer of energy and materials from one level of life to another, and from one spatial location to another, in cyclic patterns of activity. This cyclic process can be simply but incompletely described in the following steps:

1. Green plants utilize solar energy, water, and minerals from both the atmosphere and the soil to create leaves, organic tissues, and seedstock, and they give off oxygen in the process of conversion;

2. Vegetarian animals consume plant materials, expend energy, and give off heat and waste products;

3. Carnivorous animals consume the herbivorous animals, expend energy, and give off heat and

waste products;

4. Reducing agents consume leftovers and waste products at all points and convert them into soluble mineral and organic compounds that become the elements usable by the next round of green plants.

5. The heat given off at all stages is dissipated into the atmosphere for redistribution. The leftovers, organic compounds, and soluble minerals are variably transferred from one location to another as the processes run their courses.

A simple, humanly manipulated cycle may take the following form: A rural farmer grows hay crops to be fed to suburban dairy cattle to produce milk products consumed by urban residents, with conversion of wastes and leftovers being both spatially separated and cyclically disjointed.

Some transfers are made fairly directly and in a short-term cycle that is relatively complete since wastes return quickly to the local soil to be taken up again by a new growth sequence of plants and animals. In other cases the cycle involves many indirect transfers through complex routes during long periods of time involving spatial transportation. Some of the cycles involve only inorganic compounds pertaining to the purely physical world, whereas other cycles are chiefly organic cycles belonging to the biotic world. Still other cycles combine elements of the two cyclic systems.

Inorganic Cycles

As igneous materials are intruded into the crust of the earth from deep-seated sources, as volcanoes erupt, and as glacial ice scours already denuded land surfaces, chemical elements are transferred around the earth. In the inorganic cycles elements crystallize, combine, separate, and recombine. Both pressure and heat are involved in many of the inorganic cycles. Many of the common mineral deposits, either as pure minerals or as mixed compounds, result from the inorganic cycles. Some of the materials of the crust of the earth remain for long periods within the inorganic cyclic system, as in the case of eroded sediments that are transported to the oceans and stored there. Certain minerals, such as gold and silver, are elements that always remain chiefly within the inorganic cycles.

Organic Cycles

Almost one-fifth of the known chemical elements are normally involved in the organic transfer systems. Many of these elements are present only as catalysts, complements, and trace elements. Among these minor elements are boron, copper, fluorine, iodine,

iron, magnesium, manganese, molybdenum, silicon, sulfur, and zinc. The bulk of the materials transferred in organic cycles are simple-to-complex compounds of calcium, carbon, hydrogen, oxygen, nitrogen, phosphorus, potassium, and sodium. Each of the major and minor minerals has its own particular role and direct or devious path through a cyclic exchange system. The most critical are the oxygen, carbon, and nitrogen cycles that are necessary, along with the water cycle, to almost all organic forms of life. The systems of organic cycles carried on in the normal living routines of multitudes of plants and animals are far more complex than most of the inorganic cycles. The processes of the organic world are intricately interlocked and interdependent, both with one another and with those of the inorganic world.

The organic cyclic systems all start with plant photosynthesis and proceed through varying steps, each kind of life form, plant or animal, converting some material to its own uses, directly or indirectly. Widely variant groups of organic compounds are returned to the soils of the crust of the earth for some further cyclic utilization. However, large amounts of organic compounds are regularly washed into the oceans, where they eventually settle to form layers of materials. This stage in the transfer system sometimes involves long periods of storage, for it may be many millions of years until such stored compounds are raised by mountain building into positions where they may again be eroded and involved in some organic systems. In every transfer large amounts of energy are expended, the effects being widely distributed throughout many parts of the world system.

The basic production system, that of plant photosynthesis, produces an enormous annual yield of organic compounds. The plant biomass is approximately 100 *billion* metric tons for the earth's present land areas, and close to 50 *billion* metric tons for the water bodies of the earth. In past geologic periods large parts of the annual yield were put into storage, eventually to become altered into the coal beds and petroleum pools of the earth, some of which are now available at or near the crust of the earth. At the present time much smaller shares of the annual biomass product are going into storage than was true in some past geologic eras. Most of the present biomass product is going into current consumption, either by further use in plant and animal growth, as fuel supplies, or as raw materials for industrial manufacturing processes.

The Carbon Cycle. In the carbon cycle both human and nonhuman processes are active in the transfers and stages of the cycle (Fig. 11.1). All living plants utilize carbon derived chiefly from the carbon dioxide in the atmosphere, and the current total

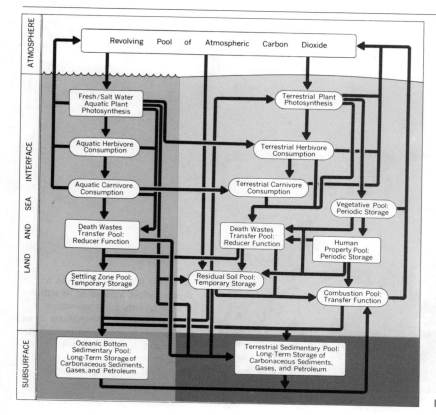

Figure 11.1 The Carbon Cycle

plant consumption of carbon is estimated at over 20 billion tons per year. A share of current "air pollution" therefore becomes a raw material for current plant growth. Any "use" of plant tissues transfers some carbon in some direction, whether it be the decaying of dead trees and annual leaf falls; the consuming of grassy and leafy materials by herbivores; the consuming of grains, fruits, and vegetables by humankind and other omnivores; the cutting of timber; the burning of wood and charcoal as fuel; the making of newsprint; or the using of plant materials in the textile and plastics industries. In most of these transfers the cyclic shift of carbon is rapid, but the carbon tied up in some of the wood used in building construction will not undergo further transfer for a long period. The burning of fossil fuels such as coal and petroleum puts back into circulation carbon that has long been stored.

Other aspects of the carbon cycle involve other kinds of transfers. Carbon dioxide from the atmosphere combines with water to form carbonic acid, which is a leaching compound in the flow of water through limestone rock formations. The great cave systems of the earth were formed partly by dilute carbonic acids in the drainage waters of such regions. The carbonates formed through leaching, carried either in solution or suspension, are transported by rivers into the oceans where part of the carbon goes into long-term storage. Forest fires con-

sume large amounts of carbon, some of which is transferred to ashes that mix into soil and other portions released into the atmosphere as smoke.

The volume of carbon dioxide present in the atmosphere has probably increased in proportion during the last two centuries, and there is some ground for believing that this increase is creating a greenhouse-warming effect on world temperatures. But carbon dioxide does not accumulate indefinitely in the atmosphere, since it precipitates with rain onto land and ocean surfaces. It is readily dissolved by water and thus passes into the oceans, where great reserves of carbon accumulate and go into storage in the shells, coral reefs, and limestones. Human action serves as but one of the many agents carrying out the total set of transfers in the carbon cycle.

Others of the elements important in the organic cycles, such as hydrogen, oxygen, and nitrogen, all have very complex cyclic systems similar to that of carbon. The transfer patterns range from simple and direct to complex, and from short-term to very long-term, with large amounts of each placed in storage at different periods in the history of the earth.

The Food Chains. The various forms of life inhabiting a region are ecologically linked together because they are sources of food to one another. There are many sequences in which the life-forms

All forest fires are disasters to those people interested in either scenic beauty or in the value of timber. However, the natural forest fire, as an agent in the recycling of carbon, is an important part of nature's ecological balance. This fire in western Oregon burned mostly Douglas fir trees.

are linked in particular ways, but any one sequence is termed a food chain, and the food chains constitute the basic structural elements of an ecosystem (Fig. 11.2). The basic producing agents, or the photosynthesizers, form the first nutritional level in any food chain, both on the land and in inhabited water bodies. These green plants are eaten by the herbivores. The carnivores consume the herbivores, and omnivorous life-forms, including mankind, consume both plant products and other animals. Scavengers consume all leftovers. Reducers who live on waste products exist at varying points along the food chain. Terminal waste products are normally soluble compounds that can be utilized by plants in another round of the food chain. Within any one region there are a large number of separate food chains, since varying life-forms have adapted to particular foods and sequences. There are varying rates of efficiency in utilization within the normal food chain, but the herbivores often operate at a rather low level. For example, the conversion efficiency of many of the higher herbivores ranges from 5 to about 14 percent, so that it takes a large amount of plant product to support a herbivore such as a horse, a cow, or a deer.

Had mankind evolved as a herbivore there would have been no need for the domestication of some of the animals, those used only for food purposes. Instead, the human species evolved as an omnivore, able to eat only certain kinds of plants, but able to eat a very wide variety of both herbivores, carnivores, and reducing agents. Additionally, the human species became dissatisfied with the natural workings of the food-chain system. The human species tampered with evolutionary and natural food-chain systems both in the plant world and among the animals, and devised ways of directing and controlling the production of both plants and animals. Wild photosynthesizers have been widely replaced by synthesizers that we call crop plants that yield organic materials more acceptable to the human taste and stomach. In the modern world the preference for meats has meant the replacement of wild animals with domesticated animals in various regional patterns. Sheep are grown not only for wool but for mutton, preferred by some culture groups. Americans prefer to raise pigs, chickens, and cattle because they like those food products. Pigs are scarce in regions of the Islamic religion, but common to non-Islamic Chinese.

The human food chain is, of course, only one of a large number that continue in force in all parts of the earth, since the many lesser forms of life continue to live within their own preferred food chain. Most forms of life, including the human species, have large numbers of items in the food chain, so that substitution of acceptable for preferred items is very common. At present, in the human food chain, much

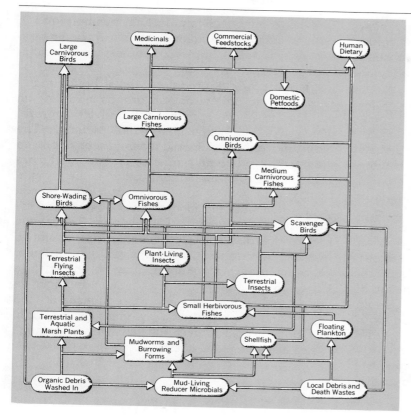

Figure 11.2 The Food Chain
Illustrated for a shore-margin
tidal estuary and its
nearby off-shore waters.

of the population of the earth lives on more completely vegetarian diets, out of economic necessity, than they might choose to do.

The Biotic Pyramid. Closely related to the food chain sequences are the structural levels of the biotic pyramid (Fig. 11.3A). Stated in the simplest of terms, it takes large numbers of plants to support a moderate population of herbivores. The herbivores can provide food for only a smaller number of carnivores, since there is a percentage loss at each stage

of the food chain. The life-forms in any food-chain sequence decline in numbers from the bottom to the top of the chain, in order of consumption. In terms of the human food chain, it takes billions of grassy plants to support hundreds of thousands of cattle to provide steaks for the thousands of prosperous human consumers who eat them.

In the natural operation of an ecosystem, the population pyramid tends to balance itself over a period of time. In a series of dry years, for example, the reproduction of herbivores may diminish because of

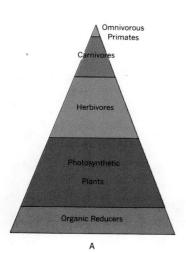

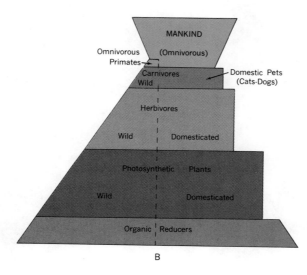

Figure 11.3

The Biotic Pyramid
A At the left is
the natural evolutionary
sequence.
B At the right,
the diagram represents
the human modification of
the pyramidal structure.

poor forage available. As the herbivore population declines under continued consumption by carnivores, the carnivore reproduction rate will decline in order to bring the carnivore population into balance with its food supply. As good grazing conditions return, the herbivore reproduction rate will increase, to be followed by an increase in the carnivore population. Other elements related to this food chain, scavengers and reducers, will also vary in numbers according to their positions in the sequence and the availability of their food stocks.

In theory, the elimination of a single life-form in any food-chain sequence may upset the balance in a population pyramid within a region. Natural physical processes, and evolutionary trends, have often interrupted sequences and upset pyramid structures. The variable natural workings of the food-chain sequence form one factor in the natural extinction of particular life-forms during geological time. Certain life-forms may retain positions in population pyramids by accepting formerly unacceptable food substitutes. There may come times, however, when diminishing sources of substitutes no longer will maintain a viable population of a particular life-form.

The human species escaped the limitations of the food-chain sequence and the population pyramid by developing several cultural technologies. Until at least the mid-Paleolithic period, the human species was still a minor life-form in total numbers and in position in most population pyramids. As *Homo sapiens* evolved (see Chapter 2), earliest modern mankind developed weapons adequate for killing off large and powerful predators who fed on humans. This ability to protect human groups made for an increase in human populations and for improvements in the positions in regional pyramids. The plant and animal domestications further improved the human food-chain sequence and enabled mankind to move to a still higher position in the population pyramid. In most parts of the earth today human beings dominate the food-chain sequence and arrange it in their own order of preference. The human species has also come to dominate the population pyramid by arranging for the population of lesser elements in such numbers as may be chosen (see Fig. 11.3B). In reality, of course, human decisions do not control all the innumerable sequences of food chains and the population pyramids of the earth; and human control over all the involved organic cycles is still far from complete.

CULTURAL INTERRUPTION OF NORMAL CHANGE

We are tempted to say that the most advanced human living systems on the earth today interrupt

Cattle feed yards, such as this one in Montfort, Colorado, are a highly productive source of meat supplies. They are a good example of one of the ways in which human beings manipulate the food chain sequence to provide for their own needs.

sequences of normal change in every way. Large parts of the earth, however, are still subject to patterns of normal change in local ecological matters, since the human interruptive impact has not become totally controlling. Expanding diffusion of technologies that are increasingly interruptive make it probable that in the not too distant future the scale of total interruption will become worldwide. This section first reviews general human interference with normal change, and then reviews historical growth of the cultural impact on the earth.

Interruption in the Geologic Norm

Soil erosion and stream-channel cutting are the most obvious and visible signs of geomorphic change, and both are normal processes in regions that have been undergoing natural uplift. Agricultural land use quickens the pace of both in such areas. Depending on the relief, condition of the regolith, climatic zonation, and kind of agriculture practiced,

the historic rates may range from 2 to 50 times the geologic norm. The alarmist writer of polemics, particularly in the United States, tends to place the whole blame on the agriculturists when this is not truly the case. It is in the dry hill-country margins that human interference with natural processes has accelerated them to the greatest degree. There are regions all over the world that are geomorphically unstable, and in all such areas there has been destructive interference with normal conditions.

Erosion and channel-cutting begin a sequential chain of abnormal developments (Fig. 11.4). Excessive soil erosion alters the whole pattern of soil development. The excessive drainage of ground water, provoked by channel-cutting, has a dehydration effect over the long-term, thereby lowering the quality of vegetative growth in affected regions. Channel cutting is often thought of as deepening of stream beds, but in many regions channel cutting tends to work horizontally. It cuts away floodplain terraces, which serve as good agricultural lands, so that

Figure 11.4 Interruption of the Geologic Norm.

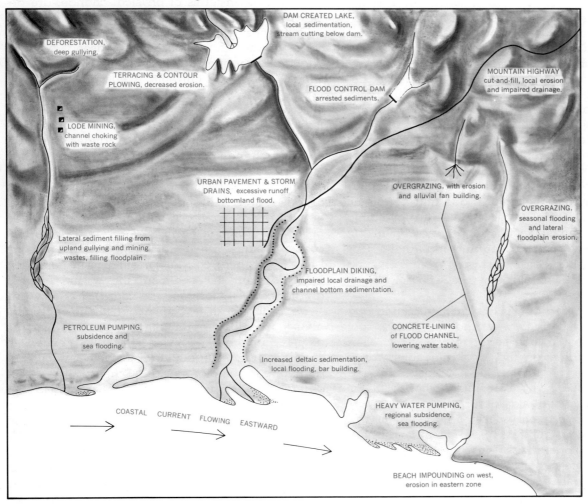

broad open river courses replace the former narrow streams and floodplain lands.

The building of dams across streams (however valuable the dams may be), results in the creation of lakes and reservoirs that serve as impounding basins. For the period it takes these basins to fill with sediment, there is created an interruptive factor on the stream course below the dam. Clear-water streams below dams often tend to erode their courses. In other cases, when the impounded water is chiefly diverted from the streams, for purposes of irrigation, wholly new situations develop along the lower channel. Lake Mead, behind Boulder Dam, may fill with sediment in a few hundred years but obviously did not do so in the 15 years predicted by alarmists. Boulder Dam, however, has created new conditions on the lower Colorado River. The new Aswan dam on the upper Nile River, has created a new situation for the Nile Basin, by eliminating the annual flood replenishment of farm lands along the stream course. The floodplain may now begin to suffer some erosion, and farmers are already having to shift to commercial fertilizers to compensate for the lack of annual flood sedimentation patterns. The Nile Delta now receives far less sediment, and in the long-term may suffer some erosion from the coastal current. In regions of potential instability, the increased surface weight-load of the impounded water is producing regional subsidence and earthquakes.

The removal of large amounts of natural gas, petroleum, and groundwater creates stress in underground strata, often causing local and regional surfaces to subside progressively. This occurs far more widely than is publicly reported, since most of these areas are in no danger of flooding by rivers, lakes, or the oceans. In some cases such subsidence is being stayed temporarily by pumping water back underground to replace the natural gas and petroleum removed. When local water supplies are available for this purpose this provides temporary solution, but when subsidence occurs after pumping water out for irrigation or urban domestic supply, there is often no replacement possible. Some of these effects will continue far into the future.

In more localized operations such as cutting and filling for highways, railways, housing tracts, and industrial plants, a long series of aftereffects often results, depending on the local situation. Water-table lowering, channel cutting, landsliding, and flood-period mudflows may result. In other situations drainage impairment may occur.

Modern mining often processes huge volumes of rock materials, in contrast to the small volumes of ancient mining operations. These huge volumes may be dumped into stream valleys where they choke drainage systems and create very different condi-

tions than were formerly present. If the waste rock material is stacked, new and unstable minor landforms are created. Strip mining, as practiced when such operations began, creates a whole series of unstable features in a landscape, often with quite unforeseen consequences.

The ancient technique of diking rivers, now very widely practiced, is beneficial in the first instance in that it prevents large-scale flooding of lowlands by rivers, but the long-term effects are troublesome. Floodplain sedimentation is prevented, channel beds fill up, lower courses of rivers and their deltas build up more rapidly, and ever higher flood levels result. The diking systems built to reclaim coastal lowlands create new landscapes that are visually impressive, but they also change natural processes in the regions affected.

When large areas are covered by buildings, paved roads, parking lots, and other such "structural" features, human interference prevents the normal patterns of physical change. Canalizing of stream courses around cities, along with the control of flood waters, creates wholly new regional situations and it is difficult often to determine whether these effects, in terms of physical change, are destructive or constructive.

Coastlines are often relatively unstable fringe zones undergoing localized or seasonal types of change according to the prevailing directions of winds, waves, and currents. A eustatic shift of sea level will of course affect them, and they may also be submerging or emerging because of local conditions. Any human attempt to control processes at particular points along these fringes normally causes a change in conditions at some other coastal location downcurrent. The attempt to create beaches at a given point by impounding sand usually results in coastal erosion at some nearby downcurrent location. Rivers may deposit enough materials eroded out of the drainage basin to aggrade downcurrent coastal fringes, but a dam built in the drainage basin may rob the coastal current of its inshore sediment. Without this sediment the waters have a greater capacity to erode the coast.

We conclude this sampling of the impacts that alter the geologic norm with an item that must be placed on the constructive side of interference with physical change. Agricultural terracing, trenching-ridging, contour plowing, and other similar procedures in many parts of the earth have worked to slow down the patterns of physical change. The total areas involved in such patterns are significant in world terms.

Human action seldom purposely alters major physical conditions and processes, and most such impacts are reaction patterns. It can be argued that,

Contoured plowing and planting is a conservation technique that reduces soil erosion and lessens the rate of physical changes of land having an irregular surface. This photo shows a farm region in Wisconsin.

aside from soil erosion, few of the humanly induced interruptions in the geologic norm are more destructive than those incidents produced through natural causes. For example, natural landslides in river canyons create dams that impound lakes behind them, and extraordinary cloudbursts produce flood discharges that strikingly rearrange stream courses and sediment patterns in valleys.

At present many remedial measures are being taken to restore natural conditions, and there are more that can be taken. Reconstitution of strip-mined areas has now reached a level of excellence. Quarries and gravel pits need not be left in their raw states, but can be turned into recreational areas by imaginative landscaping. In several of the patterns of interference mentioned above, the physical aspects are not very damaging, and it is the biotic effects that often follow that are really the significants aspects of change.

Interruption in the Biologic Norm

Biotic change is brought about by complex sets of adjustments that work toward restoring balance in a disturbed ecosystem. These disturbances have many natural causes, and they have been continuous since the early Pleistocene. Human action interferes with the adjustment process by eliminating some of the life-forms, or by changing the conditions under which the process operates. This interference creates greater instability and sometimes prevents its restoration. Climatic change over a period of time is the largest single trigger in biotic change. Either the groups of species originally inhabiting an area

Strip mining need not cause permanent physical or biotic damage. Fourteen-year-old plantings of European larch are rapidly turning this mined area in southwest West Virginia back into a forest.

adapt to the change or they die out, to be replaced by some combination of life-forms that can exist under the new conditions. Sudden natural catastrophes, if regionally significant, also trigger biotic change. Human interference in the biologic norm is often similar to the natural catastrophe in that it sets in motion impromptu and opportunistic chains of reaction. Human interference more often creates a vacuum within the biologic "web of life" by selective elimination of some particular life-form.

Plants, animals, insects, bacteria, and funguses may make their own readjustments to compensate for small-scale or short-term interferences. They can return conditions toward the earlier relative balance by colonizing a different portion of the region, by increasing or decreasing their local populations, or by finding alternatives for whatever has been eliminated from the food chain and biotic pyramid. Long-term interferences or fundamental changes, on the other hand, set up new control conditions. The result may then be a large-scale replacement by life-forms that can endure the new conditions, or it may be the reduction of the biota to a skeleton system that has fewer interrelationships and has fewer alternatives in the food-chain system.

Fire, agriculture, and the establishment of settlements have been the chief means by which wild landscapes have been changed to humanly controlled conditions. Fire may for some time eliminate certain food sources, thereby restricting occupance

of the burnt-over area by certain life-forms. Agricultural field systems replace natural habitats and markedly restrict the populations of wild life-forms. Human settlements, by creating new kinds of local environments, eliminate most wild life-forms. Fire, unless often recurrent in an area, is normally a short-term interference only. Agricultural landscapes, except those under shifting cultivation, establish controls that permit only a very reduced biota to occupy field areas. Settlements establish other controls in which human pets become predators of wild animal and bird life, and in which the environment presents opportunities for only a very restricted biotic population that can, in effect, live with human beings in a somewhat artificial and symbiotic relationship. The village and small town is normally much less restrictive than the metropolitan area. In most American settlements such life-forms as rats, mice, sparrows, mockingbirds, blue jays, doves, and finches seem quite at home, and a wide variety of weeds exist despite the efforts of gardeners. Ants, cockroaches, flies, bees, butterflies, fleas, lice, and many other forms also adapt to human settlements quite easily. In the microbial world no human settlement is without its complement of small life-forms, including some that prey on mankind directly.

The process of human beings transferring life-forms around the world continues to move forms endemic to particular areas into new environments. We are most aware of this when there occurs an introduction of a disease germ or a plant pest, but it takes place with many kinds of life-forms in an ongoing pattern. Some of these can be eliminated again, and some of them remain as minor exotics doing no great damage. Some life-forms, on the other hand, when introduced into new environments, find very few predators present and may rise to level of major significance. Normally, this sort of thing is not thought of as an element in interruption of the biologic norm, but it is very much a part of it. The accidental introduction of particular weed plants into new environments often results in a violent population explosion of the new form, causing many repercussions in the balance of plant life in the area. The introduction of the water hyacinth to the lakes of East Africa has not only altered the local plant and animal life but it is proving costly to economic activities.

The modern chemical industries, according to the alarmists, are likely to destroy almost the full range of life-forms on the surface of the earth. The human production of chemical agents destructive to some life-form is not new, for Paleolithic mankind often caught fish by suffocating them by the use of "poisons" thrown into good fishing streams. The very large numbers of highly reactive chemical compounds now being created, however, do threaten a far more serious interruption in the nature and distribution of organic life. The modern chemical wastes and chemical agents employed for specific purposes have a far more damaging impact, for they seem to affect patterns of reproduction, thereby threatening whole populations. We are not making light of this danger, however, when we point out that very little is now known about the long-term effects of the defoliants, weed killers, insect killers, rodent poisons, specialized fertilizers, and the many industrial chemicals that are termed pollutants. The fact that many species of mosquitoes have become resistant to DDT indicates only the possible evolutionary patterns of change within the insect world. Here human action may be causing abnormal evolutionary change, although it still does constitute interruption of the biologic norm.

The pollution of groundwater, rivers, lakes, and the oceans by chemicals has very complex implications in both the short run and the long run. Exactly what constitutes a total death threat for each of the life forms is not yet known. The natural conditions of areas surrounding massed bodies of water are more complex than was formerly thought, and we really know but little about long-term biotic tolerances. For example, not until 1970 was it realized that mercury is a dangerous water pollutant to some forms of life. The relatively harmless inorganic form reacts with water to become potentially dangerous methyl mercury. Minute amounts are picked up by aquatic mic-

roorganisms, accumulating along the food chain so that some of the larger fishes caught for human consumption may contain amounts of methyl mercury dangerous to man if not to the fish species themselves. How great the danger to all kinds of life forms may be is not known, for some of the marine fish eaten by human consumers have apparently contained mercuric compounds for decades. Many of the industrial and insect-killing chemicals do exact at least short-run heavy tolls on the rodent, bird, and insect populations all over the earth, and this constitutes significant interference with the biologic norm.

Various of the newly synthesized chemical compounds, when mixed together in regional environments, complicate the natural cycles of carbon, oxygen, and nitrogen, and the cumulative impact of these patterns shows up in our rivers and freshwater lakes. Here the impacts are specific, at least in the short run on the plants and animals that live in, on, and around those water bodies. We know so little about the cyclic flow patterns of these elements that we are uncertain as to what remedial action should be taken other than total cessation of production and use. In the modern industrial world, with its complex of biotic problems, cessation of use is hardly the answer needed.

Excessive erosion of any part of an occupied landscape is a physical problem with long-term biotic results. Erosion removes organic matter along with the top soil, thereby removing some of the bacteria/microbe population. Excessive erosion accompanied by channel cutting and lowering of a water table may make it impossible for some types of plants to continue occupance of the area. Drought-resistant plants may replace former varieties, but this will then determine to a considerable extent the occupance of the area by animals and birds that can exist on the new plant cover.

The building of dams to create reservoirs also has long-run biotic effects beyond the geomorphic ones. The new water bodies, and the surrounding higher water tables may create wholly new plant and animal communities in and around the waters. New species of weeds, aquatic plants, aquatic rodents, insects, and diseases often appear where they could not previously exist. Several different kinds of population "explosions" have occurred in different parts of the earth after the development of new water bodies. Those occurring in African waters have been both spectacular and somewhat frightening as to the consequences of interference with the biologic norm.

Human interference in the biotic world probably has been responsible for the extinction of what may seem like a large number of plants, animals, insects, and bacteria, although these numbers are rather slight in comparison to the totals of species in each group. Human actions may have eliminated 400 spe-

The explosive growth of "pest" plants such as the water hyacinth seen here on the Nile River above the Aswan Dam, can seriously affect the natural biotic cycle.

cies of birds, rodents, and larger animals, and brought another 400 species to the point at which they now are labeled seriously endangered species. Perhaps 50 of these species were naturally on their way to biological extinction by the end of the last glacial period, with its major climatic readjustments, in the same way that the dinosaurs had died out much earlier. In the plant realm it is possible that nearly 1,000 species (species, not varieties within species) have become extinct, and that at least 200 more species (chiefly trees with aromatic woods) have been so heavily exploited that they belong on a seriously endangered list. Among the plants it is uncertain whether normal biotic change or human action is responsible for some of the extinctions. Plant geneticists, however, are now seriously concerned over the potential loss of genetic crossbreeding material among the traditional crop plants, as peoples in the underdeveloped world modernize their agricultural economies. The replacement of traditional crop varieties by the new hybrids results in total discard of some of the old varieties that, despite their low yields, carried valuable genetic characteristics. There are no reasonably good data on the extinction of insects and bacteria, although undoubtedly some have been lost. An ominous aspect of the record is that a disproportionate share of the extinctions have occurred in the twentieth century and that the seriously endangered list is growing.

Human interference in the biotic world today is very evident in the role of modern medicine and public health. There is a self-centered human reaction, in fighting such diseases as cancer, poliomyelitis, and meningitis. The medical world hopes that the pathogens of these diseases may be one day rendered extinct, but this also constitutes interference in the biotic world. The World Health Organization hopes that it has finally and permanently wiped out smallpox, historically one of the most dangerous of the killing diseases. Is it to be regretted, then, that a life-form, the pathogen of the disease, has been rendered extinct? Among the animals both extinct and on the endangered list are some, such as the California grizzly bear, that were predators of mankind or his domestic animals in earlier time. Were those people to be blamed for trying to eliminate them? The issue of human interference is not always the simple matter the conservationists would have us believe.

The biotic world is a persistent one and highly adaptable to any kind of change, and mankind is rarely totally efficient and ruthless. Many elements of both the plant and animal world continue their ecological sequences of adjustment, adapting to cultural landscapes as they adapted to physical factors in earlier time. Domesticated plants carry on all the organic functions of wild plants, including participation in the carbon and oxygen cycles. They are the hosts for insects, bacteria, viruses, and soil microbes in the food chain, and they provide both cover and food for birds, insects, rodents, and other forms of life. In a long-cultivated landscape, a biotic complex, even though reduced in numbers of species, achieves a stability and maintains itself despite the human efforts at control. Some of this persistence is a matter of alteration of living habits, but much of it is ecological adaptation to new environmental conditions that are humanly imposed.

Notes on Earlier Interruptions

By the time *Homo sapiens* had developed in the late Paleolithic, mankind had already created several basic technologies capable of manipulating the physical/biotic environment. Paleolithic peoples carried on a number of activities already described to some extent in Chapter 2. Among these were:

1. Collecting, transporting, but also scattering, seeds of grain, fruits and nuts, legumes, berries, and plant bits reproducible by vegetative means.
2. Digging in the ground for edible roots and tubers.
3. Driving game to concentration points as well as fishing for aquatic life-forms at points of concentration.
4. Hunting the large mammals of the Pleistocene, which were then among the declining species.
5. Using poisons to catch fish in restricted bodies of freshwater.
6. Using woody materials in housing, for weapons, tools, and utensils, and as fuel for protective and cooking fires.
7. Quarrying rocky materials for the manufacture of weapons, tools, utensils, and jewelry.
8. Using fire in hunting and in improving grazing ranges.

Of these activities man's digging for roots and tubers and the quarrying of stone may have contributed slightly to erosion, but probably the many animals that rooted and dug in the earth for food caused a greater amount of erosion. Other activities of Paleolithic peoples' contributed to biotic change. But the human populations were small, and all but one of the technologies were so slight and inefficient as to have little impact on the earth. Fire was the one exception. Fire in the hands of Paleolithic peoples formed a very effective tool in acquiring food supplies, and it created a tremendous and permanent pattern of change in the biotic structure of the earth.

Paleolithic peoples must have long profited from work done by campfires escaping their bounds and by natural fires set by lightning before they started setting fires for specific purposes. Fire was used in hunting in two separate ways, to drive game into dead-end areas or over cliffs, and to remove old growth from grazing ranges. New growth of grasses and small shrubby plants attracts game animals to the burned-over areas, making them easier to hunt.

The impact of humanly set fire on the biota parallels that of naturally set fire but when often repeated is far more effective. Full-cover perennial forests are slowly reduced in area and grasses and fire-tolerant shrubs and small trees take their place. Herbivores and birds that eat seeds then flourish. There is good ground for the belief that the great grasslands of the earth were created largely because Paleolithic peoples repeatedly set fires in these regions over long periods of time. It is likely that many of the herbivore herd animals inhabiting the great grasslands and open forests increased their populations enormously only after the expansion of those grasslands.

Neolithic peoples invented many and more effective systems of controlling biotic conditions for human advantage. They thereby started another pattern of interference with the biotic world. The domestication procedures and the consequent crop growing and animal husbandry became permanent aspects of the human struggle for advancement in the biotic world. Neolithic settlement patterns began the sequence of fully controlled compact-settlement landscapes that have matured into our urban regions of today.

Land was probably first terraced and water first controlled for irrigation purposes in late Neolithic times. Both processes contributed to the growing power of the agriculturist as an agent of geomorphic change. Terracing alters the detail of landforms and affects the drainage processes. Good terracing, well maintained, probably reduces denudation below the geologic norm, as previously pointed out, but terracing systems were slow in developing effective control, and the earlier efforts probably did not maintain landscapes properly. As the wet-field terracing system was developed in Southeast Asia, terracing had to be more effective, thereby doing minimal damage.

Pastoralism in the Old World developed a maturity in grazing technology that maintained in many regions the grasslands that had been created by Paleolithic hunters who used fire as a weapon. Periodic overgrazing in some of these regions during early historic times was damaging to the biotic complexes that readapted to those grasslands.

Not all of the grasslands produced by the Paleolithic fire-hunters have been overgrazed by the cattle herds of pastoralists. In this scene, herds of zebra and gnu share a grassland range in Tanzania.

As crop growing in permanent fields developed, it began a long-term conversion process by which wild landscapes were converted into cultivated landscapes. As the plow tools became widely diffused, and agriculture replaced gardening, there began a more rapid extension of the cultivated landscapes. It is likely that this specific method of "taming the wilderness" has been the chief procedure modifying the biotic landscapes of the earth.

At the beginning of the historical era, as dike building and canal digging were developed, these formed the most significant additions to the Neolithic technologies modifying landscapes. The historical era has seen the extension of patterns of interference in the biotic systems of the earth spread farther and farther around the world as a whole, as populations grew and agricultural landscapes expanded.

Twentieth-Century Complex Pressures

When the earth's peoples were equipped only with simple tools, organic materials, and small amounts of power, only a few plants and the larger predatory animals were eliminated from the biotic world. Human impact was then often a matter of long-continued pressure that, in the end, produced major changes. With the rise of modern technologies, however, complex tools, potent chemicals, and large amounts of power, opened up an assault on sectors of the biotic world that had remained almost totally unaffected in the earlier periods. Faced with constant labor shortage, the farmers of the North American Midwest developed technologies for working ever larger acreages. Those farmers started Western culture—and eventually other culture worlds—on another phase of the Agricultural Revolution. Mechanization moved from horsepower to tractor power, freeing the land that fed the horses to grow more food, but resulting in a shift from organic operations to chemical operations, as manufactured fertilizers replaced the organic ones. Chemical pest controls and weed controls replaced the hand labor involved in earlier patterns.

By relying on complex tools operated by mechanical power sources and on the chemical fertilizers and weed-pest controls, modern agricultural practice is now threatening, to some extent at least, all aspects of biotic life on the earth. As mentioned earlier, this may be only a short-run pattern, with biological adaptation still to come, but there is the threat. World agricultural production has just barely kept ahead of modern population growth while further disturbing the organic cycles, food chains, and biotic pyramids of the wild and natural world. The pessimistic view insists that disturbance cannot go much further without major tragedy overtaking the whole biotic world, which includes the human race.

After years of effort Dr. Norman E. Borlaug developed the "miracle" high-yield strains of wheat, and received the 1970 Nobel Peace Prize for his efforts. Other work has added to these contributions so that both rice and wheat, in addition to the earlier hybrid corn, promised relief from the threat of world famine. Yields of four times earlier ones did give a little relief, but population growth around the earth has continued. There were also some side effects of the increased yields that detracted from their value. Filipinos, for example, were delighted to reap larger crops of rice, but the chemicals employed killed the fish in the rice fields, irrigation canals, and nearby streams. More rice was produced, but less protein from fish was available, unbalancing the food supply. It is not yet clear what the long-run impairment of "improved" agriculture may be to the rest of the biotic world.

The problem of maximizing food supplies for expanding populations seems to be reaching a restrictive limit, in which improvement in agriculture does not suffice. In traditional terms, when daily life became threatened, the concern for the state of the biotic world was thrust aside. This could mean the further expansion of agricultural landscapes, the increased use of mechanization, and the chemical boosters of production, all of which set in motion even more impact on the biotic world.

If the modern agriculturists of the earth are a threat to the biotic world, the urban industrial sector is an even more serious threat. Urban folk have always had difficulty in disposing of waste products, as the archaeologist and urban historian can attest, but the technological developments of modern urban life have greatly magnified the problem in the twentieth century. The gaseous pollutants released into the atmosphere from cities, and the very volume and range of liquid and solid waste products, for which no adequate recycling technologies have been developed, now threaten to infect the ecosystem of the whole earth with a new kind of "disease" that could prove disastrous to the biotic world. Chemical gases and solutions, plastic and paper packaging materials, metal and glass containers, discarded consumer goods, organic garbage, and raw and treated sewage—all the wastes of the urban-industrial world—reach far beyond the cities to infect all sections of rural regional environments.

It does little good for the urban "earth lover" to rail at the president of an industrial corporation who wishes to expand operations, or the electrical engineer who seeks to build a new power plant when that person flicks on the light switch at dark, distributes paper handbills assailing the companies, flushes an urban toilet, and stuffs a trash can with a large assortment of waste materials. No city dweller today can avoid contributing to pollution. Nor is the "simpler life" of the rural commune any real solution for the massive populations of our contemporary world. Any marked increase in the latter form of living merely spreads other varieties of environmental contamination in patches of inadequate sanitation and too-simple disposal of wastes.

Urban waste can no longer be casually disposed of, and its growing volume is taxing existing systems. One possible answer to the problem may be the wider use of the sanitary land-fill. This method is already changing the landscape of many places. In this Pennsylvania scene, a newly filled and covered terrace level lies just beyond the current disposal site.

For well over 200 years, producers of raw materials, manufacturers of commodities, distributors of goods and services, and consumers of commodities have been concerned only with producing and distributing at the lowest possible prices. All efforts were directed at production and distribution. Few in the producer/consumer cycle were concerned about the disposition of wastes, leftovers, and worn-out goods. But now that the industrial machinery of advanced societies can turn out nearly endless quantities of products, the advanced society has suddenly become aware of the enormous impact of the total wastage volume and of the impact that our industrial technology can have (Table 11.1). Recycling technologies exist, but most are still comparatively inefficient and more costly than the pricing patterns of using basic raw materials. None has been developed to operate on a scale equivalent to the problems being faced.

Unfortunately for the biotic environment, cultural habits that cause pollution and excess waste are ingrained in modern societies. Technologies for efficient conversion of wastes are still immature, and research on these problems has only begun. We know too little about the long-term effects of many aspects of the interrelations between technological society and the biotic environment. This whole complex of research needs and problems makes up a cultural "frontier" that is critical to the ecosystem of the earth, including the human race as the most active disturbing agent but also a potential victim.

SUMMARY NOTES ON THE REDUCTION OF WILD HABITATS

In the long eras before the appearance of the human race on our earth, the physical and biotic aspects of the surface were the results of natural ecological processes working in an evolutionary way. The human race was a product of those processes and, in the earliest periods of human development, the ecological processes continued to determine the ways in which growth and change occurred, including the living systems of the small populations of human beings. By the late Paleolithic era, human technologies had evolved and grown to the point that human actions began to be an important factor in the ecological sequences of change. The results were expressed in small features as areas of the earth's surface underwent different sorts of ecological adjustment to changing physical conditions. The continuing technological advances made during the Neolithic period gave regional populations the ability to set in motion significant varieties of change originating in human designs and aspirations rather than in purely natural evolutionary change. These designs involved the transformation of "wilderness" into "developed" areas, thereby reducing the wild habitats arranged by nature and increasing cultural landscapes arranged by human action (Fig. 11.5).

Table 11.1 Cumulative Human Impact on the Environment

Land Surfaces

Soil erosion/soil cycle disturbance
Gullying/water-table lowering
Stream choking with mine wastes
Deforestation/increased erosion
Coastal erosion/sedimentation
Overgrazing/increased erosion
Stream canalizing
Dam building/artificial-lake formation
Diking/floodplain restriction
Drainage/filling of wetlands
Urbanism/artificial landscapes
Roadbuilding/drainage impairment

Water Bodies

Sewage disposal/pollution
Industrial chemical disposal/pollution
Solid-waste dumping
Fertilizer infiltration/groundwater
Thermal loading/overheating
Petroleum spills
Chemical spills
Overpumping of groundwater/subsidence
Interbasin transfers

Atmosphere

Fires/Smoke pollution
Photochemicals pollution
Radioactive dissemination
Cloud seeding/storm modification
Chemical disturbances of upper atmosphere

Biotic Realm

Purposeful extinction of predators
Accidental extinction of nonpredators
Replacement of wild forms by domestics
Land clearing/disturbance of cyclic systems
Disruption of biotic pyramids
Spatial transfers of destructive exotics
Purposeful firing/disturbance of ecosystems
Disturbance of insect/predator cycles
Overfishing/disturbance of reproduction cycles
Excessive sea hunting/depletion of stocks
Overstimulation of growth cycles with organic wastes
Human overpopulation/upsetting natural systems
Excessive pet population/food wastage
Public health system/disturbing disease cycles

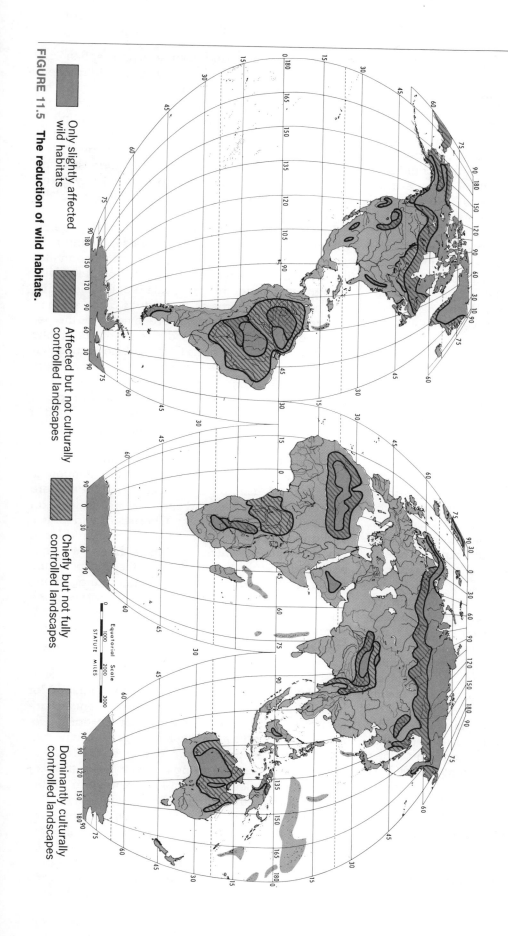

FIGURE 11.5 The reduction of wild habitats.

Only slightly affected wild habitats

Affected but not culturally controlled landscapes

Chiefly but not fully controlled landscapes

Dominantly culturally controlled landscapes

Equatorial Scale
STATUTE MILES
0   1000   2000   3000

375

Early Neolithic populations were small and the cultural landscapes produced were then minor regions on the vast surface of the earth. But the changes initiated were effective in producing marked increases in the human populations so that there began to occur the rapid expansion of those cultural landscapes and a corresponding reduction of the wild habitats. Neolithic technologies could not alter significantly the physical nature of the earth, but they could alter the biotic aspects to the greater productivity of regional environments that yielded materials useful to human beings. Neolithic people had no desires to eliminate many elements of the biotic world, but they did persevere against their predators, and they sought to restrain such biota as interfered with their productive efforts. Neolithic populations set in motion systems for transforming nature to human advantage that have been followed and improved upon by all succeeding generations of human beings.

In the two million or so years since human action became important enough to affect ecological change, human action has become increasingly a larger and larger factor in biotic change. From an interruptive agency human action has increased its power and scope until by the twentieth century it has become the chief agency directing biotic change, and it is now beginning to direct biotic evolution. Where Neolithic mankind appealed to the gods to assist the productivity of garden plots, twentieth-century mankind has become largely responsible for the biotic world. The condition of the biotic landscapes of the earth, and their future patterns of development, now rests on human shoulders rather than on the whims of the gods.

Human technology has not reached the point at which it can control the major physical processes that affect the earth. Those physical processes still control the biotic world, including all human beings. Human actions in the physical world are still at the interruptive level rather than at the directive level. The human responsibility for the condition of the biotic world, therefore, is that of a custodian who cannot act wholly independently, but who is the primary caretaker. This increases the responsibility of mankind to manipulate and manage the biotic world within the rules of the physical world. It is obviously to the human advantage, in the long-term, to maintain the biotic world in a healthy and productive state, but mankind has not fully acknowledged the responsibility taken over from nature.

Human exploitation of the physical and biotic worlds has been so successful—perhaps too successful—that the human population has outgrown its own cultural technologies and cultural systems. Human requirements for forest, marine, mineral, and agricultural resources—and simple living space—are now enormous and increasing annually. In manipulating the biotic world in satisfaction of those requirements, there continue even now the further reducing of wild habitats and the implementing of aspects of biotic change, in a desperate effort to satisfy the basic requirements of the total human population. At the present rates of change, says the pessimist, the last truly wild habitat will have vanished from the face of the earth before the end of the present century!

This does not mean, of course, that crop fields, village settlements, and the high-rise buildings of cities will soon be everywhere, but it does mean that there will be more of each. The reduction of the wild habitat involves the rearrangement of the biotic distributions and patterns on the surface of the earth in human-controlled systems. There is need for the development of careful custodianship of the biotic world, and there are signs that mankind is awakening to that need. As tree farms increase in numbers and areas, managed and more useful forests can grow where wild ones now grow. As parks replace woodlands there can be better recreational sites than before. As agriculturists recognize their problems and find better ways of maintaining productivity, there can be development within ecological rules. Dams will continue to increase the numbers of lakes in the seasonally humid and arid lands where they are most needed, and careful management can maintain them ecologically. Cherished wild plants and animals can be kept, fostered, and cared for in environments in which they may thrive. Whether each individual likes or dislikes the prospect, the whole earth is coming under human management, for good or for bad. The human race possesses the earth, and it is up to human ingenuity to operate it successfully.

At least for the near future, the largest component in the culturally ordered landscape will be those field systems providing such basic support as food and clothing. The desperate, increasing demand for larger food supplies lies behind the current expansion of the cultivated landscapes of the earth. Despite the fact that agricultural productivity has greatly increased in recent decades, the rates at which population is growing indicate that in the very near future agriculture must be extended to all the lands of the earth that can be cultivated. After that, any further increase in consumer demand will have to be met by improving the productivity of all of the lands of the earth. Either that, or humankind will have to be arbitrary in limiting the numbers of people to that total that can be more easily supported at the top of the biotic pyramid on a humanly controlled earth.

HUMANKIND THE MEASURE

Evolutionary development on the planet earth has been generally irreversible, and human developmental history is part of that pattern. In the natural sequence of events, more complex plant and animal forms have evolved and succeeded each other as the physical earth has changed. The appearance of humankind was in the line of that evolutionary development. But humanity, through cultural development, escaped the bonds of normal evolutionary trends and took over the management of the earth. Whether mankind can continue to succeed in operating the earth as a global living system depends on how well mankind meets the challenges of the next century or two to provide really constructive management of the whole earth as opposed to furthering his own immediate position in the scale of things.

As changes in environmental conditions take place, the results are good for some forms of life and bad for others. A change in environment can only be evaluated as "good" or "bad" from the viewpoint of the organism whose environment is being affected. Custodianship, control over process of change, and proper management of the biotic world, by and for people, can only be judged from the human viewpoint. The human tendency has generally grasped for obvious present good and discounted future harm. Indulgence in this tendency has led human management of the earth into activities that are proving detrimental over the long-term, but only activities or proposed changes that threaten people in the immediate period are judged "bad." People must have air, water, food, and shelter, and any environmental change that promotes the ability to obtain these is, therefore, "good," even if these may lead to future harm at some distant time. Among human values there is a very great need to distinguish between desirable and necessary. The question now facing humanity is how to grow sufficient food and fiber to feed and clothe the present population and at the same time to take these constructive steps in managing the earth so that all life may continue into the distant future. In the very near future, the necessary may not always be the desirable, and the peoples of the earth are faced by the challenge of forsaking the immediate "good" for the long-term good in order to carry on the management of the ecosystem of the earth. Can they live up to the challenge?

SUMMARY OF OBJECTIVES

After you have read this chapter, you should be able to:

Page

1. Explain the differences among the three changing human views of the earth. — 355

2. Describe what is meant by the phrases:
   a. the geologic norm of physical change — 359
   b. the biologic norm of biotic change — 359

3. Understand how human activities could possibly interfere with the natural process of the carbon cycle (Fig. 11.1). — 360

4. Describe the position of mankind in the food chain and its dependence upon the continued operation of natural systems. — 361

5. Explain why the concept of "biotic pyramid" is important for an understanding of humanity's impact upon the earth's biotic resources. — 363

6. List some ways in which people have interrupted
   a. the geologic norm of physical change — 365
   b. the biologic norm of biotic change — 367

7. Discuss the new cultural pressures upon the earth symptomatic of the twentieth century. — 372

8. Explain the meaning of the statement: "Evolutionary development on the planet earth has been generally irreversible . . . . But humankind . . . took over the management of the earth." — 377

CHAPTER 12
QUALITIES OF FUTURE CULTURAL ENVIRONMENTS

KEY CONCEPTS

A Coming Crossroads Raises Questions
Present Cultural and Human Habits Raise
    Problems
Pragmatic Practices Are Against Long-Term
    Values

Steps Toward Cultural Maturity Involve Change
    Alternatives to Success Spell Trouble
    The Way Forward Involves Improved
    *   Technology*
Responsibility for the Earth Involves Solving
    Problems

CROSSROADS AND QUESTIONS

In the last quarter of the twentieth century the human race finds itself approaching a crossroads of its own making, from which point several routes into the future lead in somewhat different but unknown directions. That crossroads may or may not be reached within the lifetimes of the young people of today's world. The human race has finally come to dominance over the biotic world, and can manipulate aspects of the physical world. The very success of the human struggle for dominance poses problems as to the use of the earth in the future. There are two alternative views in this matter. One is that some supernatural force has designed the whole structural system and controls the details of the pattern of development, with most human beings pawns in the master design and eternal salvation taking care of the chosen few. The other view is that in the general sequence of evolutionary development of the earth, humankind evolved and struggled to a dominant position in which human decision will control many aspects of the future pattern of development. If it be

assumed that the former view is correct, then little that the human race can do about biotic balance will make very much difference to the future state of the earth. If the latter view is accepted, then it becomes necessary to accept the further view that responsibility for the state of the biotic ecosystem now rests on the collective shoulders of all peoples. In this view, human decisions must be made as to which route is going to be taken beyond the crossroads. In either case, some human decision is necessary before the passage of time takes humankind beyond the crossroads and severely limits human power if not human existence.

In this volume we have taken the scientific view holding that on an evolutionary earth the appearance of *Homo sapiens sapiens* was part of an irreversible biotic trend. And we have taken the view that as humanity evolved and developed culture, human technologies began increasingly to interfere with continuing natural biotic evolution, to take daily human living systems out of the natural biotic system. The competitive struggle has brought the human race to dominance within the biotic order, though not to total control over the future development of that order. Both the physical and biotic systems of the earth continue a directional development despite the human presence and interference. If humankind can devise technologies that maintain a relative balance within the biotic world, then human dominance may continue. Unless, however, human technologies can be developed of a kind, and to a level, at which they function to maintain a relative balance, then irreversible aspects of biotic evolution may again assert themselves to take a different direction. Future development in a seriously unbalanced ecosystem could result in the eventual elimination of humankind as a life-form, along with countless other subordinate life forms subject to the same destructive influences.

To recognize that the human race is responsible for the continuing operation of the biotic earth is to pose serious and vital questions that must be solved before the crisis of the crossroads is reached. Some of these questions are:

1. What are the types and limits of more productive technologies that may be developed in the future to support humankind, while allowing the maintenance of the biotic functions of the earth at a level permitting truly long-term human occupance?

2. What population total of human beings can be supported at reasonable levels within those future technological limits?

3. How many competitive societies (e.g., national political states each claiming total autonomy) can there be on the earth in the future?

4. Can the human race continue as a set of competitive societies, each struggling for that space, body of resources, and share of the total population it can secure, to the total disregard of all other societies?

5. What cultural technologies and socioeconomic/sociopolitical institutions can be devised to prevent the struggle among societies for advantage or dominance from exceeding the limits of tolerance in the ecosystem?

6. What cultural technologies are needed to divide the resources and space of the earth in some equitable manner so that each society to be allowed a position also has a fair chance to develop a reasonable level of living and distinctive culture system?

The above questions, and others of a similar nature, focus on a further question which is: If the technologies can be devised to answer the above questions, then *who* is to make the critical decisions? The human race cannot hold a town meeting to talk it all out peacefully. The history of the occupance of the earth has thus far been a competitive one, with militant force producing the decisions of who will occupy what and how much territory. Must humankind continue in the traditional competitive pattern? If this is to be the case, then the future of the human race may not long continue, and the natural biotic world may attempt a further sequence of evolutionary developments.

THE PROBLEMS OF PRESENT CULTURAL AND HUMAN HABITS

Often heard is the speculative expression that although the present world may not be fully comfortable for all, the future world will be. Such encouraging looks into the future unroll pretty pictures of what human life will be when the human race has put into practice all the material technological procedures now known and still being learned in research laboratories. Such dream constructs often imply that no one will need to perform heavy physical labor in order to live, that quiet green spaciousness will intermingle with and surround each living community, that each such community will be air-conditioned in a temperature-humidity regime most favorable to human comfort, that recreational play and enjoyment of the finer things in life will be for all, and that people will live to be 150 years of age on the average. Other daydreams put things somewhat differently, such as that cities will be built at the bottom of the sea or under mountain ranges with elevators to the tops, and so on.

Against these interesting and perhaps pleasant views of the future of the humanized earth there stand others that reflect human failures. Much of the earth still lags seriously in production technology, and there are far too many millions suffering daily hunger in the lagging societies, and far too many even in the advanced ones. The competitive rules of modern economics work to the exploitation of the less advanced by the inventively aggressive in our present world. Political nationalisms carried to excess provoke the cruel specter of modern war. Too many societies lack the means to check the great growth in their populations, growth that threatens their own future livelihoods. To all these old cultural problems is now added the modern one of the serious danger of ecological imbalance of the earth's ecosystem.

At present the earth contains great contrasts in levels of achievement. Neolithic crop-growing procedures are still in practice, and Neolithic simple-household industry is still the rule among many peoples. Since other societies employ highly technological and highly powered systems, the gap between the most advanced and the lagging remains

Paolo Soleri conceived of this plan for a future "city". The two, 2500 foot-high pyramids, joined at their bases, would house 170,000 people. The "city center" would be placed vertically in the middle of each pyramid. Residential areas would be located at the top, while the lower parts and underground basements would be used for industry.

very wide. *In the twentieth century the earth shows a greater range of economic, technological, political, and sociopsychological development than was present in any past era.* This means that there is greater imbalance among societies today than has ever been true in the past on an earth now belonging to, and dominated by, humankind. Dominance finds the human race with almost infinite varieties of habituated living systems. It finds people still possessed of numerous instinctive animal urges and reaction patterns inherited from the far past and not significantly altered by formulation of social living systems. The processes of cultural convergence, to date, have operated to a limited degree only on the worldwide level. On the internal level in many a society, cultural convergence processes are very incomplete. The internal gap in levels of living among members of a society is as great as that between the strongest leading society and the weakest lagging society.

Although many are impatient at the slow progress of the so-called developing regions and the most underdeveloped portions of the earth, this is a short-term view. It derives chiefly from the heavy intersocietal cultural stresses that have suddenly taken shape in the articulate expression of current unhappinesses over Neolithic and post-Neolithic cultural systems. The strains produced by those stresses have been militant revolution, pillage, and

suffering. There is more of all this now because the population of the earth is larger than ever and very unequal in distribution.

PRAGMATIC PRACTICE VERSUS LONG-TERM VALUES

The more alarmed bioecologists warn that the ecological balance of the earth's ecosystem will be beyond repair by human action within about three decades unless mankind immediately ceases the multiple forms of polluting the biotic environment of the earth. They would, therefore, on a worldwide basis immediately ban all forms of industrial pollution of water; prohibit the use of all chemical forms of soil fertilization, weed killers, and insect controls; and abolish the use of all forms of fuel contributing to air pollution. They would require the full processing of sewage prior to disposal, and prohibit all indiscriminate dumping of solid wastes. And they would ban further expansion of the urban-industrial and agricultural landscapes against the wild landscapes of the earth. The pressures for such action derive from particular urban portions of the occidental world. The chiefly agricultural, underdeveloped nations of the earth, however, assert that the bans on agricultural and public-health control chemicals, as well as on agricultural expansion, would condemn them to starvation levels of existence, and they are strongly against such bans and prohibitions at present. Their position is taken because past occidental public health programs and agricultural development programs have stimulated their rapid population expansion.

In the past, basic human needs have taken precedence over policies motivated by long-term ideals, and pragmatic solutions to human demands have been sought regardless of the threats of eventual difficulty. Thus, the expansion of the agricultural landscapes of the world has continued to take place since the Neolithic beginnings of crop growing/animal husbandry. *Human practice has almost never adopted programs of current action whose payoffs are to future generations.* The immediacy of the threat to the ecological balance of the earthly ecosystem is not apparent to most of the world's peoples at the present time, in local regional terms that can be translated into primary threats to daily living systems.

Contemporary programmed change can be put into effect on a worldwide basis within a very short period of time when the item at issue is very broadly acceptable. Many of the problems of ecological preservation of, and repair to, the biotic ecosystems are in need of urgent worldwide action within a short time period, and there are few built-in handicaps. In other equally urgent aspects of ecological programs, however, such action must come to terms with the pragmatic problems of maintaining support for present generations of peoples in many parts of the earth. If the bioecologists are even remotely correct in their assessments, there is not only a crisis approaching but also the dilemma of how to achieve the long-term ideal while not drastically upsetting regional economies and bringing mass starvation to many societies.

STEPS TOWARD CULTURAL MATURITY

It may seem an unrealistic statement that the satisfactory level of living for any society anywhere on the earth no longer need be handicapped by the lack of means for production, the absence of transportation, or the ignorance of technological procedures. Problems of satisfactory levels of living for our present population now lie in the inadequacies of the cultural systems of the earth, in the socio-psycho-econo-political structuring between and among the human culture groups that exist on the earth today. Paleolithic peoples understood neither

the physical world (and how to make it produce) nor the kinds of cultural systems that could make life comfortable. Twentieth-century peoples have largely solved the primary problems of operating the physical environment to productive ends. However, twentieth-century societies have not put into operation the kinds of cultural systems throughout the earth that can make life both comfortable and peaceful for all members within each society. The steps remaining by which the human race can live satisfactorily in all parts of the earth lie entirely within human understanding. There exists, however, a basic human refusal to put into operation those procedures that can make the whole world a better place than it now is.

We describe as Paleolithic any society that has an economy based on hunting, gathering, and collecting; that is band/tribal in sociopolitical structure; that employs animistic ritualism by which to supplicate the gods of production of game and vegetable foods; that goes to war to protect its tribal territorial lands; and that does not engage in economic procedures to accumulate wealth applicable to productive ends. What is it that forces such a group to continue its Paleolithic culture system during the twentieth century? It is *not* the limitations of the physical/biotic environment. Rather it is the historic drag within the minds of the members of the culture group that causes them to cling to the old ways despite the cultural patterns that may surround them, patterns that offer innumerable avenues of cultural change by which to bring themselves toward a balanced relationship with more advanced culture groups.

We describe as feudal, backward, underdeveloped, and poverty-ridden any society that concentrates the material wealth in the hands of a small class of multiprivileged aristocrats (be it great landed estates, the ownership of mines, or the ownership of other productive elements). That society imposes a restrictive system of political management by which a few hold power, and maintain a social structure that strongly enforces stratification ranging from aristocrat-above-the-law to outcasts below having neither rights nor access to the law. Such a society does not operate an educational system that extends the privileges of cultural improvement to all members of the society to the highest possible level. And it has no rational plans or procedures by which to alleviate the poverty of its largest population component. It is indulging in Paleolithic thinking to place the blame for this compounded situation on the physical/biotic environment. The cure is available in world culture, the initiative is open to leadership of the society, and the means exist by which to lift the level of living. It is human failure that permits the continuance of the circumstances. *The human failure lies in the greed of human beings who continue the*

Territorialism between Canada and the United States has only a few technical and legal restrictions. The "border" has become little more than a symbolic line, as is indicated below by the Bechard home located on the boundary between Quebec and Maine. In contrast, the militant territorialism expressed, above, by the wall between East Berlin and West Berlin is far more common.

operation of the society in the traditional manner against collective need.

Science often holds that its rules transcend political, physiographic, and climatic boundaries, that the laws of physics or molecular biology are the same in the United States, in Patagonia, along the banks of the Congo River, or at the northern end of the Ural Mountains. The technologies of copper smelting can be operated anywhere on the earth, radio messages can be effectively relayed everywhere, and the planting of hybrid varieties of crop plants can be technologically systematized in all climatic zones except the severe microclimatic. Sociopsychologic and econopolitical rules are equally applicable to all parts of the earth, among all societies. All members of the species *Homo sapiens sapiens* possess the capacities for learning; an equitable social system needs know no physiographic, climatic, or political boundary; and industrial technology can be applied in any society. It is patent that any cultural system can be put into operation in any part of the earth.

The biological trait of exclusive territorialism may have been necessary in its elementary form to the first of the hominids, but is it any longer necessary for *Homo sapiens sapiens*? Is it beyond the capacity of civilized peoples to devise cultural systems of territorialism for orderly living that can match the one-world technology operable in research in industrial chemistry? In the failure of the League of Nations, we saw the unwillingness of mankind to integrate political management of the earth to a one-world cultural system. Do we see, in the United Nations, really revolutionary advancement equivalent to the need? In political management we see profound reluctance to take the obvious and needed steps toward one-world management of many critical problems. To a very considerable degree political nationalisms still stand in the way of progress for human society as a whole. We are not, here, denying that many United Nations programs and activities have been accomplishing a great deal.

This is not the place in which to enlarge upon the future world needs of humankind in general. We set down a simple list of the kinds of objectives that must form the remaining steps in cultural evolution necessary to the future satisfactory residence of *Homo sapiens sapiens* on the planet Earth. These are:

1. The open sharing of world natural resources.
2. The freeing of all human beings from restrictive social stratification prohibitive of individual development and the establishment of systems of international law operative throughout the world permitting peaceful settlement of problems external to individual societal boundaries.
3. The worldwide acceptance of responsibility for the rapid application of cultural procedures making for a healthy biotic environment.
4. The worldwide acceptance of population restriction to limits within world technological productive capacity.

The list is a minimal one, suggestive rather than inclusive, problem-minded rather than moralistic. The human race has come a long way since the start of the Pleistocene, and its inventiveness has brought domination of the earth, but satisfactory continuance on the earth now depends on the cultivating of human initiative, tolerance, cooperation, and willful limitation, and on the devising of cultural systems capable of dealing with huge numbers of people as residents of the planet.

The chief direction of the foregoing comments was toward external relations among human groups scattered over the earth. It is obvious, however, that within societal cultural boundaries some of the same needs apply. No human society today provides potential satisfaction for all of its members within its own sets of rules, mores, and customs. The continuance of satisfactory life, within cultural groups, depends significantly on the working out of cultural procedures satisfactory to all members of the group. No longer may a dissident group emigrate to an empty land in which to indulge its own demands for new freedoms in new systems of culture.

Alternatives to Success

Occasionally the cartoonist, the science-fiction writer, or the prophet of doom projects the human race forward through time to the odd circumstance in which culture has been reversed all the way back to the level of the Paleolithic. This "Flintstone" type of speculation is interesting but one most unlikely to come to pass. It may not be beyond the power of some government official, in a sudden fit of animal rage, to press some trigger on a technological system that could drastically reduce the numbers of mankind and poison the surface of the earth and its atmosphere for an interval. The numbers of mankind who then continue to survive might well need their maximum skills derived from training in survival living, but the skills and bodies of knowledge would be very different from those that were current in the late Paleolithic. And the biotic environments surrounding those survivors would be very lean in comparison to richness of the Paleolithic environments. Survivors most likely would carry forward assorted but quite unequal specialized patterns of knowledge derived from the cultures in which they had shared. It is conceivable that surviving remnants of our present world population could live a curious sort of life scavenging off the artifacts of the prebomb era.

The ingenuity Mrs. Fred Flintstone uses in solving her household problems is often delightful to modern urbanites, but her solutions are highly impractical in our "simplified world."

However any such living system would be unlike any that had gone before, open to impacts of a very different biotic order of life presumed also to have survived in some fractional scale.

If, instead of a bomb reducing the population to a few surviving elements, the earth's culture becomes increasingly exploitative, and total population continues to increase to the point at which the earth's resources are relatively exhausted, it is possible to speculate that declining populations might gradually and peacefully simplify their living systems by successive abandonment of industrial technologies. It is possible to invoke a future era of nonindustrial cultures perpetuated by relatively small populations struggling for survival, but such living systems would not be equivalent either to the Paleolithic or to the preindustrial era in any way. Exhaustion of resources would imply marked change in the biotic world and also quite different ecological evolutionary change unlike biotic systems of the past. Further, the assumption of peaceful simplification seems unjustified, bearing in mind the past competitiveness of mankind under pressures. Such a cultural world would evolve from present patterns but could not possibly be a return to the systems of the Paleolithic.

To presume a third directional trend, in which mankind loses control over the biotic forces struggling for some level of ecological balance on an earth suffering horribly from culturally induced environmental pollution, it is necessary to invoke such overwhelming biotic change as to make the question of human survival almost theoretical. Any human population surviving a major biotic collapse would need to invent highly specialized technologies in order to provide human support in a world entirely inimical to human success, a world very unlike that of any earlier period.

The Way Forward: Improved Technology

If we accept any element of the evolutionary system controlling life on this earth, it becomes clear that the human race cannot turn back to any past era, thereby to simplify the problems. We may turn back to a lesson learned in earlier time whose point we have perhaps recently neglected, but application of that old knowledge must be on terms that can be equated with contemporary circumstances. It is possible for a single family, or a small group of people, to go off into a wilderness and live a Neolithic life, but there are no longer sufficient wildernesses for all of the world's present population to do so. Similarly, Jeffersonian democracy, in the full sense, is simply not open by choice to the 215 million citizens of the United States. Our present environment is normal for us, and it includes such environmental components as clothing, housing, heating, modern technology, and modern medicine. The Paleolithic and Neolithic environments are now abnormal. We shall not live in those environments, and, since we are no longer fit to live in them, this may make us nostalgic but perhaps not unduly alarmed. We and our descendants are not able to dispense with the services of medicine (to maintain health) and technology (to maintain food supplies). The remedy for our dependence on technology is *more* and *improved,* rather than less, technology.

The advances in public health and sanitation in most parts of the world during the past half-century have lengthened the life of the human being remarkably, but they have also promoted the illusion that this is a one-way street and that human beings have totally conquered the whole of the biotic world. The truth is that no disease affecting human beings has yet been eliminated from the complex ecosystem, although there currently is hope that smallpox has been eliminated permanently. The biotic ecology of the surface of the earth is still an active system, which, if thrown too badly out of balance in any one direction, reacts strongly in compensation in some counterdirection. Humankind so dominates all other life-forms on the earth that lions and tigers, for example, pose no real threat to people, but the world of insects, viruses, fungi, and the like is very much alive, very evolutionary, and capable of strong pressure on the human race as a whole. Some new problems in biotic ecology have emerged out of industrialization, but most of the old problems remain in latent or merely quieted states.

It is sometimes said that human technological achievement has reached the plane on which it may be possible for science to reverse at least some of the trends in human development. Population control represents such a potential line of reversibility, but the cultural procedures by which such control may be made effective in themselves represent a further element in cultural evolution, and not a reversal. The success of population control, as a worldwide phenomenon, patently depends on advanced technologies, and it depends on the common participation of all societies on the face of the earth. Such participation in a common program would represent cultural evolution of a significant order, since it has never yet been fully achieved in any other program. Equally true is the fact that any other reversing trends will require unanimous consent and participation, furthering cultural evolution.

In our present world the human race faces a unique set of circumstances that have never before been operative. There are three primary aspects to this contemporary state of uniqueness: the total numbers of human beings; the levels of, and the

The contrasts in these two domestic scenes show the endless variety of human culture systems that have evolved in different regions of the earth. The Chinese family, below, lives in a cave home in the Loess Highlands of northwest China. They now enjoy electricity, as does the American family on the right, but the housing, clothing, furniture, eating utensils, and even the food itself are all very different.

ranges in, culture; and the present face of the earth. None of these has ever obtained in its present pattern in any earlier time. No past culture system had concepts or institutions adequately applicable to conditions now current.

From the lack of a system in the earliest of human time, groups of people developed divergent culture systems. On the maladjustment of those systems, culture groups sought for and devised new systems that grew still more divergent. Those newly successful systems developed further maladjustments, requiring still newer systems, until the modern era when convergent systems began to come into partial use among societies. It is possible to see a certain kind of cyclic pattern in the succession of problems that cause search for solutions, in solutions that make possible new levels of achievement but also create maladjustments, and in renewed searches for solutions and new breakthroughs. The Industrial Revolution has, in a sense, constituted a tremendous breakthrough, but it has in turn created a host of problems that require still newer solutions for the greatly expanded population that resulted from that breakthrough. The cultural solutions to twentieth-century problems must represent some new kinds of solutions never before achieved under patterns of divergence. The problems are common to all members of the human race in today's world, and the solution must be a solution applicable to all humanity, a convergent solution.

RESPONSIBILITY AND ITS PROBLEMS

The human race is in a peculiar position today. Most of its educated members, most of the leading societies, and a great share of the societal leaders know what mankind needs to do, but collectively they cannot bring themselves to take the necessary steps. To many people in all parts of the earth, including people in the United States, the necessary steps are at present wholly unacceptable. Attitudes of nationalism, exclusivism, and the devil-take-the-hindmost are cultural patterns out of the past that remain strongly planted in human consciousness. Each was possible on an earth inhabited by relatively small numbers of people, an earth separated by time and distance into units that could be made to accord with some physical components making up the landforms, climatic regions, vegetative associations, soil zones, and water-supply basins. Mankind went through a long series of eras during which biological and cultural divergence took place to the end that the

human race became a huge collection of differing societies practicing increasingly differing systems of culture. Most of human time on the earth is within that broad trend of divergence: 1,999,500 years out of 2,000,000 years.

Only for the last 500 years, in round terms, has the human race been on a trend of convergence; and its earlier—and larger—heritage still clings to it. Humankind may take a while to come to terms with its own past and the problems of its future (Table 12.1). Meanwhile the problems grow more urgent; they must be solved by revolutionary steps if *Homo sapiens sapiens* is to continue its mastery of the earth. Humankind now faces the responsibility of living up to the culture-creating beings it has made of itself.

The sheer diversity of humankind is thus one of the major problems facing all of humanity today. In the twentieth century people must work at the procedures needed to bring their kind back to the types of groupings that can live on the one earth. The balance between one earth and the large number of "little worlds" is tenuous, almost constantly being broken by militant competitive action so that war seems more frequent than peace. Whether this issue is viewed as biological territorialism or as militant sociopolitical regionalism makes little difference.

The solution must be worked out to the end that human control over the earth is retained, even against people or specific societies, lest mankind as a whole be threatened by upsetting the patterns of balance and by accumulating in too great numbers. The solutions must include at least three kinds of values in reaching a suitable level of balance among physical/biotic environment, human society at large, and culture at large. These three are:

1. Self-realization (identity, status) sufficient to the biological ego of individuals as members of a species,

2. Stimulus (excitement) sufficient to keep the species from dying of boredom, committing too much crime against its own kind, or refusing to participate in the obligations of human life, and

3. Security (safety, support) in terms of biological sustenance and a home territory for each member of the species.

There is not much hope of the physical/biotic environment improving markedly within the required interval, nor is there much hope that humankind can by then drain off the surplus numbers to some other planet. Without culture there is little hope that humankind can attain any solution at all. The key element in the balance is *culture* in the hands of human beings. The inquiry into the nature of the ecosystem and the means by which to maintain it in workable condition, the exploration of additional technologies,

and the search for effective means of repairing errors of commission and omission against the environment, all in the hands of science, are one phase in the achievement of that balance. The other phase must lie in the psycho-sociopolitical field. The sublimation of militant nationalism into cultural regionalism of a nonmilitant variety, the development of international institutional systems suitable to single-earth operation, and the development of social institutions encompassing all members of the species are the requisite elements on this front. The development of philosophopsychologic cultural viewpoints amenable to the present technological skills of medicine, in the extension to worldwide practice of population control, becomes the most critical element in the whole question of balance.

The finding of cultural systems sufficient to the development of the environment/mankind/culture balance is emerging as an issue of the next few years. The problems are those to be faced in the next several decades (see Table 12.1), for we have a short breathing space, still, in which to work out solutions to them. The problems are sufficient to present challenging opportunities for self-realization and stimulus in ample volume to all interested persons, since they call for the invention of some new and efficient technologies and some new culture complexes.

In any single society in the more distant past, technological change, revision in socioeconomic and sociopolitical rules and institutions, and social change usually developed in company together because the kinds and rates of change in a culture system were relatively slight at any one time and were relatively slow in occurrence. No two societies developed at exactly the same rates or underwent the same kinds of change. On the external level, between societies, rules and institutions usually developed only as a result of problems arising between two or more societies, and they often related to the different levels of cultural advancement of the societies in conflict. Those rules and institutions, therefore, followed the occurrences of problems and were patterns of response. In earlier time the lag between occurrence and response was not highly significant or critical.

Since A.D. 1500, however, technological change has taken place with increasing rapidity, greater range of application, and with greater intensity. Technological change has now affected most of the separate societies of the earth. On the internal level, within any single society, it now appears that socioeconomic and sociopolitical rules and institutions change only after technological change has created internal societal problems. On the external level, in what we now call international affairs, rules and institutions have developed only to a partial degree—and with a marked lag in time—in response to

Table 12.1 A Hierarchy of World Problems*

Priority Group	Problem or Crisis	Aspects of Problem/Crisis Requiring Solution in Next:†		
		1–5 years	5–20 years	20–50 years
1	Total Annihilation (affecting billions of people)	Nuclear or Radiological/Chemical/Biological Escalation		
2	Great Destruction or Change (physical, biological, and/or sociopolitical) (affecting hundreds of millions)	Local Wars	Famines Ecological Balance Development Failures Local Wars Rich/Poor Gap	Economic Structure and Political Theory Population Planning and Ecological Balance Patterns of Living Universal Education Communications Integration Management of World Integrative Philosophy
3	Widespread, Almost Unbearable, Tension (affecting tens of millions)	Administrative Management of Communities and Cities Slums Participation in Decision-Making Environmental Degradation Racial Conflict	Poverty Pollution Failures of Justice and Law Racial and Religious Wars Political Rigidity	
4	Large-Scale Distress (affecting millions)	Transportation Diseases Crime Neighborhood Ugliness Loss of Traditional Cultures	Communications Gap Housing Education Independence of Major Powers Control of the Seas	
5	Tension-Producing Responsive Change (affecting hundreds of thousands, but good people are working on these problems and making progress)	Regional Organization Drugs Cancer and Heart Diseases Artificial Organs Clean and Adequate Water Auto Safety Marine Resources	Education Inadequacy	
6	Other Problems, but Adequately Researched (affecting tens of thousands)	Technical Development Monetary Design Mental Illness Fusion Power		
7	Exaggerated Dangers and Hopes (affecting thousands)	Heart Transplants Mind Control	Sperm Banks Unemployment from Automation	Eugenics Melting of Ice Caps
8	Noncrisis Problems Being "Overstudied"	Man in Space Most Basic Science		

* After John Platt, *Science,* vol. 166, 28 November 1969, pp. 1115–1121.
† If unsolved, each aspect of problem or crisis will tend to advance to the next higher priority group with the passage of time (i.e. move both upward and to the right on the table).

technological change. The time lag today, however, is far more critical than it ever was in the past. Ideally, in our future world, worldwide institutions and binding regulations should lead technology rather than follow after it. At the very least, it must become the case that the formulation of those institutions and binding regulations keeps close pace with technological change, rather than trailing after. The relative integration of change and regulation may be the only way in which mankind can maintain control of the biotic ecosystem, the world cultural environment as a whole, and the future processes and patterns of human development on the planet Earth.

SUMMARY OF OBJECTIVES

After you have read this chapter, you should be able to:

1. Outline how might this volume have been different (in organization, in content) if its authors had taken the view that "a supernatural force had designed the whole struc-

GLOSSARY OF TERMS

Accretion. The increase by addition of items from outside the unit, or by accumulation of similar units over time.

Acculturation. A process of interaction between two or more culture groups through which one culture system, or one individual thereof, takes on traits of a more advanced level from the other culture system.

Adjustment. See space adjustment.

Adornment. The term having widest reference to the several ways of decorating and covering the human body. Includes scarification, hair styling, deformation, cosmetics, jewelry, and garment wearing.

Agriculture. Any system of crop production, including keeping animals, by which tilling the soil and related labor is done chiefly by nonhuman power. The focus of the activity is the preparation or treatment of whole fields. See gardening.

Agronomist. A specialist in agriculture concerned chiefly with field crops, fertilization, and soil control.

Amalgam. A mixture of different things that unifies the separate elements into one common product.

Analogue. A feature, situation, condition, or thing existing in one environment which is similar to a corresponding item in another environment.

Animism. Any unorganized and relatively simple system of belief that ascribes spiritual and supernatural power to features of the earth such as rocks, trees, or streams, and that accepts such power as controlling human destiny.

Annual. Literally once a year, but in reference to plants annual indicates plants that live only one season, completing their life cycle within that period.

Aquatic. Growing and/or living in water, or at the margins of a water body where the water table is so high that the roots are always very wet.

Artifact. Any product of human workmanship remaining from an earlier era.

Autonomy. Having the right and power to self-govern or manage, and having freedom from control by any outside power or force.

Asthenosphere. A zone of the interior of the earth from 25 to 75 miles below the surface, within which the materials are supposed to be near enough to the melting point that they will flow under heavy pressure.

Badland. Landforms in which intricate dissection, often in soft rock formations, produces many steep slopes and narrow gullies so that there is little smooth surface remaining.

Band. A territorially based small group of people that live together. The band was the smallest early socioeconomic group, and less than a tribe.

Bast. The inner bark of various shrubs and trees whose fibers can be pounded to make a pliable "cloth" material for wraparound garments.

Bauxite. A mineral high in an oxygen-rich compound of aluminum, the easiest to smelt of all the aluminum ores, and the form most commonly mined today.

Beachcombing. The procedure of searching the ocean shoreline at low tide for edible life-forms and other useful products.

Biennial. Usually used with a plant variety that ordinarily requires two years to complete a life cycle. Such plants do not flower or seed during the first year.

Biologic norm. The rate of change in plant and animal life that is produced entirely by natural and nonhuman processes and events in an evolutionary way.

Biomass. The total volume of organic plant material produced each year by all forms of plant life, and the sum total of the food supply for all herbivorous forms of animal life.

Biota. A collective term by which to refer to all the plant and animal life-forms inhabiting any space, niche, territory, or region.

Biotic pyramid. The structured population pattern in the whole of biologic life in which the primary producers of energy are at the base and most numerous; the secondary consumers must be fewer as to total consumer demand, and the tertiary consumers must be fewer still as to consumer demand; herbivores are restricted in numbers by their total consumer demand for edible plant materials, carnivores are restricted by their total consumer demand for herbivores, and so on. See food chain.

Blade. Any narrow, parallel-edged piece of stone chipped from a larger rock and used as a cutting tool.

Boreal. The northern chiefly coniferous forest sectors of the polar margins of the Holarctic Life Zone.

Brachiation. Development of the use of the hands in locomotion, particularly in climbing or swinging.

Bund. An artificial embankment constructed to restrain water, either to retain it within an enclosure or to prevent its flow into particular areas; a dike.

Burin. A specialized cutting tool of stone having a beveled point.

Carnivore. An animal life-form having primarily a meat or fish diet.

Catalyst. An agent or substance that increases the rate of positive reaction, stated in chemical terms. Applied to human affairs is any element, factor, or force that sets in motion some pattern of progressive change, usually thought of in terms of betterment of the existing situation.

Cerebellum. The two central hemispheres of the brain having to do with coordination of muscles and the maintenance of bodily equilibrium.

Charismatic. Relating to the supposedly divine or miraculously given power or gift that leaders possess that causes such awe and respect in their followers as to expect superhuman performance from those leaders.

City-state. The earliest form of the political state, in which a city formed the whole territory of the state, and included the whole state population.

Civic. Relating to affairs concerning local government, management, and good citizenship.

Civil. Pertaining to common and ordinary internal life and affairs of a society, and excluding military, naval, and religious management.

Clan. A social kinship group of families that claim descent from a common ancestor.

Climatology. The study of the long-term averages and extremes of weather conditions, including the processes and causes thereof.

Complement. The kind of feature, thing, quantity, or type of items necessary to make something complete.

Confederacy. An alliance arranged among several separate and distinct groups, clans, tribes, or nations in which all agree to combine efforts to some specific end, but in which each group retains its own independence and autonomy.

Configuration. Shape, arrangement, and form.

Continental shelf. An under-the-ocean sloping plain of variable width that forms an edge or border to most continents, usually of relatively shallow water. The shelf ends at the point where its degree of slope increases sharply and the water becomes very deep.

Continuum. Anything that has a great range in quality, quantity, or character from least to most, but in which it is very difficult to make clear distinctions in aspects very much alike at any point along the line of progression.

Convergence. A dynamic cultural process by which culture systems develop similar characteristics, institutions, and patterns; it is often held that the process produces independently developed similarities, but as used in this volume, convergence is not so strictly interpreted and at least stimulus diffusion is responsible for much of the resulting similarity.

Conversion. See resource conversion and resource extraction.

Core biface. A stone chipped on two faces to form two cutting edges, with a point narrower and sharper than the base by which it is held.

Cultigen. The ultimate stage in domestication in which the ancestral forms no longer are clearly identifiable; usually such plants no longer can maintain themselves in competition against wild plant growth, having lost characteristics of natural dissemination-reproduction.

Cultural landscape. Any landscape surface of the earth in which human action has significantly altered or replaced the natural and wild distribution of features and elements.

Cultural process. Any activity or procedure contributing to continuing change in patterns or systems of culture.

Culture. The total of historically learned human behavior and ways of doing things. See pp. 1–2.

Culture complex. A related group of traits, such as wearing "European" clothing.

Culture hearth. A region in which a particular culture system evolved; the home territory of a culture system.

Culture realm. A large area in which there are several related culture systems, each occupying a culture region, such as the African culture realm; also called culture world.

Culture region. The area within which one culture system is generally accepted and followed, such as the Japanese culture region.

Culture system. A large assemblage of culture complexes that fit together into an operating whole, such as the Chinese culture system.

Culture trait. A single item of regular practice, such as wearing cloth sandals.

Culture world. See culture realm.

Defoliant. A chemical that, when sprayed on leaf-bearing plant growth, will cause the leaves to drop off.

Deglaciation. The process of melting away glacial ice to the exhaustion of the whole of all ice masses.

Denudation. The most inclusive of the terms (such as erosion) used to refer to the removal of materials from the land surface of the earth in the ultimate lowering and smoothing of that land surface.

Desalination. The process of removing salt from sea or lake water to render the water usable in agriculture, industry, or for domestic consumption.

Determinism. The concept that all phenomena and actions result from specific causes without the possibility of alternative developments. Environmental determinism ascribes all patterns of life to causes originating in the physical environment; probabilism and possibilism are each progressively less strong interpretations allowing alternatives; cultural determinism holds that human life processes negate the influences of physical environment.

Diachronic. The approach by which events and conditions are examined in historical and evolutionary succession through time. See synchronic.

Differentiation. The process of modifying and changing culture traits and complexes so that culture systems become increasingly different from each other during a long time period.

Diffusion. The progressive spread or scattering of something. As a culture process diffusion may occur through forced acceptance, by purposeful borrowing, by dynamic spread, or by idea stimulus.

Discontinuity. A break in a regular sequence, commonly used to refer to a missing element, member, or unit in a progressive series of items.

Discovery. The act of finding something new to human experience that occurs in the natural world in its normal state, e.g., coal.

Disequilibrium. Some pattern of instability in normal ecologic conditions on the face of the earth producing change in the conditions of particular life forms. Such conditions may lead either to extinction or to increased populations if the condition continues, but a purely temporary disturbance may be overcome as conditions return toward a state of balance.

Divergence. The process of becoming more and more unlike, referring to culture systems growing apart.

Domesticable. A plant that is being worked with, or an animal that is being cared for, in which progressive changes are beginning to occur that indicate that the plant or animal is entering into the domestication process.

Domestication. The process of transforming a wild plant or a wild animal into a utilitarian plant or animal responsive to human handling; customarily the transformation includes both growing habit in plants and tameness in animals, and it also involves change in size, shape, color, build, and weight. See cultigen.

Ecological niche. The local environmental surroundings for a single life-form or closely associated group of forms; the "home area" for an animal life-form, the "growing site" for a plant form, similar to the concept of "home" for a person.

Ecology. The study of mutual interrelationships between different life-forms and their surrounding environments; often narrowed in meaning as in plant ecology, by which the study is restricted to plants and their environments. The term is often improperly used in popular language.

Ecosystem. The interaction system comprised by all the inorganic, organic, and living elements found in an environment or habitat; the earth as a whole comprises a single and complete ecosystem, but the unit of environment being examined can be reduced to a small local level.

Enculturation. The process of receiving and becoming habituated in the commonly accepted traits and complexes of a particular culture.

Endothermic. A reaction that is produced by the creation of heat.

Environment. The collective term for all of the conditions, features, and influences that affect a life-form in its homeland. For forms lesser than humankind these form the ecosystem, whereas for humankind there are both the physical ecologic elements and all the cultural ecologic elements that combine to form the total environment.

Equilibrium. In theory a state of balance between forces, pressures, and influences working in opposite or different directions. Although often thought of as a stable and "normal" condition of being "at rest," a biotic equilibrium is dynamic and constantly shifting slightly as progressive ecological change takes place in an evolutionary trend.

Estuary. Refers to a drowned river mouth in which the sea invades the lower course of the river valley, and in which tidal influences counter the usual flow of water.

Ethnic. Refers to a combination of physical and mental traits and habits in a person, or a group of people, as a product of common heredity and cultural traditions. The term is not a synonym for either race or culture system.

Eustatic. Alterations in worldwide sea level, upward or downward, according to the decrease or increase in glacial ice volumes, are termed eustatic, as opposed to the emergence or submergence of some local landform produced by deformation of the crust of the earth.

Evolution. The selective change, growth, and development of anything through small elements of change over long periods of time, usually functioning to increase complexity.

Exotic. As a noun the term refers to a plant, animal, or culture trait quite foreign to the situation in which it is found, and refers to something introduced from another region into a new setting.

Extinction. The final disappearance of a life-form from the biotic assemblage as its last members die out without reproducing their kind.

Extraction. See resource extraction.

Family. See nuclear family.

Fecundity. The physiological maximum measure of potential production of offspring during the whole bearing period. See fertility.

Fertility. The actual birth of offspring as the realized portion of the potential measure. See fecundity.

Feudalism. A contractual system of socioeconomic and sociopolitical relations imposed upon the agrarian classes of different countries by the upper classes holding politicomilitary power. Differential tenancy rights to the occupancy and cultivation of land by separately ranked agrarian classes required varying amounts of exclusive personal domestic, agricultural, or military service to be rendered to the landholding classes. These landholders in turn owed both personal services and the provision of manpower to the political ruler, who held title to all land in his domain. In the decline of feudalism, the occupancy rights became hereditary, and the personal services were replaced by fees and taxes paid in money.

Flux. When used as "in flux," refers to a condition of continuing change in shape, size, pattern, state, situation, quality, or quantity.

Foehn wind. An air-mass movement downslope which becomes warmer and drier, causing marked evaporation of moisture in its path; often termed chinook wind in northern United States and Canada.

Food chain. The linking pattern in a biological community in which each life form feeds on the form below or constitutes food to the form above (in the biotic pyramid). Plant life is the basis of the food chain, and the most powerful carnivores are at the top; the sequence of chains is the food cycle, and the food cycle controls the biotic pyramid.

Fossil fuels. Combustible products of organic origin, such as coal, petroleum, natural gas, and peat, derived from processes completed during a previous geologic era.

Friction-of-distance. The effect of lessening movement by the increasing, or greater, distance involved; alternately the greater distance may increase the cost of the item farther from the source.

Gallery forest. A strip of forest growth relatively narrow in width but scattered at length along a stream course, shore, or coast, as a fringing pattern that gives way to less than forest growth away from the line of forest.

Gardening. Any system of crop production in which tilling the soil and other needed labor is performed by human energy; the focus of activity is the care of individual plants or small patches of plantings. *See agriculture.*

Gene pool. The reservoir of genes formed by the sum total of the genetic materials of a breeding population.

Genetic drift. The result of change in gene pool composition of (usually) a small population through chance or accident so that the former range of genetic material cannot be maintained.

Geneticist. A specialist whose subject of study is the composition and variation of the genes responsible for the inherited characteristics of any life-form.

Geologic norm. The rate of change in the physical surface of the earth that is produced entirely by natural and nonhuman events and forces in an evolutionary way.

Geology. The study of the history of the earth's crust and the interior of the earth in terms of rocks, rock formations, their structures, and the processes through which change takes place.

Geomorphology. The study of the conditions and changes that occur in landforms, including origins, processes, stages, classes, and development of specific forms.

Geophysics. The study of the dynamics of the interior of the earth, including the processes that effect the gross patterns of the surface of the earth.

Glacial stage. A period marked by the accumulation of ice in heavy sheets and glaciers, accompanied by lower temperatures and, often, increased precipitation; also called pluvial stage.

Gnomonics. The measurement of time by means of sun dials.

Green Revolution. The term applied to the recent development of extremely high-yielding grain crops that are now permitting great increases in food production in a number of subtropical countries.

Grid plan. A street plan for a town or a city in which all streets run in straight lines, with one set crossing the other at right angles so as to form square blocks.

Harrow. An agricultural tool in which short, pointed teeth usually project downward from a frame that rides flat on the ground so that the teeth can break up the ridges and lumps of turned soil left by plowing.

Hatchery. A place in which eggs are incubated—kept warm until the new life-forms are born, or hatch. Usually refers to birds or fish, but may include any forms that reproduce from eggs.

Hegemony. A dominant power, control, or authority that makes possible the governing of a region.

Heliophyle. One attracted to, or adapted to, sunlight. In reference to a plant this means a preference for a sunny location in which to grow. Most crop plants, and many decorative ones, prefer full sun.

Hematite. One of the most common forms of iron ore and one of the easiest to smelt, a specific combination of iron and oxygen.

Herbaceous. A term referring to any nonwoody plants.

Also applied to grasses and herbs in broad reference.

Herbivore. An animal life-form having a primarily vegetarian diet derived from any kind of plant material. Often used to refer to the grazing animals.

Hieroglyphic. Refers to the form of writing in which drawings of picture symbols stand for certain things, sounds, ideas, numbers, or actions.

Hinduism. The native religious system developed very early in India as a comprehensive religious and social system. The system has a multitude of gods, is divided into a great many sectarian groupings, and has no single set of codes or doctrines, but generally follows similar concepts of causality, rebirth on this earth, and ultimate salvation.

Holistic. Used in reference to a viewpoint on something that takes all possible alternatives and aspects into consideration.

Holocene. On the geological time scale refers to the period since the last deglaciation, and is also termed the Recent.

Hominid. Manlike; used to refer to any humanlike form including the earliest ancestors of the human race.

Horticulture. Strictly used the term refers to the growing of tree crops. Broadly and loosely used to refer to the cultivation of vegetable gardens and small orchards of trees, usually in connection with rather simple technology.

Humanization. The altering and changing of the natural, wild earth into a pattern by which people are agents of change and is increasingly responsible for conditions and the state of affairs; the broadest term of reference for the impact of mankind on the earth, that impact expressed through culture systems practiced by culture groups.

Hybridization. The mixing of dissimilar stocks of genetic materials through interbreeding of unlike groups producing greater variety and (occasionally) greater vigor.

Hydrology. The subject matter involved in the study of water conditions in any of its aspects on the land areas of the earth, both at the surface and underground.

Imprinting. To implant firmly on the mind and fix in the memory, especially on the very young.

Incisors. The set of front teeth, both upper and lower, between the longer canine teeth on each side of the jaw; in use the incisors are the cutting teeth employed to bite off and cut materials.

Innovation. The introduction of something new into habituated customs, procedures, processes, products, or devices. Generally thought of as having broader range than invention.

Institution. Any complex of customs, relationships, procedures, agencies, or forms of action that have become so generally accepted by a culture group as to become standardized patterns to the degree that deviation from that standard is not approved by the group.

Intensification. *See space intensification.*

Interface. The boundary plane or zone between two or more situations, conditions, forces, aspects, or things that come face to face against each other. Customarily used in geography to refer to the plane of contact, or meeting, of the atmosphere against the ground/water surface of the earth. Most human activities take place along this intercontact plane.

Interglacial stage. The relatively warm and usually drier period in which deglaciation becomes quite complete and ice is mostly absent from the surface of the earth. Sometimes also called the interpluvial period.

Invention. The creation of something new to human experience that does not occur in nature, e.g., stainless steel or paper.

Invertebrate. Any animal form lacking a backbone or spinal column. By far the most numerous of the animal life-forms are invertebrates.

Isolation. Separation of culture groups, as breeding populations, for significant periods of time by some insurmountable force, so that the culture group so separated lacks significant contact with other groups.

Judaism. The term includes all sectarian groupings commonly thought of as comprising the Jewish religion.

Karst. The variety of landforms, and the processes creating them, by which solution of limestone rocks by water carries dissolved and suspended matter through underground channels rather than transporting them on the surface along stream courses. Karst landforms show irregular, nonlinear and ragged surfaces, caves, sinkholes, and have underground drainage lines.

Kitchen midden. Any refuse heap historically left on a site by its occupants. The term is often restricted to prehistoric remains of early cultures.

Landscape. As used in this volume, the general surface of the earth as occupied by, and seen by, human beings. The term cultural landscape refers to portions of the surface in which human impacts and works are significant to dominant elements in the visual scene.

Latifundia. A complex kind of system by which a few individuals control very large holdings of land on which many persons/families live, each person/family working a small area that is either rented, sharecropped, or operated under some semifeudal arrangement.

Levee. An embankment produced either by natural processes or by human procedures that serves to restrain stream flow within a channel; natural levees produced by river deposition are broad and gently inclined on their outer margins whereas humanly constructed levees are narrower, have steep banks and margins, and are sometimes called dikes.

Lineage. The smallest kinship group that traces descent from a single known ancestor.

Linkage. Process of joining different areas or groups together by transportation routes and communication systems so that there is contact, communication, and trade between formerly unconnected areas.

Littoral. The shore of an ocean, sea, or lake, including the shallow water zone and the strip of land immediately behind the shoreline.

Living system. Some specific set of patterns which an individual, or a group up to a whole society, has established, and follow, as a set of rules, customs, procedures, and activities rounding out their daily life.

Lode. An underground rock formation containing minerals of utility to man. Lode mining refers to any mining operation underground that follows particular rock formations in search of desired minerals.

Loess. The fine-grained wind-blown deposits that have accumulated in thick beds in certain areas rather distant from the source areas (loess deposits often are fringed by sandy deposits located nearer the source areas).

Longevity. Life expectancy at birth, stated as the number of years a person may expect to live.

Magic. Any combination of compulsive actions and ritual performances which attempts to control, or to influence, supernatural forces.

Mahayana. The division of the Buddhist faith that developed in the northern part of India and spread to Tibet, Mongolia, China, Vietnam, Korea, and Japan in the early centuries of the Christian Era. It became divergent from the Theravada division by incorporating alien religious concepts and practices appealing to the masses, but it developed many sectarian groupings. Sometimes called Northern Buddhism.

Mammal. The highest ranking class of life-form, warm-blooded, and having a skull enclosing the brain, a sectional backbone, and feeding the young with milk. There are a few technical distinctions that separate the earliest mammals from the later, more highly developed forms. See marsupial.

Manor. A term applicable to medieval times, and to a large unit of land owned by a lord or by the church, on which a certain number of persons/families cultivated small pieces of land, or worked at other chores, and paid rents, dues, and services. The manorial system was part of European feudalism, and the same system (the term is no longer used) applies to socioeconomic conditions in those countries in which latifundia still operates.

March site. In earlier historic eras a march site was the location of a frontier fortified post near or along the borders of a political state, established to protect the population against raids by peoples living outside the state. In time, many march sites grew into large cities, and some later became provincial capitals as territories formerly beyond the frontier were incorporated into the state.

Marsupial. Any life-form belonging to the second lowest order of mammals, such as the kangaroos and opossums, in which the female carries the newborn for a period in a pouch on the abdomen.

Megalithic. Having reference to very large stones erected in varying patterns by prehistoric culture groups according to some cultural motivation not clearly understood by modern mankind.

Megalopolis. A term coined to refer to a zone in which several large cities have grown together spatially to create an extensive urban region.

Mesolithic. The transitional age or period between the Paleolithic (Old Stone Age) and Neolithic (New Stone Age) during which plant and animal domestication began along with numerous other elements of culture growth.

Metallurgy. The extraction of minerals and the refining of metals from their ores, and the processes involved in producing usable metal products; smelting of ores by melting out the minerals in metallic form through the application of heat is the primary, but not the only, process of extraction. See smelting.

Meteorology. The study of the earth's atmosphere in terms of its short-term processes producing weather-change phenomena; the long-term study is climatology.

Metropolis. A large urban settlement, usually thought of as the largest city in a particular region; the word literally means "mother city."

Microlith. A very small chipped stone tool made from fine flakes of chipped stone, usually from flint. An arrowhead is one example.

Migration. The act or pattern of moving from one area to another by any life-form (including plants, whose movement is slow compared with that of animals). Migration may be a one-way nonrepetitive movement or, among some animals and birds, may be a repetitive and seasonal round-trip pattern. Among human beings seasonal migration of labor forces is often repetitive. See transhumance.

Mobility. In the physical sense the term refers to the ability to move from one position or place to another; dogs have mobility but trees do not. In the socioeconomic sense the term is comparative and refers to the upward or downward movement of individuals, families, clans, or tribes as compared with the status of others.

Mores. (pronounced "moreez") Those little social folkways, customs, manners, and conventional actions common to a culture group which have ethical significance and are part of a larger sociocultural system. Europeans shake right hands on greeting, whereas Chinese hold both hands folded together at chest level but do not touch each other.

Morphology. The study of structures and the forms resulting from changing evolutionary development.

Mortality. The death rate, usually stated as a proportion against the total population. Infant mortality is a special measure stated as a proportion against the total birthrate.

Mosaic. A large area or unit made up of many small and quite different pieces or units that fit together into the larger unit. A jigsaw puzzle is a special kind of mosaic.

Mountain building. The result of any one of, or a combination of, several sets of purely physical processes that produce horizontal and vertical shifts in the outermost layers of the earth's crust. Convection currents within and just below the solid crust produce displacement of crustal units (by faulting, overthrusting, folding, vulcanism, submergence, and uplift) that result in piling up masses of rocks to variable heights when measured against sea levels.

Mountain and valley breeze. A localized air movement by which cold air drifts downslope at night, and warmed air blows up-valley during the early afternoon of a warm day.

Mud way. A track along which early and simple transportation was made easier by pouring water on the track to create a slippery, muddy surface.

Mutation. A transmissible and hereditary modification in genetic materials suddenly produced by the occurrence of alterations in the cells, producing new genes or the rearrangement of genes; usually mutation is detrimental to the maintenance of the former life-form type.

Nation. A large group of people who have long been in living contact with each other, and who have quite similar values, customs, institutions, ways of living, and a sense of belonging together.

Natural landscape. A term, in theory, defining any landscape surface of the earth in which no human impact has occurred; in common usage the term is often employed to denote any wild landscape that is not permanently settled by people and in which no cultural forms are apparent.

Neolithic. The New Stone Age during which many culture complexes were invented and developed and in which production economies began to replace collecting economies.

Nuclear family. The smallest of the family types, referring to a father, mother, and immediate children; also called the stem family. Several nuclear families may live and work together, residing in the home belonging to the lineage system, and this family form is termed the extended family; since most culture groups trace descent in the male line the extended family most often consists of a set of grandparents, all the sons, the married sons' wives, and the children of the sons.

Ochre. An earthy and often impure yellow to red iron ore compound used in pigments and paints. Early peoples used red and yellow ochre for body painting, for decorating shrines, and for making cave drawings.

Omnivore. An animal life-form having an unrestricted dietary habit that includes both plant and animal materials. Such animals are termed omnivorous.

Organic cycle. In the most general terms, a complex, living production and distribution system that functions in continuous production, consumption, transfer, reduction, and storage sequences involving all levels of life in many different and often intricate ways.

Paleolithic. The Old Stone Age during which only collecting, hunting, and fishing economies were operative; refers to the earliest age of mankind.

Pastoralism. A mobile system employing animals as the basis for economic support in which the flocks/herds are moved about between seasonal grazing ranges; although nomadism is commonly used as part of the term, pastoralists normally are not nomadic (ceaseless wandering in an aimless fashion) but become nomadic when forced out of their home ranges. See transhumance.

Perception. The mental/psychological/emotional "image" of the surrounding world held by an individual or a group.

Perennial. When used with plants, refers to having a long life span covering many years, as opposed to the single-season annuals and the two-year biennials.

Permafrost. Perennially frozen ground in the polar margins, the thin upper layer of which may melt in summer.

Photosynthesis. The basic processes by which plant forms utilize solar energy, water, carbon dioxide, and some other elements to create organic compounds utilizable as food by other life-forms.

Physiography. The study of the landforms of the surface of the earth in their arrangement by classes, relief systems, and types.

Piedmont. An outward sloping zone running along the base of a mountain range.

Placental. Those high mammal groups that carry their young until birth in a saclike membrane that completely envelops the young and conveys nourishment to the young until birth.

Plankton. The very small and minute and weakly free-swimming plant and animal life-forms that live relatively near the surface of a body of water, usually the oceans. Plankton form the food supplies of many small kinds of fish.

Plasmodia. Various single-celled parasites found in red blood corpuscles that often cause disease; one variety produces malaria. *See vector.*

Pleistocene. The most recent major geologic epoch during which the last four glacial ages were operative; the period ended about 11,000 years ago, giving way to the Holocene (Recent) or present period.

Pluvial. A cooler and usually more humid period during which glacial ice usually accumulated during the Pleistocene.

Predator. Any life-form that lives by killing and eating other life-forms. Predators are usually carnivorous animals, but there are a few omnivorous plants that classify as predators. As applied to mankind, the term refers to any wild animals judged dangerous to human life.

Primate. A member of the high order of animals that includes the monkeys, the apes, and the human race.

Pristine earth. That natural and purely ecologically controlled wild earth that existed before the human race appeared and began to alter and modify natural conditions.

Process. As employed in cultural geography, any method or procedure of progressive action showing continuing application over time and serving to maintain a level of performance or to alter performance to some different level.

Pyrophyte. Any plant-form that is not killed by fire under ordinary circumstances and that has the ability to resume growth from rootstock after a fire.

Race. A regionally centered breeding population that perpetuates a distinctive body of genetic materials yielding a combination of physical traits that remain true from generation to generation.

Rain shadow. A dry zone created behind a mountain range when prevailing winds lose their excess moisture on the front, or windward, side of the range, leaving only drier air to blow over the zone behind the mountain range.

Reclamation. Operations carried out to make land economically usable again. A term often used loosely, it includes draining marsh, swamp, and too-wet lands; providing water for too-dry lands; and restoring once-used but deteriorated lands to usable state.

Redistribution. A socioeconomic process that accomplishes a degree of sharing of income, economic wealth, goods, and benefits among the less wealthy members of a society, community, village, or group. Is tied in to any local or regional system of social-status ranking in that donors achieve positions of status.

Reducer. Any of the various microorganisms (funguses, bacteria, molds, and others) that convert plant and animal remains into soluble organic compounds usable as energy producers by plant life.

Regolith. The mantle rock, weathered material, subsoils, and soil materials that comprise the surface skin covering the basic land surface of the earth.

Religion. An organized system of concepts concerning the powers believed to control human destiny on this earth and in some life hereafter, in which there is belief in a god, or gods, and in which there are sets of rules, usually accepted as divinely inspired, governing all human behavior in this life. *See animism and secularism.*

Resource. Any product, thing, or circumstance that is valued when its production, processing, and use is understood.

Resource conversion. Any of a set of processes that creates greater utility in something by rearranging it in some way.

Resource extraction. Any of a set of processes that secures raw materials thought of as resources from some situation in the environment.

Ritual. Any ceremonial set of acts, practices, or performances carried on by a person or members of a group and designed to express group participation; usually has social, political, religious, mystical, or magical significance.

Savanna. Grasslands interspersed with trees, having a winter dry season of some length but no cold period during the winter.

Sea and shore breeze. A localized wind circulation pattern along the shores of a large body of water, by which cool air blows landward in the early afternoon as the land warms up, and warm air blows offshore at night as the land becomes cooler than the water; also known as the land-and-sea-breeze system.

Secularism. The natural, temporal, nonreligious, and nonspiritual view toward the world and its events; the belief that ethical and moral standards should be formulated for life on this earth only, ignoring religious concepts concerning the gods and life in some hereafter. *See animism and religion.*

Sedentary. Staying in the same place, or remaining on the same site, for a long period of time, or over many years.

Seeding tube. A simple device such as a hollow pointed tube by which to place seed at a desired depth in the ground where it is to grow. The tube may be hand-held, or it may be rigged to be pulled by an animal. Early and simple devices planted single points (if hand-held) or a single row (if pulled by an animal), whereas the modern seed drill plants many rows at the same time.

Shaman. A specialist in mystical and religious powers who supposedly received his gift of powers from supernatural sources; roughly equivalent to medicine man or witch doctor.

Shatter belt. An area that has no cohesive characteristics, one in which many unlike sets of culture systems occur and in which politicoeconomic problems and sociocultural confrontations are frequent.

Shiah. That division of the Islamic faith that rejects the first four caliphs, claims Ali (son-in-law of Muhammad) as the rightful successor of Muhammad, and refuses to accept certain portions of the orthodox religious laws of Islam; the division is made up of several sectarian groupings, most of whom are distributed in Iraq, Iran, and West Pakistan.

Shintoism. The native religious system evolved in Japan out of a simple pattern of nature worship but which incorporated ancestor worship and various concepts taken from Buddhism; often held not to be a true religious system since it has relatively little moral content.

Sikhism. A development of religious doctrines in sixteenth-century India which attempted to take the best of both Hinduism and Islam and to bridge the conflicts between the two in a reform movement; persecuted by both Islam and Hinduism, the Sikhs developed a militant protective order whose members later were recruited as soldiers and police by the British and widely deployed throughout the British Empire, a policy that served to scatter Sikhs around the world.

Silent barter. A method of exchange by which commodities are placed at known points from which others may take them and leave different goods, there being no verbal or face-to-face contact between participants.

Site. The place or location where anything is, meaning the actual ground on which anything is placed or located. *See* situation.

Situation. The general location of anything relative to other things, features, or places on the land. *See* site.

Skid boat. A form of sled on which heavy objects were moved from place to place in early time, often on a mud way.

Slag. The leftover mixture of materials remaining after smelting operations have taken out the mineral sought after. Sometimes is used loosely to include barren rock removed from a mine or ashes removed from a furnace.

Slurry. A mixture of some hard, insoluble material in pebble-sized units with water to allow easier transportation of the material by pipeline or flume; coal, gravel, crushed rock, and mineral ores are the chief commodities so moved.

Smelting. A procedure by which mineral ores are placed in a furnace with some absorptive rock material and a fuel, in which firing the fuel melts the minerals into metal that then runs down into a pit at the base of the furnace. The traditional fuel was charcoal in the historic past. *See* metallurgy.

Society. A group of people living together in a region as a community and sharing the same general living system; a population living as a distinct and separate culture unit and possessing a specific culture system.

Space adjustment. A concept including all those procedures that change the location of things, that shorten the effective distance between places, or that shorten the time required to communicate between places distant from each other; it also includes the political, economic, and administrative means for organizing and integrating the territorial units of the earth.

Space intensification. A concept including all those procedures that lead to greater utilization within a given area, such as increased productivity and increased population density.

Speciation. The development of two or more separate species out of an ancestral form through the evolutionary growth of different characteristics among particular descendants of the ancestral species.

Splinter off. The process of separating a large group of people that lived together into two or more small groups that live apart after the separation; often used to refer to the founding of a new settlement by some former residents of an older settlement.

Spoliation. Acts that amount to plunder, robbery, destruction, ruination, and pillage.

Steppe. A region of semiarid climate in which only short grasses and small shrubby plants can be supported.

Steppe and sown. A form of reference to pastoralism (in the steppes and arid lands) and crop economies (in the humid lands) and to the frequent pattern of conflicts between the two sets of peoples practicing each system of economy.

Stratosphere. A zone from roughly 7 to about 30 miles above the earth within the atmospheric blanket that covers the earth; the stratosphere is located above the zone termed the troposphere but below the zone termed the ionosphere.

Sunni. The orthodox and conservative division of the Islamic faith that accepts the first four caliphs as rightful successors of Muhammad; the division is the largest within Islam but is itself divided into various sectarian groupings; the Sunnites are very widely distributed throughout the Islamic world.

Symbiosis. The mutual interdependence of two or more dissimilar life-forms, each finding some value of help from the other; the interdependence of social groups having dissimilar economies so that products necessary to each group can be secured from the other.

Synchronic. The approach by which events and conditions are examined as occurring during the same interval of time. *See* diachronic.

Synoptic. The general and comprehensive view of conditions and events during one interval of time; one of the approaches employed in climatology.

Taboo. An action or procedure that is forbidden by supernatural powers, whose performance is punishable. Sometimes spelled tabu.

Taoism. The native religious system developed in China in pre-Christian times out of native folk beliefs and gathered into a doctrinal body that accents conformity to the code of simplicity in social and political organization; the concepts gradually deteriorated and took on magical and mystical practices.

Technology. Any systematic body of theoretical knowledge and practical skills employed to do things, to perform and carry out operations of a purposeful sort with specific objectives in mind. In the modern era the term is often defined as "any practical art using scientific knowledge" since the term is often associated with applied science and the industrial world. The first definition, in a sense, fits the preindustrial world, but it remains applicable today in all societies.

Tectonic. Refers to the internal structural force for physical change that alters the crust of the earth in such matters as mountain building.

Teleological. The viewpoint and approach that considers all processes, events, and conditions to be determined by specific plan, usually considered to be ordered by an omniscient and omnipotent Providence. See determinism.

Temporal. Civil, political, wordly, or physical, as opposed to religious or supernatural.

Territoriality. A concept that holds that all members of all life-forms claim a territory or area as belonging to a particular individual member, and that groups defend their territory against other groups of the same species; seen most clearly among the birds and animals.

Theravada. The division of the Buddhist faith that remained faithful to the earlier concepts in which each person should participate in religious practice; its scriptures are written in Pali (an early Indian form of writing); sometimes called Hinayana Buddhism, or Southern Buddhism, it is chiefly practiced in Ceylon, southern India, Burma, Thailand, Laos, and Cambodia.

Thermonuclear. An explosive reaction created by the heat produced by nuclear fission.

Trade-transport cell. A region within which localized transportation and internal trade evolves among a culture group and excludes contact with outside regions.

Transhumance. The alternating seasonal movement between winter lowland pastures and summer highland pastures followed by many kinds of pastoralists; by extension used to describe any seasonal alternation of residence and/or productive activities. See pastoralism.

Travois. A simple device consisting of two parallel poles tied together a short distance apart and having a netting in the middle. The load was placed on the netting; one pair of ends rested on the ground; and the other pair was carried on human shoulders, or on the back of a dog. This was the only form of transport other than human portage known to the North American Indian populations. In the Old World the travois was sometimes used with cattle or water buffalo.

Tree farm. Land on which forest trees are continuously grown for timber by replanting seedling stock or by reseeding after each cutting of a specific tract.

Tribe. A social group living together, speaking a distinctive language, and possessing a distinctive culture that is different from that of other groups; the tribe may, or may not, be separately organized in political terms but it usually has a specific territory.

Tribute. Presentation of gifts by a subordinate to an acknowledged superior political leader.

Tropical forest. The tall and heavy, multilayered forest growth that develops in the equatorial regions that are wet and hot all year long when these areas are not disturbed for long periods.

Tundra. The treeless parts of the summerless Arctic zone in which plant life is chiefly sedges, mosses, and lichens.

Underdeveloped. A relative term for the not-yet industrialized parts of the earth. The term usually carries the economic meaning of having less than the world average in respect to per capita income, technology, amounts of inanimate energy available, kinds and amounts of manufacturing carried on, levels of living for most people, and consumption of goods. Usually what are now termed underdeveloped countries also have high birthrates, populations that are large compared with total national income, and an economy that is chiefly agricultural and that is not self-sufficient in manufacturing.

Ungulate. Any four-legged animal having feet that end in hoofs, such as the horse or deer.

Urbanism. The culture habit of preferring to live in a town or city; the preference for city life as opposed to rural life.

Urbanization. The increase in the proportion of a total population that resides in cities and towns as compared with the share that remains resident in rural locations.

Usufruct. The principle by which rights to claim, occupy, use, and produce from do not include individual ownership and the right to sell the site of production; in shifting cultivation the land belongs to the social group in common, but each member has rights to use land for gardening purposes; among pastoralists there is a similar right to the use of land for grazing animals.

Vector. As a biological term vector refers to the carrying-agent organism that transmits a disease from a host to a victim; the mosquito is the vector that transmits malaria. See plasmodia.

Vegetative propagation. A system of growing plants by placing some live part of the plant in the ground, as opposed to planting seeds; also termed vegetative reproduction.

Vertebrate. Any member of the higher animal orders having a backbone.

Whorl. A small rounded clay or stone disc with a small hole in it, used as a weight (as a flywheel) and inserted onto the slender rod used as a spindle by early textile-makers when spinning thread by hand. Handspun thread today usually is made by using a spindle.

Wild landscape. Any landscape surface of the earth in which man allows the natural physical/biotic processes to dominate over his own cultural forces and processes.

SELECTED REFERENCES

CONTENTS

SELECTED REFERENCES

Introducing Cultural Geography

Broek, Jan O. M. and John W. Webb
1973 *A Geography of Mankind* (2nd ed.). New York: McGraw-Hill. 571 pp.
Carter, George F.
1975 *Man and the Land; A Cultural Geography* (3rd ed.). New York: Holt, Rinehart and Winston. 532 pp.
Chisholm, Michael
1975 *Human Geography: Evolution or Revolution?* Harmondsworth, Eng.: Penguin Books. 204 pp.
Chorley, Richard J. (ed.)
1973 *Directions in Geography*. London: Methuen. 331 pp.
Chorley, Richard J. and Peter Haggett (eds.)
1968 *Socio-Economic Models in Geography* (University paperbacks, UP249). London: Methuen. 468 pp.
Dicken, Samuel N. and Forrest R. Pitts
1970 *Introduction to Cultural Geography: A Study of Man and His Environment*. Waltham, Mass.: Ginn 493 pp.
Dohrs, Fred E. and Lawrence M. Sommers (eds.)

1967 *Cultural Geography: Selected Readings*. New York: Crowell. 566 pp.
English, Paul W. and Robert C. Mayfield (eds.)
1972 *Man, Space, and Environment: Concepts in Contemporary Human Geography*. New York: Oxford University Press. 623 pp.
Haggett, Peter
1975 *Geography: A Modern Synthesis* (2nd ed.). New York: Harper & Row. 620 pp.
Jakle, John A. et al.
1976 *Human Spatial Behavior: A Social Geography*. North Scituate, Mass.: Duxbury Press. 315 pp.
Jordan, Terry G. and Lester Rowntree
1976 *The Human Mosaic: A Thematic Introduction to Cultural Geography*. San Francisco: Canfield Press. 430 pp.
Kariel, Herbert G. and Patricia E. Kariel
1972 *Explorations in Social Geography*. Reading, Mass.: Addison-Wesley. 398 pp.
Kolars, John F. and John D. Nystuen
1974 *Human Geography: Spatial Design in World Society*. New York:

McGraw-Hill. 281 pp.
Salter, Christopher L.
1971 *The Cultural Landscape*. Belmont, Calif: Duxbury Press. 281 pp.
Wagner, Philip L.
1972 *Environments and People*. Englewood Cliffs, N.J.: Prentice-Hall. 110 pp.
Wagner, Philip L. and Marvin W. Mikesell (eds.)
1962 *Readings in Cultural Geography*. Chicago: University of Chicago Press. 589 pp.

Humankind's Occupation of the Earth

Plate Tectonics

Alexander, T.
1975 "Revolution called plate tectonics has given us a whole new Earth," *Smithsonian*, Vol. 5 (Jan.), pp. 30–40.
Bird, J. M. and B. Isacks (eds.)
1972 *Plate Tectonics*. Washington, D.C.: American Geophysical Union. 951 pp.
Coulomb, Jean
1972 *Sea Floor Spreading and Continental Drift*. Dordrecht, Holland: Reidel.

184 pp.
Cox, Allan (ed.)
1973   *Plate Tectonics and Geomagnetic Reversals.* San Francisco: Freeman. 702 pp.
Dewey, John F.
1972   "Plate tectonics," *Scientific American,* Vol. 226, No. 5 (May), pp. 56–66.
Dietz, R. S. and John C. Holden
1970   "Breakup of Pangaea," *Scientific American,* Vol. 223, No. 3 (Oct.), pp. 30–41.
Hallam, A.
1975   "Alfred Wegener and the hypothesis of continental drift," *Scientific American,* Vol. 232, No. 2 (Feb.), pp. 88–89.
Heezen, B. C. and I. D. MacGregor
1973   "The evolution of the Pacific," *Scientific American,* Vol. 229, No. 5 (Nov.), pp. 102–112.
Molnar, Peter and Paul Tapponnier
1975   "Cenozoic tectonics of Asia: effects of a continental collision," *Science,* Vol. 189, No. 4201 (August 8), pp. 419–426.
Rona, P. A.
1973   "Plate tectonics and mineral resources," *Scientific American,* Vol. 229, No. 1 (July), pp. 86–95.
Sullivan, Walter
1974   *Continents in Motion: The New Earth Debate.* New York: McGraw-Hill. 399 pp.
Wilson, J. Tuzo
1975   "Earth Sciences" and "Continental Drift," *Encyclopaedia Britannica-Book of the Year 1975,* pp. 244–245.

The Pleistocene Epoch

Butzer, Karl W.
1971   *Environment and Archeology: An Ecological Approach to Prehistory* (2nd ed.). Chicago: Aldine-Atherton. 703 pp.
Flint, Richard Foster
1971   *Glacial and Quaternary Geology.* New York: Wiley. 892 pp.
Frenzel, Burkhard
1973   *Climatic Fluctuations of the Ice Age.* Cleveland, Ohio: The Press of Case Western Reserve University. 306 pp.
Kurtén, Björn
1972   *The Ice Age.* London: Hart-Davis. 178 pp.
Sparks, B. W. and R. G. West
1972   *The Ice Age in Britain.* London: Methuen. 302 pp.
Turekian, Karl K. (ed.)
1971   *The Late Cenozoic Glacial Ages.* New Haven: Yale University Press. 606 pp.
Wright, M. E., Jr. and David C. Frey (eds.)
1965   *The Quaternary of the United States.* Princeton: Princeton University Press. 922 pp.
Zeuner, Friedrich E.
1959   *The Pleistocene Period: Its Climate, Chronology, and Faunal Successions.* London: Hutchinson. 447 pp.

Plant and Animal Evolution and Distribution

Birks, H. J. B. and R. G. West (eds.)
1973   *Quaternary Plant Ecology.* New York: Wiley. 326 pp.
Colbert, Edwin H.
1969   *Evolution of the Vertebrates* (2nd ed.). New York: Wiley. 535 pp.
1973   *Wandering Lands and Animals.* New York: Dutton. 323 pp.
Darlington, C. D.
1973   *Chromosome Botany and the Origin of Cultivated Plants* (3rd ed.). London: Allen & Unwin. 237 pp.
Darlington, Philip J., Jr.
1957   *Zoogeography: The Geographical Distribution of Animals.* New York: Wiley. 675 pp.
Epstein, H.
1971   *The Origin of the Domestic Animals of Africa.* New York: Africana. 2 vols.
Gilbert, Lawrence E. and Peter H. Raven (eds.)
1975   *Co-Evolution of Animals and Plants.* Austin: University of Texas Press. 246 pp.
Good, Ronald
1974   *The Geography of the Flowering Plants* (4th ed.). London: Longman. 557 pp.
Isaac, Erich
1970   *Geography of Domestication.* Englewood Cliffs, N.J.: Prentice-Hall. 132 pp.
Keast, Allen, Frank C. Erk, and Bentley Glass (eds.)
1972   *Evolution, Mammals, and Southern Continents.* Albany, N.Y.: State University of New York Press. 543 pp.
Kurtén, Björn
1968   *Pleistocene Mammals of Europe.* Chicago: Aldine. 317 pp.
Martin, Paul S. and H. E. Wright, Jr. (eds.)
1967   *Pleistocene Extinctions: The Search for a Cause* (Proceedings of the VIIth Congress of International Association for Quaternary Research, Vol. VI). New Haven: Yale University Press. 453 pp.
Simpson, George Gaylord
1967   *The Meaning of Evolution: A Study of the History of Life and of Its Significance for Man* (2nd ed.). New Haven: Yale University Press. 368 pp.
Ucko, Peter J., and G. W. Dimbleby (eds.)
1969   *The Domestication and Exploitation of Plants and Animals.* London: Duckworth and Chicago: Aldine. 581 pp.

Primates

Clark, W. E. LeGros
1962   *The Antecedents of Man, An Introduction to the Evolution of the Primates* (2nd ed.). Edinburgh: University Press. 388 pp.
Eimerl, Sarel, Irven De Vore, and the Editors of *Life*

1965   *The Primates* (Life Nature Library). New York: Time. 200 pp.
Hulse, Frederick S.
1971   *The Human Species, An Introduction to Physical Anthropology* (2nd ed.) New York: Random House. 524 pp.

Prehistoric Peoples and Human Evolution

Berrill, Norman J.
1955   *Man's Emerging Mind.* New York: Dodd, Mead. 308 pp.
Bishop, Walter W. and J. A. Miller (eds.)
1972   *Calibration of Hominoid Evolution.* New York, London: Chatto & Windus. 487 pp.
Bordes, François
1968   *The Old Stone Age.* New York: McGraw-Hill. 255 pp.
Brace, C. L. et al.
1971   *Atlas of Fossil Man.* New York: Holt, Rinehart and Winston. 150 pp.
Buettner-Janusch, John
1973   *Physical Anthropology: A Perspective.* New York: Wiley. 572 pp.
Chard, Chester
1975   *Man in Prehistory* (2nd ed.). New York: McGraw-Hill. 406 pp.
Clark, J. Desmond
1970   *The Prehistory of Africa* (Ancient Peoples and Places). New York: Praeger. 302 pp.
Clark, J. Desmond and Eve Kemnitzer
1967   *Atlas of African Prehistory.* Chicago: University of Chicago Press. 12 maps, 32 overlays, 62 text pp.
Clark, John Grahame Douglas
1969   *World Prehistory: A New Outline* (2nd ed.). Cambridge, Eng.: University Press. 331 pp.
1970   *Aspects of Prehistory.* Berkeley: University of California Press. 161 pp.
Darlington, Cyril Dean
1969   *The Evolution of Man and Society.* London: Allen & Unwin; New York: Simon & Schuster. 753 pp.
Dubos, René J.
1965   *Man Adapting.* New Haven, Conn.: Yale University Press. 527 pp.
1972   *A God Within.* New York: Scribner's 325 pp.
Fried, Morton H.
1967   *The Evolution of Political Society: An Essay in Political Anthropology.* New York: Random House. 270 pp.
Goode, Ruth
1973   *People of the Ice Age.* New York: Crowell-Collier. 151 pp.
Howell, F. Clark and the Editors of *Life*
1965   *Early Man* (Life Nature Library). New York: Time, 200 pp.
Howells, W. W.
1973   *Evolution of the Genus Homo.* Reading, Mass.: Addison-Wesley. 188 pp.
Jacobsen, Thomas W.
1976   "17,000 years of Greek prehistory," *Scientific American,* Vol. 234, No. 6 (June), pp. 76–87.

Jennings, Jesse D.
1974 *Prehistory of North America* (2nd ed.). New York: McGraw-Hill. 436 pp.

Kennedy, Kenneth A. R.
1975 *Neandertal Man*. Minneapolis: Burgess. 106 pp.

Leakey, L. S. B. and Vanne Morris Goodall
1969 *Unveiling Man's Origins: Ten Decades of Thought About Human Evolution*. Cambridge, Mass: Schenkman 220 pp.

Lee, Richard B. and Irven DeVore (eds.)
1968 *Man the Hunter*. Chicago: Aldine. 415 pp.

Lorenz, Konrad
1966 *On Aggression*. New York: Harcourt, Brace & World. 306 pp.

Marshack, Alexander
1972 *The Roots of Civilization: The Cognitive Beginnings of Man's First Art, Symbol, and Notation*. New York: McGraw-Hill. 413 pp.

Mulvaney, D. J.
1969 *The Prehistory of Australia*. London: Thames and Hudson. 276 pp.

Mulvaney, D. J. and J. Golson (eds.)
1971 *Aboriginal Man and Environment in Australia*. Canberra: Australian National University Press. 389 pp.

Munn, Norman Leslie
1971 *The Evolution of the Human Mind*. Boston: Houghton Mifflin. 278 pp.

Neill, Wilfred
1969 *The Geography of Life*. New York: Columbia University Press. 480 pp.

Oswalt, Wendell H.
1973 *Habitat and Ecology: The Evolution of Hunting*. New York: Holt, Rinehart and Winston. 191 pp.

Pfeiffer, John E.
1969 *The Emergence of Man*. New York: Harper & Row. 477 pp.

Sahlins, Marshall
1968 *Tribesmen*. Englewood Cliffs: Prentice-Hall. 118 pp.

Sauer, Carl O.
1971 "Plants, animals, and man," pp. 34–61 in R. H. Buchanan et al. (eds.) *Man and his Habitat: Essays presented to Emyr Estyn Evans*. London: Routledge & Kegan Paul. 279 pp.

Service, Elman R.
1975 *Origins of the State and Civilization: The Process of Cultural Evolution*. New York: Norton. 361 pp.

Turney-High, Harry Holbert
1971 *Primitive War, Its Practices and Concepts* (2nd ed.). Columbia: University of South Carolina Press. 288 pp.

Ucko, Peter J. et al. (eds.)
1972 *Man, Settlement, and Urbanism*. London: Duckworth. 979 pp.

Washburn, Sherwood L. and P. J. Dolhinow
1972 *Perspectives on Human Evolution*. New York: Holt, Rinehart and Winston. 2 vols.

Differential Cultural Development

Adams, Robert McC.
1966 *The Evolution of Urban Society; Early Mesopotamia and Prehistoric Mexico*. Chicago: Aldine. 191 pp.

Ardrey, Robert
1966 *The Territorial Imperative: A Personal Inquiry into the Animal Origins of Property and Nations*. New York: Atheneum. 390 pp.
1970 *The Social Contract: A Personal Inquiry into the Evolutionary Source of Order and Disorder*. New York: Atheneum. 405 pp.

Bronowski, Jacob
1974 *The Ascent of Man*. Boston: Little, Brown. 448 pp.

Brothwell, Don, and Patricia Brothwell
1969 *Food In Antiquity: A Survey of the Diet of Early Peoples*. London: Thames and Hudson. 248 pp.

Brothwell, Don and A. T. Sandison (eds.)
1967 *Diseases in Antiquity: A Survey of the Diseases, Injuries, and Surgery of Early Populations*. Springfield, Ill.: Thomas. 766 pp.

Casson, Lionel
1971 *Ships and Seamanship in the Ancient World*. Princeton: Princeton University Press. 441 pp.

Duckham, A. N. and G. B. Masefield
1970 *Farming Systems of the World*. New York: Praeger. 542 pp.

Gailey, Alan and Alexander Fenton (eds.)
1970 *The Spade in Northern and Atlantic Europe*. Belfast: Ulster Folk Museum, Institute of Irish Studies, Queen's University. 257 pp.

Hadingham, Evan
1975 *Circles and Standing Stones: An Illustrated Exploration of Megalithic Mysteries of Early Britain*. London: Heinemann. 240 pp.

Harris, David R.
1972 "The origins of agriculture in the tropics," *American Scientist*, Vol. 60, pp. 180–193.

Heiser, Charles B.
1973 *Seed to Civilization: The Story of Man's Food*. San Francisco: Freeman. 243 pp.

Higham, Charles
1974 *The Earliest Farmers and the First Cities*. London: Cambridge University Press. 48 pp.

Hutchinson, Joseph (ed.)
1974 *Evolutionary Studies in World Crops: Diversity and Change in the Indian Sub-Continent*. Cambridge: Cambridge University Press. 175 pp.

Hyams, Edward
1971 *Plants in the Service of Man: 10,000 Years of Domestication*. Philadelphia: Lippincott. 222 pp.

Johannesson, Carl L. et al.
1970 "The domestication of maize: process or event?" *Geographical Review*, Vol. 60, No. 3 (July), pp. 393–413.

Johnson, Douglas L.
1969 *The Nature of Nomadism: A Comparative Study of Pastoral Migrations in Southwestern Asia and Northern Africa* (Research Paper No. 118). Chicago: Department of Geography, University of Chicago. 200 pp.

Krader, Lawrence
1968 *Formation of the State*. Englewood Cliffs, N.J.: Prentice-Hall, Inc. 118 pp.

Kroeber, Alfred L.
1962 *A Roster of Civilizations and Culture*. Chicago: Aldine. 96 pp.
1963 *An Anthropologist Looks at History*. Berkeley and Los Angeles: University of California Press. 213 pp.

Melko, Matthew
1969 *The Nature of Civilizations*. Boston: Sargent. 204 pp.

Rapoport, Amos
1969 *House Form and Culture*. Englewood Cliffs: Prentice-Hall. 160 pp.

Renfrew, Colin (ed.)
1973 *The Explanation of Culture Change: Models in Prehistory*. London: Duckworth; Pittsburgh: University of Pittsburgh Press. 738 pp.

Riley, Carroll L. (ed.)
1969 *The Origin of Civilization*. Carbondale: Southern Illinois University Press. 243 pp.

Sahlins, Marshall
1972 *Stone Age Economics*. Chicago: Aldine. 348 pp.

Sauer, Carl O.
1952 *Agriculture Origins and Dispersals: The Domestication of Animals and Foodstuffs*. New York: American Geographical Society. 110 pp. (Also available as *Seeds, Spades, Hearths, and Herds*. Cambridge, Mass.: MIT Press. 1969)

Service, Elman R.
1975 *Origins of the State and Civilization: The Process of Cultural Evolution*. New York: Norton. 361 pp.

Simoons, Frederick J.
1968 *A Ceremonial Ox of India: The Mithan in Nature, Culture, and History*. Madison: University of Wisconsin Press. 323 pp.

Singer, Charles, E. J. Holmyard, and A. R. Hall (eds.)
1954–1956 *A History of Technology*. London: Oxford University Press. Vol. 1, 827 pp.; Vol. 2, 802 pp.

Sopher, David E.
1967 *Geography of Religions*. Englewood Cliffs, N.J.: Prentice-Hall. 118 pp.

Spencer, J. E.
1966 *Shifting Cultivation in Southeastern Asia*. Berkeley and Los Angeles: University of California, Publications in Geography, Vol. 19, 256 pp.

Thomas, William L. Jr. (ed.)
1956 *Man's Role in Changing the Face of the Earth*. Chicago: University of Chicago Press. 1193 pp.

Ucko, Peter J., Ruth Tringham, and G. W. Dimbleby (eds.)
1972 *Man, Settlement, and Urbanism*. Cambridge, Mass.: Schenkman. 979 pp.

Waterbolk, H. T.
1968 "Food production in prehistoric Europe," *Science*, Vol. 162, No. 3858 (Dec. 6), pp. 1093–1102.
White, Lynn
1962 *Medieval Technology and Social Change*. Oxford: Clarendon Press. 194 pp.

Regional Expansion and Contacts

General

Armytage, W. H. G.
1961 *A Social History of Engineering*. London: Faber & Faber. 378 pp.
Belshaw, Cyril S.
1965 *Traditional Exchange and Modern Markets*. Englewood Cliffs, N.J.: Prentice-Hall. 149 pp.
Daumas, Maurice (ed.)
1969 *A History of Technology and Invention*. New York: Crown. 2 vols.
Hornell, James
1970 *Water Transport, Origins and Early Evolution*. London: David & Charles. 307 pp.
Landstrom, Björn
1967 *The Ship, An Illustrated History*. Garden City, N.Y.: Doubleday. 318 pp.
Nash, Manning
1966 *Primitive and Peasant Economic Systems*. San Francisco: Chandler. 166 pp.
Rostow, W. W.
1975 *How It All Began: Origins of the Modern Economy*. New York: McGraw-Hill. 264 pp.
Starr, Chester G. et al.
1974 *A History of the Ancient World* (2nd ed.). Oxford: Oxford University Press. 742 pp.
Wolf, E. R.
1966 *Peasants*. Englewood Cliffs, N.J.: Prentice-Hall. 116 pp.

Mesopotamia

Adams, Robert McC.
1965 *Land Behind Baghdad*. Chicago: University of Chicago Press. 187 pp.
Fairservis, Walter A.
1964 *Mesopotamia, The Civilization that Arose out of Clay*. New York: Macmillan. 126 pp.
Kramer, Samuel N.
1969 *Cradle of Civilization* (Time-Life Books). New York: Time. 183 pp.
Margueron, Jean Claude
1965 *Mesopotamia*. Cleveland: World. 211 pp.
Oppenheim, A. L.
1964 *Ancient Mesopotamia*. Chicago: University of Chicago Press. 433 pp.
Roux, Georges
1964 *Ancient Iraq*. London: George Allen & Unwin. 431 pp.

Egypt

Aldred, Cyril

1965 *Egypt to the End of the Old Kingdom*. London: Thames and Hudson. 143 pp.
Casson, Lionel
1965 *Ancient Egypt* (Time-Life Books). New York: Time. 192 pp.
Fairservis, Walter A.
1963 *Egypt, Gift of the Nile*. New York: Macmillan. 146 pp.
Harris, John Richard (ed.)
1971 *The Legacy of Egypt* (2nd ed.). Oxford: Clarendon Press. 510 pp.
Hayes, William C.
1965 *Most Ancient Egypt*. Chicago: University of Chicago Press. 160 pp.
White, J. E. M.
1970 *Ancient Egypt: Its Culture and History* (2nd ed.). London: Allen & Unwin. 217 pp.

India

Basham, A. L.
1968 *The Wonder that was India* (3rd ed.). New York: Taplinger. 572 pp.
Coedes, Georges
1966 *The Making of South East Asia*. Berkeley: University of California Press. 268 pp.
1967 *The Indianized States of Southeast Asia*. Honolulu: East-West Center Press. 403 pp.
Fairservis, Walter A.
1971 *The Roots of Ancient India: The Archaeology of Early Indian Civilization*. New York: Macmillan. 482 pp.
Majumdar, R. C.
1968 *Ancient India* (5th ed.). Delhi: Mutilal Banarsidass. 538 pp.
Malik, S. C.
1968 *Indian Civilization: The Formative Period*. Simla: Indian Institute of Advanced Study. 204 pp.
Piggott, Stuart
1962 *Prehistoric India to 1000 B.C.* London: Cassell. 294 pp.
Wheeler, Mortimer
1968 *Early India and Pakistan: To Ashoka* (rev. ed.). London: Thames and Hudson. 241 pp.

China

Boulnois, Luce
1966 *The Silk Road*. New York: Dutton. 250 pp.
Chang Kwang-chih
1968 *The Archaeology of Ancient China* (rev. ed.). New Haven, Conn.: Yale University Press. 483 pp.
Eberhard, Wolfram
1968 *The Local Cultures of South and East China*. Leiden: Brill. 520 pp.
Hay, John
1973 *Ancient China*. London: Bodley Head. 128 pp.
Needham, Joseph
1954–1976 *Science and Civilization in China*. Cambridge, England: Cambridge University Press. To be completed in 7 volumes, with Vol. 5, Part 3

having appeared in 1976.
Schafer, Edward H.
1967 *Ancient China* (Time-Life Books). New York: Time. 191 pp.
Stover, Leon E.
1974 *The Cultural Ecology of Chinese Civilization: Peasants and Elites in the Last of the Agrarian States*. New York: Pica Press. 305 pp.
Treistman, Judith M.
1972 *The Prehistory of China: An Archeological Exploration*. New York: Natural History Press. 156 pp.
Tuan Yi-fu
1970 *China* (The Making of the World's Landscapes). London: Longmans. 225 pp.
Watson, William
1966 *Early Civilization in China*. New York: McGraw-Hill. 143 pp.
Wheatley, Paul
1971 *The Pivot of the Four Quarters: A Preliminary Inquiry into the Origins and Character of the Ancient Chinese City*. Chicago: Aldine. 602 pp.

The New World

Driver, Harold E.
1969 *Indians of North America* (2nd ed.). Chicago: University of Chicago Press. 632 pp.
Gorenstein, Shirley et al.
1974 *Prehispanic America*. New York: St. Martin's Press. 165 pp.
Jennings, Jesse D.
1974 *Prehistory of North America* (2nd ed.). New York: McGraw-Hill. 436 pp.
Josephy, Alvin M. Jr.
1968 *The Indian Heritage of America*. New York: Knopf. 384 pp.
Lanning, Edward P.
1967 *Peru Before the Incas*. Englewood Cliffs, N.J.: Prentice-Hall. 216 pp.
Riley, Carroll L. et al. (eds.)
1971 *Man Across the Sea: Problems of Pre-Columbian Contacts*. Austin: University of Texas Press. 552 pp.
Sanders, William T. and Barbara J. Price
1968 *Mesoamerica: The Evolution of a Civilization*. New York: Random House. 264 pp.
von Hagen, Victor Wolfgang
1961 *The Ancient Sun Kingdoms of the Americas: Aztec, Maya, Inca*. Cleveland: World. 617 pp.
Willey, Gordon R.
1966 *An Introduction to American Archaeology*. Vol. 1: *North and Middle America*. Vol. 2: *South America*. New York: Prentice-Hall. 2 vols.
Willey, Gordon R. and Jeremy A. Sabloff
1974 *A History of American Archaeology*. San Francisco: Freeman. 252 pp.

Africa

Bohannon, Paul and Philip Curtin
1971 *Africa and Africans* (rev. ed.). Garden City, N.Y., Natural History Press. 391 pp.

Davidson, Basil
1966 *African Kingdoms* (Time-Life Books). New York: Time. 192 pp.

Gann, Lewis H. and Peter Duignan
1972 *Africa and the World: A History of Sub-Saharan Africa from Antiquity to 1840.* San Francisco: Chandler. 522 pp.

Hiernaux, Jean
1974 *The People of Africa.* London: Weidenfeld & Nicolson. 217 pp.

McEwan, P. J. M. (ed.)
1968 *Africa from Early Times to 1800.* London: Oxford University Press. 436 pp.

Mitchison, Naomi
1970 *The Africans.* London: Blond. 232 pp.

Murdock, George Peter
1959 *Africa, Its Peoples and Their Culture History.* New York: McGraw-Hill. 456 pp.

Paden, John N. and Edward W. Soja (eds.)
1970 *The African Experience.* Evanston: Northwestern University Press. 2 vols.

Shinnie, Margaret
1971 *The African Iron Age.* Oxford: Clarendon Press. 281 pp.

Welch, Galbraith
1965 *Africa Before They Came: The Continent North, South, East, and West Preceding the Colonial Powers.* New York: Morrow. 396 pp.

Islam

Dunlop, D. M.
1971 *Arab Civilization to A. D. 1500.* New York: Praeger. 368 pp.

Gibb, H. A. R.
1962 *Studies on the Civilization of Islam.* London: Routledge & Kegan Paul. 369 pp.

Hitti, Philip K.
1968 *The Arabs: A Short History* (5th ed.). New York: St. Martin's Press. 211 pp.

Hottinger, A.
1963 *The Arabs, Their History, Culture and Place in the Modern World.* Berkeley: University of California Press. 344 pp.

Planhol, Xavier de
1959 *The World of Islam.* Ithaca, N.Y.: Cornell University Press (English edition). 142 pp.

Roolvink, R. et al. (comps.)
1957 *Historical Atlas of the Muslim Peoples.* Cambridge, Mass.: Harvard University Press. 40 pp.

Watt, W. M.
1974 *The Majesty that was Islam: The Islamic World, 661–1100.* London: Sidgwick & Jackson. 424 pp.

Mongols

Charol, Michael
1967 *The Mongol Empire: Its Rise and Legacy.* New York: Free Press. 581 pp.

Grousset, René
1966 *Conqueror of the World. The Life of Chingis-Khan.* New York: Orion Press. 300 pp.

1970 *The Empire of the Steppes: A History of Central Asia.* New Brunswick, N.J.: Rutgers University Press. 687 pp.

Phillips, Eustace D.
1969 *The Mongols.* London: Thames and Hudson. 208 pp.

Saunders, John J.
1971 *The History of the Mongol Conquests.* London: Routledge & Kegan Paul. 275 pp.

Spuler, Bertold
1972 *History of the Mongols.* Berkeley: University of California Press. 221 pp.

Europe

Duby, Georges (trans. Howard B. Clarke)
1974 *The Early Growth of the European Economy: Warriors and Peasants from the Seventh to the Twelfth Centuries.* London: Weidenfeld & Nicolson; Ithaca: Cornell University Press. 292 pp.

East, W. Gordon
1966 *An Historical Geography of Europe* (5th ed.). London: Methuen. 492 pp.

Franklin, S. H.
1969 *The European Peasantry: The Final Phase.* London: Methuen. 256 pp.

Glacken, Clarence J.
1967 *Traces on the Rhodian Shore: Nature and Culture in Western Thought from Ancient Times to the End of the Eighteenth Century.* Berkeley and Los Angeles: University of California Press. 763 pp.

Hale, John R.
1965 *Renaissance* (Time-Life Books). New York: Time. 192 pp.

Lambert, Audrey M.
1971 *The Making of the Dutch Landscape: An Historical Geography of the Netherlands.* London: Seminar Press. 412 pp.

Lewis, A. R.
1958 *The Northern Seas: Shipping and Commerce in Northern Europe, A.D. 300–1100.* Princeton, N.J.: Princeton University Press. 498 pp.

1970 *The Islamic World and the West, A.D. 622–1492.* New York: Wiley. 146 pp.

Renfrew, Colin
1972 *The Emergence of Civilization: The Cyclades and the Aegean in the Third Millennium B.C.* London: Methuen. 595 pp.

Timmer, C. Peter
1969 "The turnip, the new husbandry, and the English agricultural revolution," *Quarterly Journal of Economics,* Vol. 83, No. 3 (August), pp. 375–395.

Europe's Expansion Overseas

Bell, Christopher
1974 *Portugal and the Quest for the Indies.* New York: Barnes & Noble. 247 pp.

Boxer, Charles R.
1965 *The Dutch Seaborne Empire, 1600–1800.* New York: Knopf. 326 pp.

1967 *The Christian Century in Japan, 1549–1650.* Berkeley: University of California Press. 535 pp.

1969a *The Portuguese Seaborne Empire, 1415–1825.* New York: Knopf. 315 pp.

1969b *Four Centuries of Portuguese Expansion, 1415–1825; A Succinct Survey.* Berkeley: University of California Press. 102 pp.

Chamberlain, Muriel E.
1974 *The Scramble for Africa.* London: Longman. 163 pp.

Crone, G. R.
1969 *The Discovery of America.* New York: Weybright and Talley. 224 pp.

1972 *The Discovery of the East.* London: Hamilton. 178 pp.

Cumming, W. P. et al.
1974 *The Exploration of North America, 1630–1776.* New York: Putnam's. 272 pp.

Davies, Kenneth G.
1974 *The North Atlantic World in the Seventeenth Century.* Minneapolis: University of Minnesota Press. 366 pp.

Dunmore, John
1965–1969 *French Explorers in the Pacific.* Oxford: Clarendon Press. 2 vols.

Easton, Stewart C.
1964 *The Rise and Fall of Western Colonialism.* New York: Praeger. 402 pp.

Fieldhouse, D. K.
1966 *The Colonial Empires: A Comparative Survey from the Eighteenth Century.* New York: Delacorte Press. 450 pp.

Friis, Herman R. (ed.)
1967 *The Pacific Basin: A History of Its Geographical Exploration.* New York: American Geographical Society. 457 pp.

Gann, Lewis H. and Peter Duignan (eds.)
1969–1973 *Colonialism in Africa, 1870–1960.* London: Cambridge University Press. 4 vols.

Hale, John R.
1966 *Age of Exploration.* New York: Time. 192 pp.

Lach, D. F.
1965 *Asia in the Making of Europe, Vol. 1, The Century of Discovery.* Chicago: University of Chicago Press. 1040 pp.

1970 *Asia in the Making of Europe, Vol. 2, A Century of Wonder.* Chicago: University of Chicago Press.

Lach, D. F., and C. Floumenhaft (eds.)
1965 *Asia on the Eve of Europe's Expansion.* New York: Prentice-Hall. 211 pp.

McDermott, J. F. (ed.)
1974 *The Spanish in the Mississippi Valley, 1762–1804.* Urbana: University of Illinois Press. 421 pp.

Morison, Samuel Eliot
1971 *The European Discovery of America: The Northern Voyages, A.D. 500–1600.* New York: Oxford University Press. 712 pp.

1974 *The European Discovery of America: The Southern Voyages, 1492–1616.*

New York: Oxford University Press. 758 pp.

Osei, G. K.
1968 *Europe's Gift to Africa.* London: African Publication Society. 120 pp.

Parry, J. H.
1961 *The Establishment of the European Hegemony: 1415–1715, Trade and Exploration in the Age of the Renaissance.* New York: Harper & Row. 202 pp.
1967 *The Spanish Seaborne Empire* (2nd ed.). London: Hutchinson. 416 pp.
1971 *Trade and Dominion: The European Overseas Empires in the Eighteenth Century.* London: Weidenfeld & Nicolson. 409 pp.
1974 *The Discovery of the Sea.* New York: Dial Press. 302 pp.

Parsons, James J.
1968 *Antioqueño Colonization in Western Colombia* (rev. ed.). Berkeley: University of California Press. 233 pp.

Penrose, B.
1952 *Travel and Discovery in the Renaissance, 1420–1620.* Cambridge, Mass.: Harvard University Press (reprinted as an Atheneum paperback, 1962). 463 pp.

Rotberg, Robert I. (ed.)
1970 *Africa and Its Explorers: Motives, Methods, and Impact.* Cambridge, Mass.: Harvard University Press. 351 pp.

Sauer, Carl O.
1966 *The Early Spanish Main.* Berkeley and Los Angeles: University of California Press. 306 pp.
1968 *Northern Mists.* Berkeley and Los Angeles: University of California Press. 204 pp.
1971 *Sixteenth-Century North America: The Land and People as Seen by Europeans.* Berkeley: University of California Press. 319 pp.

Severin, Timothy
1973 *The African Adventure: A History of Africa's Explorers.* London: Hamilton. 288 pp.

Uzoigwe, G. N.
1974 *Britain and the Conquest of Africa: The Age of Salisbury.* Ann Arbor: University of Michigan Press. 403 pp.

Wright, Louis B.
1970 *Gold, Glory, and Gospel: The Adventurous Lives and Times of the Renaissance Explorers.* New York: Atheneum. 362 pp.

Wright, Louis B. and Elaine W. Fowler (comps.)
1971 *West and by North: North America seen through the Eyes of Its Seafaring Discoverers.* New York: Delacorte Press. 389 pp.
1972 *The Moving Frontier: North America seen through the Eyes of Its Pioneer Discoverers.* New York: Delacorte Press. 348 pp.

Recent Convergence of Humanity

Anonymous
1968 *Treaties and Alliances of the World: An International Survey covering Treaties in Force and Communities of States.* New York: Scribner's. 214 pp.

Bock, Philip K. (ed.)
1969 *Peasants in the Modern World.* Albuquerque: University of New Mexico Press. 173 pp.

Boulding, Kenneth E.
1964 *The Meaning of the Twentieth Century: The Great Transition.* New York: Harper & Row. 199 pp.

Brockington, Colin F.
1975 *World Health* (3rd ed.). New York: Longman. 345 pp.

Brunner, Ronald D. and Gerry D. Brewer
1971 *Organized Complexity. Empirical Theories of Political Development.* New York: Free Press. 190 pp.

Cartwright, Frederick F. and M. D. Biddiss
1972 *Disease and History.* New York: Crowell. 247 pp.

Claude, Inis L.
1971 *Swords into Plowshares: The Problems and Progress of International Organization.* New York: Random House. 514 pp.

Cobb, Roger W. and Charles Elder
1970 *International Community: A Regional and Global Study.* New York: Holt, Rinehart and Winston. 160 pp.

Coon, Carleton S.
1971 *The Hunting Peoples.* Boston: Little, Brown. 413 pp.

Crosby, Alfred W.
1973 *The Columbian Exchange: Biological and Cultural Consequences of 1492.* Westport, Conn.: Greenwood Press. 268 pp.

Desmonde, William H.
1962 *Magic, Myth, and Money: The Origin of Money in Religious Ritual.* New York: Free Press. 208 pp.

Dirks, Lee E.
1969 *The Ecumenical Movement.* New York: Public Affairs Committee. 28 pp.

Fischer, Eric
1948 *The Passing of the European Age: A Study of the Transfer of Civilization and Its Renewal in Other Continents* (rev. ed.). Cambridge, Mass.: Harvard University Press. 228 pp.

Frazier, E. Franklin
1957 *Race and Culture Contacts in the Modern World.* New York: Knopf. 338 pp.

Goldman, Marshall I.
1970 "The convergence of environmental disruptions," *Science,* Vol. 170, No. 3953, (Oct. 2), pp. 37–42.

Goodman, Elliot R.
1975 *The Fate of the Atlantic Community.* New York: Praeger. 583 pp.

Haskell, P. T.
1971 "International locust research and control," *Journal of the Royal Society of Arts,* Vol. 119, No. 5176 (March), pp. 249–263.

Hilton, Ronald (ed.)
1969 *The Movement Toward Latin American Unity.* New York: Praeger. 561 pp.

Honnold, John (ed.)
1964 *The Life of the Law: Readings on the Growth of Legal Institutions.* Glencoe, Ill.: Free Press. 581 pp.

Irving, Keith
1970 *The Rise of the Colored Races.* New York: Norton. 646 pp.

Mathis, F. John and Walter Krause
1970 *Latin America and Economic Integration: Regional Planning for Development.* Iowa City: University of Iowa Press. 105 pp.

National Geographic Society
1968 *Vanishing Peoples of the Earth* (Special Publication). Washington, D.C.: National Geographic Society. 207 pp.
1971 *Great Religions of the World* (Special Publication). Washington, D.C.: National Geographic Society. 420 pp.

Nettleship, Martin A. et al. (eds.)
1975 *War: Its Causes and Correlates.* Chicago: Aldine. 813 pp.

Noss, John B.
1969 *Man's Religions* (4th ed.). New York: Macmillan. 598 pp.

Pachter, Henry M.
1975 *The Fall and Rise of Europe: A Political, Social, and Cultural History of the Twentieth Century.* New York: Praeger. 481 pp.

Patton, Clyde P.
1969 "The origins and diffusion of the European universities," *Yearbook,* Association of Pacific Coast Geographers, Vol. 31, pp. 7–26.

Pei, Mario A.
1965 *The Story of Language* (rev. ed.). Philadelphia. Lippincott. 491 pp.

Price, A. Grenfell
1963 *The Western Invasions of the Pacific and Its Continents. A Study of Moving Frontiers and Changing Landscapes. 1513–1958.* Oxford: Clarendon Press. 236 pp.

Pye, Lucien W. (ed.)
1963 *Communications and Political Development.* Princeton: Princeton University Press. 381 pp.

Ribeiro, Darcy (Transl., and with a foreword by Betty J. Meggers)
1968 *The Civilization Process.* Washington, D.C.: Smithsonian Institution Press. 201 pp.

Rodnick, David
1966 *An Introduction to Man and His Development.* New York: Appleton-Century-Crofts, 520 pp.

Said, Abdul Aziz and Luiz R. Simmons
1975 *The New Sovereigns: Multinational Corporations as World Powers.* Englewood Cliffs, N.J.: Prentice-Hall. 186 pp.

Sopher, David E.
1967 *Geography of Religions.* Engle-

wood Cliffs, N.J.: Prentice-Hall. 118 pp.

Sopher, David E. and Ismail Raqi A. al Faruqi
1974 *Historical Atlas of the Religions of the World.* New York: Macmillan. 346 pp.

Swadesh, Morris
1971 *The Origin and Diversification of Language.* Chicago: Aldine-Atherton. 350 pp.

Thut, I. N., and Don Adams
1964 *Educational Patterns in Contemporary Societies.* New York: McGraw-Hill. 494 pp.

Wagar, W. Warren
1963 *The City of Man: Prophecies of a World Civilization in Twentieth Century Thought.* Boston. Houghton Mifflin. 310 pp.
1971 *Building the City of Man: Outlines of a World Civilization.* New York: Grossman. 180 pp.

Wallerstein, I. M.
1974 *The Modern World-System: Capitalist Agriculture and the Origins of the European World-Economy in the Sixteenth Century.* New York: Academic Press. 410 pp.

Wilkins, Mira
1974 *The Maturing of Multinational Enterprise: American Business Abroad from 1914 to 1970.* Cambridge, Mass.: Harvard University Press. 590 pp.

The Population Spiral

Beaver, Steven E.
1975 *Demographic Transition Theory Reinterpreted: An Application to Recent Natality Trends in Latin America.* Lexington, Mass.: Lexington Books. 177 pp.

Beyer, G.
1970 "International and national migratory movements," *International Migration,* Vol. 8, No. 3, pp. 93–109.

Brown, Lester R.
1974 *In the Human Interest: A Strategy to Stabilize World Population.* New York: Norton. 190 pp.

Cipolla, Carlo
1974 *The Economic History of World Population* (6th ed.) Harmondsworth Eng.: Penguin Books. 154 pp.

Clarke, John I.
1972 *Population Geography* (2nd ed.) Oxford: Pergamon Press. 176 pp.

Cragg, J. B. (ed.)
1955 *The Numbers of Man and Animals.* Edinburgh: Oliver & Boyd. 152 pp.

Demko, George J. et al. (eds.)
1970 *Population Geography: A Reader.* New York: McGraw-Hill. 526 pp.

Desmond, Annabelle
1962 "How many people have ever lived on Earth?" *Population Bulletin,* Vol. 18, No. 1 (Feb.). Reprinted in *Smithsonian Institution Annual Report 1962,* pp. 545–565.

George, Pierre
1969 *Population et Peuplement.* Paris:

Presses Universitaires de France. 212 pp.

Hartley, Shirley F.
1972 *Population Quantity vs. Quality: A Sociological Examination of the Causes and Consequences of the Population Explosion.* Englewood Cliffs, N.J.: Prentice-Hall. 343 pp.

Heer, David M.
1975 *Society and Population* (2nd ed.). Englewood Cliffs, N.J.: Prentice-Hall. 155 pp.

Kuroda, Toshio
1973 *Japan's Changing Population Structure.* Tokyo: Ministry of Foreign Affairs. 95 pp.

Petersen, William
1970 *Population* (3rd ed.). New York: Macmillan. 735 pp.

Polgar, Stephen (ed.)
1975 *Population, Ecology, and Social Evolution.* Chicago: Aldine. 368 pp.

Scientific American
1974 *The Human Population.* San Francisco: Freeman. 147 pp.

Thomas, Robert N. (ed.)
1973 *Population Dynamics of Latin America: A Review and a Bibliography* (Publication Series, Vol. 2). East Lansing, Mich.: Conference of Latin Americanist Geographers. 200 pp.

Trewartha, Glenn T.
1969 *A Geography of Population: World Patterns.* New York: Wiley. 186 pp.
1972 *The Less-Developed Realm: A Geography of Its Population.* New York: Wiley. 449 pp.

Wrigley, E. A.
1969 *Population and History.* London: World University Library. 256 pp.

Zelinsky, Wilbur
1966 *A Prologue to Population Geography.* Englewood Cliffs, N.J.: Prentice-Hall. 150 pp.

Resource Extraction

Adelman, M. A.
1972 *The World Petroleum Market.* Baltimore: Johns Hopkins University Press. 438 pp.

Bardach, John
1972 *Aquaculture: The Farming and Husbandry of Fresh Water and Marine Organisms.* New York: Wiley. 868 pp.

Barrons, Keith C.
1975 *The Food in Your Future: Steps to Abundance.* New York: Van Nostrand Reinhold. 180 pp.

Boesch, Hans H.
1974 *A Geography of World Economy* (2nd ed.). New York: Wiley. 303 pp.

Clawson, Marion
1964 *Natural Resources and International Development.* Baltimore: Johns Hopkins Press, for Resources for the Future, 462 pp.
1972 *America: Land and Its Uses.* Baltimore: Johns Hopkins University Press. 166 pp.

1975 *Forests for Whom and for What?* Baltimore: Johns Hopkins University Press. 175 pp.

Deju, Raul A.
1974 *Extraction of Minerals and Energy: Today's Dilemma.* Ann Arbor, Mich.: Ann Arbor Science. 301 pp.

Firth, Raymond, and B. S. Yamey (eds.)
1964 *Capital, Savings and Credit in Peasant Societies: Studies from Asia, Oceania, The Caribbean, and Middle America.* Chicago: Aldine. 399 pp.

Food and Agriculture Organization
1967 *Wood: World Trends and Prospects.* Rome: Food and Agriculture Organization, United Nations. 130 pp.

Ginsburg, Norton D.
1961 *Atlas of Economic Development.* Chicago: University of Chicago Press. 119 pp.

Gregor, Howard F.
1970 *Geography of Agriculture: Themes in Research.* Englewood Cliffs, N.J.: Prentice-Hall. 181 pp.

Grigg, David
1970 *The Harsh Lands: A Study in Agricultural Development.* London: Macmillan. 321 pp.
1974 *Agricultural Systems of the World: An Evolutionary Approach.* London: Cambridge University Press.

Hodder, B. W. and Roger Lee
1974 *Economic Geography.* New York: St. Martin's Press. 207 pp.

Iversen, E. S.
1968 *Farming the Edge of the Sea.* London: Fishing News. 301 pp.

Manners, Gerald
1971 *The Geography of Energy* (2nd ed.). London: Hutchinson University Library. 222 pp.
1974 *Minerals and Man.* Baltimore: Johns Hopkins University Press. 175 pp.

McKee, Alexander
1969 *Farming the Sea.* New York: Crowell. 198 pp.

National Research Council, Committee on Resources and Man
1969 *Resources and Man: A Study and Recommendations.* San Francisco: Freeman. 259 pp.

Odell, Peter R.
1974 *Oil and World Power* (3rd ed.). Baltimore: Penguin Books. 245 pp.

Schultz, Theodore W.
1972 *Human Resources.* New York: National Bureau of Economic Research, Columbia University Press. 97 pp.

Schurr, S. H. and B. C. Netschert
1972 *Energy, Economic Growth, and the Environment.* Baltimore: Johns Hopkins University Press. 232 pp.

Skinner, Brian J. and K. K. Turekian
1973 *Man and the Ocean.* Englewood Cliffs, N.J.: Prentice-Hall. 149 pp.

Warren, Kenneth
1973 *Mineral Resources.* New York: Wiley. 272 pp.

Waters, John F.
1970 *The Sea Farmers*. New York: Hastings House. 120 pp.

Wharton, Clifton R. (ed.)
1969 *Subsistence Agriculture and Economic Development*. Chicago: Aldine. 481 pp.

Resource Conversion

Alexandersson, Gunnar
1967 *The Geography of Manufacturing*. New York: Prentice-Hall. 154 pp.

Becht, J. Edwin
1975 *World Resource Management: Key to Civilization and Social Achievement*. Englewood Cliffs, N.J.: Prentice-Hall. 329 pp.

Boyce, Ronald R.
1974 *The Bases of Economic Geography*. New York: Holt, Rinehart and Winston. 358 pp.

Brookfield, Harold C.
1975 *Interdependent Development*. Pittsburgh: University of Pittsburgh Press. 234 pp.

De Garmo, Ernest Paul
1974 *Materials and Processes in Manufacturing* (4th ed.). New York: Macmillan. 982 pp.

deSchweinitz, Karl, Jr.
1964 *Industrialization and Democracy: Economic Necessities and Political Possibilities*. New York: Free Press. 309 pp.

Guyol, N. B.
1969 *The World Electric Power Industry*. Berkeley: University California Press. 366 pp.
1971 *Energy in the Perspective of Geography*. Englewood Cliffs, N.J.: Prentice-Hall. 156 pp.

Hagen, Everett E.
1975 *The Economics of Development* (rev. ed.). Homewood, Ill.: Irwin. 563 pp.

Heady, Earl O.
1971 *Economic Models and Quantitative Methods for Decisions and Planning*. Ames, Iowa: Iowa State University Press. 518 pp.
1972 *Agricultural and Water Politics and the Environment: An Analysis of National Alternatives in Natural Resource Use, Food Supply Capacity, and Environmental Quality*. Ames, Iowa: Iowa State University Press. 295 pp.

Heady, Earl O. and Leroy L. Blakeslee
1973 *World Food Production: Demand and Trade*. Ames, Iowa: Iowa State University Press. 417 pp.

Herfindahl, O. C. and A. V. Kneese
1974 *Economic Theory of Natural Resources*. Columbus, Ohio: Merrill. 405 pp.

Highsmith, Richard M. and Ray M. Northam
1968 *World Economic Activities: A Geographic Analysis*. New York: Harcourt, Brace & World. 526 pp.

Hull, Oswald
1968 *A Geography of Production*. London: Macmillan; New York: St. Martin's Press. 344 pp.

Jarrett, H. R.
1969 *A Geography of Manufacturing*. London: Macdonald & Evans. 349 pp.

Martindale, Don
1962 *Social Life and Cultural Change*. Princeton, N.J.: Van Nostrand. 528 pp.

Mountjoy, Alan B.
1966 *Industrialization and Underdeveloped Countries* (2nd ed.) London: Hutchinson: and Chicago: Aldine. 200 pp.

Nef, John U.
1964 *The Conquest of the Material World*. Chicago: University of Chicago Press. 408 pp.

O'Riordan, Timothy
1971 *Perspectives on Resource Management*. London: Pion. 183 pp.

Packer, D. W.
1964 *Resource Acquisition in Corporate Growth*. Cambridge, Mass.: MIT Press. 118 pp.

Parker, Geoffrey
1972 *The Geography of Economics: A World Survey* (2nd ed.). London: Longman. 258 pp.

Parker, J. E. S.
1974 *The Economics of Innovation: The National and Multinational Enterprise in Technological Change*. London: Longman. 297 pp.

Rogers, Everett M. and F. Floyd Shoemaner
1971 *Communication of Innovations: A Cross-Cultural Approach* (2nd ed.). New York: Free Press. 476 pp.

Samuelson, Paul A.
1973 *Economics* (9th ed.). New York: McGraw-Hill. 917 pp.

Shaw, J. H. and Emery, J. S.
1968 *Cities and Industries*. Melbourne: Jacaranda Press. 220 pp.

Taylor, Theodore B. and C. C. Humpstone
1973 *The Restoration of the Earth*. New York: Harper & Row. 166 pp.

Thoman, Richard S. and Peter Corbin
1974 *The Geography of Economic Activity* (3rd ed.) New York: McGraw-Hill. 420 pp.

Wagstaff, H. Reid
1974 *A Geography of Energy*. Dubuque, Iowa: Brown. 122 pp.

Warren, Kenneth
1975 *World Steel: An Economic Geography*. New York: Crane, Russak. 335 pp.

Space Adjustment

Travel and Transportation

Alexandersson, Gunnar, and Göran Norström
1963 *World Shipping, An Economic Geography of Ports and Seaborne Trade*. New York: Wiley. 507 pp.

Bird, James
1971 *Seaports and Seaport Terminals*. London: Hutchinson University Library. 240 pp.

Cantor, Leonard M.
1970 *A World Geography of Irrigation*. New York: Praeger. 252 pp.

Colby, Charles C.
1966 *North Atlantic Arena: Water Transport in the World Order*. Carbondale: Southern Illinois University Press. 253 pp.

Eliot-Hurst, Michael E. (ed.)
1974 *Transportation Geography*. New York: McGraw-Hill. 528 pp.

Fabre, Maurice
1963 *A History of Land Transportation* (The New Illustrated Library of Science and Invention, Vol. 7). New York: Hawthorn Books. 112 pp.

Heady, Earl O.
1975 *An Interregional Analysis of United States Domestic Grain Transportation*. Ames, Iowa: Iowa State University Press. 199 pp.

Hoyle, B. S. (ed.)
1973 *Transport and Development*. New York: Barnes & Noble. 230 pp.

Kraft, Gerald et al.
1971 *The Role of Transportation in Regional Economic Development*. Lexington, Mass.: Lexington Books. 129 pp.

Lowe, John C. and S. Moryadas
1975 *The Geography of Movement*. Boston: Houghton Mifflin. 333 pp.

Marlowe, John
1964 *World Ditch, The Making of the Suez Canal*. New York: Macmillan. 294 pp.

O'Dell, Andrew C. and Peter S. Richards
1971 *Railways and Geography*, (2nd ed.). London: Hutchinson University Library. 248 pp.

Sealy, Kenneth R.
1968 *The Geography of Air Transport* (2nd ed.). Chicago: Aldine. 198 pp.

Seckler, David (ed.)
1971 *California Water: A Study in Resource Management*. Berkeley: University of California Press. 348 pp.

Straszheim, Mahlon R.
1969 *The International Airline Industry*. Washington, D.C.: Brookings Institution. 297 pp.

Taaffe, Edward J. and Howard L. Gauthier
1973 *Geography of Transportation*. Englewood Cliffs, N.J.: Prentice-Hall. 226 pp.

White, Gilbert F.
1971 *Strategies of American Water Management* (Ann Arbor Paperback 176). Ann Arbor: University Michigan Press. 170 pp.

Williams, Ernest W. (ed.)
1971 *The Future of American Transportation*. Englewood Cliffs, N.J.: Prentice-Hall. 211 pp.

Wolfe, Roy I.
1963 *Transportation and Politics* (Searchlight Book #18). Princeton, N.J.: Van Nostrand. 136 pp.

Trade

Belshaw, Cyril S.
1965 *Traditional Exchange and Modern Markets.* Englewood Cliffs, N.J.: Prentice-Hall. 149 pp.

Berry, Brian J. L.
1967 *Geography of Market Centers and Retail Distribution.* Englewood Cliffs, N.J.: Prentice-Hall. 146 pp.

Berry, Brian J. L. and Duane F. Marble
1967 *Spatial Analysis: A Reader in Statistical Geography.* Englewood Cliffs, N.J.: Prentice-Hall. 512 pp.

Berry, Brian J. L. et al.
1976 *The Geography of Economic Systems.* Englewood Cliffs, N.J.: Prentice-Hall. 529 pp.

Fryer, Donald W.
1965 *World Economic Development.* New York: McGraw-Hill. 627 pp.

Lewis, W. Arthur
1968 "World trade since the war," *Proceedings of the American Philosophical Society,* Vol. 112, No. 6 (Dec.), pp. 362–366.

McCreary, Edward A.
1964 *The Americanization of Europe.* Garden City, N.Y.: Doubleday. 295 pp.

Mulvihill, Donald F. and Ruth C. Mulvihill
1970 *Geography, Marketing, and Urban Growth.* New York: Van Nostrand Reinhold. 190 pp.

Parker, Geoffrey
1969 *An Economic Geography of the Common Market.* New York: Praeger. 178 pp.
1972 *The Geography of Economics: A World Survey* (2nd ed.). London: Longman. 258 pp.

Polanyi, K., C. M. Arensberg, and H. W. Pearson
1957 *Trade and Market in Early Empires.* Glencoe, Ill.: Free Press. 382 pp.

Thoman, Richard S., and Edgar C. Conkling
1967 *Geography of International Trade.* Englewood Cliffs, N.J.: Prentice-Hall. 190 pp.

Thorbecke, Erik
1960 *The Tendency Towards Regionalization in International Trade, 1928–1956.* The Hague: Nijhoff. 223 pp.

Vance, James E.
1970 *The Merchant's World: The Geography of Wholesaling.* Englewood Cliffs, N.J.: Prentice-Hall. 167 pp.

Communications

American Telephone and Telegraph Company
1967 *The World's Telephones, 1967.* New York: A T & T, Business Research Division. 24 pp.

Cherry, Colin
1971 *World Communication, A Threat or Promise? A Socio-Technical Approach.* New York: Wiley. 229 pp.

Cherry, Colin (ed.)

1974 *Pragmatic Aspects of Human Communication.* Boston: Reidel. 178 pp.

Davison, W. Phillips
1974a *Mass Communication and Conflict Resolution: The Role of the Information Media in the Advancement of International Understanding.* New York: Praeger. 155 pp.
1974b *Mass Communication Research and Future Directions.* New York: Praeger. 246 pp.

Fabre, Maurice
1963 *A History of Communications* (The New Illustrated Library of Science and Invention, Vol. 9). New York: Hawthorn Books. 112 pp.

Jeffries, Charles
1967 *Illiteracy: A World Problem.* New York: Praeger. 204 pp.

Zilliacus, Lawrin
1965 *From Pillar to Post, the Troubled History of the Mail.* London: Heinemann. 217 pp.

Political Territories and Spatial Organization

Abler, Ronald et al.
1971 *Spatial Organization: The Geographer's View of the World.* Englewood Cliffs, N.J.: Prentice-Hall. 587 pp.

Anonymous
1968 *Treaties and Alliances of the World: An International Survey Covering Treaties in Force and Communities of States.* New York: Scribner's. 214 pp.

Brunn, Stanley D.
1974 *Geography and Politics in America.* New York: Harper & Row. 443 pp.

East, W. Gordon, O. H. K. Spate, and Charles A. Fisher
1971 *The Changing Map of Asia, A Political Geography* (5th ed.). London: Methuen. 678 pp.

Gottmann, Jean
1973 *The Significance of Territory.* Charlottesville: University Press of Virginia. 169 pp.

Jacob, Philip E., and James V. Toscano (eds.)
1964 *The Integration of Political Communities.* Philadelphia: Lippincott. 314 pp.

Johnson, E. J.
1970 *The Organization of Space in Developing Countries.* Cambridge, Mass.: Harvard University Press. 452 pp.

Massam, Bryan H.
1972 *The Spatial Structure of Administrative Systems* (Resource Paper No. 12). Washington, D.C.: Commission on College Geography, Association of American Geographers. 38 pp.

Morrill, Richard L.
1974 *The Spatial Organization of Society* (2nd ed.). North Scituate, Mass.: Duxbury Press. 267 pp.

Muir, Richard
1975 *Modern Political Geography.* New York: Wiley. 262 pp.

Nader, Laura (ed.)
1969 *Law in Culture and Society.* Chicago: Aldine. 454 pp.

Pounds, Norman J. G.
1972 *Political Geography* (2nd ed.). New York: McGraw-Hill. 453 pp.

Prescott, John R. V.
1972 *Political Geography.* London: Methuen. 124 pp.
1975 *The Political Geography of the Oceans.* Newton Abbot Eng.: David & Charles. 247 pp.

Soja, Edward W.
1971 *The Political Organization of Space* (Resource Paper No. 8). Washington, D.C.: Commission on College Geography, Association of American Geographers. 54 pp.

Wainhouse, David W.
1964 *Remnants of Empire. The United Nations and the End of Colonialism.* New York: Harper. 153 pp.

Space Intensification

Evolution of Cities

Adams, R. McC.
1966 *The Evolution of Urban Society: Early Mesopotamia and Precolumbian Mexico.* Chicago: Aldine. 191 pp.

Bellan, Ruben C.
1971 *The Evolving City.* Vancouver: Copp Clark. 420 pp.

Chandler, Tertius and Gerald Fox
1974 *3000 Years of Urban Growth.* New York: Academic Press. 431 pp.

Davis, Kingsley (comp.)
1973 *Cities: Their Origin, Growth, and Human Impact.* San Francisco: Freeman. 297 pp.

Green, Constance M.
1965 *The Rise of Urban America.* New York: Harper & Row. 208 pp.

Hammond, Mason
1972 *The City in the Ancient World.* Cambridge, Mass.: Harvard University Press. 617 pp.

McKelvey, Blake
1963 *The Urbanization of America, 1860–1915.* New Brunswick, N.J.: Rutgers University Press. 370 pp.
1968 *The Emergence of Metropolitan America, 1915–1966.* New Brunswick, N.J.: Rutgers University Press. 311 pp.
1973 *American Urbanization: A Comparative History.* Glenview, Ill.: Scott, Foresman. 166 pp.

Mayer, Harold M.
1969 *The Spatial Expression of Urban Growth* (Resource Paper No. 7). Washington, D.C.: Commission on College Geography, Association of American Geographers. 57 pp.

Mumford, Lewis
1961 *The City in History: Its Origins, Its Transformations, and Its Prospects.* New York: Harcourt, Brace & World. 657 pp.

Shaw, J. H. and J. E. Emery
1968 *Cities and Industries.* Melbourne:

Jacaranda Press. 220 pp.

Sjoberg, Gideon
1960 *The Preindustrial City: Past and Present.* Glencoe, Ill.: Free Press. 353 pp.

Urban Regions

Berry, Brian J. L.
1973 *Growth Centers in the American Urban System.* Cambridge, Mass.: Bollinger. Two vols. 195 and 181 pp.

Berry, Brian J. L. and Frank E. Horton
1970 *Geographic Perspectives on Urban Systems, with Integrated Readings.* Englewood Cliffs, N.J.: Prentice-Hall. 564 pp.

Carter, Harold
1972 *The Study of Urban Geography.* London: Edward Arnold. 346 pp.

Chapin, F. Stuart, Jr.
1965 *Urban Land Use Planning* (2nd ed.). Urbana: University of Illinois Press. 498 pp.
1974 *Human Activity Patterns in the City: Things People Do in Time and Space.* New York: Wiley. 272 pp.

Clawson, Marion and Peter Hall
1973 *Planning and Urban Growth: An Anglo-American Comparison.* Baltimore: Johns Hopkins University Press. 300 pp.

Detwyler, Thomas R. and Melvin G. Marcus
1972 *Urbanization and Environment: The Physical Geography of the City.* Belmont, Calif.: Duxbury Press. 287 pp.

Dickinson, Robert E.
1964 *City and Region: A Geographical Interpretation.* London: Routledge & K. Paul. 588 pp.

Doxiadis, K. A. and J. G. Papaioannou
1974 *Ecumenopolis: The Inevitable City of the Future.* Athens: Athens Center of Ekistics. 469 pp.

Gottmann, Jean
1961 *Megalopolis: The Urbanized Northeastern Seaboard of the United States.* New York: Twentieth Century Fund. 810 pp.
1974 *The Evolution of Urban Centrality: Orientations for Research.* Oxford: School of Geography, University of Oxford. 44 pp.

Gottmann, Jean and Robert A. Harper (eds.)
1967 *Metropolis on the Move: Geographers Look at Urban Sprawl.* New York: Wiley. 203 pp.

Hall, Peter G.
1966 *The World Cities* (World University Library). New York: McGraw-Hill. 256 pp.
1969 *London 2000* (rev. ed.). New York: Praeger. 287 pp.
1975 *Urban and Regional Planning.* New York: Wiley. 312 pp.

Hall, Peter G. et al.
1973 *The Containment of Urban England.* London: Allen & Unwin. 2 Vols.

Hauser, Philip M., and Leo F. Schnore

1965 *The Study of Urbanization.* New York: Wiley. 554 pp.

Herbert, David
1973 *Urban Geography: A Social Perspective.* New York: Praeger. 320 pp.

Hicks, U. K. W.
1974 *The Large City: A World Problem.* New York: Wiley. 270 pp.

Johnson, James H.
1972 *Urban Geography* (2nd ed.). New York: Pergamon Press. 203 pp.

Mayer, Harold M., and Clyde F. Kohn (eds.)
1959 *Readings in Urban Geography.* Chicago: University of Chicago Press. 625 pp.

Murphy, Raymond E.
1972 *The Central Business District.* Chicago: Aldine-Atherton. 193 pp.
1974 *The American City: An Urban Geography* (2nd ed.). New York: McGraw-Hill. 556 pp.

Northam, Ray M.
1975 *Urban Geography.* New York: Wiley. 270 pp.

Steiss, Alan W.
1974 *Urban Systems Dynamics.* Lexington, Mass.: Lexington Books. 323 pp.

Swatridge, L. A.
1971 *The Bosnywash Megalopolis: A Region of Great Cities.* Toronto: McGraw-Hill 106 pp.

Von Eckardt, Wolf
1964 *The Challenge of Megalopolis.* New York: Macmillan. 126 pp.

Yeates, Maurice H.
1975 *Main Street: Windsor to Quebec City.* Toronto: Macmillan. 431 pp.

Yeates, Maurice H. and Barry J. Garner
1971 *The North American City.* New York: Harper & Row. 536 pp.

Vertical Intensification and Internal Complexity

Andrews, Richard B.
1971 *Urban Land Economics and Public Policy.* New York: Free Press. 159 pp.

Fitch, Lyle C., et al.
1964 *Urban Transportation and Public Policy.* San Francisco: Chandler. 279 pp.

Fitch, Lyle C. and A. H. Walsh (eds.)
1970 *Agenda for a City: Issues Confronting New York.* Beverly Hills: Sage. 718 pp.

Lang, Albert S., and Richard M. Soberman
1964 *Urban Rail Transit: Its Economics and Technology.* Cambridge, Mass.: MIT Press. 139 pp.

Schaeffer, K. H. and Elliott Sclar
1975 *Access for All: Transportation and Urban Growth.* Harmondsworth Eng.: Penguin Books. 182 pp.

Spreiregen, Paul D.
1965 *Urban Design: The Architecture of Towns and Cities.* New York: McGraw-Hill. 243 pp.

Penalties of Human Concentration

Berry, B. J. L.
1973 *The Human Consequences of Urbanization.* New York: St. Martin's Press. 205 pp.

Berry, Brian J. L. and Frank E. Horton
1974 *Urban Environmental Management: Planning for Pollution Control.* Englewood Cliffs, N.J.: Prentice-Hall. 425 pp.

Berry, Brian J. L., and Jack Meltzer (eds.)
1967 *Goals for Urban America.* Englewood Cliffs, N.J.: Prentice-Hall. 152 pp.

Clawson, Marion
1973 *Modernizing Urban Land Policy.* Baltimore: Johns Hopkins University Press. 248 pp.

Doxiades, K. A.
1966 *Urban Renewal and the Future of the American City.* Chicago: Public Administration Service. 174 pp.

Doxiades, K. A. and J. G. Papaioannou
1974 *Ecumenopolis: The Inevitable City of the Future.* Athens: Athens Center of Ekistics. 469 pp.

Gappert, Gary and Harold M. Rose (eds.)
1975 *The Social Economy of Cities.* Beverly Hills: Sage. 640 pp.

Greer, Scott A.
1972 *The Urbane View: Life and Politics in Metropolitan America.* New York: Oxford University Press. 355 pp.

Harvey, David
1973 *Social Justice and the City.* Baltimore: Johns Hopkins University Press. 336 pp.

Jacobs, Jane
1961 *The Death and Life of Great American Cities* (Vintage Book, V-241). New York: Vintage Books. 458 pp.
1969 *The Economy of Cities.* New York: Random House. 268 pp.

Levin, Melvin R. (ed.)
1971 *Exploring Urban Problems.* Boston: Urban Press. 667 pp.

Levin, Melvin R. and Norman A. Abend
1971 *Bureaucrats in Collision: Case Studies in Area Transportation Planning.* Cambridge: MIT Press. 295 pp.

Meyer, John R. et al.
1965 *The Urban Transportation Problem.* Cambridge, Mass.: Harvard University Press. 427 pp.

Mumford, Lewis
1968 *The Urban Prospect.* New York: Harcourt, Brace & World. 255 pp.

Rose, Harold M.
1971 *The Black Ghetto: A Spatial Behavioral Perspective* (Problem series in geography). New York: McGraw-Hill. 147 pp.

Rose, Harold M. (ed.)
1972 *Geography of the Ghetto: Perceptions, Problems, and Alternatives.* DeKalb: Northern Illinois University Press. 273 pp.

Smerk, George M.
1974 *Urban Mass Transportation: A Dozen Years of Federal Policy.* Bloom-

ington: Indiana University Press. 388 pp.

Swatridge, L. A.
1972 *Problems in the Bosnywash Megalopolis: Pollution, Transportation, Sprawl, Social Problems.* Toronto: McGraw-Hill Ryerson. 113 pp.

Weaver, Robert C.
1964 *The Urban Complex: Human Values in Urban Life.* Garden City. N.Y. Doubleday. 297 pp.
1965 *Dilemmas of Urban America.* Cambridge, Mass.: Harvard University Press. 138 pp.

Wilson, James Q. (ed.)
1966 *Urban Renewal: The Record and the Controversy.* Cambridge, Mass.: MIT Press. 683 pp.

Urbanization as a Worldwide Process

Arango, Jorge
1970 *The Urbanization of the Earth.* Boston: Beacon Press. 175 pp.

Beyer, Glenn H. (ed.)
1967 *The Urban Explosion in Latin America: A Continent in Process of Modernization.* Ithaca: Cornell University Press. 360 pp.

Breese, Gerald W.
1966 *Urbanization in Newly Developing Countries.* Englewood Cliffs, N.J.: Prentice-Hall. 151 pp.

Breese, Gerald W. (ed.)
1969 *The City in Newly Developing Countries: Readings on Urbanism and Urbanization.* Englewood Cliffs, N.J.: Prentice-Hall. 556 pp.

Davis, Kingsley
1969–1972 *World Urbanization 1950–1970* (Population Monograph Series, Nos. 4 and 9). Berkeley: University of California Press. 2 vols.

DuToit, Brian M. and Helen I. Safa
1975 *Migration and Urbanization: Models and Adaptive Strategies.* Chicago: Aldine. 305 pp.

Dwyer, D. J. (ed.)
1974 *The City in the Third World.* New York: Barnes & Noble. 253 pp.

Hance, William A.
1970 *Population, Migration, and Urbanization in Africa.* New York: Columbia University Press. 450 pp.

Juppenlatz, Morris
1970 *Cities in Transformation: The Urban Squatter Problem of the Developing World.* St. Lucia, Queensland: University of Queensland Press. 257 pp.

Mabogunje, Akin L.
1968 *Urbanization in Nigeria.* New York: Africana. 353 pp.
1972 *Regional Mobility and Resource Development in West Africa.* Montreal: McGill-Queen's University Press. 154 pp.

Meier, Richard L.
1974 *Planning for an Urban World: The Design of Resource-Conserving Cities.* Cambridge: MIT Press. 515 pp.

Merlin, Pierre
1971 *New Towns: Regional Planning and Development.* London: Methuen. 276 pp.

Palen, J. John
1975 *The Urban World.* New York: McGraw-Hill. 480 pp.

Rugg, Dean S.
1972 *Spatial Foundations of Urbanism.* Dubuque, Iowa: Brown. 313 pp.

Safa, Helen I. and Brian M. DuToit (eds.)
1975 *Migration and Development: Implications for Ethnic Identity and Political Conflict.* Chicago: Aldine. 360 pp.

Scientific American
1965 *Cities.* New York: Knopf. 211 pp.

Sovani, N. V.
1966 *Urbanization and Urban India.* New York: Asia Publishing House. 160 pp.

Strong, Ann Louise
1971 *Planned Urban Environments: Sweden, Finland, Israel, The Netherlands, France.* Baltimore: Johns Hopkins Press. 406 pp.

Tilly, Charles
1974 *An Urban World.* Boston: Little, Brown. 487 pp.

People in Transformation

Bain, G. S., D. Coates, and V. Ellis
1973 *Social Stratification and Trade Unionism.* London: Heinemann. 174 pp.

Berelson, Bernard (ed.)
1974 *Population Policy in Developed Countries.* New York: McGraw-Hill. 793 pp.

Bird, Caroline
1972 *The Crowding Syndrome: Learning to Live with Too Much and Too Many.* New York: McKay. 337 pp.

Bock, Philip K. (ed.)
1969 *Peasants in the Modern World.* Albuquerque: University of New Mexico Press. 173 pp.
1970 *Culture Shock.* New York: Knopf. 379 pp.

Boot, Johannes C. G.
1974 *Common Globe or Global Commons: Population Regulation and Income Distribution.* New York: Dekker. 139 pp.

Brown, Lester R.
1970 *Seeds of Change: The Green Revolution and Development in the 1970's.* New York: Praeger. 205 pp.
1971 *The Social Impact of the Green Revolution.* New York: Carnegie Endowment for International Peace. 61 pp.
1972a *By Bread Alone.* New York: Praeger. 272 pp.
1972b *Man and His Environment: Food.* New York: Harper & Row. 208 pp.

Clarke, John I.
1971 *Population Geography and the Developing Countries.* Oxford, New York: Pergamon Press. 282 pp.
1972 *Population Geography* (2nd ed.). New York: Pergamon Press. 176 pp.

Cosgrove, Isobel and Richard Jackson
1972 *The Geography of Recreation and Leisure.* London: Hutchinson University Library. 168 pp.

Dalton, George (ed.)
1971 *Economic Anthropology: Essays on Tribal and Peasant Economies.* New York: Basic Books. 386 pp.

Finkle, Jason L. and Richard W. Gable
1971 *Political Development and Social Change* (2nd ed.). New York: Wiley. 685 pp.

Foster, George M.
1973 *Traditional Societies and Technological Change* (2nd ed.). New York: Harper & Row. 286 pp.

Fraser, Dean
1971 *The People Problem: What You Should Know about Growing Population and Vanishing Resources.* Bloomington: Indiana University Press. 248 pp.

Freedman, Ronald (ed.)
1964 *Population: The Vital Revolution.* Garden City, N.Y.: Anchor Books. 274 pp.

Frejka, Tomas
1973 *The Future of Population Growth: Alternative Paths to Equilibrium.* New York: Wiley. 268 pp.

Goldschneider, Calvin
1971 *Population, Modernization, and Social Structure.* Boston: Little, Brown. 345 pp.

Griffin, Keith B.
1974 *The Political Economy of Agrarian Change: An Essay on the Green Revolution.* London: Macmillan. 264 pp.

Hagen, Everett E.
1975 *The Economics of Development* (rev. ed.). Homewood, Ill.: Irwin. 563 pp.

Halacy, Daniel Stephen
1972 *The Geometry of Hunger.* New York: Harper & Row. 280 pp.

Hardin, Garrett (ed.)
1969 *Population, Evolution, and Birth Control: A Collage of Controversial Ideas* (2nd ed.). San Francisco: Freeman. 386 pp.
1972 *Exploring New Ethics for Survival: The Voyage of the Spaceship Beagle.* New York: Viking Press. 273 pp.

Hauser, Philip M. (ed.)
1969 *The Population Dilemma* (2nd ed.). Englewood Cliffs, N.J.: Prentice-Hall. 211 pp.

Hayami, Yujiro and Vernon W. Ruttan
1971 *Agricultural Development: An International Perspective.* Baltimore: Johns Hopkins Press. 367 pp.

Huxley, Julian S.
1974 *Evolution: The Modern Synthesis* (3rd ed.). London: Allen & Unwin. 705 pp.

International Bank for Reconstruction and Development
1974 *Population Policies and Economic Development.* Baltimore: Johns Hopkins University Press. 214 pp.

Jacoby, Erich H. and Charlotte F. Jacoby

Selected References

1971 *Man and Land: The Fundamental Issue in Development*. London: Deutsch. 400 pp.

Kurihara, Kenneth K.
1971 *The Growth Potential of the Japanese Economy*. Baltimore: Johns Hopkins Press. 148 pp.

Mead, Margaret
1964 *Continuities in Cultural Evolution*. New Haven, Conn.: Yale University Press, 471 pp.
1972 *Twentieth-Century Faith: Hope and Survival*. New York: Harper & Row. 172 pp.

Mesarovic, Milajlo D.
1974 *Mankind at the Turning Point*. New York: Dutton. 210 pp.

Morrill, Richard L. and Ernest H. Wohlenberg
1971 *The Geography of Poverty in the United States* (Problem Series in Geography). New York: McGraw-Hill. 148 pp.

Myrdal, Gunnar
1968 *Asian Drama: An Enquiry into the Poverty of Nations*. New York: Pantheon. 3 vols.; 2284 pp.
1970 *The Challenge of World Poverty*. New York: Pantheon. 518 pp.

Ng, L. K. Y. and Stuart Mudd
1966 *The Population Crisis: Implications and Plans for Action*. Bloomington: Indiana University Press. 364 pp.

Peach, Ceri
1968 *West Indian Migration to Britain: A Social Geography*. London: Oxford University Press. 122 pp.

Pitchford, J. D.
1974 *The Economics of Population: An Introduction*. Canberra: Australian National University Press. 100 pp.

Population Reference Bureau
1972 *The World Population Dilemma*. Washington, D.C.: Columbia Books. 79 pp.

Rosenzweig, Michael L.
1974 *And Replenish the Earth: The Evolution, Consequences, and Prevention of Overpopulation*. New York: Harper & Row. 304 pp.

Spengler, Joseph J.
1974 *Population Change, Modernization, and Welfare*. Englewood Cliffs, N.J.: Prentice-Hall. 182 pp.

Stanford, Quentin H. (ed.)
1972 *The World's Population: Problems of Growth*. New York: Oxford University Press. 346 pp.

Waldron, Ingrid and Robert E. Ricklefs
1973 *Environment and Population: Problems and Solutions*. New York: Holt, Rinehart and Winston. 232 pp.

Wharton, Clifton R. (ed.)
1969 *Subsistence Agriculture and Economic Development*. Chicago: Aldine. 481 pp.

Wolf, Eric R.
1969 *Peasant Wars of the Twentieth Century*. New York: Harper & Row. 328 pp.

Zelinsky, Wilbur
1970 "Beyond the exponentials; the role of geography in the great transition," *Economic Geography*, Vol. 46, No. 3 (July), pp. 498–535.
1971 "The hypothesis of the mobility transition," *Geographical Review*, Vol. 61, No. 2 (April), pp. 219–249.

Zelinsky, Wilbur et al. (eds.)
1969 *A Symposium on Population Pressures upon Physical and Social Resources in the Developing Lands*. University Park: Dept. of Geography, Pennsylvania State University 132 pp.
1970 *Geography and a Crowding World. A Symposium on Population Pressures upon Physical and Social Resources in the Developing Lands*. New York: Oxford University Press. 601 pp.

Humankind's Dominance of the Earth

Bach, Wilfrid
1972 *Atmospheric Pollution* (Problem Series in Geography). New York: McGraw-Hill. 144 pp.

Balon, E. K. and A. G. Coche (eds.)
1974 *Lake Kariba: A Man-made Tropical Ecosystem in Central Africa*. The Hague: Junk. 767 pp.

Becht, J. Edwin and L. D. Belzung
1975 *World Resource Management: Key to Civilizations and Social Achievement*. Englewood Cliffs, N.J.: Prentice-Hall. 329 pp.

Benarde, Melvin A.
1970 *Our Precarious Habitat*. New York: Norton. 362 pp.
1971 *The Chemicals We Eat*. New York: American Heritage Press. 208 pp.

Bennett, Charles F.
1968 *Human Influences on the Zoogeography of Panama* (Ibero-Americana, No. 51). Berkeley and Los Angeles: University of California Press. 112 pp.

Bigger, J. W., and R. B. Corey
1969 "Eutrophication: causes, consequences, corrections." *Proceedings of the Symposium*, pp. 404–445. Washington, D.C.: National Academy of Science.

Borgstrom, Georg
1973a *Focal Points: A Global Food Strategy*. New York: Macmillan. 320 pp.
1973b *The Food and People Dilemma*. Belmont, Calif.: Duxbury Press. 140 pp.
1973c *Harvesting the Earth*. New York: Abelard-Schuman. 237 pp.

Boulding, Kenneth E.
1970 *Economics of Pollution*. New York: New York University Press. 158 pp.

Brockman, C. Frank and Lawrence C. Merriam
1973 *Recreational Use of Wild Lands* (2nd ed.). New York: McGraw-Hill. 329 pp.

Brown, Theodore L.
1971 *Energy and the Environment*. Columbus, Ohio: Merrill. 141 pp.

Buchanan, Keith
1970 *The Transformation of the Chinese Earth*. New York: Praeger. 336 pp.

Carr, Donald E.
1971 *Death of the Sweet Waters*. New York: Norton. 257 pp.

Cornwall, I. W.
1968 *Prehistoric Animals and Their Hunters*. London: Faber & Faber. 214 pp.

Curry-Lindahl, Kai
1972 *Let Them Live: A Worldwide Survey of Animals Threatened with Extinction*. New York: Morrow. 394 pp.

Darling, Frank Fraser
1970 *Wilderness and Plenty: The Reith Lectures, 1969*. Boston: Houghton Mifflin. 84 pp.

Darling, Frank Fraser and Raymond F. Dasmann
1969 "The ecosystem view of human society," *Impact of Science on Society*, Vol. 19, No. 2, pp. 109–121.

De Bach, P. (ed.)
1964 *Biological Control of Insect Pests and Weeds*. New York: Reinhold. 844 pp.

Detwyler, Thomas R. (ed.)
1971 *Man's Impact on Environment*. New York: McGraw-Hill. 731 pp.

Ehrenfeld, David W.
1972 *Conserving Life on Earth*. New York: Oxford University Press. 360 pp.

Ehrlich, Paul R. and A. H. Ehrlich
1972 *Population/Resources/Environment* (2nd ed.). San Francisco: Freeman. 509 pp.
1974 *The End of Affluence: A Blueprint for Your Future*. New York: Ballantine Books. 307 pp.

Fisher, James, Noel Simon, and Jack Vincent (eds.)
1969 *Wildlife in Danger*. New York: Viking Press. 368 pp.

Flanagan, Dennis (ed.)
1970 *The Biosphere*. San Francisco: Freeman. 134 pp. (Book version of the September, 1969, *Scientific American* issue.)
1971 *Energy and Power*. San Francisco: Freeman. 144 pp. (Book version of the September, 1971, *Scientific American* issue.)

Flawn, P. T.
1970 *Environmental Geology; Conservation, Land Use, and Resource Management*. New York: Harper & Row. 313 pp.

Freeman, A. Myrick III et al.
1973 *The Economics of Environmental Policy*. New York: Wiley. 184 pp.

Furon, Raymond (trans. Paul Barnes)
1967 *The Problem of Water: A World Study*. New York: American Elsevier. 208 pp.

George, Pierre
1971 *L'environnement*. Paris: Presses Universitaires de France. 127 pp.
1973 *Géographie de l'électricité*. Paris: Presses Universitaires de France. 192 pp.

Gerasimov, I. P. et al. (eds.)
1971 *Natural Resources of the Soviet Union: Their Use and Renewal.* San Francisco: Freeman. 349 pp.

Graham, Frank
1971 *Man's Dominion: The Story of Conservation in America.* New York: Evans. 339 pp.

Greenwood, Ned H. and J. M. B. Edwards
1973 *Human Environments and Natural Systems.* North Scituate, Mass.: Duxbury Press. 429 pp.

Guggisberg, C. A. W.
1970 *Man and Wildlife.* New York: Arco. 224 pp.

Helfrich, H. W. (ed.)
1970 *The Environmental Crisis.* New Haven: Yale University Press. 187 pp.

Higbee, Edward
1970 *A Question of Priorities: New Strategies for Our Urbanized World.* New York: Morrow. 214 pp.

Hines, Lawrence G.
1973 *Environmental Issues: Population, Pollution, and Economics.* New York: Norton. 339 pp.

Hoult, D. P. (ed.)
1969 *Oil on the Sea.* New York: Plenum Press. 114 pp.

Kneese, Allen V. et al. (eds.)
1971 *Managing the Environment: International Economic Cooperation for Pollution Control.* New York: Praeger. 356 pp.

Kneese, Allen V. and Orris C. Herfindahl
1974 *Economic Theory of Natural Resources.* Columbus, Ohio: Merrill. 405 pp.
1975 *Pollution, Prices, and Public Policy.* Washington, D.C.: Brookings Institution. 125 pp.

Kaplan, D. J. and E. Kivy (eds.)
1973 *Ecology and the Quality of Life.* Springfield, Ill.: Thomas. 296 pp.

Kormondy, E. J.
1969 *Concepts of Ecology.* Englewood Cliffs, N.J.: Prentice-Hall. 209 pp.

Krantz, Grover S.
1970 "Human activities and megafaunal extinctions," *American Scientist,* Vol. 58, No. 2 (March-April), pp. 164–170.

Lowe-McConnel, R. H. (ed.)
1966 *Man-Made Lakes: Proceedings.* New York: Academic Press. 218 pp.

McKenzie, Garry D., and Russel O. Utgard (eds.)
1975 *Man and His Physical Environment: Readings in Environmental Geology* (2nd ed.). Minneapolis: Burgess. 388 pp.

Manners, Ian and Marvin W. Mikesell (eds.)
1974 *Perspectives on Environment.* Washington, D.C.: Association of American Geographers. 395 pp.

Martin, P. S. and H. E. Wright, Jr.
1967 *Pleistocene Extinctions; The Search for a Cause.* New Haven: Yale University Press. 453 pp.

Marx, W.

1967 *The Frail Ocean.* New York: Coward-McCann. 248 pp.
1971 *Man and his Environment: Waste.* New York: Harper & Row. 179 pp.

Matthews, W. H., F. E. Smith, and E. D. Goldberg (eds.)
1971 *Man's Impact on Terrestrial and Oceanic Ecosystems.* Cambridge, Mass.: MIT Press. 540 pp.

Meinig, D. W.
1968 *The Great Columbia Plain: A Historical Geography, 1805–1910.* Seattle: University of Washington Press. 576 pp.

Murdoch, William W. (ed.)
1975 *Environment, Resources, Pollution, and Society* (2nd ed.). Sunderland, Mass.: Sinauer 488 pp.

Nash, Roderick
1973 *Wilderness and the American Mind* (Rev. ed.). New Haven: Yale University Press. 300 pp.
1976 *The American Environment: Readings in the History of Conservation* (2nd ed.). Reading, Mass.: Addison-Wesley. 364 pp.

Nicholson, Max
1970 *The Environmental Revolution; A Guide for the New Masters of the World.* New York: McGraw-Hill. 366 pp.
1973 *The Big Change: After the Environmental Revolution.* New York: McGraw-Hill. 288 pp.

Nicol, Hugh
1967 *The Limits of Man, An Enquiry into the Scientific Bases of Human Population.* London: Constable. 283 pp.

Odum, Howard T.
1971 *Environment, Power, and Society.* New York: Wiley-Interscience. 331 pp.

Passmore, John
1974 *Man's Responsibility for Nature: Ecological Problems and Western Traditions.* New York: Scribner's. 213 pp.

Report of the Study of Man's Impact on Climate (SMIC)
1972 *Inadvertent Climate Modification.* Cambridge, Mass: MIT Press. 308 pp.

Revelle, Roger and Hans H. Landsberg
1970 *America's Changing Environment.* Boston: Houghton Mifflin. 314 pp.

Rubin, Neville and William M. Warren (eds.)
1968 *Dams in Africa; An Interdisciplinary Study of Man-made Lakes in Africa.* New York: Kelley. 188 pp.

Shepard, Paul
1967 *Man in the Landscape: A Historic View of the Esthetics of Nature.* New York: Knopf. 290 pp.

Simmons, Ian G.
1974 *The Ecology of Natural Resources.* New York: Halstead Press. 424 pp.

Singer, S. Fred (ed.)
1970 *Global Effects of Environmental Pollution.* New York: Springer-Verlag. 218 pp.
1971 *Is There an Optimum Level of Population?* New York: McGraw-Hill. 426 pp.

1975 *The Changing Global Environment.* Boston: Reidel. 423 pp.

Small, William E.
1971 *Third Pollution: The National Problems of Solid Waste Disposal.* New York: Praeger. 173 pp.

Stern, A. C. et al. (eds.)
1973 *Fundamentals of Air Pollution.* New York: Academic Press. 492 pp.

Strong, Maurice F. (ed.)
1973 *Who Speaks for Earth?* New York: Norton. 173 pp.

Taft, C. E.
1965 *Water and Algae: World Problems.* Chicago: Educational Publishers. 236 pp.

Thomas, William L. (ed.)
1956 *Man's Role in Changing the Face of the Earth.* Chicago: University of Chicago. 1193 pp.
1959 *Man, Time, and Space in Southern California* (Supplement to *Annals,* Association of American Geographers, Vol. 49, No. 3, Part 2). Washington, D.C.: Association of American Geographers. 120 pp.

Treshow, M.
1970 *Environment and Plant Response.* New York: McGraw-Hill. 422 pp.

Ucko, P. J., and C. W. Dimbleby (eds.)
1969 *The Domestication and Exploitation of Plants and Animals.* London: Duckworth and Chicago: Aldine. 581 pp.

Van Dyne, G. M. (ed.)
1969 *The Ecosystem Concept in Natural Resource Management.* New York: Academic Press. 386 pp.

Vesey-Fitzgerald, Brian
1969 *The Vanishing Wild Life of Britain.* London: MacGibbon & Kee. 159 pp.

Vos, Antoon de
1975 *Africa, The Devastated Continent? Man's Impact on the Ecology of Africa.* The Hague: Junk. 236 pp.

Wagner, Richard H.
1974 *Environment and Man* (2nd ed.). New York: Norton. 528 pp.

White, Philip L. and Diane Robins (eds.)
1974 *Environmental Quality and Food Supply.* Mount Kisco, N.Y.: Futura. 248 pp.

Whiteside, Thomas
1970 *Defoliation.* New York: Ballantine Books. 168 pp.

Wiens, John A. (ed.)
1972 *Ecosystem Structure and Function.* Proceedings of the 31st Annual Biology Colloquium, Corvallis, Oregon, April, 1970. Corvallis: Oregon State University Press. 176 pp.

Wilson, B. R. (ed.)
1968 *Environmental Problems: Pesticides, Thermal Pollution, and Environmental Synergisms.* Philadelphia: Lippincott. 183 pp.

Wilson, Thomas W.
1971 *International Environmental Action: A Global Survey.* New York: Dunellen. 364 pp.

414

Selected References

Future Environments and Their Quality

Abler, Ronald et al. (eds.)
1975 *Human Geography in a Shrinking World.* North Scituate, Mass.: Duxbury Press. 307 pp.

Bell, Daniel
1973 *The Coming of Post-Industrial Society.* New York: Basic Books. 507 pp.

Boughey, Arthur S.
1971 *Man and Environment: An Introduction to Human Ecology and Evolution.* New York: Macmillan. 472 pp.

Brown, Harrison S.
1954 *The Challenge of Man's Future; An Inquiry Concerning the Conditions of Man During the Years that Lie Ahead.* New York: Viking Press. 290 pp.

Brown, Harrison S. (ed.)
1972 *Are Our Descendants Doomed? Technological Change and Population Growth.* New York: Viking Press. 377 pp.

Brown, Lester R.
1972 *World Without Borders.* New York: Random House. 395 pp.

Brubaker, Stirling
1975 *In Command of Tomorrow: Resource and Environmental Strategies for Americans.* Baltimore: Johns Hopkins University Press. 177 pp.

Brzezinski, Zbigniew
1970 *Between Two Ages: America's Role in the Technetronic Era.* New York: Viking Press. 334 pp.
1972 *The Fragile Blossom: Crisis and Change in Japan.* New York: Harper & Row. 153 pp.

Burton, Ian and Kenneth Hewitt
1971 *The Hazardousness of a Place: A Regional Ecology of Damaging Events.* Toronto: University of Toronto Press. 154 pp.

Clarke, Arthur C.
1973 *Profiles of the Future: An Inquiry into the Limits of the Possible* (rev. ed.). New York: Harper & Row. 237 pp.

Dansereau, Pierre (ed).
1970 *Challenge for Survival: Land, Air, and Water for Man in Megalopolis.* New York: Columbia University Press. 235 pp.
1971 *Dimensions of Environmental Quality.* Montreal: Institut d'Urbanisme, Université de Montreal. 109 pp.
1975 *Inscape and Landscape: The Human Perception of Environment.* New York: Columbia University Press. 118 pp.

Darling, J. Fraser, and John P. Milton (eds.)
1966 *Future Environments of North America.* Garden City, N.Y.: Natural History Press. 767 pp.

Dasmann, Raymond F.
1968 *A Different Kind of Country.* New York: Macmillan. 276 pp.
1972 *Environmental Conservation* (3rd ed.). New York: Wiley. 473 pp.
1975 *The Conservation Alternative.* New York: Wiley. 164 pp.

Dasmann, Raymond F. et al.
1973 *Ecological Principles for Economic Development.* New York: Wiley. 252 pp.

Edwards, Gordon
1969 *Land, People, and Policy: The Problems and Techniques of Assembling Land for the Urbanization of 100 Million New Americans.* West Trenton, N.J.: Chandler-Davis. 159 pp.

Ehrlich, Paul R., and Anne H. Ehrlich
1971 *The Population Bomb* (rev. ed.). New York: Ballantine Books. 201 pp.
1972 *Population, Resources, Environment* (2nd ed.). San Francisco: Freeman. 509 pp.
1974 *The End of Affluence: A Blueprint for your Future.* New York: Ballantine Books. 307 pp.

Ehrlich, Paul R. and John P. Holdren
1973 *Human Ecology: Problems and Solutions.* San Francisco: Freeman. 304 pp.

Ehrlich, Paul R. and Dennis Pirages
1974 *Ark II: Social Response to Environmental Imperatives.* San Francisco: Freeman. 344 pp.

Fairbrother, Nan
1970 *New Lives, New Landscapes: planning for the 21st century.* New York: Knopf. 397 pp.

Falk, Richard A.
1971 *This Endangered Planet: Prospects and Proposals for Human Survival.* New York: Random House. 498 pp.
1975 *A Study of Future Worlds.* New York: Free Press. 506 pp.

Falk, Richard A. and Saul Mendkovitz (comps.)
1973 *Regional Politics and World Order.* San Francisco: Freeman. 475 pp.

Feinberg, Gerald
1968 *The Prometheus Project: Mankind's Search for Long-Range Goals.* Garden City, N.J.: Doubleday. 216 pp.

Ferkiss, Victor
1974 *The Future of Technological Civilization.* New York: Braziller. 369 pp.

Franklin, Jerry et al. (eds.)
1975 *Productivity of World Ecosystems.* Washington, D.C.: National Academy of Sciences. 166 pp.

Gordon, Theodore J.
1965 *The Future.* New York: St Martin's Press. 184 pp.

Graubard, Stephen R. (ed.)
1967 "The year 2000—the trajectory of an idea," *Daedalus,* Vol. 96, No. 3, pp. 639–997, of the Proceedings of the American Academy of Arts and Sciences.

Hellman, Hal
1973 *Energy in the World of the Future.* New York: Evans. 240 pp.

Hutchinson, Joseph
1975 *The Challenge of the Third World.* Cambridge: Cambridge University Press.

Jarrett, Henry
1966 *Environmental Quality in a Growing Economy.* Baltimore: Johns Hopkins Press, for Resources for the Future.

173 pp.

Kahn, Herman, and Anthony J. Wiener
1967 *The Year 2000: A Framework for Speculation on the Next Thirty-Three Years.* New York: Macmillan. 431 pp.

Kahn, Herman and Bruce Briggs
1972 *Things to Come: Thinking about the Seventies and Eighties.* New York: Macmillan. 262 pp.

Landsberg, Hans H., L. L. Fischman, and J. L. Fisher
1963 *Resources in America's Future; Patterns of Requirements and Availabilities, 1960–2000.* Baltimore: Johns Hopkins Press, for Resources for the Future. 1017 pp.

McHale, John
1971 *The Future of the Future* (2nd ed.). New York: Braziller. 370 pp.

Mead, Margaret
1975 *World Enough: Rethinking the Future.* Boston: Little, Brown. 218 pp.

Meier, Richard L.
1966 *Science and Economic Development: New Patterns for Living* (2nd ed.). Cambridge, Mass.: MIT Press. 273 pp.
1971 *Organized Responses to Communications Stress in the Future Urban Environment.* Berkeley: Institute of Urban and Regional Development, University of California. 33 pp.
1975 *Studies on the Futures of Asian Cities.* Berkeley: Institute of Urban and Regional Development. 110 pp.

Perloff, Harvey S. (ed.)
1969 *The Quality of the Urban Environment.* Washington, D.C.: Resources for the Future. 332 pp.

Perloff, Harvey S. and Neil C. Sandberg (eds.)
1971 *New Towns—Why and for Whom?* New York: Praeger. 250 pp.

Platt, John R.
1966 *The Step to Man.* New York: Wiley. 216 pp.
1970 *Perception and Change: Projections for Survival* (2nd ed.). Ann Arbor: University of Michigan Press. 178 pp.

Saarinen, Thomas F.
1976 *Environmental Planning: Perception and Behavior.* Boston: Houghton Mifflin. 262 pp.

Ward, Barbara
1966 *Spaceship Earth.* New York: Columbia University Press. 152 pp.

Ward, Barbara, and René Dubos
1972 *Only One Earth: The Care and Maintenance of a Small Planet.* New York: Norton. 225 pp.

Wolstenholme, Gordon (ed.)
1963 *Man and His Future.* London: Churchill. 410 pp.

Toward a Geography of Cultures

Barth, Fredrik (ed.)
1969 *Ethnic Groups and Boundaries: The Social Organization of Culture Difference.* Boston: Little, Brown. 153 pp.

Bjorklund, Elaine M.
1964 "Ideology and culture exemplified in Southwestern Michigan," *Annals, Association of American Geographers*, Vol. 54 (June), pp. 227–241.

Brookfield, Harold C.
1972 *Colonialism, Development, and Independence: The Case of the Melanesian Islands in the South Pacific.* Cambridge: Cambridge University Press. 225 pp.
1975 *Interdependent Development.* London: Methuen. 234 pp.

Brookfield, Harold C. (ed.)
1973 *The Pacific in Transition: Geographical Perspectives on Adaptation and Change.* London: Edward Arnold. 332 pp.

Brookfield, Harold C. and Paula Brown
1963 *Struggle for Land: Agriculture and Group Territories among the Chimbu of the New Guinea Highlands.* New York: Oxford University Press. 193 pp.

Brookfield, H. C. and Doreen Hart
1971 *Melanesia: A Geographical Interpretation of an Island World.* London: Methuen. 464 pp.

Bruk, S. I., and V. S. Apenchenko (eds.)
1964 *Atlas Naradov Mira* (Atlas of the Peoples of the World). Moscow: Akademii Nauk SSSR. 184 pp. [106 colored plates detail the location of 910 ethnic groups.]

Clarke, William C.
1971 *Place and People: An Ecology of a New Guinean Community.* Berkeley: University of California Press. 265 pp.

Evans, E. Estyn
1973 *The Personality of Ireland: Habitat, Heritage, and History.* Cambridge: Cambridge University Press. 123 pp.

Gastil, Raymond D.
1975 *Cultural Regions of the United States.* Seattle: University of Washington Press. 366 pp.

Gottmann, Jean
1969 *A Geography of Europe* (4th ed.). New York: Holt, Rinehart, and Winston. 866 pp.

Hance, William A.
1964 *The Geography of Modern Africa.* New York: Columbia University Press. 653 pp.

Hodder, B. W. and D. R. Harris (eds.)
1967 *Africa in Transition; Geographical Essays.* London: Methuen. 378 pp.

Houston, J. M.
1968 *A Social Geography of Europe* (rev. ed.). New York: Praeger. 271 pp.

Jackson, James C.
1970 *Chinese in the West Bornean Goldfields: A Study in Cultural Geography.* Hull: University of Hull Publications. 88 pp.

Jakle, John A.
1973 *Ethnic and Racial Minorities in North America: A Selected Bibliography of the Geographical Literature.* Monticello, Ill.: Council of Planning Librarians. 71 pp.

Jakle, John A. and James O. Wheeler
1969 "The changing residential structure of the Dutch population in Kalamazoo, Michigan," *Annals*, Association of American Geographers, Vol. 59, No. 3 (September), pp. 441–460.

Jordan, Terry G.
1973 *The European Culture Area: A Systematic Geography.* New York: Harper & Row. 381 pp.

Kolb, Albert
1971 *East Asia: China, Japan, Korea, Vietnam; A Geography of a Cultural Region.* London: Methuen. 591 pp.

Longrigg, Stephen H.
1970 *The Middle East: A Social Geography* (2nd ed.). Chicago: Aldine. 291 pp.

Lowenthal, David
1972 *West Indian Societies.* London: Oxford University Press. 385 pp.

Lowenthal, David and Lambras Comitas
1973 *Slaves, Free Men, Citizens: West Indian Perspectives.* New York: Anchor Books. 340 pp.

Meinig, Donald W.
1971 *Southwest: Three Peoples in Geographical Change, 1600–1970.* New York: Oxford University Press. 151 pp.

Mikesell, Marvin W.
1961 *Northern Morocco* (Publications in Geography, Vol. 14), Berkeley: University of California. 135 pp.

Mutton, Alice F. A.
1968 *Central Europe, A Regional and Human Geography* (2nd ed.). New York: Praeger. 488 pp.

Pounds, Norman J. G.
1969 *Eastern Europe.* Chicago: Aldine. 912 pp.
1973 *An Historical Geography of Europe: 450 B.C.–A.D. 1330.* Cambridge: Cambridge University Press. 475 pp.

Pritchard, J. M.
1971 *Africa: The Geography of a Changing Continent.* New York: African Publishing Corporation. 248 pp.

Prothero, R. Mansell (ed.)
1969 *A Geography of Africa: Regional Essays on Fundamental Characteristics, Issues, and Problems.* New York: Praeger. 480 pp.
1972 *People and Land in Africa South of the Sahara: Readings in Social Geography.* New York: Oxford University Press. 344 pp.

Prothero, R. Mansell and Wilbur Zelinsky (eds.)
1970 *Geography and a Crowding World.* New York: Oxford University Press. 601 pp.

Sandhu, Kernial Singh
1969 *Indians in Malaya: Some Aspects of their Immigration and Settlement (1786–1957).* Cambridge Eng.: University Press. 345 pp.

Simoons, Fred
1960 *Northwest Ethiopia, Peoples and Economy.* Madison: University of Wisconsin Press. 250 pp.

Spate, O. H. K. and A. T. A. Learmonth
1967 *India and Pakistan; A general and Regional Geography* (3rd ed.). London: Methuen. 877 pp.

Spate, O. H. K.: Charles A. Fisher: and W. Gordon East (eds.)
1971 *The Changing Map of Asia: A Political Geography.* London: Methuen. 678 pp.

Spencer, J. E. and William L. Thomas
1971 *Asia, East by South: A Cultural Geography* (2nd ed.). New York: Wiley. 669 pp.

Tregear, T. R.
1965 *A Geography of China.* Chicago: Aldine. 342 pp.

Watson, J. Wreford
1967 *North America; Its Countries and Regions* (rev. ed.). New York: Praeger. 881 pp.

Zelinsky, Wilbur
1973 *The Cultural Geography of the United States.* Englewood Cliffs, N.J.: Prentice-Hall. 164 pp.

PHOTO CREDITS

INDEX

Longevity, modern increase in human, 266–268

Malaria, transfer from Old to New World, 150
Malthus, Thomas R., on population growth, 261–263
Mammals, evolution of, 25
Mankind, as agent responsible for ecological
  balance in future, 377
  alternatives facing, in future, 384–385
  ancestors of, 29–36
  brain development in early, 40
  chart of Pleistocene development of, 31
  culture as evolutionary complement for, 37–38
  evolution of, chart of, 35
  genetic variations in, 344–345
  mobility of early, 41
  as now responsible for earth, 379–380
  occupation of whole early by early, 41
  as omnivorous consumer, 37–38, 49–50
  as powerful agent in landscape change, 365–367
  problems facing, on future earth, 379–380
  problems facing, table of, 389
  as single species, 344
  skull forms of, 32
  steps toward cultural maturity for, 382–384
  table of chronology of development of, 35
  territoriality of, 45–46
  tropical origin of, 37–38
Manorial system, of Medieval Europe, 136–137
Manufacturing, economic and administrative
  elements in, 203–204
  locational factors in, 206
  regions, maps of, 206–208
  as resource conversion, 201–202
  terminology of, 201–202
  worldwide assembly patterns of, 209
Marco Polo, as Chinese official, 133
Markets, regionalism of, 235–237
Mediterranean Basin, eastern, as minor culture
  hearth, 111
Megalopolis, growth of, 245–246
Meso-America, as New World culture hearth, 129
Mesolithic, characteristics of, for mankind, 41
  as era of plant and animal domestication, 55
  as start of Recent period, 41
Mesopotamia, as primary culture hearth, 107–108
Metals, mining and smelting of ores of, 92–93
Metropolis, maps of typical cases, 248–249, 340
Migration, of Europeans, 1500–1900, map of, 141
  table of types of, 90
  to towns, earliest, 88
Midlatitude grasslands, opening of, 148–150
Minerals, cyclic mining of, 187–188
  earth as stockpile of, 187–188
  involved in organic cycles, 360
  table of critical, 200
Mining, effects of, on landscape, 366
  evolution of, 92–93

nature of cycle of, 188
Mining and smelting, map of spread of, 92
Minor culture hearths, Crete and the Eastern
  Mediterranean, 110–111
Miracle grains, recent development of, 372–373
Mississippi River, changes in delta of, 12–15, 359
Modernization, as cultural process, 294–296
  example of Japan, 295–296
  as related to colonialism, 329–330
Mongol culture realm, cavalry tactics used in, 85
  extent of, 132
  map of, 131
Monuments to cultural systems, 297–311
Moscow, map of metropolitan, 249
Mountain building and climatic change, 26

Nationalism, spread of political, 330
Nation-state, characteristics of, 323–324
  growth of, 155–156
Natural resources, table of critical, 200
Neandertaloids, emergence of, 34
  penetration of Europe by, 34
Neolithic, expansion of housing in, 64
  impact on environment during, 371
  meaning of term, 55
  origin of warfare in, 48
  outer limit of cropping in, map of, 58–59
  period of plant and animal domestication, 50–62
  processing of raw materials during, 84
  view of the earth, 355–356
New World, culture hearths in, 128–130
New York, changing area of, map of, 340
Nile Valley, as primary culture hearth, 108–109
Nomadism, as opposed to pastoralism, 84
North China, as early culture hearth, 110
North America, settlement of, map of, 317

Ocean transportation, 232–233
Organic cycle, nature of, 360

Pakistan, population pyramid for, 273
Paleolithic, cave art of, 36–37
  food resources of, 37–38
  impact on environments, 371
  population of the earth in, 260–261
  tools, 33–34
  view of the earth, 355–356
Paris, map of metropolitan, 248
Pastoralism, as alternative to crop-growing, 63
  invasion of agricultural regions by peoples
    following, 84
  lack of, in pre-Columbian New World, 63
  maturing of, 84–85
  as opposed to nomadism, 84
  and transhumance, 84
Peace, infrequency of, 47–48
Peasantry, emergence of, in early political state, 107

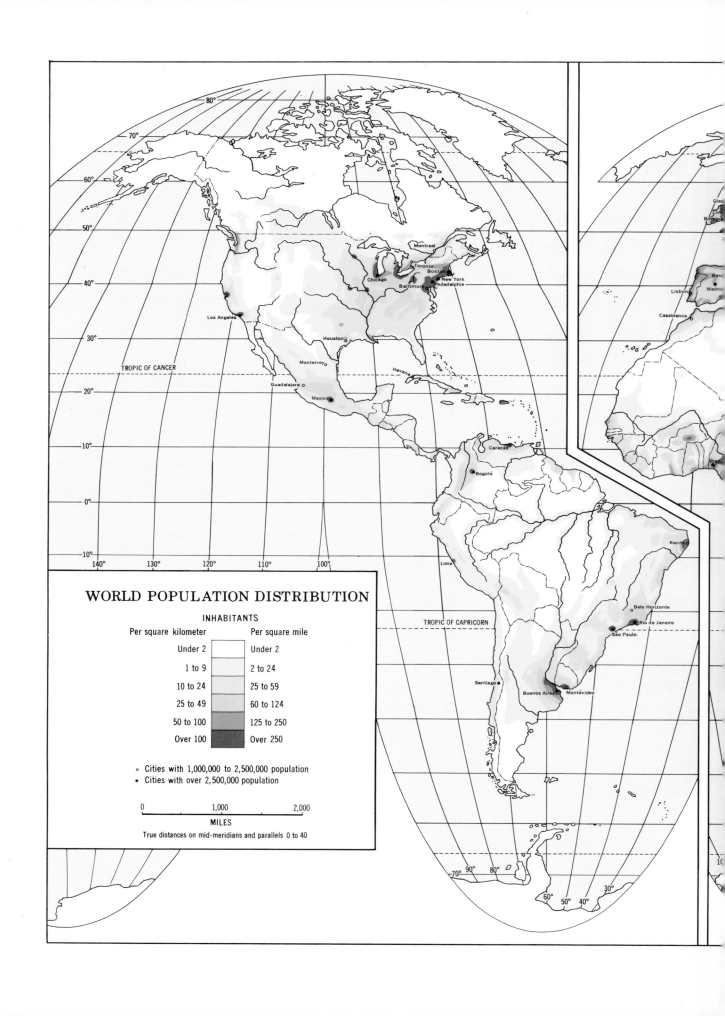

WORLD POPULATION DISTRIBUTION

INHABITANTS

Per square kilometer	Per square mile
Under 2	Under 2
1 to 9	2 to 24
10 to 24	25 to 59
25 to 49	60 to 124
50 to 100	125 to 250
Over 100	Over 250

○ Cities with 1,000,000 to 2,500,000 population
● Cities with over 2,500,000 population

```
0              1,000              2,000
```
MILES

True distances on mid-meridians and parallels 0 to 40